Biogeochemical Cycle of Mercury in Reservoir Systems in Wujiang River Basin, Southwest China

Xinbin Feng · Bo Meng · Haiyu Yan
Xuewu Fu · Heng Yao · Lihai Shang

Biogeochemical Cycle of Mercury in Reservoir Systems in Wujiang River Basin, Southwest China

Science Press
Beijing

Springer

Xinbin Feng
Institute of Geochemistry
Chinese Academy of Sciences
Guiyang, Guizhou
China

Xuewu Fu
Institute of Geochemistry
Chinese Academy of Sciences
Guiyang, Guizhou
China

Bo Meng
Institute of Geochemistry
Chinese Academy of Sciences
Guiyang, Guizhou
China

Heng Yao
Institute of Geochemistry
Chinese Academy of Sciences
Guiyang, Guizhou
China

Haiyu Yan
Institute of Geochemistry
Chinese Academy of Sciences
Guiyang, Guizhou
China

Lihai Shang
Institute of Geochemistry
Chinese Academy of Sciences
Guiyang, Guizhou
China

ISBN 978-981-10-6718-1 ISBN 978-981-10-6719-8 (eBook)
https://doi.org/10.1007/978-981-10-6719-8

Jointly published with Science Press, Beijing, China

The print edition is not for sale in China Mainland. Customers from China Mainland please order the print book from: Science Press.

Library of Congress Control Number: 2017955243

Printed on acid-free paper

This Springer imprint is published by Springer Nature
The registered company is Springer Nature Singapore Pte Ltd.
The registered company address is: 152 Beach Road, #21-01/04 Gateway East, Singapore 189721, Singapore

Contents

Chapter 1
Introduction

Abstract Newly built reservoirs favor mercury (Hg) methylation and then accumulation of methylmercury (MeHg) in food chains and therefore were regarded as "Hg sensitive ecosystems". Construction of dams in rivers will have an effect on the biogeochemistry and eco-environmental evolution of river system. Wujiang River Basin is a typical impounded river system with cascade reservoirs in China. The background of the Hg biogeochemical cycle research in Wujiang River Basin was elaborated in this chapter.

Keywords Wujiang river · Reservoir · Mercury · Methylmercury

1.1 Reservoirs Are Typical "Mercury Sensitive Ecosystems"

According to the International Association of Dams, there are at least 36,000 medium or large-sized dams built throughout the world in 1990s, and these dams control over 20% of freshwater flows globally (Vörösmarty et al. 1997). In China, there are only 2693 lakes with surface areas greater than 1.0 km^2, but as many as 86,353 reservoirs have been built until 2007. The number of reservoirs in China is expected to increase with the implementation of the "Go West" policy. In fact, almost all rivers in China were impounded at different degrees that impounded rivers are the most significant and universal feature of rivers in China. Due to the fact that rapid development of dams in river systems around the world, the scientific community is paying closer attention to the impacts of dams built on the biogeochemical cycling of mercury (Hg) in river systems (Heyes et al. 2000; Kelly et al. 1997).

It is a key question for scientific communities and governments to evaluate the health risks of exposure to Hg, the environmental behavior and the environmental effect of Hg in sensitive ecosystems. Numerous studies showed that newly built reservoirs are hotspot of Hg methylation, and MeHg accumulation in food chains (Lucotte et al. 1999), and therefore are regarded as "Hg sensitive ecosystems" (Wiener et al. 2003). In reservoir ecosystems, MeHg concentrations in fish at the

X. Feng et al., *Biogeochemical Cycle of Mercury in Reservoir Systems in Wujiang River Basin, Southwest China*, https://doi.org/10.1007/978-981-10-6719-8_1

top of food chain can be high enough to pose health risk to human. Since the impoundment of rivers could affect the biogeochemical cycling of Hg on both global and local scales, an increasing number of scientists have started to investigate the ecological risk of Hg in newly built reservoirs (Wiener et al. 2003).

In early 1970s, it was reported that MeHg concentrations in fish from newly built reservoirs in the US were generally higher compared to those same fish species from the nearby natural lakes (Abernathy and Cumbie 1977; Cox et al. 1979; Meister et al. 1979; Smith et al. 1974), and similar phenomenon was observed in other countries in North America, South America, and Europe (Anderson et al. 1995; Bodaly et al. 1984; Johnston et al. 1991; Louchouarn et al. 1993; Lucotte et al. 1999). These studies demonstrated that newly built reservoirs could be Hg sensitive ecosystems. Submersed soils are sites for Hg methylation, and the formed MeHg is subsequently bioaccumulated and biomagnified in food chains (Hecky et al. 1987, 1986; Jackson 1988). With the growth of the age of reservoir, organic matters in submersed soils are decomposed, and MeHg production will therefore decline in a few years after the built of the dam. During the past few decades, many studies have been devoted to understand the mechanism of Hg methylation in newly built reservoirs (Anderson et al. 1995; Bodaly et al. 1984; Johnston et al. 1991; Louchouarn et al. 1993; Porvari 1998).

In 1970s and early 1980s, scientists from China investigated the cycling of Hg in Hg-contaminated rivers such as the Songhua River and the Jiyun River, and high total Hg and MeHg concentrations were found in waters, sediments, and fish (Lin et al. 1983; Wang et al. 1986; Zhang et al. 1985). These studies confirmed that sediments were the main source of MeHg, and MeHg fluxes from sediments to waters were also estimated, which provided important scientific and technological foundations for Hg remediation in the future. In recent decades, Hg in fish from the Three Gorges Dam in Yangtze River were investigated (Jin and Xu 1997; Xu et al. 1998), but the biogeochemical cycling and ecological risk of Hg in newly built reservoirs in China were rarely studied.

1.2 Characteristics of Reservoir Systems

Reservoirs (or manmade impoundments) have the most significant and common disruptive effects on rivers and their associate ecosystems. Reservoirs play an important role in flood control, irrigation, water supply, electricity generation, and fish farming. Dams can not only regulate the water flow of the river, but affect the biogeochemistry of the river ecosystem (Naiman et al. 1987).

After a dam is built, the water depth, hydrodynamics, water residence time can change significantly. Human adjustment of water flows changes both the flood fluctuation of water and the environmental conditions generated by the periodical flood event (Bayley 1991). When a dam is built in a river, a complex accumulation effect will be associated. Impoundment construction could change the natural features and processes of the river. For instance, sedimentation is accelerated, and assimilation by aquatic biota is therefore enhanced, and temporal and spatial fluxes, and chemical composition of substances are also significantly changed (Hamilton and Schladow 1997; Menshutkin et al. 1998).

The natural features of rivers can be strongly affected by enhanced human activities in the river catchment (Degens et al. 1982; Nilsson and Dynesius 1994; Vörösmarty et al. 1997), and therefore the biogeochemical cycling and eco-environmental issues related to impounded river systems are becoming a hot topic of environmental studies globally.

1.3 Why Study Reservoirs in Wujiang River Basin?

1.3.1 Wujiang River Basin Is a Typical Impounded River System with Cascade Reservoirs Built in China

Wujiang River originating from Wumeng Mountain in Western Guizhou Province is the longest branch in the upstream of Yangtze River on the south bank, and it flows through the central and northeastern part of Guizhou Province, then though Chongqing and ends in Yangtze River at Fuling (Fig. 1.1). The total length of Wujiang River is 1037 km with a total catchment area of 88,267 km^2 and the total catchment area in Guizhou Province is 67,500 km^2. It is one of the most important hydropower generating rivers in China, and 13 reservoirs have been planned to be built on the mainstream of Wujiang River. Since 1970s, a number of reservoirs have been built including Wujiangdu Reservoir (WJD, built in 1979), Dongfeng Reservoir (DF, 1994), Puding Reservoir (PD, 1994), Yinzidu Reservoir (YZD, 2003), Suofengying Reservoir (SFY, 2003), Hongjiadu Reservoir (HJD, 2004), and Goupitan Reservoir (GPT, 2004). Silin Reservoir (SL), Shatuo Reservoir (ST), and Pengshui Reservoir (PS). In addition, a number of reservoirs were also built in the branches of Wujiang River, such as Baihua Reservoir (BH, 1966), Hongfeng Reservoir (HF, 1966), and Aha Reservoir (AH, 1960). It is obvious that Wujiang River is a typical impounded river with so many reservoirs built both on the mainstream and on the branches, and therefore it is an ideal site to investigate Hg biogeochemcial cycling of Hg in reservoirs in China.

1.3.2 Wujiang River Basin Can Be a Model of Yangtze River Catchment and Rivers of Southwestern China

After completion of the Three Gorges Dam, the impact of dam construction on the eco-environment of the aquatic system is a big concern to the scientific community. Meanwhile, other rivers in China are facing the same issue. For example, the upstream of Yangtze River, Jinsha River, Yalong River, Minjiang River, Daduhe

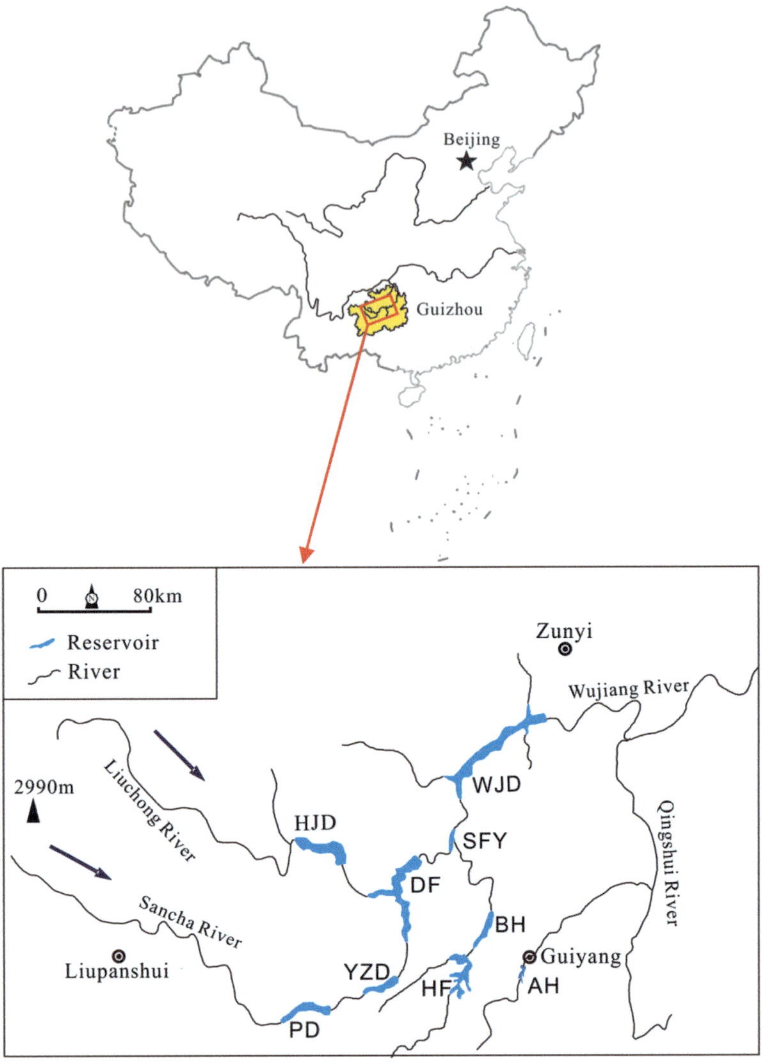

Fig. 1.1 Map of Wujiang River Basin

River, the middle and upstream of Pearl River, Lantsang River have been impounded by cascade reservoirs. It is definitely an extremely difficult task to investigate the biogeochemical cycling of Hg in Three Gorge Reservoir because this reservoir system is so complex and enormously gigantic, and the boundary conditions are so complicated as well. Cascade reservoirs were built in Wujiang River, where the environmental and the background conditions such as the geographic conditions, the shapes of the reservoirs, the hydrological conditions, and the climate are very much similar to those of Three Gorges Dam. Therefore, it will be much easier to understand the biogeochemical cycling of Hg in small reservoirs in Wujiang River. Meanwhile, Wujiang River is one of the important branches of Yangtze, and its large spatial scale can represent many important hydrological processes of different reservoirs. The outcomes of studying the Hg biogeochemical cycling in reservoirs of Wujiang River Basin can be definitely applied to other large branch catchments of Yangtze Basin, such as the mainstream of up and middle Yangtze River, Jinsha River, Yalong River, Miniiang River, Dadu River, the middle and upstream of Pearl River, Lantsang River, Hongshui River, and Nujiang River.

1.3.3 Reservoirs with Different Ages in Wujiang River Provide the Possibility to Investigate Mercury Biogeochemical Cycling Characteristics with the Evolution of Reservoirs

A number of reservoirs at different ages were built in Wujiang River, and this provided ideal study sites to investigate Hg biogeochemical cycling with the evolution of reservoirs. The total load and accumulation rates of biogenic elements in reservoirs might be controlled by the characteristics and evolution of reservoir ecosystems (Vollenweider 1976). After impoundment of a reservoir, the biological, physical, and chemical processes of the aquatic system will change subsequently. The Three Gorges Dam started to impound in 2003, and it was difficult to predict the environmental and ecological evolution after its impoundment. On a small spatial scale of Wujiang River Basin, a number of reservoirs were built at different time, which resulted in the fact that we can find reservoirs in different eutrophication stages and in different Hg pollution status in a small area. This provided important opportunities to investigate the changes of Hg biogeochemical cycling with the evolution of reservoirs.

1.3.4 The Scientific and Special Significance of Studying the Biogeochemical Cycling of Mercury in Reservoirs in Wujiang River Basin

Due to the geological background of Wujiang River Basin, it is of great scientific importance to investigate Hg biogeochemical cycling in Wujiang River Basin. Wujiang River Basin is located in the Circum-Pacific mercuriferous belt, and a number of Hg ore deposits were discovered in the basin. Due to the special geological background, Hg concentrations in soil of the basin were elevated against other regions in China. Meanwhile, Hg mining also resulted in serious Hg contamination in the environment. For example, Hg mining and retorting activities in

Wuchuan Hg mine resulted in serious Hg contamination on the Hongdu River, one of the branches of Wujiang River (Qiu et al. 2006); artisanal Zn smelting in Hezhang also gave rise to Hg contamination in the surface water of the upstream of Wujiang River (Feng et al. 2004a). Wujiang River Basin is also one of important coal producing regions in Guizhou Province, and due to special geological background, Hg concentrations in coal in this region were much elevated compared to other areas in China (Feng and Hong 1999; Feng et al. 2002). Hg emission from coal burning in Guiyang resulted in the annual average total gaseous Hg concentration in ambient air reached 8.4 ng m^{-3}, which was much higher than the values (1.5–2.0 ng m^{-3}) observed at background sites in Northern hemisphere (Feng et al. 2004b, 2003; Tang et al. 2003). It was estimated that the annual Hg emission from Guizhou Province reached about 20 t (Feng et al. 1997). Hg emission from coal burning is another important Hg pollution source in Wujiang River Basin (Feng and Hong 1999; Feng et al. 2002). Coal-fired power plants are one of the most important industries in Guizhou Province, and the total coal consumption reached 60–70 million t in 2010. Therefore, both dry and wet deposition are important sources of Hg pollution to Wujinag River Basin. Besides, many energy, chemical engineering, mechanical, metallurgy, and textile and other light industries are distributed in the catchments of HF, BH, and AH, these industries discharged Hg directly into the aquatic systems, aggravating Hg pollution.

References

Abernathy AR, Cumbie PM (1977) Mercury accumulation by largemouth bass (Micropterus-Salmoides) in recently impounded reservoirs. Bull Environ Contam Toxicol 17:595–602

Anderson MR, Scruton DA, Williams UP, Payne JF (1995) Mercury in fish in the smallwood reservoir, Labrador, 21 one years after impoundment. Water Air Soil Pollut 80:927–930

Bayley PB (1991) The flood pulse advantage and the restoration of river-floodplain systems. Regulated Rivers Res Manag 6:75–86

Bodaly RA, Hecky RE, Fudge RJP (1984) Increases in fish mercury levels in lakes flooded by the Churchill river diversion, Northern Manitoba. Can J Fish Aquat Sci 41:682–691

Cox JA, Carnahan J, Dinunzio J, McCoy J, Meister J (1979) Source of mercury in fish in new impoundments. Bull Environ Contam Toxicol 23:779–783

Degens ET, Herrera R, Kempe S, Soliman H (1982) Transport of carbon and minerals in major world rivers (No. 52). Selbstverlag des Geologisch-Paläontologischen Institutes der Universität Hamburg

Feng XB, Hong YT (1999) Modes of occurrence of mercury in coals from Guizhou, People's Republic of China. Fuel 78:1181–1188

Feng XB, Chen YC, Zhu WG (1997) Vertical fluxes of volatile mercury over soil surfaces in Guizhou Province, China. J Environ Sci 9:241–245 (In Chinese, with English abstract)

Feng XB, Sommar J, Lindqvist O, Hong YT (2002) Occurrence, emissions and deposition of mercury during coal combustion in the Province Guizhou China. Water Air Soil Pollut 139:311–324

Feng XB, Tang SL, Shang LH, Yan HY, Sommar J, Lindqvist O (2003) Total gaseous mercury in the atmosphere of Guiyang, PR China. Sci Total Environ 304:61–72

Feng XB, Li GH, Qiu GL (2004a) A preliminary study on mercury contamination to the environment from artisanal zinc smelting using indigenous methods in Hezhang county,

Guizhou, China—Part 1: mercury emission from zinc smelting and its influences on the surface waters. Atmos Environ 38:6223–6230

Feng XB, Shang LH, Wang SF, Tang SL, Zheng W (2004b) Temporal variation of total gaseous mercury in the air of Guiyang, China. J Geophys Res Atmos 109. doi: 10.1029/2003JD004159

Hamilton DP, Schladow SG (1997) Prediction of water quality in lakes and reservoirs 1 model description. Ecol Modell 96:91–110

Hecky RE, Bodaly RA, Ramsey DJ, Strange NE. (1986) Enhancement of mercury bioaccumulation in fish by flooded terrestrial materials in experimental ecosystems. In: Canada-Manitoba agreement on the study and monitoring of mercury in the Churchill River diversion, appendix

Hecky RE, Bodaly RA, Ramsey DJ, Ramlal PS, Strange NE (1987) Evolution of limnological conditions, microbial methylation of mercury and mercury concentrations in fish in reservoirs of Northern Manitoba. In: Technical appendices to the summary report of the Canada-Manitoba agreement on the study and monitoring of mercury in the Churchill River diversion, vol 3. pp 1–7

Heyes A, Moore TR, Rudd JWM, Dugoua JJ (2000) Methyl mercury in pristine and impounded boreal peatlands, experimental Lakes Area, Ontario. Can J Fish Aquat Sci 57:2211–2222

Jackson TA (1988) The mercury problem in recently formed reservoirs of Northern Manitoba (Canada)—effects of impoundment and other factors on the production of Methyl Mercury by microorganisms in sediments. Can J Fish Aquat Sci 45:97–121

Jin LJ, Xu XQ (1997) Methylmercury distribution in surface water and fish in the three Gorge reservoir area. Resour Environ Yangtze Valley 6:324–328 (In Chinese, with English abstract)

Johnston TA, Bodaly RA, Mathias JA (1991) Predicting fish mercury levels from physical characteristics of boreal reservoirs. Can J Fish Aquat Sci 48:1468–1475

Kelly CA, Rudd JWM, Bodaly RA, Roulet NP, StLouis VL, Heyes A, Moore TR, Schiff S, Aravena R, Scott KJ, Dyck B, Harris R, Warner B, Edwards G (1997) Increases in fluxes of greenhouse gases and methyl mercury following flooding of an experimental reservoir. Environ Sci Technol 31:1334–1344

Lin YH, Kang MD, Liu JY (1983) Speciation of mercury in sediment of Jiyun River. Envion Chem 2:10–19 (in Chinese, with English abstract)

Louchouarn P, Lucotte M, Mucci A, Pichet P (1993) Geochemistry of mercury in 2 hydroelectric reservoirs in Quebec, Canada. Can J Fish Aquat Sci 50:269–281

Lucotte M, Montgomery S, Bégin M (1999) Mercury dynamics at the flooded soil-water interface in reservoirs of Northern Québec: in situ observations. In: Lucotte M, Schetagne R, Thérien N, Langlois C, Tremblay A (eds) Mercury in the biogeochemical cycle: natural environments and hydroelectric reservoirs of Northern Québec (Canada). Springer, Berlin Heidelberg, Berlin, Heidelberg, pp 165–189

Meister JF, Dinunzio J, Cox JA (1979) Source and level of mercury in a new impoundment. J Am Water Works Assoc 71:574–576

Menshutkin VV, Astrakhantsev GP, Yegorova NB, Rukhovets LA, Simo TL, Petrova NA (1998) Mathematical modeling of the evolution and current conditions of the Ladoga Lake ecosystem. Ecol Model 107:1–24

Naiman RJ, Melillo JM, Lock MA, Ford TE, Reice SR (1987) Longitudinal patterns of ecosystem processes and community structure in a Sub-Arctic River Continuum. Ecology 68:1139–1156

Nilsson C, Dynesius M (1994) Ecological effects of river regulation on mammals and birds—a review. Regulated Rivers Res Manage 9:45–53

Porvari P (1998) Development of fish mercury concentrations in Finnish reservoirs from 1979 to 1994. Sci Total Environ 213:279–290

Qiu GL, Feng XB, Wang SF, Shang LH (2006) Environmental contamination of mercury from Hg-mining areas in Wuchuan, northeastern Guizhou China. Environ Pollut 142:549–558

Smith FA, Sharma RP, Lynn RI, Low JB (1974) Mercury and selected pesticide levels in fish and wildlife of Utah. 1. Levels of mercury, DDT, DDE, Dieldrin and PCB in fish. Bull Environ Contam Toxicol 12:218–223

Tang SL, Feng XB, Shang LH, Yan HY, Hou YM (2003) Mercury speciation in the flue gas of a small-scale coal-fired boiler in Guiyang PR China. J De Physique Iv 107:1287–1290

Vollenweider RA (1976) Advances in defining critical loading levels for phosphorus in lake eutrophication. Memorie dell'Istituto Italiano di Idrobiologia, Dott. Marco de Marchi Verbania Pallanza

Vörösmarty CJ, Sharma KP, Fekete BM, Copeland AH, Holden J, Marble J, Lough JA (1997) The storage and aging of continental runoff in large reservoir systems of the world. Ambio 26:210–219

Wang SH, Wang NH, Wang QC (1986) Evaluation standard of methylmercury pollution in aquatic system and evaluation of methylmercury pollution in the second Songhua River. J Environ Sci 7:73–76 (in Chinese)

Wiener JG, Krabbenhoft DP, Heinz GH, Scheuhammer AM (2003) Ecotoxicology of mercury. Handb Ecotoxicol 2:409–463

Xu XQ, Qiu CQ, Deng GH, Hui JY, Zhang XH (1998) Prediction of fish mercury bioaccumulation in reservoirs. Acta Hydrobiologica Sinica 22:244–250 (in Chinese, with English abstract)

Zhang LT, Guang JK, Zhang YS, Ma XS (1985) Speciation of mercury in sediment from Hadawan to Shaokuo of the second Songhua River. Envion Chem 4:40–46 (in Chinese)

Chapter 2
Analysis of Mercury Species in the Environmental Samples

Abstract Accurate analysis of mercury (Hg) species is the basis of understanding fate and behavior of Hg in aquatic environment. In this chapter, different Hg speciation systems were introduced according to analytical theory or research aim. In this chapter, we elucidate the detailed procedures, including the sample collection, preparation, and analytical methods in water, sediment, plankton, and sediment employed in this study.

Keywords Mercury · Species · Measurement · Water · Sediment · Plankton · Fish

2.1 Mercury Speciation in Water

2.1.1 Definition of Mercury Species

Mercury (Hg) widely exists in nature, including the lithosphere, pedosphere, hydrosphere, biosphere, and atmosphere. Hg has three chemical states (0, +1, +2) and exists as elemental, inorganic, and organic Hg in the environment. Concentrations of Hg usually represent much lower levels than other heavy metals in the natural environment, like Cu, Cd, Pb. etc. It is a great challenge to accurately analyze the trace level Hg species in samples, such as uncontaminated water samples, which hampered the understanding of the Hg biogeochemical cycle and the estimation of the potential Hg exposure in aquatic food chain. With the improvement of sensitive, specific, and precise Hg species analysis method, it appears that the data of Hg in natural waters before the 1980s are not reliable. High concentration of Hg in water samples was caused by contamination during the sample collection or pretreatment. Furthermore, trace level Hg species in natural water sample could not be measured since the Lowest Detection Limit of Method for Hg is not good enough (Bloom and Fitzgerald 1988; Yan et al. 2005a).

Biogeochemical cycling of Hg in aquatic ecosystems involves the distribution, transportation, and transformation of Hg in sediment, water, sediment/water

X. Feng et al., *Biogeochemical Cycle of Mercury in Reservoir Systems in Wujiang River Basin, Southwest China*, https://doi.org/10.1007/978-981-10-6719-8_2

interface, water/air interface, phytoplankton, zooplankton, shellfish, fish, etc. Elemental Hg (Hg^0), divalent Hg (Hg^{2+}), and methylmercury (MeHg) are the main concerned Hg species in aquatic ecosystem. Hg^0 mainly exists in the atmosphere due to its high volatility, and accounts for more than 95% of Hg in the atmosphere. The Hg^0 exchange between water/air interface is a key transportation process between the two large Hg pools, including both emission and deposition process. The deposition of Hg from the atmosphere to the water includes wet deposition and dry deposition. The deposition can directly input into water surface or input from the watershed to the water body as runoff after it deposited on the land. Divalent Hg (Hg^{2+}) is the main fraction of Hg in water, which is regarded as high activity, named as reactive Hg (RHg). MeHg is the most concerned species due to its high toxicity, bioaccumulation, and biomagnification through the food chains. Generally, MeHg is the predominant form of Hg in fish tissues. MeHg is formed from inorganic Hg (such as Hg^0, Hg^{2+}, etc.) via methylation usually involving with bacteria. Sediment is the pool of Hg in aquatic system and its anaerobic environment favors the methylation of Hg. The diffusion of Hg^{2+} and MeHg between sediment/water interface is the key transportation process from sediment to water body.

In chemistry, speciation analysis refers to the analytical activities of identifying and/or measuring the quantities of one or more individual chemical species in a sample (IUPAC 1997). For Hg, the speciation analysis includes both chemical defined species, like MeHg, ethylmercury (EtHg), etc., and operationally defined species, like dissolved Hg (DHg), reactive Hg (RHg), etc. In general, the analytic species are first extracted from the sample matrix, then following separation of Hg species, and detected by an appropriate detector. With regard to different sample media, acid or alkaline digestion technique will be processed before organic solvent extraction. Chromatographic technique is often applied for the separation of different chemical forms, such as high-performance liquid chromatography (HPLC), gas chromatography (GC), supercritical Fluid chromatography (SFE), or capillary zone electrophoresis (CZE). A number of detection methods are available, but its sensitivity, multielemental capability, and the possibility of isotopic information make inductively coupled plasma mass spectrometry (ICP-MS) the detector of first choice. Hg detectors usually include cold vapor atomic absorption spectroscopy (CVAAS), ICP-MS, cold vapor atomic fluorescence spectrometry (CVAFS), and atomic emission spectroscopy (AES), in which CVAFS is most popular due to its high sensitivity and low cost.

In this chapter, the analysis of Hg species in water, sediment, and biota is reviewed, including the cleaning procedure of sampling vessels or equipment, sample collection, storage, pretreatment, analytical methods, etc.

There are various Hg species (Fig. 2.1) in natural water in river, lake, reservoir, etc. In chemistry, Hg can be categorized into inorganic Hg and organic Hg. Inorganic Hg includes elemental Hg, divalent Hg (Hg^{2+}), monovalent Hg, which is not stable and easily transforms into elemental Hg and divalent Hg through disproportionation reaction. Organic Hg includes methylmercury (MeHg), dimethylmercury (DMeHg), ethylmercury (EtHg), phenylmercury (PhHg), etc.

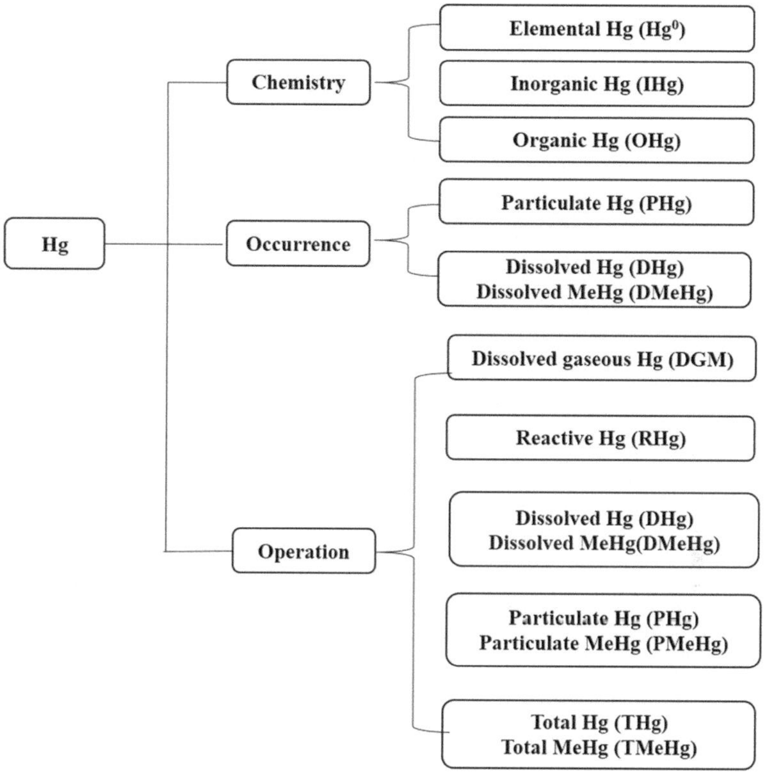

Fig. 2.1 Hg speciation in natural water

In water, the dissolved Hg^0 is defined as the Hg trapped by N_2 bubbling, named as dissolved gaseous mercury (DGM). Hg^{2+} is operationally defined as $SnCl_2$ reduced Hg, named reactive mercury (RHg) (Lindqvist et al. 1991). MeHg is the most concerned species in water, which is easily bioaccumulated in food chain. The concentration of DMeHg, EtHg, and PhHg in water is much lower than MeHg, and they are not measured so popular and only reported in a few references in deep sea and wetland systems (Cai et al. 1996; Cossa et al. 1994; Wallschlager et al. 1995).

According to the operation definition of filtration, Hg in water was operationally defined as total mercury (THg), soluble Hg (dissolved Hg, DHg), and particulate Hg (PHg). Accordingly, MeHg in water is categorized into total methylmercury (TMeHg), dissolved methylmercury (DMeHg), and particulate methylmercury (PMeHg).

2.1.2 Sample Collection Program in River and Reservoirs

1. Cleaning procedure of vessels and equipment

Due to the extremely low mercury concentration (ppt and sub-ppt level) in natural water, quality control of every step including blank of sampling vessels, sampling equipment, sample collection, storage, and determination is vital to get accurate measurement of Hg. It is quite essential to do ultra-clean treatment during the preparation of sampling bottle and sampling equipment.

For Hg measurement, collection of water samples should be bottled with fluoropolymer or borosilicate glass. For water sample filtering, borosilicate glass filter device was employed. Great caution is needed during the cleaning of fluoropolymer bottle or glass bottle to avoid any kind of contamination.

All of the borosilicate glass sampling bottle or filtration device are subjected to a strict ultra-clean treatment before sampling. The processes are as follows:

(1) Cleaning: cleaning the bottle with tap water with detergent.
(2) Acid soaking: soaking the bottle in 10% HNO_3 (V/V) at least 24 h.
(3) Rinsing: first rinsing the bottle or device from the acid with tap water, and then rinsing with double-distilled water (DDW).
(4) Baking: Putting the bottle or device into muffle furnace, keeping the temperature at 500 °C at least 1 h.
(5) Storage: Sealing the bottle or device with double-layer polyethylene bags.

If fluoropolymer bottle is used, the cleaning procedure should follow EPA Method 1631E.

(1) Keep the bottle in hot 4 N HCl for more than 48 h.
(2) After cooling and rinsed with DDW, the bottle was filled with 1% HCl in an oven at 60–70 °C overnight.
(3) After cooling and rinsed with DDW, the bottle was filled with 0.4% (V/V) HCl.

After cleaning, the bottle is tightly capped (with a wrench), double bagged in new polyethylene zip bags and stored in wooden or plastic boxes.

2. Sampling method for atmospheric deposition

A bulk precipitation sampler was designed based on the version of the collector used by European countries (Commission OaP 1997). It is demonstrated that there is no significant difference (at the 10% level) between co-located bulk and wet deposition samples for THg (Landing et al. 1998).

To reduce the Hg volatilization in the precipitation samples, the precipitation sample should avoid the exposure to air and sunlight as less as possible. The sampling train consisted of three borosilicate glass components (Fig. 2.2), a funnel (~15 cm diameter), a connecting tube (~2 cm diameter), and a sampling bottle (~1 L volume). The connecting tube plays a role of capillary to prevent the volatilization of Hg in the precipitation sample. The connecting tube and

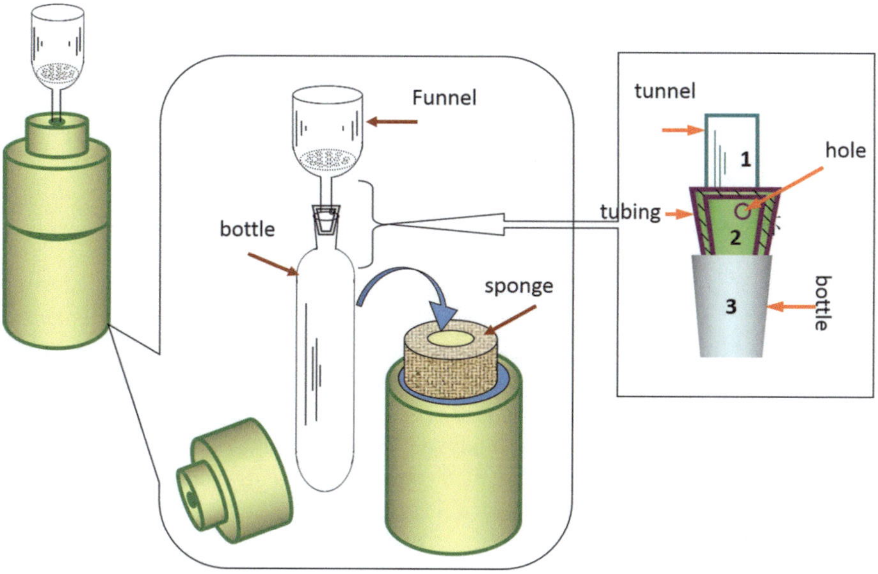

Fig. 2.2 Sampling equipment for wet deposition

the sampling bottle were placed inside a PVC column which was filled with sponge to be shielded from sunlight.

The cleaning of the sampling train components follows the cleaning procedure for borosilicate glass bottles. The sample collector was mounted on the roof of houses approximately 1.5 m above the ground to avoid contamination from soil particles by splashing during heavy rainfall.

3. Water sample sampling method

Water samples of river reservoirs system were collected using an acid-washed, Teflon-coated, 10 L Niskin sampler (Model 1010X series, General Oceanics Inc. U.S.A.). The Niskin bottle was positioned in the upstream direction relative to the operator. All samples were collected following ultra-clean sample handling protocols. Water samples were transferred from the sampler into pre-cleaned bottles. Each bottle was rinsed three times with reservoir/river water before sample collection.

Unfiltered samples were packed into double layer of zip bags for the analysis of THg, RHg, TMeHg, and total suspended solid (TSS) in lab. Water samples was filtered on site using a 0.45 μm filter (Millipore) to get dissolved water samples for analysis of DHg and DMeHg. All the water samples were acidified on site to 0.4% (v/v) with ultra-pure concentrated hydrochloric acid (HCl). The sample bottles were then capped, sealed with Parafilm, and stored in a refrigerator at 3–4 °C in the dark. The analysis of concentrations of Hg species in water samples was conducted within 28 d after sampling.

2.1.3 Analysis Total Mercury in Water Samples

Basically, the analysis of THg and inorganic mercury in water samples followed US EPA Method 1631 (US EPA 2001b). For RHg analysis, take an aliquot of unfiltered water samples into bubbler with 50 mL DDW, 0.1 mL HNO_3 and 0.1 mL of 20% $SnCl_2$. Then purge with 300 mL·min^{-1} N_2 for 20 min to collect RHg by gold trap. Finally, connect the gold trap with analysis system, RHg will be analyzed by due amalgamation combined with CVAFS.

Analysis of DHg and THg in water followed a same method. In 100 mL of water samples, add 0.5 mL BrCl at room temperature for 24 h. Just prior to analysis, 0.2 mL 20% NH_2OHHCl were added into the sample bottle to remove the excess amount of BrCl. The sample was further reduced with 0.1 mL 20% $SnCl_2$, purged with N_2, the formed elemental Hg vapor (Hg^0) was trapped by gold amalgamation, and finally thermally desorbed and analyzed by dual amalgamation combined with CVAFS les was following the same method with THg. PHg was calculated as the difference between the concentration of THg and DHg in the sample.

2.1.4 Analysis of Methylmercury in Water Samples

The analysis of MeHg in water followed the distillation-$NaBEt_4$ ethylation-Tenax trap-GC-CVAFS as EPA Method 1630 (US EPA 2001a). MeHg analysis in water samples based on distillation of CH_3HgCl, which is formed by adding 0.4% HCl into water samples during the sample collection. For distillation, 45 mL of water sample and 200 μL of 1% APDC was placed into a fluoropolymer distillation vial. For the distillate receiving vial, 5.0 mL of reagent water and 200 μL of acetate buffer was added. The distillation vial was in an Al-heating block at 125 with purging with 60 mL·min^{-1} N_2. The receiving vial was kept at about 4 °C in a water-ice bath.

Distilled water samples were added into bubblers for ethylation by adding 50 μL 1% $NaBEt_4$. MeHg were collected onto Tenax trap by purging the solution with 300 mL·min^{-1} N_2. Then Tenax trap was heated to release MeHg into the carrier gas and MeHg finally was analyzed by GC-CVAFS. Analysis of TMeHg and DMeHg in water samples was following the same method. PMeHg was calculated as the difference between the concentrations of TMeHg and DMeHg in the sample.

2.2 Analysis of Mercury Species in Sediment Samples

2.2.1 Sample Collection

Sediment samples were collected by a gravity corer with a 6.3 cm diameter 64 cm long plexiglass tube. The overlying water in the core tube was collected into acid-cleaned borosilicate glass bottle by siphoning down to about 10–20 cm above the sediment surface. Sediment cores were sectioned every 1 or 2 cm under N_2 atmosphere in a glove bag. Sediment samples were placed in centrifuge tubes, capped and sealed with parafilm.

In the lab, sediment samples were centrifuged at 3500 rpm for 25 min at 4 °C. After being filtered through 0.45 μm membrane filter (Millipore), pore water samples were preserved with 0.4% (V/V) HCl in acid clean borosilicate glass tubes and stored in a refrigerator at 3–4 °C in the dark until DMeHg and DHg analysis. Field blanks were also prepared by adding Milli-Q water in sampling tubes. Subsequently, the freeze-dried sediment samples were ground and homogenized to a size of 150 meshes per inch with a mortar for solid-phase THg, MeHg, and organic matter (OM) concentration analysis.

2.2.2 Total Mercury Analysis in Sediment Samples

THg measurement in sediments was performed by reduction with $SnCl_2$ following oxidation by acid digestion (Li et al. 2005). Sediment samples were freeze-dried and ground before analysis. The digestion procedure required approximately a 0.20 g dry weight sample. The sample was placed inside a 25 mL glass tube covered with a glass ball, and 5 mL DDW and 5 mL aqua regia were added in turn, digested at 95 °C for 5 min. After that, 1 mL BrCl solution was added continuously to digest at 95 °C for 30 min. After cooling, the sample reacted for an additional 24 h and 0.2 mL NH_2OHHCl solution (25 g of reagent grade NH_2OHHCl is dissolved in 100 mL DDW) was added in the sample. Then, THg was determined by CVAFS. The lowest limit of detection was 0.01 ng g^{-1} for THg analysis, which was calculated based on 3 times the standard deviation of blank measurements. Quality control for the THg determinations was addressed with method blanks, blank spikes, matrix spikes, certified reference materials, and blind duplicates.

2.2.3 Methylmercury Analysis in Sediment Samples

MeHg in sediment samples were analyzed following the method of He et al. (2004). Approximately 0.3 g of sediment was placed into a 50 mL centrifuge tube. 1.5 mL of 1 M $CuSO_4$, 7.5 mL of 3 M HNO_3, and 10 mL of CH_2Cl_2 were added. The tube was closed and shaken for 30 min. 5 mL of the CH_2Cl_2 layer was pipetted into another 50 mL centrifuge tube after the tube was centrifuged at 3000 rpm for 30 min. About 40 mL of double-deionized water was added to the tube. The tube was heated at 45 °C in a water bath until no visible solvent was left in the tube and the remaining liquid was then purged with nitrogen for 8 min in a water bath at 80 °C to remove solvent residue. The sample was brought to 50 mL with double-deionized water before an appropriate volume (generally 15 mL) of the sample was transferred to a borosilicate bubbler for MeHg analysis following the procedure described previously. Quality control for the MeHg determinations was addressed with method blanks, blank spikes, matrix spikes, certified reference materials, and blind duplicates.

2.3 Analysis of Mercury Species in Plankton and Fish

Hg transport and accumulation in aquatic ecosystems is mainly through the food chain. Due to the biomagnification effect of Hg in the food chain, Hg levels are low at the bottom of the food chain and in aquatic organisms of relatively small age. As the trophic level increases, aquatic organisms absorb Hg into the body through ingestion, mainly MeHg, and progressively with trophic levels. In order to study the enrichment characteristics of Hg in foodstuffs in the reservoirs of different evolution stages, the dominant aquatic species of different trophic levels were collected from upper to lower reaches of the Wujiang River for Hg and MeHg analysis.

2.3.1 Sample Collection

1. Plankton

Phytoplankton is an ecological concept that refers to the life of small plants floating in the water, usually refers to the phytoplankton, mainly including Cyanophyta, Bacillariophyta, Chrysophyta, Xanthophyta, Pyrrophyta, Cryptophyta, Euglenophyta, and Chlorophyta. Phytoplankton is in the first trophic level in the aquatic food chain.

The phytoplankton was collected by plankton net with different pore size, and finally classified according to their size. Phytoplankton was fixed with formalin 3–5% in the field and identified and enumerated (random fields) under the microscope using the settling technique in the laboratory. In addition, the cells colonies and filaments were enumerated to at least 300 specimens of the combined specie (Li et al. 2014).

According to size difference, zooplankton can be divided into macrozooplankton, mesozooplankton, and microzooplankton. Zooplankton has a wide range of feeding behavior: filter feeding, predation, and symbiosis with autotrophic phytoplankton as seen in corals. Zooplankton feed on bacterioplankton, phytoplankton, other zooplankton, detritus and even nektonic organisms.

Zooplankton samples were collected using a conical net of 64 μm mesh. The organisms were rinsed with filtered (0.45 μm porosity) reservoir water to remove adhered particulate matter and filtered through Nitex sieves of 610, 216, 108, and 38 μm mesh size to obtain four size fractions: 38–108, 108–216, 216–610, and >610 μm. Samples collected for zooplankton identification as well as for Hg and MeHg determination were separated into four fractions as described above. Samples for zooplankton identification were stored in vials and fixed in 4% formalin solution, while samples for THg and MeHg analyses were kept frozen until lyophilization, then again stored frozen for analysis (Wang et al. 2011).

They were taken back to the laboratory for freeze-drying and used to determine THg, MeHg, carbon (C) and nitrogen (N) isotopic composition. Samples of C and N isotopes were freeze-dried and ground through a 60 mesh nylon sieve. The samples were wrapped in aluminum foil and packed in a centrifuge tube and sealed in a dry dish.

2. Fish

The fish was collected from the fisherman in each reservoir. The number of fish species and the number of specimens sampled for each species were limited by availability of fish during the sampling campaign. All fish were visually inspected for fin and body deformations to avoid farmed fish in the samples. The collected fish were stored alive in barrels with water and air purge, or dead on ice in freeze boxes until sampling in the lab. The fish collected may not represent all species in the reservoirs, but the most abundant ones should be included.

The weight and length of the fish were measured before a sample of the dorsal muscle was removed and stored frozen for Hg analysis. The length and weight of the fish were recorded before collecting the fish muscle. About 20 g of muscle in the dorsal muscle of the fish was collected and wrapped in a tin foil and stored in a zip bag at −18 °C. Fish scales are used to identify their age, fish without scales, or partial scales, and their otoliths are age-matched.

2.3.2 *Total Mercury Analysis in Plankton and Fish Samples*

All samples were analyzed for THg following the method of acid digestion, $SnCl_2$ reduction, gold trap collection, and CVAFS. About 0.5–1.0 g of fresh muscle tissue was added into a 25 mL glass tube covered with a glass ball, and 10 mL of HNO_3: H_2SO_4 (8:2, V/V) were added in turn, digested at 95 °C for 3 h. After the samples cooled down, add appropriate volume of DDW to 25 mL. Then 0.5 mL BrCl were added for 24 h. Before analysis, 0.2 mL 25% $NH_2OH{\cdot}HCl$ were added to remove excess of BrCl. An aliquot of the digestate were taken for Hg analysis by $SnCl_2$, gold trap collection, and CVAFS following EPA Method 1631 (US EPA 2001b).

Quality control consisted of duplicates, method blanks, and standard reference material. Blank spikes and duplicates were taken regularly (>10% of samples) throughout each sampling process. The analysis of THg in plankton is following the similar method with fish THg analysis, but 0.10–0.20 g of dry samples was taken.

2.3.3 *Methylmercury Analysis in Plankton and Fish Samples*

About 0.5–1.0 g of fresh fish samples and 5 mL of 20% KOH were added into 25 mL fluoropolymer vials. Then the fluoropolymer bottles were kept at 75 °C in a water bath for 3 h. An aliquot of fish sample digestate were taken for MeHg analysis by aqueous ethylation, Tenax trap, GC-CVAFS following US EPA (2001a) and Yan et al. (2005b). Analysis of MeHg in plankton is similar with fish samples, but only about 5–10 mg of dry sample was taken for digestion. Quality control consisted of duplicates, method blanks, and standard reference material. Blank spikes and duplicates were taken regularly (>10% of samples) throughout each sampling process.

References

Bloom N, Fitzgerald WF (1988) Determination of volatile mercury species at the picogram level by low-temperature gas-chromatography with cold-vapor atomic fluorescence detection. Anal Chim Acta 208:151–161

Cai Y, Jaffe R, Alli A, Jones RD (1996) Determination of organomercury compounds in aqueous samples by capillary gas chromatography atomic fluorescence spectrometry following solid-phase extraction. Anal Chim Acta 334:251–259

Commission OaP (1997) JAMP guidelines for the sampling and analysis of mercury in air and precipitation. OSPAR Commission, Paris, pp 1–20

Cossa D, Martin J-M, Sanjuan J (1994) Dimethylmercury formation in the alboran sea. Mar Pollut Bull 28:381

He TR, Feng XB, Dai QJ, Qiu GL, Shang LH, Jiang HM, Liang L (2004) Determination of methylmercury in sediments and soils by GC-CVAFS after aqueous phase ethyltion. Earth Environ 32:83–86 (In Chinese, with English abstract)

IUPAC (1997) Compendium of Chemical Terminology, 2nd edn (the “Gold Book”). In: McNaught AD, Wilkinson A (eds) Blackwell Scientific Publications, Oxford

Landing WM, Guentzel JL, Gill GA, Pollman CD (1998) Methods for measuring mercury in rainfall and aerosols in Florida. Atmos Environ 32:909–918

Li QH, Shang LH, Gao TJ, Zhang L, Ou T, Huang GJ, Chen C, Li CX (2014) Use of principal component scores in multiple linear regression models for simulation of chlorophyll-a and phytoplankton abundance at a karst deep reservoir, Southwest of China. Acta Ecologica Sinica 34:72–78

Li ZG, Feng XB, He TR, Yan HY, Liang L (2005) Determination of total mercury in soil and sediment by aqua regia digestion in the water bath coupled with cold vapor atom fluorescence spectrometry detection. Bull Mineral Peterol Geochem 24:140–143 (In Chinese, with English abstract)

Lindqvist O, Johansson K, Aastrup M, Andersson A, Bringmark L, Hovsenius G, Hakanson L, Iverfeldt A, Meili M, Timm B (1991) Mercury in the Swedish environment—recent research on causes, consequences and corrective methods. Water Air Soil Pollut 55:1–261

US EPA. Method 1630: Methylmercury in water by distillation, aqueous ethylation, purge and trap, and CVAFS (2001a) U.S. Environmental Protection Agency, Washington, DC, pp 1–41

US EPA. Method 1631, Revision E: mercury in water by oxidation, purge & trap, and cold vapor atomic fluorescence spectrometry (2001b) U.S. Environmental Protection Agency, Wangshington, DC, pp 1–41

Wallschlager D, Hintelmann H, Evans RD, Wilken RD (1995) Volatilization of dimethylmercury and elemental mercury from river elbe floodplain soils. Water Air Soil Pollut 80:1325–1329

Wang Q, Feng XB, Yang YF, Yan HY (2011) Spatial and temporal variations of total and methylmercury concentrations in plankton from a mercury-contaminated and eutrophic reservoir in Guizhou Province, China. Environ Toxicol Chem 30:2739–2747

Yan HY, Feng XB, Li ZG, Jiang HM, He TR (2005a) A methodological development in measuring total mercury in fish using semi-closed digestion and CVAFS. Shanghai Environ Sci 33:89–92 (In Chinese, with English abstract)

Yan HY, Feng XB, Liang L, Shang LH, Jiang HM (2005b) Determination Of methylmercury in fish using GC-CVAFS. J Instrmental Anal 24:78–80 (In Chinese, with English abstract)

Chapter 3
Wet Deposition Flux of Total Mercury and Methylmercury in Wujiang River Basin

Abstract Wet deposition is an important pathway for the removal of mercury (Hg) from the atmosphere and the loading of Hg to the terrestrial and aquatic ecosystem systems. In this chapter, we (1) measured total mercury (THg) and methylmercury (MeHg) in bulk precipitation at five rural sites in Wujiang River Basin; (2) studied the monthly and annual deposition fluxes of THg and MeHg at the sampling sites in Wujiang River Basin and the underlying factors, and (3) estimated the input of THg and MeHg to the studied reservoirs in Wujiang River Basin via wet deposition.

Keywords Total mercury · Methylmercury · Wet deposition flux

3.1 Sampling Location and Sample Collection

3.1.1 Sampling Location

Wujiang River is the largest tributary in the upstream of the Yangtze River basin, mainly flowing in a Karst environment in Guizhou Province, southwestern China. Wujiang River Basin belongs to a subtropical monsoon humid climate, and the mean annual air temperature is 14.6 °C. In the studied area, rain is the dominant pathway of wet deposition, and snow is rare throughout the calendar year. The multi-year average annual rainfall is approximately 1100 mm. Precipitation samples were collected at five sites in the studied area: Puding Reservoir (PD, 26°22′N, 105°48′E), Yinzidu Reservoir (YZD, 26°34′N, 106°07′E), Hongjiadu Reservoir (HJD, 26°53′N, 105°51′E), Dongfeng Reservoir (DF, 26°51′N, 106°08′E), and Wujiangdu Reservoir (WJD, 27°19′N, 106°46′E) (Fig. 3.1). In general, the surrounding areas of the sampling sites were rural and there were no nearby sources at the sampling sites. However, there were several settlements within 5 km of the DF and WJD which might be related to anthropogenic Hg emissions because of coal burning in house heating activities.

X. Feng et al., *Biogeochemical Cycle of Mercury in Reservoir Systems in Wujiang River Basin, Southwest China*, https://doi.org/10.1007/978-981-10-6719-8_3

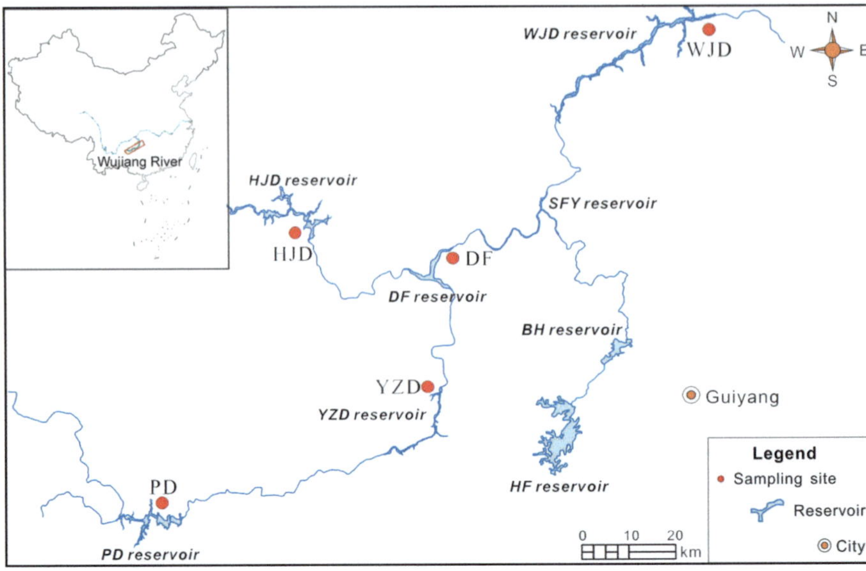

Fig. 3.1 Sampling locations (WJD, DF, YZD, HJD, and PD) of precipitation in Wujiang River (redrawn from Guo et al. (2008), with permission from Elsevier Ltd.)

3.1.2 Collection, Processing, and Analysis of Mercury in Precipitation

From January to December in 2006, precipitation samples were collected monthly at the sampling sites using a bulk precipitation collector constructed following the designation of the collector developed by European countries (Oslo 1998). This sampling train consists of three parts: (1) a funnel-type beaker (inner diameter of 15 cm), (2) a connecting tube, acting as capillary tube, furthermore, it can prevent the deposition of Hg into the precipitation sample as well as the evaporation of Hg from sample, and (3) a sampling bottle (volume: 1 L) (more detail see Sect. 2.1). These three components were made of borosilicate glass and pre-cleaned rigorously before filed sampling in the laboratory to prevent possible contamination. The connecting tube and the sampling bottle were placed inside a PVC column which was filled with sponge to be shielded from sunlight. The sample collector was fixed on an iron shelf and elevated by 1.5 m above the ground surface.

Cleaning procedures were utilized using trace metal clean protocols. All funnels, tubes, and bottles were cleaned rigorously by soaking in 10% HNO_3 solution for 24 h, rinsing with ultrapure deionized water (18 MΩ cm) and baking for one hour in a muffle furnace at 500 °C. Then the sampling components were double-bagged, stored in a plastic box before filed sampling. Just prior to sampling of precipitation, 5 mL of trace metal grade HCl (12 N) was added into the sampling bottle to minimize the adsorption of Hg in precipitation to the wall of borosilicate glass and evaporation of Hg from precipitation. The precipitation collector was replaced monthly using a new pre-cleaned collector to avoid possible cross contaminations. At the end of each month during the sampling period, precipitation collected were transferred carefully to two pre-cleaned borosilicate glass bottles (volume: 100 mL), individually sealed into three successive polyethylene bags and kept in a refrigerator at 4 °C until Hg analysis.

THg and MeHg concentrations in precipitation were determined following the U.S. EPA Method 1631 (USEPA 2002) and Method 1630 (USEPA 1998), respectively. Briefly, Hg in precipitation was purged from solution in a Hg-free nitrogen stream and concentrated onto a gold-coated sand trap after oxidation by BrCl followed by addition $NH_2OH{\cdot}HCl$ to discharge the excess BrCl and reduction of divalent Hg by $SnCl_2$ to Hg^0. The trapped Hg is then thermally desorbed from the gold trap into an inert gas stream and quantified using a dual amalgamation technique followed by cold vapor atomic fluorescence spectrometry (CVAFS) (USEPA 2002). MeHg concentration in precipitation was analyzed via distillation, aqueous ethylation, purge and trap, and detected by CVAFS (USEPA 1998). A 45-mL sample aliquot was added into a Teflon® distillation vessel, and then placed in an aluminum heating pan and distilled at 125 °C for 3–4 h. The distillated samples then underwent aqueous phase ethylation with the addition of 0.2 mL 2 M sodium acetate and 0.1 mL 1% sodium tetraethylborate in sequence, and followed by purging with N_2 onto an Tenax® trap. The MeHg was then desorbed with heat onto an isothermal GC column for peak separation and analyzed by CVAFS (Tekran model 2500).

A previous study suggested that there were no statistical differences in precipitation Hg concentrations between the wet-only and the bulk precipitation samples (Landis and Keeler 1997). However, given the highly elevated atmospheric particulate bound mercury (PBM) concentrations in most areas of China (Fu et al. 2015), using bulk precipitation collectors in the present study would probably overestimate the concentrations and deposition fluxes of Hg in the studied areas because of the incorporation of dry deposition of atmospheric Hg (especially the dry deposition of atmospheric PBM).

Quality control for the analysis of THg and MeHg concentrations was conducted using method blank, blank spikes, matrix spikes, and duplicates. The method detection limits (MDL), based on three times the standard deviation of duplicated measurements of the blanks, were 0.02 ng L^{-1} for THg and 0.01 ng L^{-1} for MeHg. The method blanks were found to be lower than the detection limits in all cases. The average relative standard deviation on precision test for the duplicate analysis of THg and MeHg was 4.5 and 5.4%, respectively. Spike recoveries for THg and MeHg were 93–110% and 88–108%, respectively.

3.1.3 Estimation of Wet Deposition Fluxes of Total Mercury and Methylmercury

Volume-weighted mean (VWM) Hg concentration at each sampling site was calculated using Eq. (3.1):

$$VWM = \frac{\sum_i^n Hg_i \times PD_i}{\sum_i^n PD_i}, \quad (3.1)$$

where VWM is the volume-weighted mean of precipitation Hg concentrations in ng L^{-1}, Hg_i, and PD_i are the Hg concentration (ng L^{-1}) and precipitation depth (mm) of a monthly precipitation i, respectively.

The precipitation depth calculated using collected sample volumes was not significantly different from that observed at nearby meteorological stations. Therefore, annual wet deposition fluxes of THg and MeHg were calculated using the VWM THg and MeHg concentrations and the corresponding annual precipitation depth.

3.2 Concentrations of Total Mercury and Methylmercury in Precipitation

Concentrations of THg in precipitation collected at the five sampling sites in Wujiang River ranged from 7.5 to 149.1 ng L^{-1}, with an overall VWM concentration of 36.7 ng L^{-1} (Fig. 3.2). During the whole year, the highest monthly mean THg concentration of the five sampling sites, 117.0 ng L^{-1}, was observed in January, which was approximately 10 times higher than the lowest monthly mean THg concentration in September (Fig. 3.2). To examine the seasonal variations in THg, we compared the precipitation THg concentrations at the five sampling sites in spring (March, April, May), summer (June, July, August), fall (September, October, November), and winter (December, January, February). The seasonal variation of THg in precipitation samples was significant ($p < 0.05$). Average THg concentrations in spring and winter samples (62.8 and 85.5 ng L^{-1}, respectively) were 2.3–4.0 times higher than those sampled in summer and fall (21.2 and 27.1 ng L^{-1}, respectively). This finding was in contrast to the observations in North America which showed increased THg concentration during summer months (Landis et al. 2002). In Wujiang River Basin, most of the Hg in precipitation was found to be associated with particulate matters, which accounted for 67.6–96.1% of the total Hg in precipitation, indicating that particulate bound Hg was the dominant form of Hg in precipitation. Increasing THg concentrations in precipitation during winter could be attributed to increasing emissions of Hg to the atmosphere from

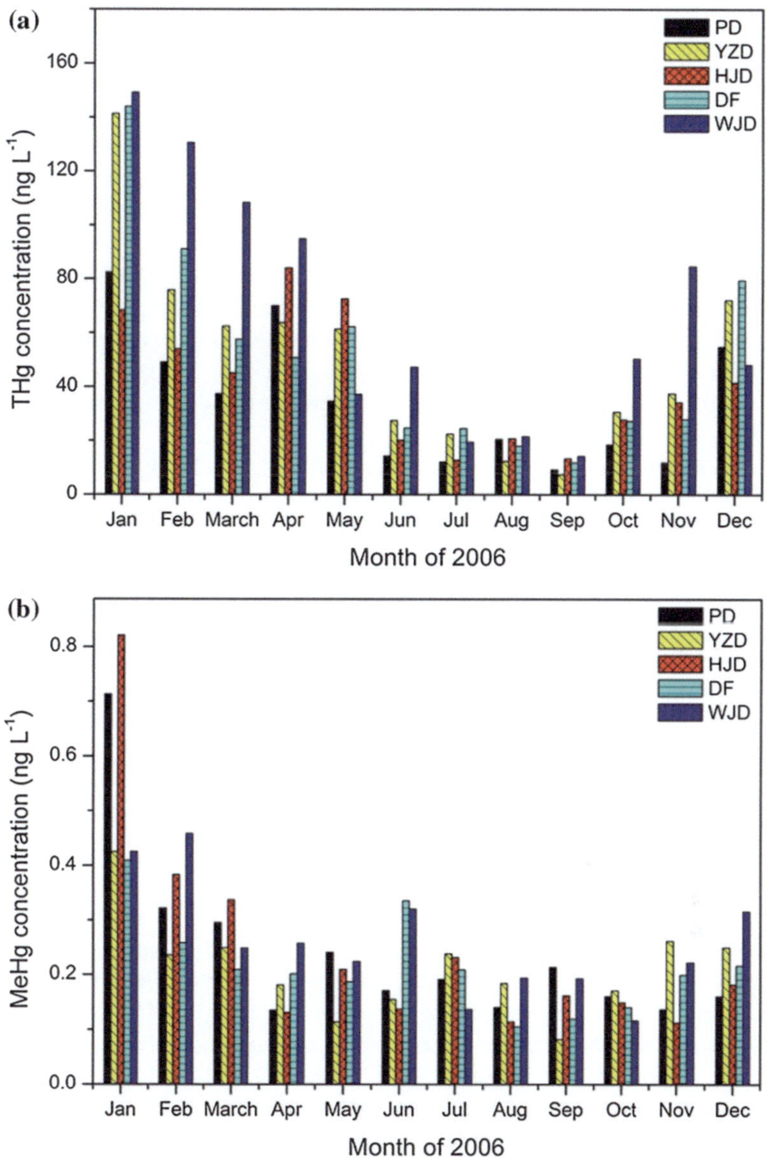

Fig. 3.2 **a** Total mercury (THg) and **b** methylmercury (MeHg) concentrations (ng L^{-1}) in precipitation at the five sites from January to December in 2006 (reprinted from Guo et al. (2008), with permission from Elsevier Ltd.)

coal combustion in domestic house heating activities (Feng et al. 2004; Fu et al. 2008, 2011), which could be incorporated into precipitation in cloud (i.e., rainout) and below cloud (i.e., washout). Additionally, there is generally a lack of precipitation events during winter, which would probably result in an accumulation of atmospheric Hg in ambient air due to the lack of washout processes and drive an increase of Hg in precipitation.

MeHg concentrations in precipitation samples at the five sites were in the range of 0.08–0.82 ng L^{-1} (VWM = 0.19 ng L^{-1}) (Fig. 3.2), representing 0.2–2.3% (mean = 0.7%) of the THg in precipitation. The ratio of MeHg to THg is comparable to those reported by previous studies. For example, Lamborg et al. (1995) and Mason et al. (2000) observed the MeHg/THg ratios in precipitation were 0.5–1.3% in northern Wisconsin and Chesapeake Bay. Studies in Europe observed the fraction of MeHg in THg in precipitation were in the range of 0.5–3.0% in Sweden (Munthe et al. 1995), and 0.25–1.48% in the Lake Balaton, Hungary (Nguyen et al. 2005). Similar to THg concentrations, the highest mean MeHg concentrations at the five sites were also observed in January (mean = 0.56 ng L^{-1}), which was significantly higher than the mean MeHg concentrations in the remaining months ($p < 0.01$).

There is a clear spatial distribution patterns in THg and MeHg in precipitation in Wujiang River. The highest VWM THg concentration (57.1 ng L^{-1}) was observed in WJD, followed by HJD (39.4 ng L^{-1}), DF (37.4 ng L^{-1}), YZD (35.7 ng L^{-1}) and PD (20.6 ng L^{-1}). The highest VWM MeHg was also observed in WJD (0.24 ng L^{-1}), followed by DF (0.20 ng L^{-1}), PD (0.18 ng L^{-1}), HJD (0.18 ng L^{-1}), and YZD (0.18 ng L^{-1}). The WJD site situated closely to several settlements which might be related to anthropogenic Hg emissions, which could be partly responsible for the elevated THg and MeHg concentrations at WJD.

3.3 Wet Deposition Fluxes of Total Mercury and Methylmercury

The monthly wet deposition fluxes of THg and MeHg as well as rain depth at the five sites in Wujiang River were shown in Fig. 3.3. Monthly wet deposition fluxes of THg and MeHg at each site varied with the rain depth. The THg and MeHg deposition fluxes were relatively higher in rainy season (from May to October) than in dry season (from November to April). The average monthly deposition fluxes during wet season were 3.6 μg m^{-2} mon^{-1} for THg and 23.3 ng m^{-2} mon^{-1} for MeHg, which were 1.64 and 3.1 times higher than that (THg flux: 2.2 μg m^{-2} mon^{-1}, MeHg flux: 7.5 ng m^{-2} mon^{-1}) during dry season, respectively. During wet season (from May to October), the averaged total wet deposition fluxes for THg and MeHg at the five sites were 21.5 μg m^{-2} and 0.14 ng m^{-2}, respectively, which accounted for 63% and 75% of the annual THg and MeHg wet depositions, respectively. Higher wet deposition fluxes during wet season could be explained by the intensive precipitation events. It is observed that the accumulative rain depth during wet season was 780 mm, contributing approximately 80% of the annual total rain depth in the study area. Correlation analysis was performed to evaluate the importance of rain depth and Hg concentration to wet deposition flux. The analysis

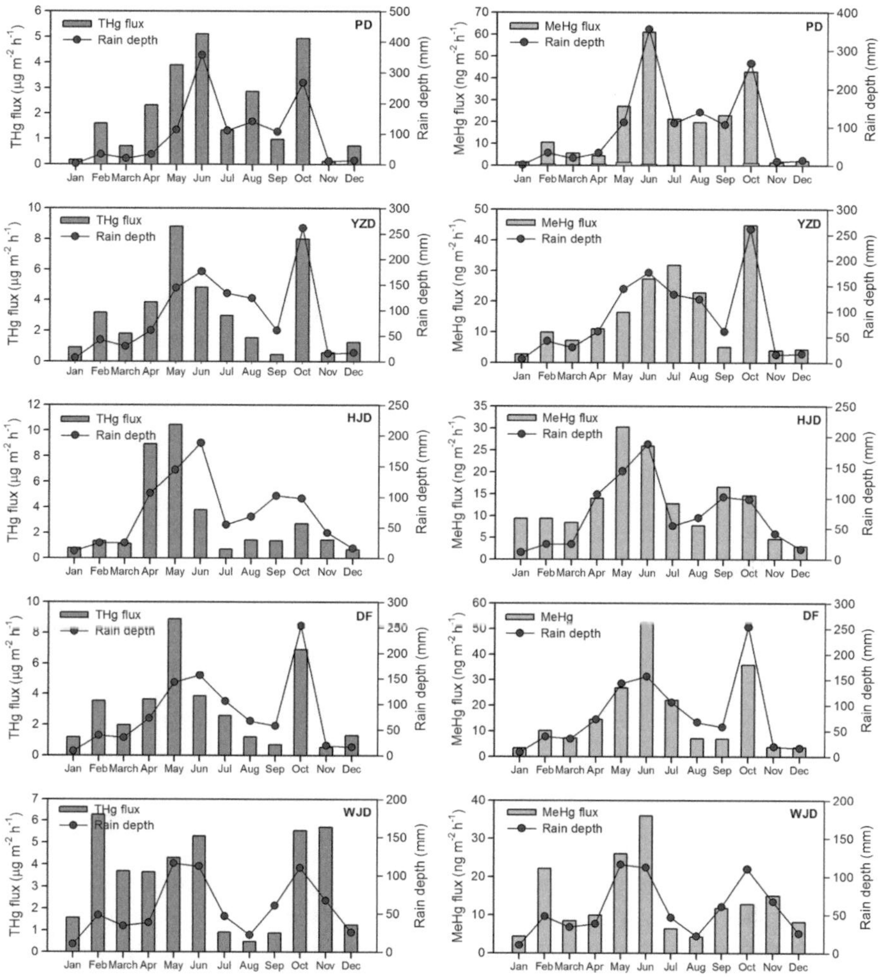

Fig. 3.3 Monthly variations of THg and MeHg fluxes as well as rain depth at the five sampling sites in Wujiang River (reprinted from Guo et al. (2008), with permission from Elsevier Ltd.)

revealed that the monthly rain depth was strongly correlated with monthly wet deposition fluxes of THg ($r^2 = 0.39$, $p < 0.01$, $n = 60$) and MeHg ($r^2 = 0.83$, $p < 0.01$, $n = 60$), while no significant correlations were observed between THg and MeHg concentrations and THg and MeHg wet deposition fluxes, respectively ($p > 0.05$, $n = 60$). Thus, about 40 and 80% of the seasonal variance of THg and MeHg wet depositions were caused by precipitation depth, respectively. The remaining variance should be due to local and regional sources of PHg and RGM and also meteorological conditions. We also note that MeHg deposition fluxes correlated stronger with the rain depth than THg, suggesting that MeHg in wet deposition tend to be less affected by local sources (an major source of atmospheric

Hg in the planetary boundary layer that could be incorporated into wet deposition via washout processes below cloud) than THg.

The annual wet deposition fluxes of THg and MeHg at the five sites were summarized in Table 3.1. The wet deposition fluxes in Wujiang River ranged from 24.8 to 39.6 μg m^{-2} $year^{-1}$ (mean = 34.7 ± 5.8 μg m^{-2} $year^{-1}$) for THg and from 0.16 to 0.22 μg m^{-2} $year^{-1}$ (mean = 0.18 ± 0.03 μg m^{-2} $year^{-1}$) for MeHg.

3.4 Comparison with Observations in China and in Other Regions Worldwide

A comparison of precipitation THg and MeHg concentrations and wet deposition fluxes between the Wujiang River and other areas worldwide is given in Table 3.2. THg and MeHg concentrations in precipitation of this study were much higher than those reported in the remaining rural areas of China, North America, and Europe, and well in the range of precipitation THg and MeHg concentrations in urban areas of China. The mean annual wet deposition flux of THg (34.7 μg m^{-2} $year^{-1}$) in Wujiang was higher than that (9.3 μg m^{-2} $year^{-1}$) at the MDN sites (Prestbo and Gay 2009). This is mainly because the VWM THg concentration in precipitation (36.7 ng L^{-1}) in Wujiang was much higher than that (9.5 ng L^{-1}) at the MDN sites even though the mean annual rain depth (963 mm) in Wujiang was similar to that (1014 mm) at the MDN sites (Prestbo and Gay 2009). The highly elevated precipitation THg concentrations in Wujiang River Basin resulted in a considerably greater wet deposition flux that was almost four times higher than the value in MDN sites. The elevated THg concentrations in precipitation in Wujiang River were partly due to the incorporation of dry deposition during sampling. Additionally, it should be noted the great local and regional anthropogenic Hg emissions also played an important role. Here, we proposed three possible sources. (1) Coal combustion in industrial activities: the average mercury content in coal from Guizhou (0.53 ppm) is substantially higher than the average value in China (0.19 ppm) and also in U.S. (0.17 ppm) (Feng et al. 2002). The highly elevated Hg

Table 3.1 Annual wet deposition of THg and MeHg and annual rain depths at five sites in Wujiang River in 2006

Site	Wet deposition flux (μg m^{-2} $year^{-1}$)		Rain depth (mm)
	THg	MeHg	
PD	24.8	0.22	1203
YZD	38.1	0.19	1068
HJD	34.7	0.17	881
DF	36.3	0.19	970
WJD	39.6	0.17	693
Mean ± 1SD	34.7 ± 5.8	0.18 ± 0.03	963 ± 193

Reprinted from Guo et al. (2008), with permission from Elsevier Ltd concentrations in raw coal and a great demand of coal consumption in industrial activities in Guizhou province resulted in significant Hg emissions to the atmosphere. It is reported that the annual total Hg emissions from industrial coal combustion was 4.54 tons (Zhang et al. 2015). The great emissions resulted in highly elevated atmospheric PBM and GOM concentrations in urban areas (Fu et al. 2011), which are important sources of precipitation Hg and contributed inevitably to the elevated wet deposition fluxes in Wujiang River. (2) Residential coal combustion induced Hg emissions: coal is a major energy source in rural area in Guizhou, and 16% of the total coal consumption in Guizhou province came from domestic coal combustion for cooking and house heating activities (Feng et al. 2004). The residential use of coal for house cooking and heating, which is lacking emission control devices, is expected to release a large amount of mercury to the atmosphere and consequently result in an increase of mercury concentrations in ambient air, especially in cold season (Feng et al. 2004; Fu et al. 2011). The Hg emissions from residential coal combustion are widely distributed in rural areas of China and could probably cause elevate wet depositions of Hg. (3) Hg emissions from artisanal Hg and Zn production: Hg emission from artisanal Hg production is an important sources of atmospheric Hg in Guizhou province. It releases approximately 1.97 ton annually and accounts for 14.8% of the total anthropogenic Hg emissions in Guizhou province.

3.5 Estimates of Wet Deposition of Mercury in Wujiang River

Inputs of atmospheric Hg to the water body of Wujiang River via wet deposition were calculated by multiplying the measured wet deposition flux at a single reservoir by the corresponding area of the reservoir. In this study, measurements of wet deposition were not conducted at the SFY reservoir, and the mean of wet deposition fluxes at the WJD and DF reservoir, which were located closely to the downstream and upstream of the SFY reservoir, was adopted to calculated the wet deposition of Hg to the SFY reservoir.

Wet deposition of THg and MeHg to the reservoirs in Wujiang River was shown in Table 3.3. The annual wet deposition to the five reservoirs ranged from 217 to 2790 g for THg and from 1.0 to 13.7 g for MeHg, which varied with both the wet deposition fluxes and reservoir areas. The total wet deposition of THg and MeHg in Wujiang River was 6.64 kg and 33.5 g per year, respectively. We caution that, due to the incorporation of dry deposition in wet deposition, our estimates would probably overestimate the wet deposition of Hg to the reservoirs. Unfortunately, we are not able to accurately quantify the contributions of dry deposition to the estimates of wet depositions.

Table 3.2 Comparison of precipitation Hg concentrations and wet deposition fluxes in this study and literatures

Site	Time	Classification	Precipitation amount (mm)	VWM concentration (ng L^{-1})		Wet deposition		(μg m^{-2} $year^{-1}$)
THg	MeHg	THg	MeHg					Reference
Wujiang River, China[a]	2006	Rural	963	36.7	0.19	34.7	0.18	This study
Guiyang, China	September 2012–August 2013	Urban	1057	11.9		12.6		(Fu et al. 2016)
Lhasa, China	2010	Urban	331	24.8		8.2		(Huang et al. 2013)
Chongqing, China	July 2010–June 2011	Urban	935	30.7	0.31	28.7	0.29	(Wang et al. 2012)
Nanjing, China	June 2011–February 2012	Urban	1068	52.9		56.5		(Zhu et al. 2014)
Xiamen, China	June 2012–May 2013	Urban	1138	12.3	0.05	14.0	0.06	(Xu et al. 2014)
Mt. Changbai, China	August 2011–August 2014	Rural	757	7.4		5.6		(Fu et al. 2010, 2016)
Mt. Damei, China	August 2012–August 2014	Rural	1621	3.7		6.0		
Mt. Leigong, China	May 2008–May 2009	Rural	1533	4.0	0.04	6.1	0.06	
Mt. Ailao, China	May 2011–May 2014	Rural	1931	3.7		7.2		
Mt. Waliguan, China	May 2012–August 2014	Rural	290	6.9		2.0		
Bayinbuluk, China	December 2013–December 2014	Rural	266	7.7		2.0		
Nam Co, China	July 2009–July 2011	Rural	375	4.8	0.04	1.8	0.01	(Huang et al. 2012)
Set station, China	May 2010–October 2012	Rural	975	4.0	0.11	3.9	0.11	(Huang et al. 2015)
Mt. Simian, China	February 2012–February 2013	Rural	1413	10.9	0.24	15.4	0.26	(Ma et al. 2016)
MDN, North America	1996–2005	Rural, Urban		9.5		9.3		(Prestbo and Gay 2009)
EMEP, Europe	–	Rural, Urban				6.8		(EMEP)

[a]Deposition pathway: bulk deposition, while the literatures are related to wet-only deposition

Table 3.3 Estimates of wet deposition input of THg and MeHg to the reservoirs in Wujiang River

Site	Wet deposition flux ($\mu g\ m^{-2}\ year^{-1}$)		Reservoir areas ($\times\ 10^6\ m^2$)	Wet Hg deposition (g)	
	THg	MeHg		THg	MeHg
PD	24.8	0.22	19.2	476	4.2
YZD	38.1	0.19	15.0	572	2.9
HJD	34.7	0.17	80.5	2790	13.7
DF	36.3	0.19	19.1	693	3.6
SFY	38.0	0.18	5.7	217	1.0
WJD	39.6	0.17	47.8	1890	8.1
Total				6640	33.5

References

EMEP. European Monitoring and Evaluation Programme (EMEP) Website. http://www.emep.int/

Feng XB, Sommar J, Lindqvist O, Hong YT (2002) Occurrence, emissions and deposition of mercury during coal combustion in the Province Guizhou, China. Water Air Soil Pollut 139:311–324

Feng XB, Shang LH, Wang SF, Tang SL, Zheng W (2004) Temporal variation of total gaseous mercury in the air of Guiyang, China. J Geophys Res Atmos 109, D03303, doi:10.1029/2003JD004159

Fu XW, Feng XB, Zhu WZ, Zheng W, Wang SF, Lu JY (2008) Total particulate and reactive gaseous mercury in ambient air on the eastern slope of the Mt. Gongga area, China. Appl Geochem 23:408–418

Fu XW, Feng X, Dong ZQ, Yin RS, Wang JX, Yang ZR et al (2010) Atmospheric gaseous elemental mercury (GEM) concentrations and mercury depositions at a high-altitude mountain peak in south China. Atmos Chem Phys 10:2425–2437

Fu XW, Feng XB, Qiu GL, Shang LH, Zhang H (2011) Speciated atmospheric mercury and its potential source in Guiyang, China. Atmos Environ 45:4205–4212

Fu XW, Zhang H, Yu B, Wang X, Lin CJ, Feng XB (2015) Observations of atmospheric mercury in China: a critical review. Atmos Chem Phys 15:9455–9476

Fu X, Yang X, Lang X, Zhou J, Zhang H, Yu B et al (2016) Atmospheric wet and litterfall mercury deposition at urban and rural sites in China. Atmos Chem Phys 16:11547–11562

Guo YN, Feng XB, Li ZG, He TR, Yan HY, Meng B et al (2008) Distribution and wet deposition fluxes of total and methyl mercury in Wujiang River Basin, Guizhou, China. Atmos Environ 42:7096–7103

Huang J, Kang SC, Zhang QG, Yan HY, Guo JM, Jenkins MG et al (2012) Wet deposition of mercury at a remote site in the Tibetan Plateau: Concentrations, speciation, and fluxes. Atmos Environ 62:540–550

Huang J, Kang SC, Wang SX, Wang L, Zhang QG, Guo JM et al (2013) Wet deposition of mercury at Lhasa, the capital city of Tibet. Sci Total Environ 447:123–132

Huang J, Kang SC, Zhang QG, Guo JM, Sillanpaa M, Wang YJ et al (2015) Characterizations of wet mercury deposition on a remote high-elevation site in the southeastern Tibetan Plateau. Environ Pollut 206:518–526

Lamborg CH, Fitzgerald WF, Vandal GM, Rolfhus KR (1995) Atmospheric mercury in northern Wisconsin—Sources and species. Water Air Soil Pollut 80:189–198

Landis MS, Keeler GJ (1997) Critical evaluation of a modified automatic wet-only precipitation collector for mercury and trace element determinations. Environ Sci Technol 31:2610–2615

Landis MS, Stevens RK, Schaedlich F, Prestbo EM (2002) Development and characterization of an annular denuder methodology for the measurement of divalent inorganic reactive gaseous mercury in ambient air. Environ Sci Technol 36:3000–3009

Ma M, Wang DY, Du HX, Sun T, Zhao Z, Wang YM et al (2016) Mercury dynamics and mass balance in a subtropical forest, southwestern China. Atmos Chem Phys 16:4529–4537

Mason RP, Lawson NM, Sheu GR (2000) Annual and seasonal trends in mercury deposition in Maryland. Atmos Environ 34:1691–1701

Munthe J, Hultberg H, Iverfeldt A (1995) Mechanisms of deposition of methylmercury and mercury to coniferous forests. Water Air Soil Pollut 80:363–371

Nguyen HL, Leermakers M, Kurunczi S, Bozo L, Baeyens W (2005) Mercury distribution and speciation in Lake Balaton, Hungary. Sci Total Environ 340:231–246

Oslo PC (1998) JAMP guidelines for the sampling and analysis of mercury in air and precipitation. Joint Assessment and Monitoring Programme. pp 1–20

Prestbo E, Gay DA (2009) Wet deposition of mercury in the U.S. and Canada, 1996–2005: results and analysis of the NADP mercury deposition network (MDN). Atmos Environ 43:4223–4233

USEPA (1998) Method 1630: methyl mercury in water by distillation, aqueous ethylation, purge and trap, and cold vapor atomic fluorescence spectrometry, USA, pp 1–55

USEPA (2002) Method 1631, Revision E: mercury in water by oxidation, purge and trap, and cold vapor atomic fluorescence spectrometry. United States Environmental Protection Agency, USA, pp 10–46

Wang YM, Wang DY, Meng B, Peng YL, Zhao L, Zhu JS (2012) Spatial and temporal distributions of total and methyl mercury in precipitation in core urban areas, Chongqing, China. Atmos Chem Phys 12:9417–9426

Xu LL, Chen JS, Yang LM, Yin LQ, Yu JS, Qiu TX et al (2014) Characteristics of total and methyl mercury in wet deposition in a coastal city, Xiamen, China: concentrations, fluxes and influencing factors on Hg distribution in precipitation. Atmos Environ 99:10–16

Zhang L, Wang SX, Wang L, Wu Y, Duan L, Wu QR et al (2015) Updated emission inventories for speciated atmospheric mercury from anthropogenic sources in China. Environ Sci Technol 49:3185–3194

Zhu J, Wang T, Talbot R, Mao H, Yang X, Fu C et al (2014) Characteristics of atmospheric mercury deposition and size-fractionated particulate mercury in urban Nanjing, China. Atmos Chem Phys 14:2233–2244

Chapter 4
Water/Air Mercury Flux in Reservoirs

Abstract Dynamics of mercury (Hg) in aquatic ecosystem is of particular concern because Hg delivered to watersheds can be converted to methylmercury (MeHg), a highly toxic, persistent, and bioaccumulative species that poses a serious threat to human health and wildlife by consuming fish. Emission of Hg from water is an important source of global atmospheric Hg budget and would also reduce the Hg burden in the water bodies which in turn decreases the MeHg production and bioaccumulation in aquatic ecosystem. Therefore, exchange of Hg between water and atmosphere plays a crucial role in cycling of Hg in aquatic ecosystem and the atmosphere. In this chapter, we (1) measured water/air exchange flux at eight reservoirs in Wujiang River Basin; (2) studied the diurnal and seasonal variations in water/air Hg flux and related factors; and (3) developed water/air Hg fluxes empirical models and estimated Hg emissions from Wujiang River.

Keywords Water/air Hg flux · Total gaseous mercury (TGM) · Factor · Emission

4.1 Sampling Sites and Sampling Techniques

4.1.1 Studied Reservoirs and Sampling Sites

The field experiments were carried out at eight reservoirs in the Wujiang River, which is one of the largest branches of the upper Yangtze River (Fig. 4.1). Some relevant information of the studied reservoirs are shown in Table 4.1. The eight reservoirs are all situated on the Yunnan-Guizhou Plateau with the altitude of 700–1300 m above sea level. The study area belongs to a subtropical moist and warm climate region, with distinct rainy (April to October) and dry (November to March) seasons. The annual average air temperatures for the eight reservoirs are in the range of 12–16 °C (Table 4.1).

Most of the studied reservoirs were surrounding by natural preserved forests. For the eight reservoirs, two reservoirs, named HF and BH, were frequently impacted by discharges of Hg enriched wastewaters from surrounding urbanized and

X. Feng et al., *Biogeochemical Cycle of Mercury in Reservoir Systems in Wujiang River Basin, Southwest China*, https://doi.org/10.1007/978-981-10-6719-8_4

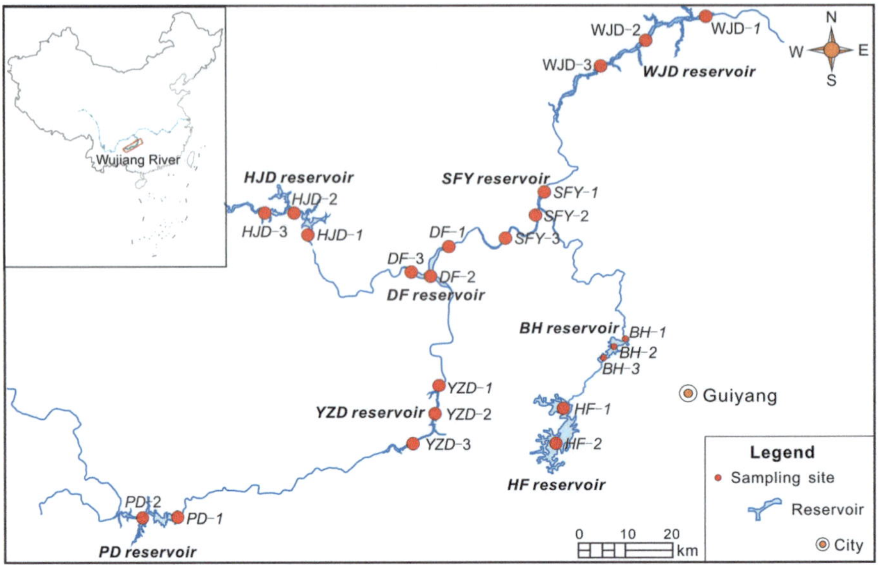

Fig. 4.1 The study area and sampling sites of water/air mercury flux in Wujiang River

industrialized regions as well as wet and dry deposition of atmospheric Hg emitted by industrial sources in Guiyang City, southwestern China. The six remaining reservoirs are isolated from industrial sources and populated regions. As shown in Table 4.1, the total Hg (THg) concentrations in water in HF and BH reservoirs were in the range of 6.9–22.8 ng L^{-1}, which are 3.0–10.8 times higher than that in the six remaining reservoirs (mean THg: 2.1–2.3 ng L^{-1}).

At each reservoir, two to three sampling sites were selected to study the spatial patterns of water/air exchange flux of Hg, which were located close to the dam, at the middle, and at the upper reach of the reservoirs, respectively (Fig. 4.1). At each of the sampling site in reservoirs, measurements of water/air exchange flux of Hg were conducted in both warm season (July to October) and cold season (November, December, and January). All the sampling sites were shown in Fig. 4.1.

4.1.2 Measurements of Water/Air Exchange Flux of Mercury

Exchange flux of Hg between water and atmosphere was measured by using a dynamic flux chamber method (Fig. 4.2) (Carpi and Lindberg 1998; Poissant and Casimir 1998; Xiao et al. 1991). A semicylinder, open-bottom quartz glass chamber (Ø20 × 60 cm, $V = 0.009\ m^3$) was placed on a bottom-open floating boat at the water/air interface. The quartz glass chamber is characterized by low blanks and

Table 4.1 Basic parameters of the studied reservoirs

Reservoirs	HF	BH	WJD	SFY	DF	YZD	PD	HJD
Operation time	1960	1966	1979	2005	1994	2003	1993	2004
Immersed area (m^2)	57.2×10^6	14.5×10^6	47.8×10^6	5.7×10^6	19.1×10^6	15.0×10^6	19.2×10^6	80.5×10^6
Storage capacity (m^3)	6.0×10^8	1.9×10^8	13.5×10^8	0.7×10^8	4.9×10^8	3.2×10^8	2.5×10^8	33.6×10^8
Max water depth (m)	45	45	120	90	110	100	45	110
Water residence time (day)	118	72	30	7	33	37	27	380
Altitude (m, a.s.l.)	1240	1230	740	837	970	1088	1145	1130
Precipitation depth (mm)	1200	1200	693	976	970	1068	1203	881
Mean air temperature (°C)	14.1	13.8	15.3	14.0	14.0	14.1	14.1	12.8
Water Hg conc. (ng L^{-1})	6.9	22.8	2.7	2.2	2.3	2.1	2.3	2.1
PH	8.4	8.2	7.8	7.9	8.0	8.0	8.1	8.1

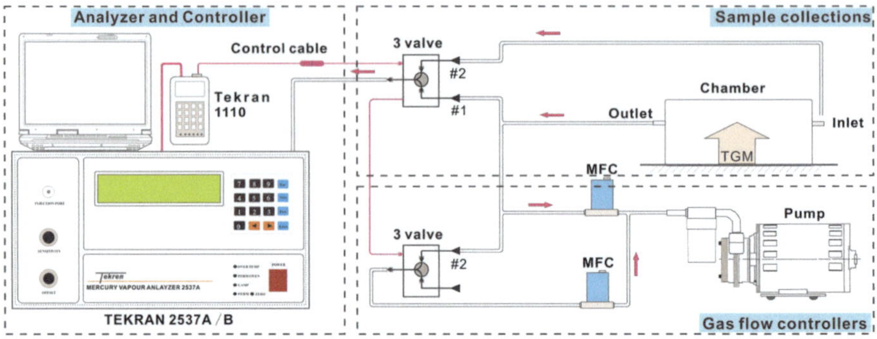

Fig. 4.2 Setup of the dynamic flux chamber for measuring mercury flux over water surface

high transparency, and could be precleaned easily and used repeatedly. The chamber is completely exposed to the ambient conditions and the inlet holes of the chamber were exposed to ambient wind. In order to prevent the wind and water movement factors causing rapid pressure fluctuations inside the chamber, three inlet holes with 8 mm diameter are applied to the chamber.

Water/air Hg flux was obtained via measuring the difference in atmospheric Hg concentrations inside and outside the flux chamber and exchange flux of air mass in the chamber. Water/air exchange flux of Hg was calculated using Eq. (4.1):

$$F = (C_o - C_i) \times Q/A, \tag{4.1}$$

where F is Hg flux in ng m^{-2} h^{-1}; C_o and C_i are total gaseous mercury (TGM) concentrations of the outlet and inlet air stream (ng m^{-3}), respectively; A is the enclosed water area (0.12 m^2); and Q is the flushing flow rate through the chamber (8–15 L min^{-1}).

TGM concentrations of the inlet and outlet air were measured using an automated Hg vapor analyzer (Tekran 2537A/B). This analyzer has been used extensively for atmospheric TGM measurements worldwide (Ebinghaus et al. 1999; Fu et al. 2012; Munthe et al. 2001). It combines the pre-concentration of TGM onto gold traps, thermal desorption, and cold vapor atomic fluorescence spectrometry detection of GEM. The analyzer has two gold cartridges working in parallel. While one cartridge is collecting TGM, the other one is performing analysis of the collected TGM. The function of the cartridges is then reversed, allowing continuous sampling of ambient air. Air particulate matters were removed by using two 45-mm diameter Teflon filters (pore size 0.2 μm), which were installed at the inlets of the sampling Teflon tube and analyzer, respectively. The analyzer was programmed to measure atmospheric TGM at the time resolution of 5 min with a volumetric sampling flow rate of ~1.0 L min^{-1}. The data quality of the analyzer was controlled by periodic (every 25 h) automatic permeation source injections. Alternate measurements of air TGM concentrations from the inlet to the outlet of the chamber every 10 min were achieved by using a magnetic three-way valve (Tekran 1110).

It was suggested that flushing flow rate and associated chamber turnover time (TOT) have a crucial influence on Hg flux from soil (Eckley et al. 2010). To the best of our knowledge, studies on the role of flushing flow rate in Hg flux from waters are very limited. The study on soil/air exchange flux of Hg by Eckley et al. (2010) suggested that a TOT of 0.84 min could be adopted for substrates with low Hg concentrations. The chamber TOTs in the present study were in the range of 0.7–1.2 min, which are consistent with that proposed for low Hg substrates by Eckley et al. (2010). Blanks of the chamber were measured with an ultra-clean quartz glass plate and in the range of −0.1 to 0.3 ng m^{-2} h^{-1} (mean = 0.1 ± 0.2 ng m^{-2} h^{-1}). The detection limit (based on three times the standard deviation of blanks) of the DFC system was estimated to be 0.5 ng m^{-2} h^{-1}. The blanks were small and we did not make blank correction for measured fluxes.

4.1.3 Meteorological and Water Quality Parameters

Meteorological parameters were continuously monitored at the sampling sites using a portable weather station (Puhui, Wuhan, China). Air temperature (accuracy ± 0.1 °C), air relative humidity (accuracy ± 1%), solar radiation (accuracy ± 5%), wind speed (accuracy ± 0.3 m s^{-1}), and wind direction (accuracy ± 1%) were measured at 2 m height above the water surface on the reservoir bank. Water temperature was measured near the sampling site using a temperature probe (Puhui, Wuhan, China, accuracy ± 0.1 °C). 100 mL unfiltered water samples were collected from surface layer of reservoirs and preserved in precleaned Teflon bottles with addition of trace metal grade HCl (to 5‰ of total sample volume). Teflon bottles with samples were individually sealed into three successive polyethylene bags and rapidly brought to the laboratory and stored in a refrigerator until analysis. THg concentrations in surface waters were determined in the laboratory using the standard method of BrCl oxidation followed by $SnCl_2$ reduction, purge and trap, and dual amalgamation combined with CVAFS detection (USEPA 2002).

4.2 Atmospheric Total Gaseous Mercury Concentrations Over Water of Reservoirs

Averaged atmospheric TGM concentrations over waters of reservoirs in Wujiang River are shown in Table 4.2 and Fig. 4.3. The averaged atmospheric TGM concentrations in Wujiang River ranged from 2.71 to 15.1 ng m^{-3}, with an overall mean of 5.81 ± 3.25. This value is relatively lower than the annual mean TGM concentration (8.40 ± 4.87 ng m^{-3}) in Guiyang city, which is one the largest city in southwestern China and impacted by strong local anthropogenic emission

Table 4.2 Water/air Hg flux, atmospheric total gaseous mercury (TGM) concentrations, water total mercury (THg) concentrations, and meteorological parameters at the filed sampling sites in Wujiang River

Reservoir	Site	Time	Hg flux (ng m^{-2} h^{-1})	TGM (ng m^{-3})	THg (ng L^{-1})	Air Temperature (°C)	Solar radiation (w m^{-2})	Relative humidity (%)
HF	HF-1	January 2003	6.5 ± 8.5	5.92 ± 2.92		15.3	78.3	91.0
	HF-1	March 2003	5.1 ± 8.4	5.48 ± 1.57		6.1	68	80.0
	HF-1	March 2004	4.8 ± 9.0	9.76 ± 3.23	6.1	13.1	162	69.2
	HF-1	November 2004	2.0 ± 3.9	8.10 ± 4.10	5.3	11.6		
	HF-2	March 2003	1.8 ± 4.4	8.34 ± 3.13		5.2	16	95.1
	HF-2	March 2004	4.0 ± 5.2	9.09 ± 1.94	9.0	8.1	66	79.6
	HF-2	November 2004	2.8 ± 3.4	8.10 ± 4.10	4.9	12.3	25	97.1
BH	BH-1	April 2003	7.2 ± 16.0	9.96 ± 7.90	29.5	10.1	60	
	BH-2	April 2003	10.1 ± 10.0	4.40 ± 2.30	42.6	21.7	192	
	BH-2	November 2001	3.0 ± 2.8	7.47 ± 2.46	33.2	13.1	18.4	
	BH-2	May 2002	8.1 + 5.4	4.60 ± 1.10	12.1	15.4	103	
	BH-3	April 2003	9.9 + 10.0	8.09 ± 2.41	37.8	12.2	70	
WJD	WJD-1	May 2006	19.7 ± 11.9	6.79 ± 1.80	2.68	22.4		76.1
	WJD-2	May 2006	15.6 ± 6.1	6.98 ± 2.51	2.13	19.4		82.1
	WJD-3	May 2006	10.1 ± 4.2	3.75 ± 0.64	2.04	14.0		82.3
	WJD-1	February 2007	4.7 ± 9.7	11.5 ± 6.85	1.82	10.7	72	82.8
	WJD-2	February 2007	4.7 ± 7.1	9.80 ± 2.76	1.25	12	103	75.1
	WJD-3	February 2007	4.3 ± 5.7	9.20 ± 2.21	1.28	15.1	115	72.9

(continued)

Table 4.2 (continued)

Reservoir	Site	Time	Hg flux ($ng\ m^{-2}\ h^{-1}$)	TGM ($ng\ m^{-3}$)	THg ($ng\ L^{-1}$)	Air Temperature (°C)	Solar radiation ($w\ m^{-2}$)	Relative humidity (%)
SFY	SFY-1	October 2006	3.81 ± 5.5	15.1 ± 3.78	1.4	21.8	47	86
	SFY-2	October 2006	2.3 ± 3.1	5.57 ± 3.30	1.4	24.2	133	86
	SFY-3	October 2006	4.3 ± 5.3	6.71 ± 1.99	2.1	23.5	114	86
	SFY-1	January 2007	1.3 ± 2.2	6.08 ± 1.41	1.1	5.8	6	73.4
	SFY-2	January 2007	1.2 ± 2.7	8.84 ± 1.13	0.9	9.7	61	63.3
	SFY-3	January 2007	1.3 ± 2.4	9.32 ± 1.52	1.2	9.2	15	76.1
DF	DF-1	July 2006	3.6 ± 4.6	3.08 ± 0.95	2.2	25.9	134	76.8
	DF-2	July 2006	4.3 ± 4.6	3.66 ± 0.68	2.5	25.0	125	81.9
	DF-3	July 2006	3.3 ± 2.2	4.09 ± 0.87	2.2	24.2	118	82.5
	DF-1	December 2007	0.7 ± 1.2	6.94 ± 1.89	1.5	8.9	22	71.7
	DF-3	December 2007	0.9 ± 1.0	4.69 ± 1.60	1.6	8.3	26	48.5
YZD	YZD-1	October 2006	4.0 ± 6.1	3.73 ± 1.46	2.9	23.7	163	83.6
	YZD-2	October 2006	3.7 ± 4.7	4.06 ± 1.70	2.9	22.0	84	86.7
	YZD-3	October 2006	3.0 ± 1.6	5.02 ± 1.20	3.2	17.4	26	94.2
	YZD-1	December 2007	0.1 ± 0.9	14.3 ± 4.06	0.6	9.0	9	79.6
	YZD-2	December 2007	0.4 ± 0.7	9.59 ± 1.89	0.9	9.2	8	78.8
	YZD-3	December 2007	1.0 ± 1.1	4.84 ± 0.99	1.1	16.2	157	53.3

(continued)

Table 4.2 (continued)

Reservoir	Site	Time	Hg flux (ng m^{-2} h^{-1})	TGM (ng m^{-3})	THg (ng L^{-1})	Air Temperature (°C)	Solar radiation (w m^{-2})	Relative humidity (%)
PD	PD-1	September 2008	2.3 ± 1.5	3.40 ± 1.10	1.8	22.1	60	95.1
	PD-2	September 2008	4.8 ± 5.9	3.43 ± 1.38	2.1	25.2	90	60.1
	PD-1	December 2008	0.2 ± 0.62	4.78 ± 1.54	1.1	14.0	132	69.2
	PD-2	December 2008	0.1 ± 0.8	3.63 ± 0.88	1.3	3.6	10	72.1
HJD	HJD-1	September 2008	4.4 ± 3.9	4.32 ± 2.20	2.0	27.1	209	65.9
	HJD-3	September 2008	4.9 ± 3.3	2.71 ± 1.60	2.3	27.2	251	69.5
	HJD-1	November 2007	3.8 ± 5.0	6.58 ± 1.11	1.7	12.8	124	66.4
	HJD-2	November 2007	2.4 ± 3.3	6.79 ± 2.39	1.8	13.4	76	60.0
	HJD-3	November 2007	2.1 ± 2.6	7.86 ± 2.65	1.4	11.9	115	67.9

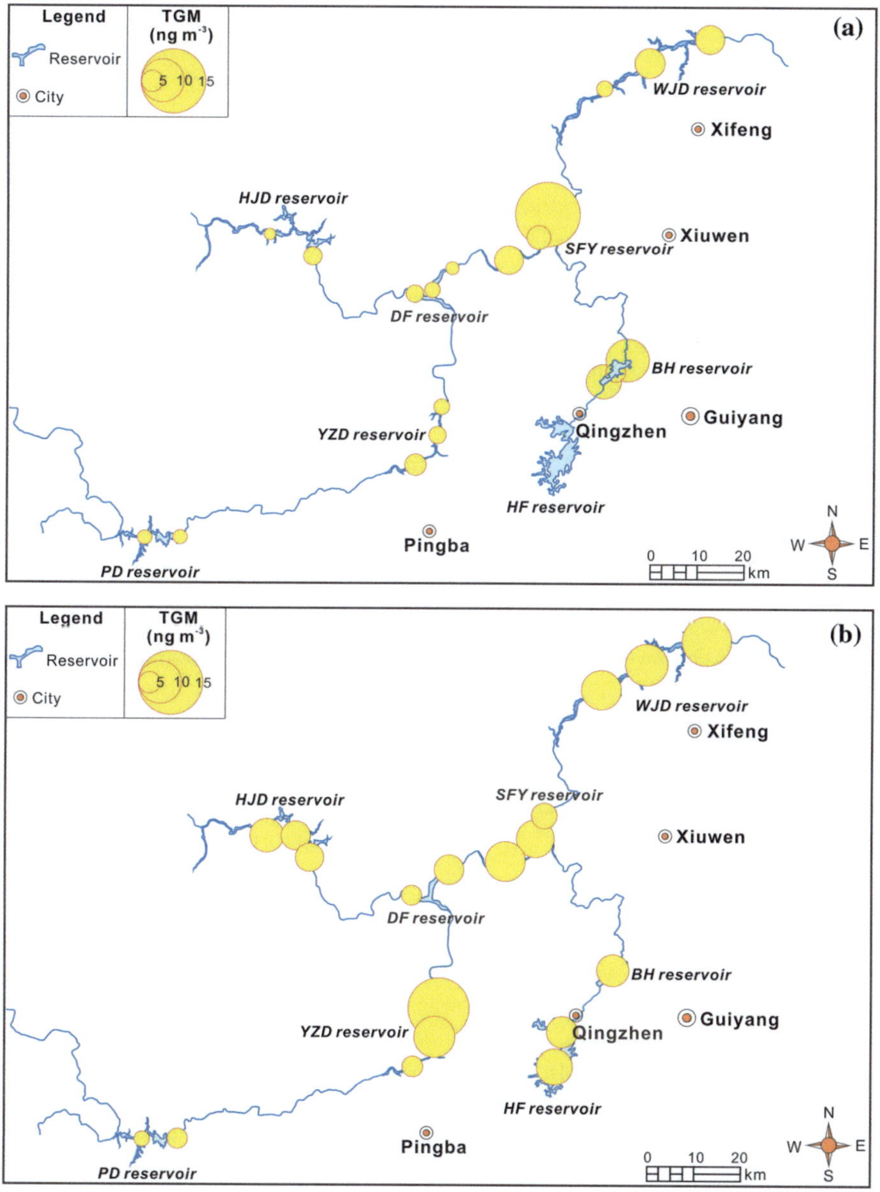

Fig. 4.3 Atmospheric TGM concentrations over waters in **a** warm season (April to October) and **b** cold season (November to March) in Wujiang River

sources (Feng et al. 2004a; Fu et al. 2011). On the other hand, atmospheric TGM concentrations in Wujiang River were significantly elevated compared to the values (1.32–1.93 ng m^{-3}) observed in remote areas North America and Europe

(Kellerhals et al. 2003; Lan et al. 2012; Munthe et al. 2003; Sprovieri et al. 2010). Guizhou is one of the largest anthropogenic sources of atmospheric Hg in China. Due to the great Hg emissions from coal combustion in coal-fired power plants, industrial and residential activates as well as non-ferrous metal smelting, artisanal mercury production, cement production, the total anthropogenic Hg emission in Guizhou in 2003 reached 55.5 tons, which was the largest anthropogenic source region over China (Wu et al. 2006). Additionally, natural emissions of Hg from soil and water bodies were also found to be higher in Guizhou province, due to the fact that the study area is located in the Circum-Pacific Global Mercuriferous Belt (Feng et al. 2005; Wang et al. 2007a, b, 2005). Hence, elevated atmospheric TGM concentrations in Wujiang River would be probably a result of significant anthropogenic and natural emission in Guizhou province.

A clear spatial variation in atmospheric TGM concentrations was observed in Wujiang River (Fig. 4.3). Averaged TGM concentrations in cold and warm season showed the highest (8.60 ± 3.52 ng m^{-3}) in SFY reservoir, followed by the WJD reservoir (8.00 ± 2.74 ng m^{-3}), HF reservoir (7.83 ± 1.58 ng m^{-3}), YZD reservoir (6.92 ± 4.19 ng m^{-3}), BH reservoir (6.90 ± 2.38 ng m^{-3}), HJD reservoir (5.65 ± 2.09 ng m^{-3}), DF reservoir (4.49 ± 1.49 ng m^{-3}), and PD reservoir (3.81 ± 0.65 ng m^{-3}). No significant correlation was observed between atmospheric TGM concentrations and water/air Hg fluxes in Wujiang river ($r^2 = 0.01$, p value > 0.05), suggesting that emission of Hg from water in Wujiang River was not responsible for the spatial variation in atmospheric TGM concentrations in Wujiang River. The SFY, WJD, HF, and YZD reservoirs were located close to industrial and urbanized centers in Guizhou province, such as Guiyang, Qingzhen, Xifeng, Xiuwen, and Pingba (Fig. 4.3), the anthropogenic Hg emissions in which would probably contribute to the elevated TGM concentrations in these reservoirs. On the other hand, the DF, PD, and HJD reservoirs were located relatively further from these anthropogenic source regions.

Within each reservoir, TGM concentrations in the atmosphere, in most case, showed the highest value near the dam, where some residential centers and industrial point sources were generally located (Fig. 4.3). TGM concentrations in the atmosphere also showed a clear season pattern with higher concentrations during cold season and lower concentrations during warm season with the exception of BH reservoir (Fig. 4.4). This pattern is generally consistent with observations in rural and urban areas in mainland China (Feng et al. 2004a; Fu et al. 2010a, 2008), but in contrast with the seasonal variations in water/air Hg flux (see Sect. 4.4). This could be partly due to the enhanced coal and biomass burning in regional domestic heating activities during cold seasons (Feng et al. 2004a; Fu et al. 2008).

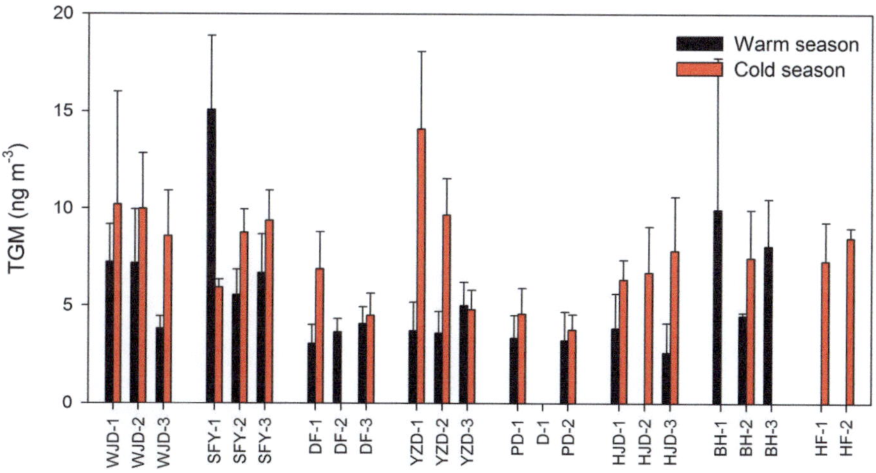

Fig. 4.4 Comparison of average atmospheric TGM concentrations at sampling sites in warm season (April to October) and cold season (November to March) in Wujiang River

4.3 Overall Characteristics of Water/Air Mercury Flux

A statistical summary of water/air exchange flux of Hg at the eight reservoirs in Wujiang River is shown in Table 4.2. The average water/air Hg flux at the eight reservoirs in Wujiang River ranged from 2.3 to 19.7 ng m^{-1} h^{-1} in warm season, and ranged from 0.1 to 6.5 ng m^{-1} h^{-1} in cold season.

The water/air Hg fluxes in Wujiang River was significantly higher than that measured from freshwater bodies in the North America, but comparable to that measured from river in Sweden (Table 4.3). Due to the discharge of wastewaters from surrounding urbanized and industrialized regions, water THg concentrations in the BH and HF reservoirs during the study period ranged from 5.3 to 42.6 ng L^{-1}, which were significantly higher than the values in the North America and Sweden. Apart from the BH and HF reservoirs, water THg concentrations in other six reservoirs in Wujiang River ranged from 0.6 to 3.2 ng L^{-1}, which were comparable to the values obtained in the North America and Sweden (Tables 4.2 and 4.3).

A clear variation in water/air Hg flux was observed among the eight studied reservoirs in Wujiang River. Figure 4.5 shows the mean water/air Hg flux at the eight reservoirs during warm and cold seasons. In this study, the HF, BH, and WJD reservoirs are defined as polluted reservoirs, while the five remaining reservoirs are defined as non-polluted reservoirs. The HF and BH reservoirs were impacted by discharge of wastewaters from surrounding urbanized and industrialized regions and had elevated water THg concentrations (5.3–42.6 ng L^{-1}). Due the prevalent cage aquaculture activities, the WJD reservoir was characterized by highly elevated contents of TN, TP and chlorophyll II and regarded as a hyper-eutrophic reservoir

Table 4.3 Comparison of water/air Hg flux between this study and previous studies worldwide

Sites	Time	Hg flux (ng m^{-2} h^{-1})	THg concentration (ng L^{-1})	Reference
Wujiang River	Warm season	2.3–19.7	1.4–42.6	This study
	Cold season	0.05–6.5	0.6–33.2	
St. Lawrence River, Canada	July 1995	−0.02	0.52	Poissant and Casimir (1998)
Kejimkujik National Park, Canada	July and August 1997	2.3	2.6	Boudala et al. (2000)
Lake Ontario, Canada	July 1998	0.8	0.08	Schroeder et al. (2005)
Quebec, Canda	August 1998	0.3	0.31	
Nova Scotia, Canada	Summer 1997, 1999, and 2000	0.6–2.2	0.7–2.8	
St. Francois, Canada	June 2003	0.67	1.2–2.0	Zhang et al. (2006)
Borrsjön Lake, Sweden	August 1999	11.1	2.36	Gårdfeldt et al. (2001)

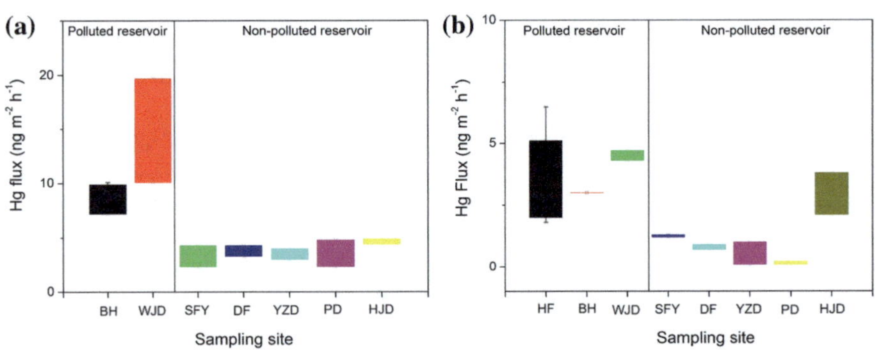

Fig. 4.5 Water/air Hg fluxes in polluted reservoirs and non-polluted reservoirs in **a** warm season and **b** cold season

in the study areas (Fu et al. 2010b). As shown in Fig. 4.5, water/air Hg fluxes at the polluted reservoirs were significantly elevated compared to non-polluted reservoirs in both warm and cold seasons. Higher water/air Hg fluxes in the HF and BH reservoirs were possibly related to elevated water THg concentrations, which could be transformed to dissolve gaseous mercury (DGM) via photochemical and microbial reduction and then preferentially emitted to the atmosphere. The hyper-eutrophic WJD reservoir had relatively higher DOC and Fe(III) contents and stronger microbial activities (Fu et al. 2010b), which would contribute to the increased water/air Hg flux at the WJD reservoir (Garcia et al. 2005; Peters et al. 2007; Poulain et al. 2007; Wollenberg and Peters 2009). Mean water/air Hg fluxes were comparable at the SFY, DF, YZD, PD, and HJD reservoirs in warm season (Fig. 4.5a), but the mean water/air Hg flux at HJD reservoir in cold season was relatively higher than the DF, YZD, PD, and HJD reservoirs (Fig. 4.5b), which could be due to the higher air temperature (mean = 11.9–13.4 °C) and solar radiation (mean = 76–124 w m^{-2}) (Table 4.2).

Spatial variations in water/air Hg flux within each reservoir were different. The water/air Hg fluxes at WJD reservoir showed the highest values near the dam in both warm and cold season, whereas no consistent spatial variations in water/air Hg fluxes were observed in the remaining reservoirs (Table 4.2). This suggests water/air Hg fluxes were affected by a combination of many water chemical and meteorological parameters.

4.4 Diurnal and Seasonal Patterns of Water/Air Mercury Flux

The time series of water/air Hg fluxes and solar radiation in the eight reservoirs in both warm and cold seasons were shown in Figs. 4.6–4.21. The water/air Hg fluxes at all the sampling sites in Wujiang River showed clear diurnal trend especially during sunny days. In general, water/air Hg fluxes displayed the minimum values in the early morning (from 4:00 to 6:00), then the values increased throughout the morning and forenoon and reached the highest around noon (from 12:00 to 14:00), after that the fluxes decreased consistently from afternoon to the early morning. This pattern was strongly correlated with the variations of solar radiation and air

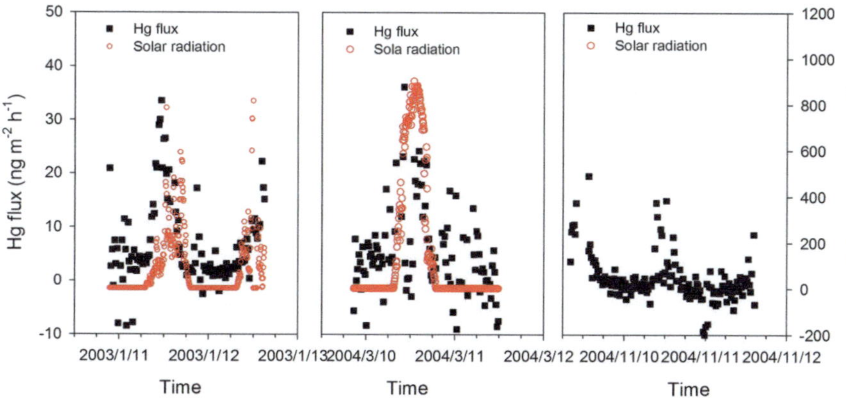

Fig. 4.6 Water/air Hg flux and solar radiation at HF-1 in Hongfeng (HF) reservoir (reprinted from Feng et al. (2008), with permission from John Wiley & Sons, Inc.)

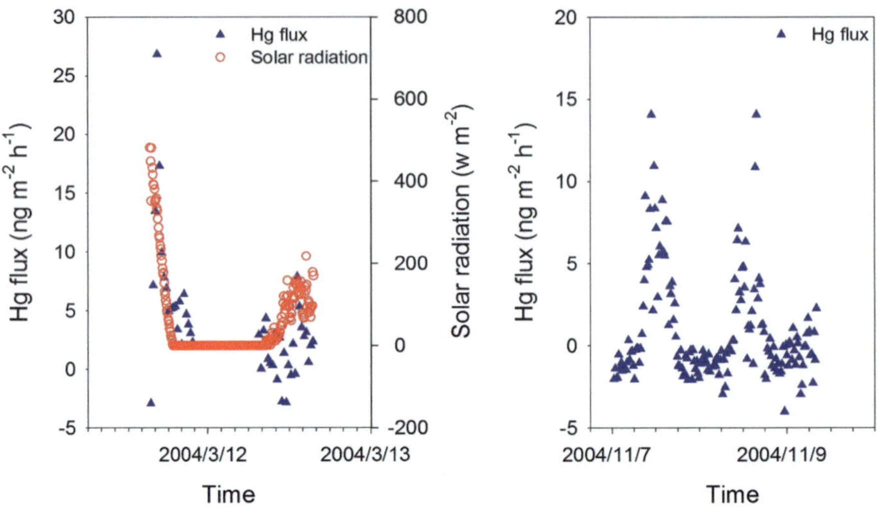

Fig. 4.7 Water/air Hg flux and solar radiation at HF-2 in Hongfeng (HF) reservoir (reprinted from Feng et al. (2008), with permission from John Wiley & Sons, Inc.)

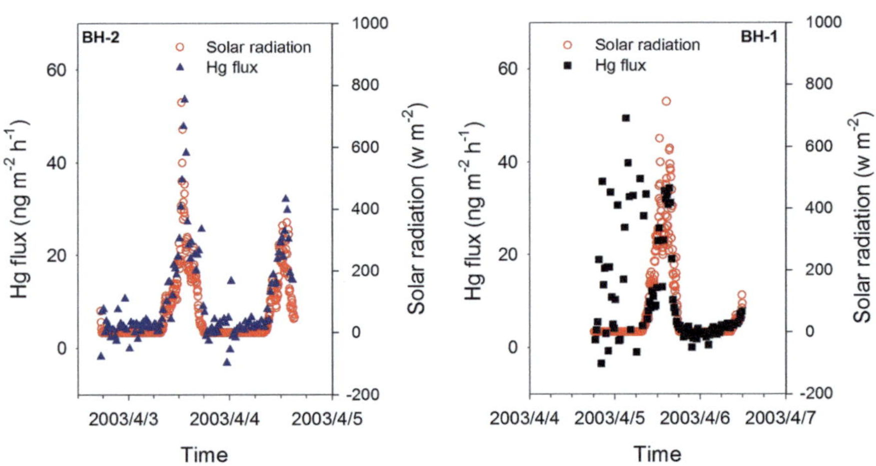

Fig. 4.8 Water/air Hg flux and solar radiation at BH-1 and BH-2 in Baihua (BH) reservoir (reprinted from Feng et al. (2004b), with permission from Elsevier Ltd.)

temperature at all the sampling sites, suggesting that solar radiation and air temperature might play an important role in water/air Hg fluxes.

Water/air Hg fluxes also displayed clear seasonal distribution pattern in Wujiang River (Fig. 4.22). The water/air Hg fluxes in warm season were found to be elevated compared to that in cold season in all the reservoirs in Wujiang River. The difference of water/air Hg fluxes between warm and cold season showed the largest

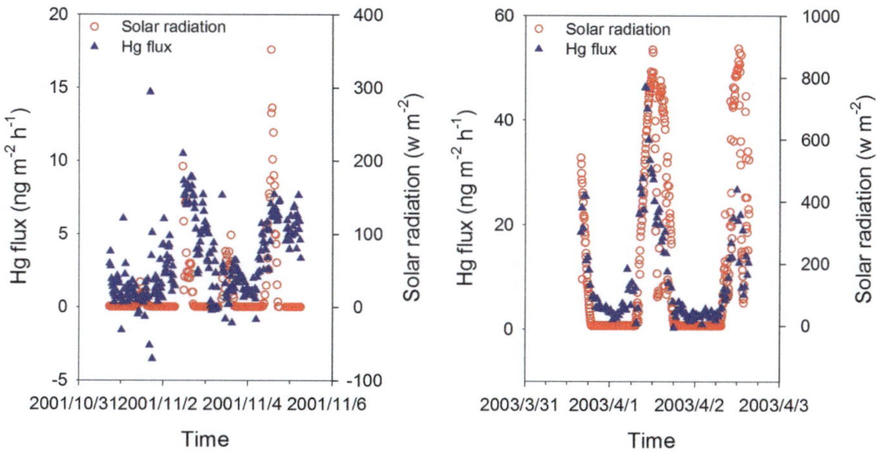

Fig. 4.9 Water/air Hg flux and solar radiation at BH-2 in Baihua (BH) reservoir (reprinted from Feng et al. (2004b), with permission from Elsevier Ltd.)

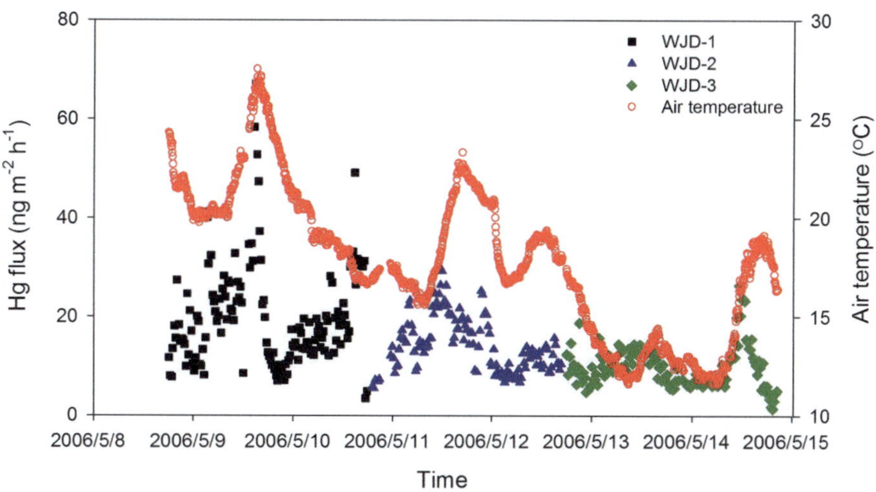

Fig. 4.10 Water/air Hg flux and air temperature in Wujiangdu (WJD) reservoir in warm season (reprinted from Fu et al. (2010b), with permission from Elsevier Ltd.)

in PD reservoir, with the mean water/air Hg flux in warm season 24 times higher than that in cold season. The elevated factor of water/air Hg fluxes between warm and cold season showed the lowest in HJD reservoir, with the mean water/air Hg fluxes in warm season 1.7 times higher than that in cold season. The seasonal distribution pattern in water/air Hg fluxes was probably related to seasonal variations in meteorological parameters. As shown in Table 4.2, solar radiation and air temperature at all the sampling sites in Wujiang River in warm season were

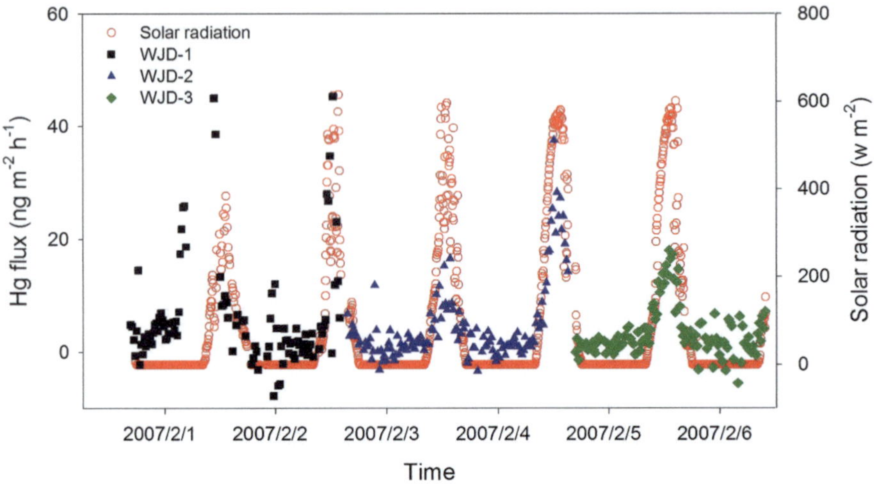

Fig. 4.11 Water/air Hg flux and solar radiation in Wujiangdu (WJD) reservoir in cold season (reprinted from Fu et al. (2010b), with permission from Elsevier Ltd.)

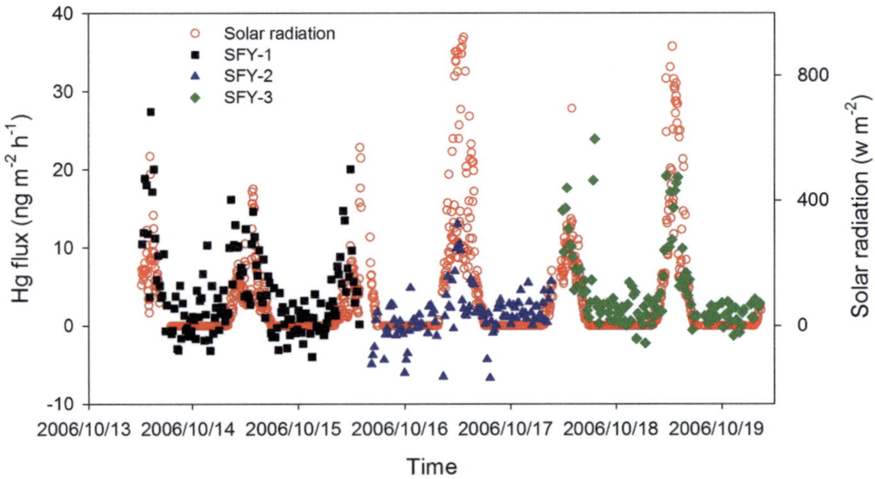

Fig. 4.12 Water/air Hg flux and solar radiation in Suofengying (SFY) reservoir in warm season (reprinted from Fu et al. (2010b), with permission from Elsevier Ltd.)

relatively higher than that in cold season. The observations in the present and previous studies have shown that water/air Hg fluxes were generally positively correlated with solar radiation and air temperature, which suggested that increasing solar radiation and air temperature could facilitate the emissions of Hg from water bodies (Boudala et al. 2000; Fu et al. 2013a; Poissant and Casimir 1998). Additionally, water THg concentrations in the studied reservoirs were relatively

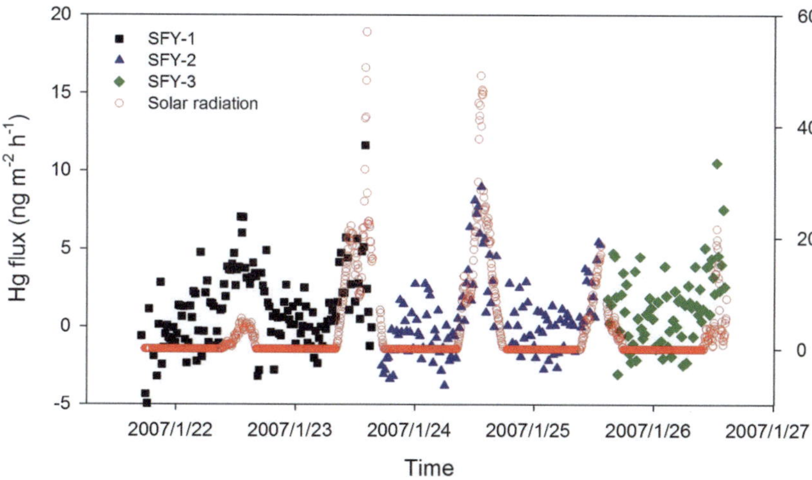

Fig. 4.13 Water/air Hg flux and solar radiation in Suofengying (SFY) reservoir in cold season (reprinted from Fu et al. (2010b), with permission from Elsevier Ltd.)

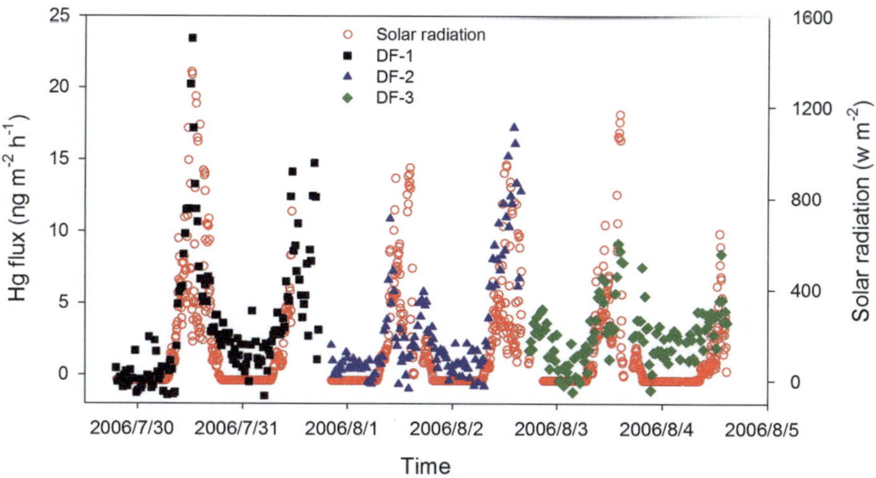

Fig. 4.14 Water/air Hg flux and solar radiation in Dongfeng (DF) reservoir in warm season

higher in warm season than in cold season, and this could be due to the increased loading of Hg via wet deposition and runoff in warm season (Guo et al. 2008). Phytoplankton and microbial activities were generally stronger in warm season than in cold season. The factors mentioned above were thought to accelerate the production of dissolve gaseous mercury (DGM) in warm season, which consequently increased Hg emission from water bodies (Amyot et al. 1997; Fu et al. 2013a; Mason et al. 1995; Siciliano et al. 2002).

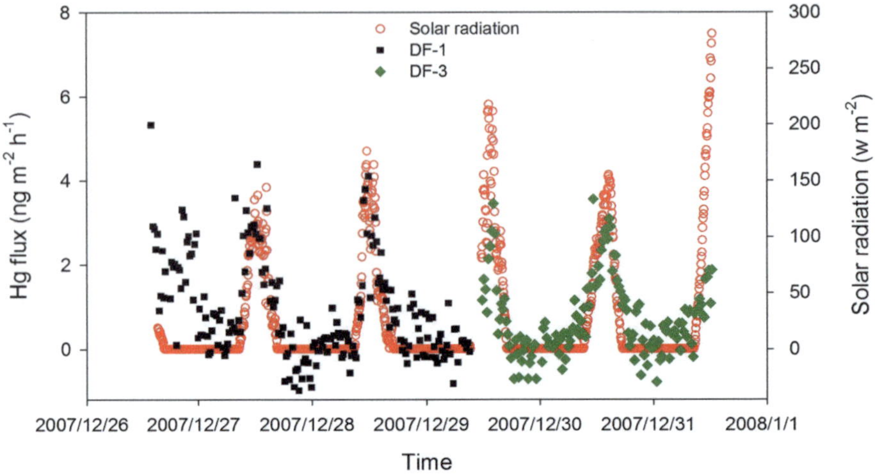

Fig. 4.15 Water/air Hg flux and solar radiation in Dongfeng (DF) reservoir in cold season

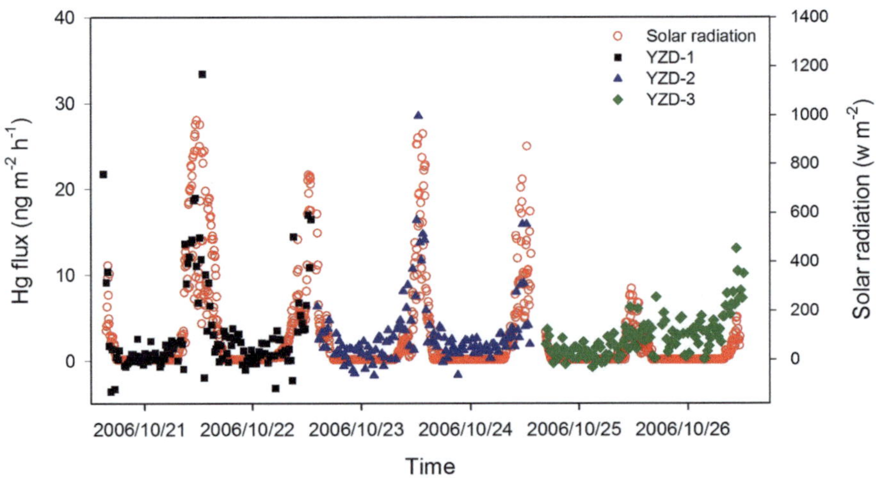

Fig. 4.16 Water/air Hg flux and solar radiation in Yinzidu (YZD) reservoir in warm season

4.5 Factors Influencing Water/Air Mercury Flux

Water/air Hg flux is a complicated process and influenced by many environmental factors. It consists of a thermal dynamic equilibrium process and a chemical kinetics process. The thermal dynamic equilibrium process is generally depicted as the thin film gas exchange model which demonstrates that Hg flux from water is mainly driven by the diffusion of DGM from water to the atmosphere. On the other hand, the chemical kinetics process at the water/air interface could be also important

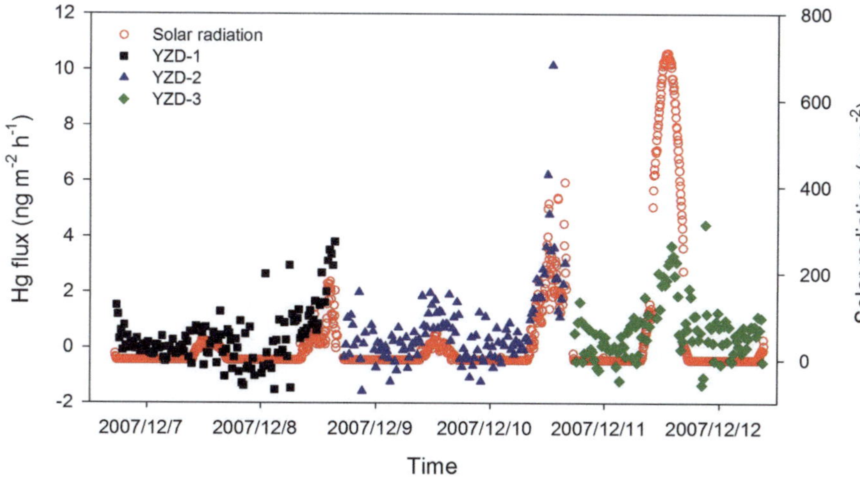

Fig. 4.17 Water/air Hg flux and solar radiation in Yinzidu (YZD) reservoir in cold season

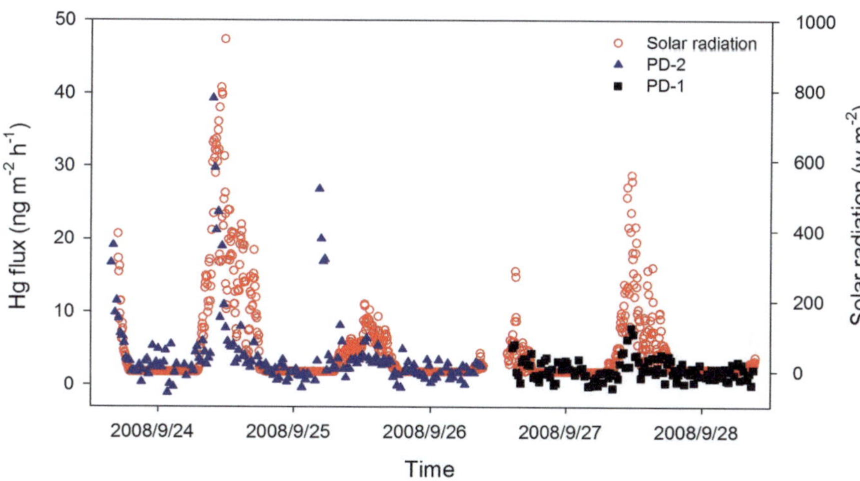

Fig. 4.18 Water/air Hg flux and solar radiation in Puding (PD) reservoir in warm season (reprinted from Fu et al. (2013b), with permission from John Wiley & Sons, Inc.)

for water/air Hg flux. The DGM produced at the water/air interface by abiotic and biotic induced reductions of other Hg species could be released to the atmosphere directly, which might be not controlled by the thin film gas exchange model. The above two processes could be further controlled by a combination of many water chemical and meteorological parameters including concentrations and forms of Hg, DOC content, reducing substance, microbial activities in water, solar radiation,

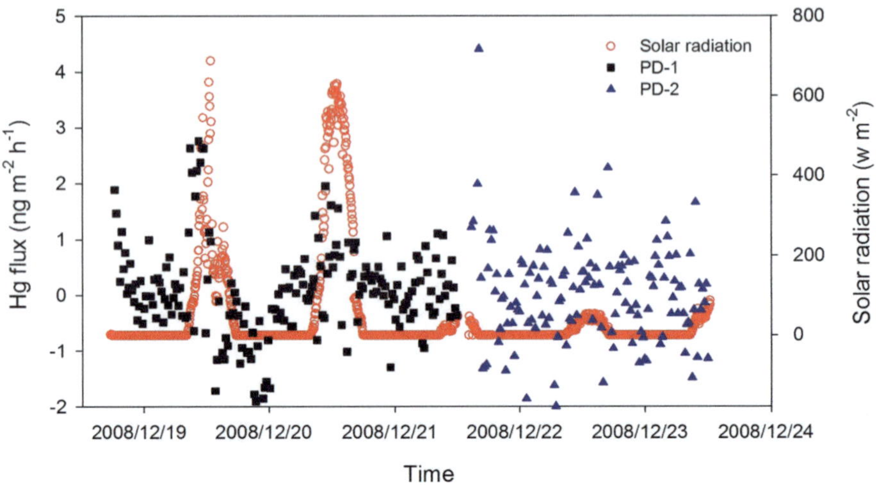

Fig. 4.19 Water/air Hg flux and solar radiation in Puding (PD) reservoir in cold season (reprinted from Fu et al. (2013b), with permission from John Wiley & Sons, Inc.)

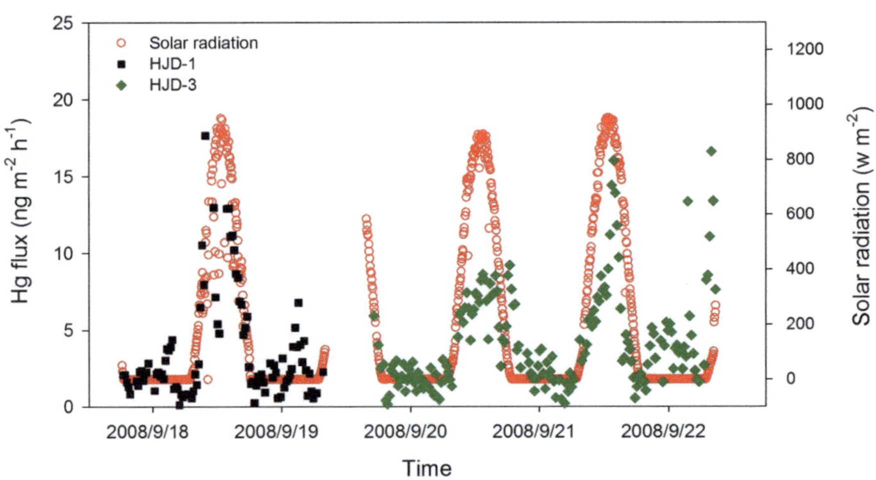

Fig. 4.20 Water/air Hg flux and solar radiation in Hongjiadu (HJD) reservoir in warm season

water and air temperature, and wind speed (Amyot et al. 1997; Feng et al. 2004b; Mason et al. 1995; Poissant and Casimir 1998; Schroeder et al. 2005).

In this study, it was observed that THg concentration in surface water played an important role in water/air Hg fluxes. As shown in Table 4.2, concentrations of THg in surface water of the HF and BH reservoirs (5.3–42.6 ng L^{-1}) were much higher than other reservoirs, and this is corresponding to elevated water/air Hg fluxes in these two polluted reservoirs. For the six remaining unpolluted reservoirs,

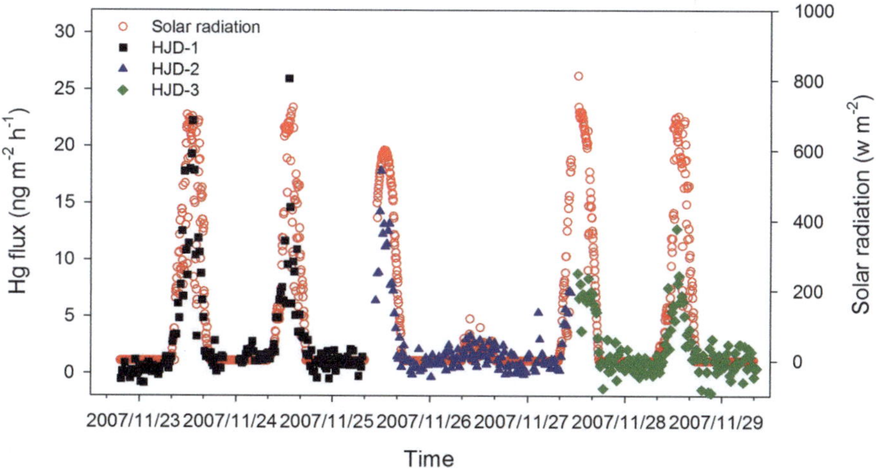

Fig. 4.21 Water/air Hg flux and solar radiation in Hongjiadu (HJD) reservoir in cold season

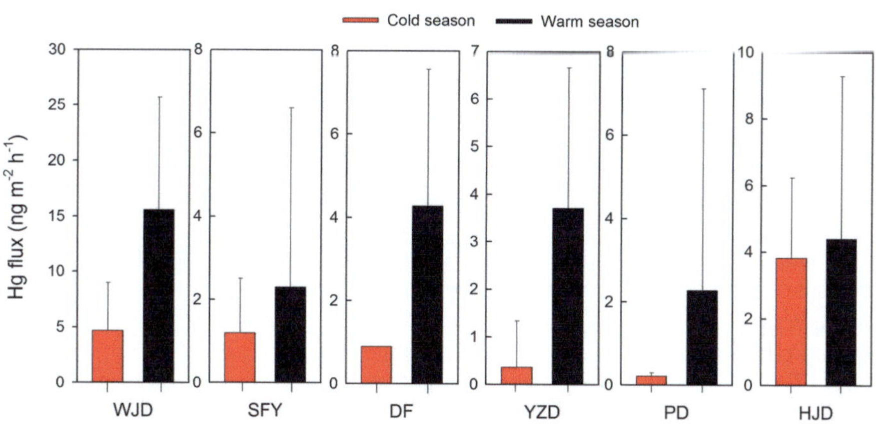

Fig. 4.22 Comparisons of water/air Hg flux in the six reservoirs in Wujiang River between cold and warm seasons

water/air Hg fluxes were also significantly correlated with THg concentrations in surface water (Fig. 4.23). It is noteworthy that the influence of water THg concentrations on water/air Hg flux was different among reservoirs, suggesting that the transformation of THg to DGM in water was different among reservoirs, and this could be partly due to the variations of physicochemical and microbial processes in different reservoirs. As shown in Fig. 4.23, the correlation slope between water/air Hg fluxes and THg concentrations in surface water in WJD reservoir (slope = 10.9) was approximately 5.6 time greater than that (slope = 1.95) in the remaining five reservoirs. This could be due to the fact that the prevalent cage aquaculture

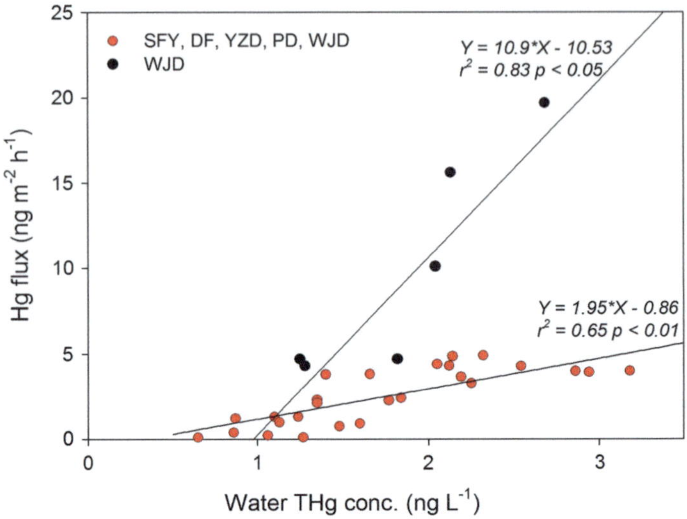

Fig. 4.23 Correlations between water/air Hg fluxes and water THg concentrations in the six reservoirs in Wujiang River

activities in WJD reservoir increased the phytoplankton and microbial activities. The remaining five reservoirs were rarely impacted anthropogenic activities and had similar water physicochemical parameters, which may explain the strong correlation between water/air Hg fluxes and water THg concentrations in these five reservoirs.

Meteorological parameter also played important roles in water/air Hg fluxes. The correlations between water/air Hg fluxes and meteorological parameters and atmospheric TGM concentrations at all the sampling sites in Wujiang River are shown in Table 4.4. In general, water/air Hg fluxes were positively correlated with solar radiation, water and air temperature and inversely correlated with atmospheric relative humidity at most of the sampling sites, whereas no consistent correlation was observed between water/air Hg fluxes and wind speed and atmospheric TGM concentrations. The correlation analysis reveals that water/air Hg fluxes were most significantly correlated with solar radiation, indicating that the solar radiation might play the most important role in the variation of Hg fluxes in Wujaing River among all the meteorological parameters. Sunlight-induced production of DGM in surface water is an important source of DGM in water and consequently drives Hg emissions from water, and this process is generally triggered by dissolved organic carbon (DOC) and oxidized iron (e.g., Fe(III)) in water (Amyot et al. 1997; Zhang and Lindberg 2001).

There have been debates as to whether air and water temperature could significantly affect water/air Hg fluxes. A previous study by Gårdfeldt et al. (2001) on Hg fluxes over river water observed a significant Arrhenius relationship between water/air Hg fluxes and water temperature, suggesting that water and air

Table 4.4 Statistical summary of the correlations between water/air Hg fluxes and meteorological parameters and atmospheric TGM concentrations in Wujiang River

Reservoir	Site	Season	Solar radiation	Air temperature	Water temperature	Relative humidity	Wind speed	TGM
WJD	WJD-1	Warm season	–	0.50**	0.21	−0.45**	0.56**	−0.16
		Cold season	0.78**	0.34*	–	−0.22	0.09	0.67
	WJD-2	Warm season	–	0.36**	-0.17	−0.26*	0.57**	0.75**
		Cold season	0.84**	0.54**	–	−0.39**	−0.01	0.52**
	WJD-3	Warm season	–	0.213*	0.29**	−0.20*	0.06	0.27**
		Cold season	0.86**	0.28*	–	−0.48*	0.20	−0.15
SFY	SFY-1	Warm season	0.73**	0.45**	–	−0.36**	–	−0.34*
		Cold season	0.62**	0.04	–	−0.24*	0.09	−0.42**
	SFY-2	Warm season	0.53**	0.39**	–	−0.38**	–	0.05
		Cold season	0.78**	0.16	–	−0.53**	0.21	-0.19
	SFY-3	Warm season	0. 88	0.68**	–	-0.60**	–	0.24
		Cold season	0.57**	0.10	–	–	–	−0.36**
DF	DF-1	Warm season	0.82**	0.78**	–	−0.74**	0.13	0.05
		Cold season	0.84**	0.47**	–	−0.33**	−0.11	0.04
	DF-2	Warm season	0.722**	0.87**	0.38**	−0.89**	0.16	0.79**
	DF-3	Warm season	0.68**	0.70**	0.51**	−0.66**	0.04	0.32**
		Cold season	0.86**	0.55**	–	−0.25*	−0.26*	0.64**
YZD	YZD-1	Warm season	0.73**	0.60**	–	-0.54**	–	0.61**
		Cold season	0.1	0.16	–	−0.14	0.08	−0.31*
	YZD-2	Warm season	0.81**	0.53**	–	−0.41**	–	0.64**
		Cold season	0.59**	−0.14	–	−0.26*	−0.21	−0.27*
	YZD-3	Warm season	0.59*	–	–	-0.27*	–	-0.30*
		Cold season	0.67**	0.51**	–	−0.50**	−0.04	−0.15

(continued)

Table 4.4 (continued)

Reservoir	Site	Season	Solar radiation	Air temperature	Water temperature	Relative humidity	Wind speed	TGM
PD	PD-1	Warm season	0.77**	0.51**	0.27	−0.39*	0.11	0.12
		Cold season	0.34**	0.33**	–	−0.34	0.38**	0.32
	PD-2	Warm season	0.64**	0.14	0.02	−0.16	0.08	0.06
		Cold season	0.20	0.07	0.16	0.06	0.06	0.06
HJD	HJD-1	Warm season	0.80**	0.54**	0.65**	−0.27*	0.05	0.79**
		Cold season	0.89**	0.53**	–	−0.52**	0.05	0.60**
	HJD-2	Cold season	0.92**	0.41**	–	−0.24**	−0.17**	0.43**
	HJD-3	Warm season	0.62**	0.42**	0.17	−0.30**	0.27**	0.53**
		Cold season	0.82**	0.39**	–	−0.33**	0.08	−0.34**

Pearson correlation coefficient: $^*p < 0.05$; $^{**}p < 0.01$; no star $p > 0.05$

temperature might play an important role in water/air Hg fluxes. On the other hand, Feng et al. (2004b) observed both negative and positive correlations between water/air Hg fluxes and water and air temperature in the BH reservoir. This implies that air and water temperatures were not the driving forces of Hg exchange between water and the air. In the present study, water/air Hg fluxes were mostly positively correlated with water and air temperature at most of the sampling sites in Wujiang River, whereas no significant correlation or negative correlation can be established at the remaining several sites (Table 4.4). This is overall consistent with previous studies (Boudala et al. 2000; Feng et al. 2004b; Gårdfeldt et al. 2001; Poissant and Casimir 1998). It should be pointed out that some of the positive correlations between water/air Hg fluxes and water and air temperature were obtained during sunny days, during which the variations of water/air Hg fluxes were significantly linked to solar radiation. This implies that, for a short sampling period, some of the strong correlations between water/air Hg fluxes and water and air temperature might be a reflection of sunlight-induced evasion of Hg from water bodies. However, this does not necessarily rule out the possible effect of water and air temperature in water/air Hg fluxes. The water and air temperature is an important variable affecting production of microorganism and phytoplankton in water, which have been verified to play an important role in water/air Hg fluxes (Poulain et al. 2004; Siciliano et al. 2002). Additionally, water and air temperature is also a crucial factor in the biogeochemical transformation of Hg in water (Lin and Pehkonen 1999). Therefore, water and air temperature could probably influence water/air Hg fluxes over a extended period (e.g., seasonal time scale). We used the forward stepwise multiple regression to determine the relative importance of water THg concentration, solar radiation, air temperature, relative humidity, atmospheric TGM concentrations, etc., to water/air Hg fluxes during warm and cold seasons. The result revealed that the water THg concentrations and air temperature were the two most important factors in water/air Hg fluxes.

To establish the empirical models predicting water/air Hg fluxes in the polluted and non-polluted reservoirs, we used the forward stepwise multiple regression to determine the relationships between water/air Hg fluxes and water THg concentrations and air temperature in the polluted and non-polluted reservoirs. The result revealed that the empirical models in the polluted and non-polluted reservoirs were different.

The empirical model for the polluted reservoirs of BH and HF reservoirs is:

$$F = 0.11 * \mathrm{THg} + 0.23 * T_{\mathrm{air}} + 0.53\, r^2 = 0.53, \quad p < 0.01 \tag{4.2}$$

The empirical model for the hyper-eutrophic WJD reservoir is:

$$F = 6.09 * \mathrm{THg} + 0.80 * T_{\mathrm{air}} - 14.0\, r^2 = 0.97, \quad p < 0.01 \tag{4.3}$$

And the empirical model for the non-polluted reservoirs (SFY, DF, YZD, PD, and HJD) is:

$$F = 1.08 * THg + 0.12 * T_{air} - 1.34 \quad r^2 = 0.81, \quad p < 0.01 \tag{4.4}$$

where F is the water/air Hg flux (μg m^{-2} h^{-1}), THg is the total Hg concentration in surface water (ng L^{-1}), and T_{air} is the air temperature.

The simulated Hg fluxes using above equations showed a good agreement with measured values in the study area (Figs. 4.24, 4.25, and 4.26).

There are no consistent relationships between water/air Hg fluxes and wind speed. Strong positive correlations between water/air Hg fluxes and wind speed were only observed at the WJD-1 and WJD-2 in warm season, PD-1 in cold season, and HJD-3 in warm season, whereas the water/air Hg fluxes were weakly or negatively correlated with wind speed (Table 4.4). This is quite different from the thin film gas exchange model. According to the thin film gas exchange model, evasion of Hg from water is controlled by the gradient of Hg between water and ambient air, wind speed, water temperature, Hg diffusion coefficient in water, and kinematic viscosity of water (Gardfeldt et al. 2003; Wangberg et al. 2001; Wanninkhof 1992), as depicted in Eq. (4.5).

$$F = K_w \times (\mathrm{DGM} - \mathrm{GEM}/H'(T)), \tag{4.5}$$

where F is the water/air Hg flux, DGM and GEM are the concentrations of Hg^0 in water, $H'(T)$ is the dimensionless partitioning coefficient for Hg^0 between freshwater and air as known as the Henry's law constant and was calculated according to Eq. (4.6) (Andersson et al. 2008), K_w is the gas transfer velocity of Hg^0 in the water/air surface (cm h^{-1}) and was calculated according to Eq. (4.7) (Wanninkhof 1992):

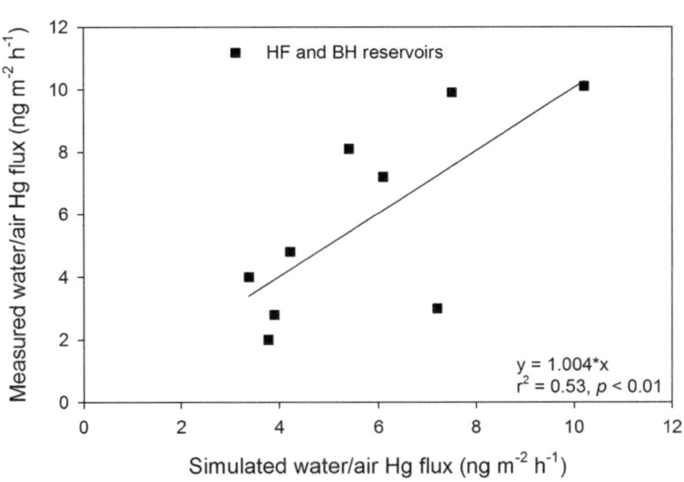

Fig. 4.24 Comparison between the simulated and measured water/air Hg fluxes in the HF and BH reservoirs

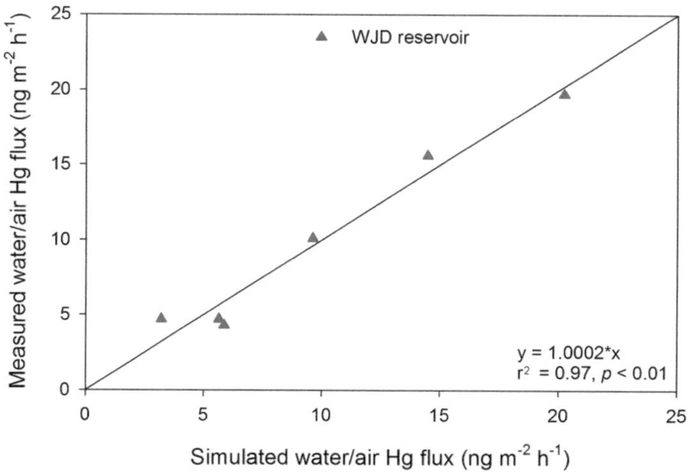

Fig. 4.25 Comparison between the simulated and measured water/air Hg fluxes in the WJD reservoir

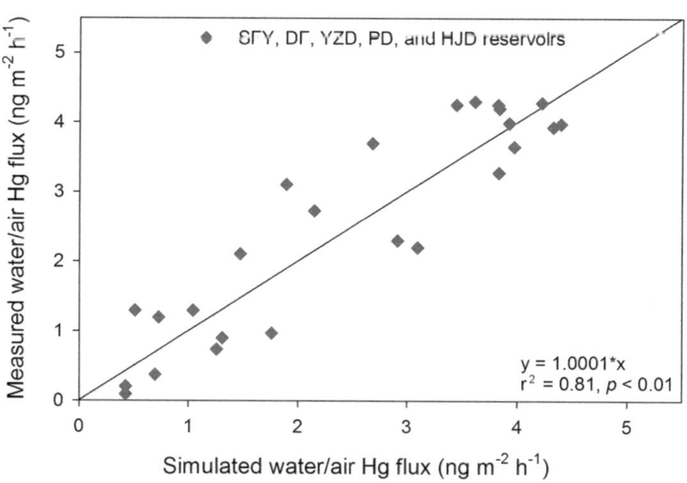

Fig. 4.26 Comparison between the simulated and measured water/air Hg fluxes in the SFY, DF, YZD, PD, and HJD reservoirs

$$H'(T) = \exp(-2404.3/T + 6.92) \tag{4.6}$$

$$K_w = 0.31 \times U_{10}^2 \times \left(\mathrm{Sc}_{\mathrm{Hg}}/600\right)^{-0.5}, \tag{4.7}$$

where U_{10} is the wind speed normalized to 10 m above water surface, and $\mathrm{Sc}_{\mathrm{Hg}}$, the Schmidt number for Hg, is defined as:

Table 4.5 Estimated monthly and annual Hg flux from water in the eight reservoirs in Wujiang River

Reservoir	Monthly mean Hg flux($\mu g\ m^{-2}\ mon^{-1}$)												Annual mean ($\mu g\ m^{-2}\ yr^{-1}$)
	January	February	March	April	May	June	July	August	September	October	November	December	
PD	1.8	1.6	1.8	2.1	2.2	2.7	2.9	2.7	2.4	3.0	2.0	2.0	27.0 ± 5.6
HJD	1.7	0.7	1.7	2.3	2.1	2.5	2.6	2.4	3.2	2.7	1.5	1.3	24.6 ± 8.4
YZD	1.2	1.2	1.8	2.0	2.3	2.3	2.6	2.7	2.4	2.7	1.4	1.0	23.7 ± 7.5
DF	1.2	1.7	1.7	2.6	2.8	2.4	2.8	2.5	2.5	2.6	1.7	1.6	26.0 ± 6.7
SFY	1.3	1.5	2.1	2.6	2.6	2.5	2.5	2.4	2.8	2.4	1.6	0.9	25.2 ± 7.4
WJD	7.5	3.3	10.8	12.0	16.1	12.9	13.4	12.9	11.2	13.5	8.6	7.6	129.8 ± 42.1
BH	5.4	5.8	6.8	8.0	8.7	9.3	9.7	9.6	8.9	7.9	7.0	6.0	92.9 ± 18.2
HF	2.3	2.7	3.6	4.9	5.7	6.2	6.6	6.1	5.5	4.5	3.6	2.7	54.4 ± 18.1

$$Sc_{Hg} = v/D_{Hg}, \tag{4.8}$$

where v is the kinematic viscosity ($cm^2\ s^{-1}$) of freshwater, and D_{Hg} is the Hg diffusion coefficient ($cm^2\ s^{-1}$) in freshwater, which was calculated by the molecular dynamics simulation, as described by Kuss et al. (2009).

The thin film gas exchange model demonstrates that wind speed is an important factor in water/air Hg flux, however, we did not observe clear relationships between water/air Hg fluxes and wind speed. This is probably attributed to (1) the implementation of dynamic chamber that minimizes the effect of wind speed on water/air Hg flux; and (2) the low speed during the whole study period, which would significantly decreased the gas transfer velocity of Hg^0 in the water/air surface (O'Driscoll et al. 2003; Poissant et al. 2000).

There existed both strong negative and positive correlations between water/air Hg fluxes and atmospheric TGM concentrations (Table 4.4). The positive correlations between water/air Hg fluxes and atmospheric TGM concentrations suggested that emission of Hg from water was an important source of atmospheric TGM over reservoir water. On the other hand, the significant negative relationships between water/air Hg fluxes and atmospheric TGM concentrations at SFY-1, SFY-3, YZD-1, YZD-2, and HJD-3 in cold season and YZD-3 in warm season suggested the increasing atmospheric TGM concentrations could inhibit evasions of Hg from water. This phenomenon could be explained by the Eq. (4.5), that is water/air Hg flux increases with DGM in water and decreases with atmospheric TGM concentration above surface water.

4.6 Estimates of Hg Emission for Water in Wujiang River

As discussed in Sect. 4.5, water/air Hg fluxes in Wujiang River were found to be significantly correlated with water THg concentration and air temperature. To estimate the annual Hg emissions from the studied reservoirs, monthly THg concentrations in surface waters and monthly mean air temperature of the eight reservoirs were employed as inputs in the empirical models (Eqs. 4.2, 4.3, and 4.4). The estimated monthly evasion of Hg from water in Wujiang River is listed in Table 4.5.

Estimated annual mean Hg fluxes at the eight reservoirs in Wujiang River range from 23.7 to 129.8 $\mu g\ m^{-2}\ yr^{-1}$. With the exception WJD, BH, and HF reservoirs, Hg emission fluxes were relatively lower than the wet deposition fluxes (24.8–38.1 $\mu g\ m^{-2}\ yr^{-1}$) in Wujiang River (see Chap. 3). This indicates exchange of Hg between water and the atmosphere in WJD, BH, and HF reservoirs were probably the net sources of atmospheric Hg, whereas they were net sinks of atmospheric Hg in SFY, DF, YZD, PD, and HJD reservoirs. As discussed earlier, the WJD, BH, and HF reservoirs were impacted by discharges of Hg enriched wastewaters from surrounding urbanized and industrialized regions and hyper-eutrophic issues. The great loading of Hg to these reservoirs probably drove an increase of reemission of Hg from water bodies.

Figure 4.27 shows the monthly variations in water/air Hg flux in the eight reservoirs in Wujiang River. Water/air Hg fluxes showed a clear seasonal trend in all the studies reservoirs with higher Hg fluxes in warm season and lower Hg fluxes in cold season. Water THg concentrations in reservoirs in Wujiang River did not vary significantly among different seasons, and therefore the seasonal trend in water/air Hg fluxes could be function of variation in air temperature.

Output of Hg from each of the reservoir in Wujiang River via evaporation of Hg from water bodies were estimated by multiplying the annual mean Hg flux by the corresponding area of the reservoir. The estimated Hg emissions from the eight reservoirs in Wujiang River are shown in Table 4.6. Emission of Hg from the reservoirs ranged from 144 to 6204 g/year, with a total emission of 14,158 g/year. Overall, the total emission of Hg (9699 g/year) from the WJD, SFY, DF, YZD, PD, and HJD reservoirs was relatively higher than the wet deposition (6640 g/year).

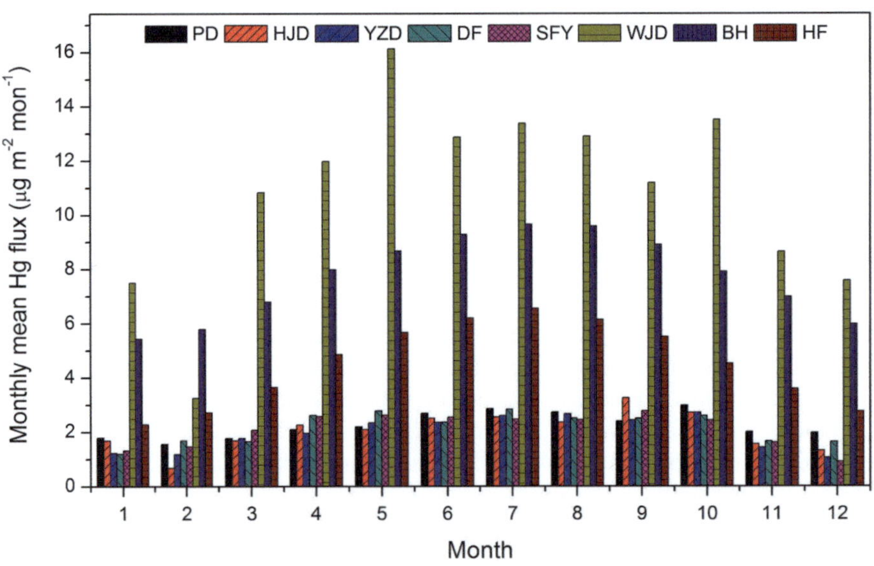

Fig. 4.27 Monthly variations in water/air Hg flux in the eight reservoirs in Wujiang River

Table 4.6 Estimates of Hg emissions from the reservoirs in Wujiang River

Site	Hg emission flux (μg m^{-2} yr^{-1})	Reservoir areas ($\times$ 10^6 m^2)	Hg emission (g)
	27.0	19.2	518
HJD	24.6	80.5	1980
YZD	23.7	15.0	356
DF	26.0	19.1	497
SFY	25.2	5.7	144
WJD	129.8	47.8	6204
BH	92.9	14.5	1347
HF	54.4	57.2	3112
Total			14,158

References

Amyot M, Gill GA, Morel FMM (1997) Production and loss of dissolved gaseous mercury in coastal seawater. Environ Sci Technol 31:3606–3611

Andersson ME, Gardfeldt K, Wangberg I, Stromberg D (2008) Determination of Henry's law constant for elemental mercury. Chemosphere 73:587–592

Boudala FS, Folkins I, Beauchamp S, Tordon R, Neima J, Johnson B (2000) Mercury flux measurements over air and water in Kejimkujik National Park, Nova Scotia. Water Air Soil Pollut 122:183–202

Carpi A, Lindberg SE (1998) Application of a Teflon (TM) dynamic flux chamber for quantifying soil mercury flux: tests and results over background soil. Atmos Environ 32:873–882

Ebinghaus R, Jennings SG, Schroeder WH, Berg T, Donaghy T, Guentzel J et al (1999) International field intercomparison measurements of atmospheric mercury species at Mace Head, Ireland. Atmos Environ 33:3063–3073

Eckley CS, Gustin M, Lin CJ, Li X, Miller MB (2010) The influence of dynamic chamber design and operating parameters on calculated surface-to-air mercury fluxes. Atmos Environ 44: 194–203

Feng XB, Shang LH, Wang SF, Tang SL, Zheng W (2004a) Temporal variation of total gaseous mercury in the air of Guiyang, China. J Geophys Res Atmos 109

Feng XB, Yan HY, Wang SF, Qiu GL, Tang SL, Shang LH et al (2004b) Seasonal variation of gaseous mercury exchange rate between air and water surface over Baihua reservoir, Guizhou, China. Atmos Environ 38:4721–4732

Feng XB, Wang SF, Qiu GA, Hou YM, Tang SL (2005) Total gaseous mercury emissions from soil in Guiyang, Guizhou, China. J Geophy Res Atmos 110

Feng X, Wang S, Qiu G, He T, Li G, Li Z et al (2008) Total gaseous mercury exchange between water and air during cloudy weather conditions over Hongfeng Reservoir, Guizhou, China. J Geophys Res Atmos 113:n/a–n/a

Fu XW, Feng XB, Zhu WZ, Wang SF, Lu JL (2008) Total gaseous mercury concentrations in ambient air in the eastern slope of Mt. Gongga, South-Eastern fringe of the Tibetan plateau, China. Atmos Environ 42:970–979

Fu XW, Feng X, Dong ZQ, Yin RS, Wang JX, Yang ZR et al (2010a) Atmospheric gaseous elemental mercury (GEM) concentrations and mercury depositions at a high-altitude mountain peak in south China. Atmos Chem Phys 10:2425–2437

Fu XW, Feng XB, Wan Q, Meng B, Yan HY, Guo YN (2010b) Probing Hg evasion from surface waters of two Chinese hyper/meso-eutrophic reservoirs. Sci Total Environ 408:5887–5896

Fu XW, Feng XB, Qiu GL, Shang LH, Zhang H (2011) Speciated atmospheric mercury and its potential source in Guiyang, China. Atmos Environ 45:4205–4212

Fu XW, Feng XB, Sommar J, Wang SF (2012) A review of studies on atmospheric mercury in China. Sci Total Environ 421:73–81

Fu XW, Feng XB, Guo YN, Meng B, Yin RS, Yao H (2013a) Distribution and production of reactive mercury and dissolved gaseous mercury in surface waters and water/air mercury flux in reservoirs on Wujiang River, Southwest China. J Geophys Res Atmos 118:3905–3917

Fu XW, Feng XB, Yin RS, Zhang H (2013b) Diurnal variations of total mercury, reactive mercury, and dissolved gaseous mercury concentrations and water/air mercury flux in warm and cold seasons from freshwaters of southwestern China. Environ Toxicol Chem 32:2256–2265

Garcia E, Amyot M, Ariya PA (2005) Relationship between DOC photochemistry and mercury redox transformations in temperate lakes and wetlands. Geochim Cosmochim Acta 69:1917–1924

Gardfeldt K, Sommar J, Ferrara R, Ceccarini C, Lanzillotta E, Munthe J et al (2003) Evasion of mercury from coastal and open waters of the Atlantic Ocean and the Mediterranean Sea. Atmos Environ 37:S73–S84

Gårdfeldt K, Feng XB, Sommar J, Lindqvist O (2001) Total gaseous mercury exchange between air and water at river and sea surfaces in Swedish coastal regions. Atmos Environ 35:3027–3038

Guo YN, Feng XB, Li ZG, He TR, Yan HY, Meng B et al (2008) Distribution and wet deposition fluxes of total and methyl mercury in Wujiang River Basin, Guizhou, China. Atmos Environ 42:7096–7103

Kellerhals M, Beauchamp S, Belzer W, Blanchard P, Froude F, Harvey B et al (2003) Temporal and spatial variability of total gaseous mercury in Canada: results from the Canadian Atmospheric Mercury Measurement Network (CAMNet). Atmos Environ 37:1003–1011

Kuss J, Holzmann J, Ludwig R (2009) An elemental mercury diffusion coefficient for natural waters determined by molecular dynamics simulation. Environ Sci Technol 43:3183–3186

Lan X, Talbot R, Castro M, Perry K, Luke W (2012) Seasonal and diurnal variations of atmospheric mercury across the US determined from AMNet monitoring data. Atmos Chem Phys 12:10569–10582

Lin CJ, Pehkonen SO (1999) The chemistry of atmospheric mercury: a review. Atmos Environ 33:2067–2079

Mason RP, Morel FMM, Hemond HF (1995) The role of microorganisms in elemental mercury formation in natural-waters. Water Air Soil Pollut 80:775–787

Munthe J, Wangberg I, Pirrone N, Iverfeldt A, Ferrara R, Ebinghaus R et al (2001) Intercomparison of methods for sampling and analysis of atmospheric mercury species. Atmos Environ 35:3007–3017

Munthe J, Wangberg I, Iverfeldt A, Lindqvist O, Stromberg D, Sommar J et al (2003) Distribution of atmospheric mercury species in Northern Europe: final results from the MOE project. Atmos Environ 37:S9–S20

O'Driscoll NJ, Beauchamp S, Siciliano SD, Rencz AN, Lean DRS (2003) Continuous analysis of dissolved gaseous mercury (DGM) and mercury flux in two freshwater lakes in Kejimkujik Park, Nova Scotia: evaluating mercury flux models with quantitative data. Environ Sci Technol 37:2226–2235

Peters SC, Wollenberg JL, Morris DP, Porter JA (2007) Mercury emission to the atmosphere from experimental manipulation of DOC and UVR in mesoscale field chambers in a freshwater lake. Environ Sci Technol 41:7356–7362

Poissant L, Casimir A (1998) Water-air and soil-air exchange rate of total gaseous mercury measured at background sites. Atmos Environ 32:883–893

Poissant L, Amyot M, Pilote M, Lean D (2000) Mercury water-air exchange over the Upper St. Lawrence River and Lake Ontario. Environ Sci Technol 34:3069–3078

Poulain AJ, Lalonde JD, Amyot M, Shead JA, Raofie F, Ariya PA (2004) Redox transformations of mercury in an Arctic snowpack at springtime. Atmos Environ 38:6763–6774

Poulain AJ, Ni Chadhain SM, Ariya PA, Amyot M, Garcia E, Campbell PGC et al (2007) Potential for mercury reduction by microbes in the high arctic. Appl Environ Microbiol 73:2230–2238

Schroeder WH, Beauchamp S, Edwards G, Poissant L, Rasmussen P, Tordon R et al (2005) Gaseous mercury emissions from natural sources in Canadian landscapes. J Geophys Res Atmos 110

Siciliano SD, O'Driscoll NJ, Lean DRS (2002) Microbial reduction and oxidation of mercury in freshwater lakes. Environ Sci Technol 36:3064–3068

Sprovieri F, Pirrone N, Ebinghaus R, Kock H, Dommergue A (2010) A review of worldwide atmospheric mercury measurements. Atmos Chem Phys 10:8245–8265

USEPA (2002) Method 1631, Revision E: mercury in water by oxidation, purge and trap, and cold vapor atomic fluorescence spectrometry. United States Environmental Protection Agency, Washington DC, USA, pp 10–46

Wang SF, Feng XB, Qiu GL, Wei ZQ, Xiao TF (2005) Mercury emission to atmosphere from Lanmuchang Hg–Tl mining area, Southwestern Guizhou, China. Atmos Environ 39:7459–7473

Wang SF, Feng XB, Qiu GG, Shang LH, Li P, Wei ZQ (2007a) Mercury concentrations and air/soil fluxes in Wuchuan mercury mining district, Guizhou province, China. Atmos Environ 41:5984–5993

Wang SF, Feng XB, Qiu GL, Fu XW, Wei ZQ (2007b) Characteristics of mercury exchange flux between soil and air in the heavily air-polluted area, eastern Guizhou, China. Atmos Environ 41:5584–5594

Wangberg I, Schmolke S, Schager P, Munthe J, Ebinghaus R, Iverfeldt A (2001) Estimates of air-sea exchange of mercury in the Baltic Sea. Atmos Environ 35:5477–5484

Wanninkhof R (1992) Relationship between wind-speed and gas-exchange over the ocean. J Geophys Res Oceans 97:7373–7382

Wollenberg JL, Peters SC (2009) Diminished mercury emission from waters with duckweed cover. J Geophys Res Biogeosciences 114

Wu Y, Wang SX, Streets DG, Hao JM, Chan M, Jiang JK (2006) Trends in anthropogenic mercury emissions in China from 1995 to 2003. Environ Sci Technol 40:5312–5318

Xiao ZF, Munthe J, Schroeder WH, Lindqvist O (1991) Vertical Fluxes of Volatile Mercury over Forest Soil and Lake Surfaces in Sweden. Tellus Ser B Chem Phys Meteorol 43:267–279

Zhang H, Lindberg SE (2001) Sunlight and iron(III)-induced photochemical production of dissolved gaseous mercury in freshwater. Environ Sci Technol 35:928–935

Zhang HH, Poissant L, Xu XH, Pilote M, Beauvais C, Amyot M et al (2006) Air-water gas exchange of mercury in the Bay Saint Francois wetlands: observation and model parameterization. J Geophys Res Atmos 111

Chapter 5
Mercury in Inflow/Outflow Rivers of Reservoirs

Abstract Rivers are major pathways for a variety of materials to flow in and out of reservoirs. Impounding rivers changes the characteristics of a water body from "rivers" to "reservoirs", affecting not only their hydrology but also their physical, chemical, and biological characteristics. To understand the influence of damming on the distribution and methylation of Hg within a river-reservoir ecosystem, this chapter illustrates the spatial and temporal distributions of Hg species in the inflow–outflow rivers of the six cascade reservoirs in Wujiang River Basin, Southwest China. Furthermore, the influence of cascade reservoirs on the distributions of Hg species in rivers of Wujiang River Basin is elucidated as well.

Keywords River · Inflow · Outflow · Mercury · Reservoir

5.1 Sampling Sites Description, Sample Collection, Sample Analyses, Analytical Methods, and QA/QC

Eighteen sampling stations recognized as the inflow and outflow of the six reservoirs were chosen, as depicted in Fig. 5.1 and Table 5.1. The six selected cascade reservoirs, including the Puding Reservoir (PD), Hongjiadu Reservoir (HJD), Yinzidu Reservoir (YZD), Suofengying Reservoir (SFY), Dongfeng Reservoir (DF), and Wujiangdu Reservoir (WJD), are located in the Wujiang River Basin, southwestern China (Fig. 5.1). Basic parameters of six selected reservoirs in Wujiang River Basin are shown in Table 5.2.

On the basis of the trophic state of the specific reservoir, the six selected reservoirs in Wujiang River Basin are classified as the oligotrophic–mesotrophic stage (YZD, SFY, and HJD), the mesotrophic–eutrophic stage (PD and DF), and the hyper-eutrophic stage (WJD) (Table 5.2). In detail, the primary productivity in newly constructed reservoirs (YZD, SFY, and HJD) is currently represented as oligotrophic–mesotrophic. The major source of organic matter in the newly constructed reservoirs is mainly derived from the watershed input with little autochthonous contribution because of the low primary productivity (Meng et al. 2010,

X. Feng et al., *Biogeochemical Cycle of Mercury in Reservoir Systems in Wujiang River Basin, Southwest China*, https://doi.org/10.1007/978-981-10-6719-8_5

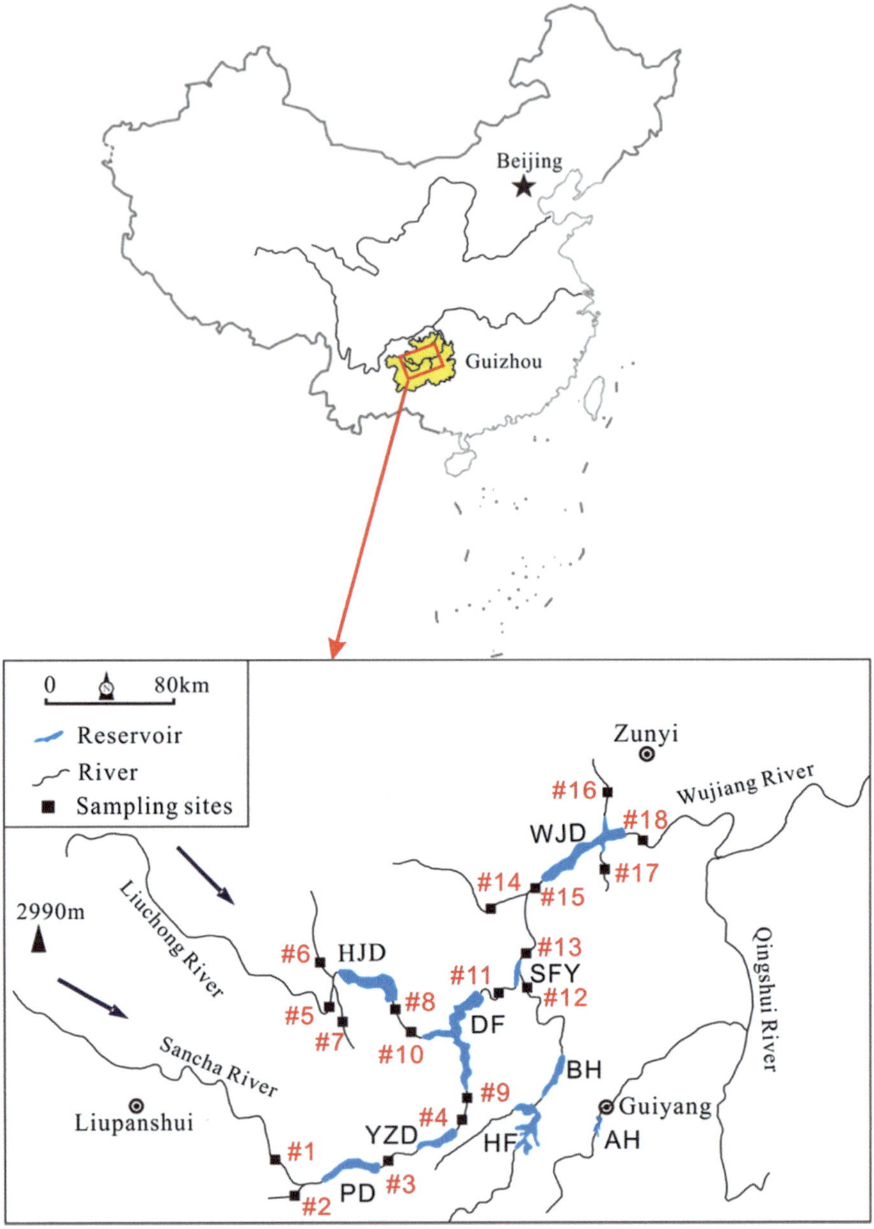

Fig. 5.1 Map of the study area and sampling sites of inflow–outflow rivers of reservoirs in Wujiang River Basin, Guizhou Province, China (*HJD* Hongjiadu Reservoir; *YZD* Yinzidu Reservoir; *SFY* Suofengying Reservoir; *PD* Puding Reservoir; *DF* Dongfeng Reservoir; *WJD* Wujiangdu Reservoir; *AH* Aha Reservoir; *HF* Hongfeng Reservoir; *BH* Baihua Reservoir) (redrawn from Zhao et al. 2017, with permission from Elsevier Ltd.)

Table 5.1 Summary of inflow–outflow rivers to reservoirs in Wujiang River Basin, Guizhou province, China (reprinted from Zhao et al. 2017, with permission from Elsevier Ltd.)

Reservoir	Inflow into reservoir	Outflow of reservoir
PD	SC (#1) and BY (#2)	Discharge water (#3)
YZD	Outflow of PD (#3)	Discharge water (#4)
HJD	LCH (#5), LJ (#6), MK (#7)	Discharge water (#8)
DF	mainstream (#10) and YZD outflow (#9)	Discharge water (#11)
SFY	Outflow of DF (#11), MT river (#12)	Discharge water (#13)
WJD	YJ river (#14), Mainstream of WJD (#15), PY river (#16), XF river (#17)	Discharge water (#18)

PD Puding Reservoir; *YZD* Yinzidu Reservoir; *HJD* Hongjiadu Reservoir; *DF* Dongfeng Reservoir; *SFY* Suofengying Reservoir; *WJD* Wujiangdu Reservoir

2016; Yao et al. 2011). With the continuous evolution of reservoirs, PD and DF are mesotrophic–eutrophic, which indicates a much higher level of primary productivity compared to the newly constructed reservoirs in the Wujiang River Basin. The organic matter content in the sediment of PD and DF was significantly elevated as a result of the continuously increasing autochthonous material (Feng et al. 2009a; Zhang 2009). Over a long-term evolution of the reservoir, the primary productivity of WJD was significantly elevated (hyper state of eutrophication) (Meng et al. 2010, 2016). The high-intensity cage aquaculture activity in WJD contributed to the higher primary productivity compared to the newly constructed reservoirs (SFY, HJD, and YZD). Phytoplankton-derived organic matter and the fish feeds and feces were potentially significant sources of organic matter input to the surface sediments of WJD (Meng et al. 2010, 2016).

Briefly, the Liuchong River (LC, #5) and Sancha River (SC, #1) as the main sources of the Wujiang River, are the predominant inflows into HJD and PD, respectively, while the contribution from the Luojiao River (LJ, #6), Mukong River (MK, #7) and Boyu River (BY, #2) to these two reservoirs is less pronounced. The inflow into YZD is limited to the outflow of PD (#3). The outflow both from HJD (#10) and YZD (#9) contribute to the inflow into DF. The inflows into SFY are from the outflow of DF (#11) and the Maotiao River (MT, #12). There are four inflows into WJD, and the largest inflow is from the outflow of SFY (#13). The Yeji River (YJ, #14), Pianyan River (PY, #16), and Xifeng River (XF, #17) are the other three rivers that enter into WJD. Since PD and YZD are adjacent, we set up one sampling site (#3) to represent the outflow of PF and the inflow of YZD. Similarly, the sampling sites of #11 represent the outflow of DF and the inflow of SFY.

In this study, all the outflow sites of the reservoirs were located within 1 km downstream from the dams. There was no tributary or any other inflow river between dam and outflow site throughout the six reservoirs. Therefore, the water samples collected from outflow sites were mainly from discharged water from hypolimnion of the reservoirs. However, the exact depth of the discharged water from dam of the reservoir was unavailable in this study. It should be noted that XF

Table 5.2 Basic parameters of six selected reservoirs in Wujiang River Basin, Guizhou province, China (reprinted from Zhao et al. (2017), with permission from Elsevier Ltd.)

Parameters	HJD[a]	YZD[a]	SFY[a]	PD[a]	DF[a]	WJD[a]
Construction time/year	2004	2003	2005	1993	1994	1979
Height of dam/m	180	136	122	75	162	165
Total water volume/10^8 m^3	49.47	4.55	2.01	4.2	10.3	23.0
Surface area of the reservoir/km^2	80.5	15.0	5.7	19.3	19.1	47.8
Annual runoff volume/10^8 m^3	47.0	44.2	124.6	33.8	109	158
Normal water level/m	1140	1086	837	1145	970	760
Distance from river source/km	268	289.9	372	238.4	333	443
Average annual flow/m^3 s^{-1}	155	141	395	123	355	502
Water residence time/day[b]	380	44	7	45	33	53
Regulation mode	Annual	Seasonal	Daily	Seasonal	Seasonal	Seasonal
Total nitrogen/mg L^{-1c}	5.21	4.38	7.57	6.25	8.81	8.46
Total phosphorus/mg L^{-1c}	0.019	0.003	<0.001	0.019	0.035	0.468
Chl.a/μg L^{-1c}	2.25	3.73	1.34	10.36	3.35	8.99
Trophic level index[c]	49	42	40	57	56	72
Trophic stage[c]	I	I	I	II	II	III

[a]Hongjiadu Reservoir, HJD; Yinzidu Reservoir, YZD; Suofengying Reservoir, SFY; Puding Reservoir, PD; Dongfeng Reservoir, DF; Wujiangdu Reservoir, WJD
[b]Data obtained from Guo (2008)
[c]Data obtained from Yu (2008)
I oligotrophic–mesotrophic; *II* mesotrophic–eutrophic; *III* hyper-eutrophic

(#17) was an Hg-polluted tributary of WJD by Xifeng Phosphorus and Calcium Factory; but other than that, no more point source of Hg within Wujiang River Basin was found during the entire sampling period.

Surface water samples (20-cm depth) from each sampling site were collected once a month during a one-year period from January to December in 2006 using an acid-washed, Teflon coated, 10 L Niskin bottle water sampler (Model 1010X series, General Oceanics Inc. USA.). The Niskin bottle was positioned in the upstream direction relative to the operator. All samples were collected following ultra-clean sample-handling protocols. Water samples were transferred from the sampler into acid-cleaned borosilicate glass bottles. These bottles had been rigorously cleaned prior to use by immersion in dilute acid (10% HNO_3), followed by rinsing three times with ultrapure deionized water (18.2 MΩ Milli-Q) and heating for 1 h in a muffle furnace at 500 °C; they were then double-bagged and stored in a wooden box before sampling. Each bottle was rinsed three times with reservoir water before sample collection. Filtered samples were conducted within 24 h using 0.45 μm pore-size, 47 mm (diameter) polycarbonate membrane filter (Millipore), and subsequently analyzed for dissolved Hg (DHg), dissolved MeHg (DMeHg), and dissolved organic carbon (DOC). Total Hg (THg), reactive Hg (RHg), total MeHg (TMeHg), and total suspended solids (TSS) were analyzed in each of the unfiltered samples.

All water samples used for Hg species analysis were acidified on-site to 0.5% (*v/v*) using concentrated hydrochloric acid (HCl); the sample bottles were then sealed, double-bagged, and transported to the lab on ice within 24 h. They were stored in a refrigerator at +4 °C in the dark and analyzed in three weeks. The TSS content was determined gravimetrically by filtering an aliquot of water (typically 1500 mL) through a pre-weighed 0.45 μm pore-size, 47 mm (diameter) polycarbonate membrane filter. Water quality parameters, such as pH, temperature (T), and dissolved oxygen (DO), were measured in situ using a portable analyzer (Pioneer 65 Portable Multiparameter Instrument, France). The DOC was qualified in the laboratory using a high-temperature combustion technique with a TOC analyzer (Elementar High TOC II).

Quality control for the THg and MeHg determinations was conducted by field blanks, method blank, matrix spikes, and duplicate samples. The method detection limit, based on three times the standard deviation of replicate measurements of blank solution ($3 \times \sigma$), was 0.02 ng L^{-1} for THg and 0.01 ng L^{-1} for MeHg. The method blank in each case was obviously less than the detection limit. Field blanks were all lower than 20% of the lowest concentrations of Hg species in samples. Accordingly, the determined Hg concentrations in water samples were not corrected for either method blank or field blank. The relative standard deviations for duplicate sample analyses were 4.5 and 5.4% for THg and MeHg. Recoveries for matrix spikes (within the range of samples concentrations) ranged from 88 to 108% for the MeHg analysis, and from 93 to 110% for the THg analysis.

Statistical evaluation was performed using SPSS 11.5 software. Relationships between covariant sets of data were analyzed by regression analysis. Correlation coefficients (r) and significance probabilities (p) were computed for the linear regression fits. T-test or AVOVA test was employed to compare the significant difference between paired or unpaired samples. In addition, Kolmogorov–Smirnov (K–S) and Kruskal–Wallis (K–W) tests were performed to compare significant differences between two or more independent datasets. Significant differences were all declared at $P < 0.05$. To reveal any seasonal variations of geochemical characteristics and Hg species in the surface water in Wujiang River Basin, all the data including water quality parameters and concentrations of Hg species were compared among spring (March, April, and May), summer (June, July, and August), fall (September, October, and November), and winter (December, January, and February). It should be noted that the mean concentrations of Hg species in inflow/outflow rivers were calculated through the arithmetic averaged concentrations of multiple inflow/outflow rivers throughout all the sampling campaigns.

5.2 General Water Quality Characteristics

The distribution of monthly pH, T, DO, DOC, TSS, and conductivity in inflow and outflow rivers of six studied reservoirs in Wujiang River Basin are shown in Tables 5.3 and 5.4.

The surface water temperature exhibited expected seasonal patterns, with the maximum and minimum values occurring in the summer and winter, respectively. However, the seasonal trend of surface water pH was not clear, though pH in the surface water was slightly alkaline in all of the samples (annual average value of 8.15 ± 0.36). The annual average concentration of DOC in the surface water in the Wujiang River was 0.91 ± 0.42 mg L^{-1}, which was lower than that observed in the Yangtze River, as well as in most rivers globally (Duan 2000). The high pH and low DOC in surface water can be explained by the karstic geology in Wujiang River Basin. Moreover, DOC showed obvious seasonal patterns, with the highest average DOC values in the summer and lowest values in the winter, which may be caused by the inner source of natural organic matter (NOM), produced in situ (phytoplankton) to the DOC.

Slightly elevated values of TSS in the surface water of Wujiang River Basin were observed in May, June, July, and October. Seasonal patterns of DO were pronounced and well correlated with TSS and T. Elevated DO in surface water during the period from late May to August compared to the remaining months of the year were explained by algal blooms in some sections of reservoirs, such as the downstream of WJD (Meng et al. 2010; Zhu 2005). Conductivity showed slightly seasonal distributions that were opposite to that of DOC and TSS, with a decreased average conductivity value in rainy seasons (during the period from August to October). This was believed to be the result of dilution of the river water by rainfall during the rainy season.

5.3 Spatial and Temporal Variations of Mercury Species in River Water

Temporal distributions of THg, DHg, PHg, and RHg concentrations in surface water in inflow and outflow rivers of the six studied reservoirs are shown in Figs. 5.2, 5.3, 5.5 and 5.6, and summary data are shown in Tables 5.5 and 5.6.

5.3.1 Total Mercury

The annual mean concentration of THg in inflow–outflow rivers of reservoirs in Wujiang River Basin was 3.41 ± 1.98 ng L^{-1} (ranging from 1.43 ng L^{-1} to

Table 5.3 Monthly values (mean ± SD) of pH, temperature (T), dissolved oxygen (DO), dissolved organic carbon (DOC), total suspended solid (TSS), and conductivity in inflow rivers of six reservoirs in Wujiang River Basin, Guizhou Province, China (January–December 2006) (reprinted from Zhao et al. 2017, with permission from Elsevier Ltd.)

Month	T (°C)	pH	DO ($mg\ L^{-1}$)	DOC ($mg\ L^{-1}$)	TSS ($mg\ L^{-1}$)	Conductivity ($\mu s\ cm^{-2}$)
January	11.7 ± 1.16	8.10 ± 0.16	9.07 ± 0.47	0.78 ± 0.12	1.37 ± 0.83	406 ± 31
February	11.6 ± 0.78	8.35 ± 0.24	8.86 ± 0.35	0.78 ± 0.34	1.66 ± 0.71	408 ± 22
March	14.6 ± 0.65	8.26 ± 0.23	8.51 ± 0.33	0.88 ± 0.35	1.84 ± 0.83	412 ± 14
April	18.0 ± 3.2	8.67 ± 0.17	8.59 ± 0.32	0.91 ± 0.37	2.38 ± 1.35	407 ± 39
May	21.2 ± 2.2	8.40 ± 0.28	9.59 ± 0.83	1.25 ± 0.52	3.25 ± 1.95	398 ± 54
June	21.6 ± 2.5	8.11 ± 0.15	9.65 ± 1.14	1.33 ± 0.42	3.51 ± 1.88	417 ± 53
July	27.2 ± 2.7	8.22 ± 0.26	9.80 ± 1.14	1.21 ± 0.34	2.78 ± 1.38	428 ± 35
August	27.7 ± 1.3	8.38 ± 0.21	9.39 ± 0.81	1.06 ± 0.46	1.71 ± 0.87	404 ± 43
October	21.2 ± 2.3	7.98 ± 0.31	8.92 ± 0.77	0.89 ± 0.26	2.76 ± 1.33	369 ± 39
September	20.7 ± 1.5	8.23 ± 0.24	8.43 ± 1.18	0.92 ± 0.32	3.54 ± 2.60	381 ± 46
November	17.8 ± 1.7	8.01 ± 0.31	8.95 ± 0.96	0.81 ± 0.39	2.07 ± 1.09	400 ± 53
December	14.6 ± 1.7	8.14 ± 0.35	9.23 ± 1.07	0.84 ± 0.54	2.14 ± 1.37	375 ± 40

Table 5.4 Monthly values (mean ± SD) of pH, temperature (T), dissolved oxygen (DO), dissolved organic carbon (DOC), total suspended solid (TSS), and conductivity in outflow rivers of six reservoirs in Wujiang River Basin, Guizhou Province, China (January–December 2006) (reprinted from Zhao et al. 2017, with permission from Elsevier Ltd.)

Month	T (°C)	pH	DO ($mg\ L^{-1}$)	DOC ($mg\ L^{-1}$)	TSS ($mg\ L^{-1}$)	Conductivity ($\mu s\ cm^{-2}$)
January	11.5 ± 1.81	8.19 ± 0.14	8.39 ± 0.40	0.75 ± 0.14	1.93 ± 0.83	395 ± 41
February	11.6 ± 0.93	8.28 ± 0.22	8.28 ± 0.18	0.64 ± 0.10	1.66 ± 0.71	414 ± 14
March	13.1 ± 0.45	8.23 ± 0.24	7.88 ± 0.18	0.71 ± 0.14	1.84 ± 0.83	406 ± 5.0
April	14.1 ± 1.58	8.50 ± 0.06	8.17 ± 0.35	0.70 ± 0.10	2.38 ± 1.35	397 ± 28
May	15.8 ± 2.01	8.47 ± 0.26	9.82 ± 1.57	0.88 ± 0.14	3.25 ± 1.95	406 ± 45
June	17.6 ± 1.75	8.02 ± 0.08	8.16 ± 0.42	1.42 ± 0.83	3.51 ± 1.88	407 ± 31
July	20.9 ± 1.29	7.66 ± 0.19	8.75 ± 0.76	1.30 ± 0.15	2.78 ± 1.38	381 ± 31
August	21.9 ± 1.79	7.49 ± 0.37	8.35 ± 0.62	1.01 ± 0.25	1.71 ± 0.87	386 ± 44
October	21.6 ± 0.87	7.71 ± 0.22	8.06 ± 0.72	0.93 ± 0.17	2.76 ± 1.33	369 ± 41
September	20.2 ± 1.95	7.56 ± 0.20	7.81 ± 0.40	0.79 ± 0.30	3.54 ± 2.60	355 ± 17
November	18.6 ± 1.36	7.69 ± 0.23	8.32 ± 0.74	0.71 ± 0.09	2.07 ± 1.09	371 ± 31
December	15.0 ± 1.56	7.94 ± 0.38	8.26 ± 1.03	0.57 ± 0.13	2.14 ± 1.37	352 ± 41

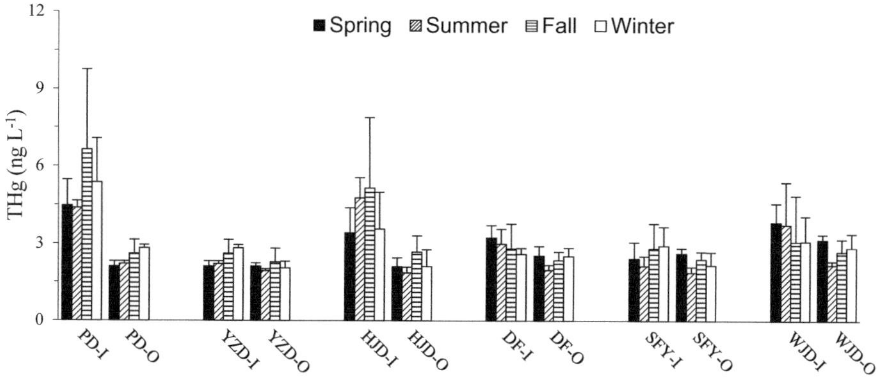

Fig. 5.2 Annual mean concentrations of total mercury (THg) in inflow and outflow rivers of six reservoirs in Wujiang River Basin, Guizhou Province, China (PD-I, inflow of PD; PD-O, outflow of PD; YZD-I, inflow of YZD; YZD-O, outflow of YZD; HJD-I, inflow of HJD; HJD-O, outflow of HJD; DF-I, inflow of DF; DF-O, outflow of DF; SFY-I, inflow of SFY; SFY-O, outflow of SFY; WJD-I, inflow of WJD; WJD-O, outflow of WJD) (reprinted from Zhao et al. 2017, with permission from Elsevier Ltd.)

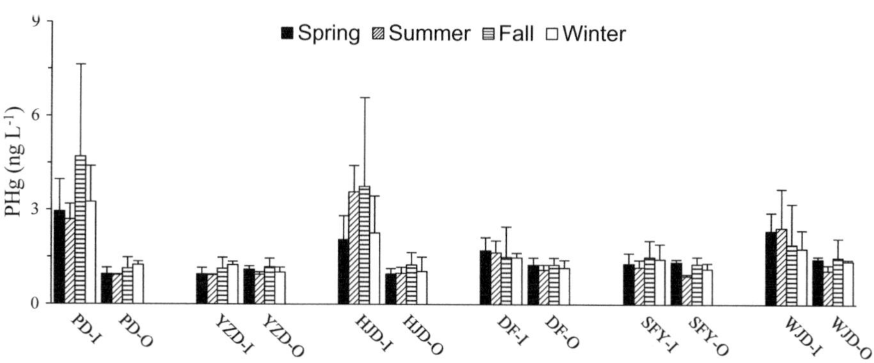

Fig. 5.3 Annual mean concentrations of particulate mercury (PHg) in inflow and outflow rivers of six reservoirs in Wujiang River Basin, Guizhou Province, China (PD-I, inflow of PD; PD-O, outflow of PD; YZD-I, inflow of YZD; YZD-O, outflow of YZD; HJD-I, inflow of HJD; HJD-O, outflow of HJD; DF-I, inflow of DF; DF-O, outflow of DF; SFY-I, inflow of SFY; SFY-O, outflow of SFY; WJD-I, inflow of WJD; WJD-O, outflow of WJD) (reprinted from Zhao et al. 2017, with permission from Elsevier Ltd.)

19.04 ng L^{-1}) throughout all the sampling sites. The annual average concentration of THg in the inflow rivers of the reservoirs was 3.90 ± 2.21 ng L^{-1}, with the highest value observed in XF (#17) and the lowest seen in the inflow river of SFY (Outflow of DF, #11). The elevated levels of THg in XF (#17) could be attributed to the point source of Hg from local factories (Xifeng Phosphorus and Calcium

Table 5.5 Annual mean concentrations (mean ± SD) of Hg species in inflow rivers of reservoirs in Wujiang River Basin, Guizhou Province, China (ng L^{-1}) (reprinted from Zhao et al. 2017, with permission from Elsevier Ltd.)

Reservoirs	THg[b]	DHg[b]	PHg[b]	RHg[b]	MeHg[b]	DMeHg[b]
PD[a]	5.21 ± 1.96	1.81 ± 0.48	3.40 ± 1.75	0.27 ± 0.09	0.12 ± 0.02	0.06 ± 0.01
HJD[a]	4.22 ± 1.73	1.30 ± 0.29	2.92 ± 1.71	0.28 ± 0.11	0.13 ± 0.04	0.06 ± 0.02
YZD[a]	2.43 ± 0.40	1.37 ± 0.23	1.07 ± 0.23	0.24 ± 0.06	0.18 ± 0.05	0.11 ± 0.03
DF[a]	2.90 ± 0.63	1.28 ± 0.26	1.70 ± 0.55	0.21 ± 0.11	0.11 ± 0.02	0.05 ± 0.01
SFY[a]	2.56 ± 0.74	1.21 ± 0.36	1.35 ± 0.41	0.20 ± 0.07	0.15 ± 0.04	0.08 ± 0.03
WJD[a]	4.59 ± 2.99	1.54 ± 0.60	3.05 ± 2.64	0.28 ± 0.15	0.16 ± 0.10	0.07 ± 0.03

[a]*PD* Puding Reservoir; *HJD* Hongjiadu Reservoir; *YZD* Yinzidu Reservoir; *DF* Dongfeng Reservoir; *SFY* Suofengying Reservoir; *WJD* Wujiangdu Reservoir

[b]*THg* total mercury; *DHg* dissolved mercury; *PHg* particulate mercury; *RHg* reactive mercury; *TMeHg* total methylmercury; *DMeHg* dissolved methylmercury

Table 5.6 Annual mean concentrations (mean ± SD) of Hg species in outflow rivers of reservoirs in Wujiang River Basin, Guizhou Province, China (ng L^{-1}) (reprinted from Zhao et al. 2017, with permission from Elsevier Ltd.)

Reservoirs	THg[b]	DHg[b]	PHg[b]	RHg[b]	TMeHg[b]	DMeHg[b]
PD[a]	2.43 ± 0.40	1.37 ± 0.23	1.07 ± 0.23	0.24 ± 0.06	0.18 ± 0.05	0.11 ± 0.03
HJD[a]	2.18 ± 0.53	1.11 ± 0.26	1.06 ± 0.30	0.23 ± 0.08	0.14 ± 0.03	0.07 ± 0.02
YZD[a]	2.09 ± 0.29	1.02 ± 0.16	1.06 ± 0.17	0.25 ± 0.09	0.14 ± 0.02	0.07 ± 0.01
DF[a]	2.33 ± 0.36	1.14 ± 0.21	1.19 ± 0.20	0.17 ± 0.05	0.16 ± 0.05	0.09 ± 0.03
SFY[a]	2.25 ± 0.40	1.09 ± 0.21	1.16 ± 0.22	0.20 ± 0.09	0.14 ± 0.02	0.07 ± 0.01
WJD[a]	2.70 ± 0.50	1.36 ± 0.37	1.35 ± 0.32	0.25 ± 0.09	0.25 ± 0.05	0.16 ± 0.06

[a]*PD* Puding Reservoir; *HJD* Hongjiadu Reservoir; *YZD* Yinzidu Reservoir; *DF* Dongfeng Reservoir; *SFY* Suofengying Reservoir; *WJD* Wujiangdu Reservoir

[b]*THg* total mercury; *DHg* dissolved mercury; *PHg* particulate mercury; *RHg* reactive mercury; *TMeHg* total methylmercury; *DMeHg* dissolved methylmercury

Factory). The average concentration of THg in surface water in Wujiang River Basin is comparable with the values that are recognized as uncontaminated rivers of North American and Europe (<5 ng L^{-1}) (Ullrich et al. 2001). Moreover, the mean THg concentration in surface water in Wujiang River Basin is significantly below the Chinese surface water standard of 50 ng L^{-1} (Ministry of Environmental Protection 2002), and also below the 12 ng L^{-1} standard for THg recommended by the USEPA to protect against adverse chronic effects on aquatic life (USEPA 1992).

The seasonal distribution patterns and mean concentrations of THg in inflow and outflow rivers of reservoirs in Wujiang River Basin were illustrated in Fig. 5.2. No discernable difference in the THg concentration in inflow and outflow rivers of reservoirs was observed among the spring, summer, fall, and winter (K–W test, $p > 0.05$, $n = 39$), with the inflow river water in PD and HJD as exceptions. The mean concentrations of THg in inflow rivers in PD (5.21 ± 1.96 ng L^{-1}) and HJD (4.22 ± 1.73 ng L^{-1}) were significantly higher than those in the four other reservoirs (2-way ANOVA, $p < 0.01$). The statistical analysis further observed that concentrations of THg in inflow river of PD and HJD in rainy seasons (summer and fall) were significantly higher than those in the dry season (spring and winter) (T-test, $p < 0.01$).

As shown in Fig. 5.1, PD and HJD are located at the upper end of Wujiang River Basin. With the increase of rainfall and rainfall intensity during the rainy season, more Hg in surface runoff and sediment re-suspension entered the rivers (Guo 2008), which resulted in the elevated THg concentration compared with the dry season. When the river water flows into the reservoir, the water flow rate decreases significantly. Therefore, concentrations of TSS in outflow river of PD (2.05 ± 0.81 mg L^{-1}) and HJD (1.28 ± 0.84 mg L^{-1}) were also significantly decreased when compared with that in inflow rivers of PD (3.37 ± 1.12 mg L^{-1}) and HJD (3.14 ± 1.47 mg L^{-1}) (T-test, $p < 0.01$, both for PD and HJD). Consequently, a considerable amount of Hg bound to particulate matter was intercepted by way of sedimentation (Guo 2008), which explains the lower THg concentrations in outflow rivers of reservoirs compared with that in inflow rivers in PD and HJD. As shown in Fig. 5.2, THg concentrations in outflow rivers of reservoirs were lower than that in the corresponding inflow river throughout the six reservoirs (T-test, $p < 0.05$), which was primarily caused by the sedimentation of PHg after the construction of reservoirs. Previous studies observed that the construction of the reservoir reduced the concentration of Hg in river water (Bonzongo et al. 1996; Lawson et al. 2001; Quemerais et al. 1999), which is consistent with our results.

The statistical analysis showed that significant positive correlations were observed between PHg and TSS in inflow rivers throughout the six reservoirs during the sampling periods (Table 5.7), indicating that the Hg bound to particulate matter from runoff (riverbank and riverbed) was the primary source of PHg in inflow rivers. Furthermore, concentrations of THg in inflow rivers were significantly correlated with PHg and TSS, with correlation coefficients of 0.98 and 0.62, respectively, indicating that the level of THg in inflow rivers was controlled by PHg and TSS. However, no correlation was observed between TSS and THg, PHg in outflow rivers of reservoirs during the sampling periods (TSS *vs.* THg: $r = -0.04$, $p = 0.71$, $n = 72$; TSS *vs.* PHg: $r = -0.07$, $p = 0.58$, $n = 72$), which could be explained by the dissolution of PHg in reservoir and contribution of planktonic derived NOM to TSS. Our results were consistent with observations from previous studies (Lawson et al. 2001; Guentzel et al. 2007; Paraquetti et al. 2004).

Table 5.7 Pearson's Correlation Matrix, giving the Linear Correlation Coefficients (r) among the concentrations of Hg species and total suspended solid (TSS) in inflow rivers of reservoirs in Wujiang River Basin, Guizhou Province, China (n = 168) (reprinted from Zhao et al. 2017, with permission from Elsevier Ltd.)

	THg	PHg	DHg	RHg	TMeHg	PMeHg	DMeHg	TSS
THg	1							
PHg	0.98^{b}	1						
DHg	0.61^{b}	0.45^{b}	1					
RHg	0.49^{b}	0.50^{b}	0.23^{b}	1				
TMeHg	0.48^{b}	0.50^{b}	0.17^{a}	0.26^{b}	1			
PMeHg	0.58^{b}	0.61^{b}	0.16^{a}	0.37^{b}	0.93^{b}	1		
DMeHg	0.15	0.14	0.14	0.02	0.76^{b}	0.49^{b}	1	
TSS	0.62^{b}	0.62^{b}	0.31^{b}	0.31^{b}	0.47^{b}	0.53^{b}	0.21^{b}	1

THg total mercury; *PHg* particulate mercury; *DHg* dissolved mercury; *RHg* reactive mercury; *TMeHg* total methylmercury; *PMeHg* particulate methylmercury; *DMeHg* dissolved methylmercury; *TSS* total suspended solid

[a]Correlation is significant at the 0.05 level (2-tailed)

[b]Correlation is significant at the 0.01 level (2-tailed)

5.3.2 *Particulate Mercury*

The annual mean concentration of PHg in rivers (inflow and outflow rivers) of the reservoirs in Wujiang River Basin was 2.05 ± 1.73 ng L^{-1} (ranging from 0.62 ng L^{-1} to 17.1 ng L^{-1}) throughout all the sampling sites. The annual mean concentrations of PHg in inflow river and outflow rivers were 2.46 ± 1.95 ng L^{-1} and 1.15 ± 0.26 ng L^{-1}, respectively. As shown in Fig. 5.3, concentrations of PHg in inflow rivers in the rainy seasons (summer and fall) were generally higher than those in the dry seasons (winter and spring) (T-test, $p < 0.05$), especially at PD and HJD, which was similar to the seasonal trend of THg. Furthermore, the PHg concentration in outflow rivers was significantly lower than that in inflow rivers throughout the six reservoirs during the sampling periods (T-test, $p < 0.05$).

Due to the relatively low primary productivity (Chlorophyll a) of the inflow rivers of reservoirs in Wujiang River Basin (Zhu 2005), the adsorption of Hg by phytoplankton was very limited. Our previous study showed that concentrations of PHg in the bottom water from upstream to downstream of WJD were all significantly higher than the corresponding overlying water, confirming the input of particulate Hg from sediment re-suspension (Meng et al. 2010). When compared to the reservoir, inflow river represents a relatively strong hydrodynamic condition. Hence, it is reasonable to speculate that the PHg in inflow river water of Wujiang River Basin was primarily from runoff together with sediment re-suspension (Hurley et al. 1998; Meng et al. 2010).

PHg represented 56.1 ± 10.1% of THg in the river water. The statistical analysis showed that the ratio of PHg to THg (PHg/THg) in inflow rivers (59%) was significantly higher than that in outflow rivers (49%) throughout the six reservoirs

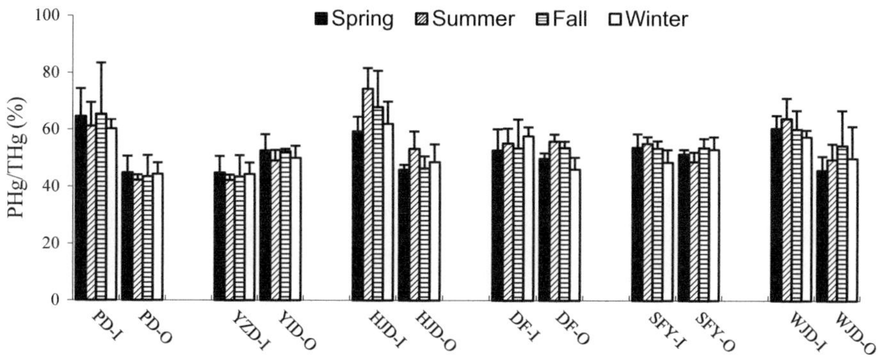

Fig. 5.4 Annual ratios of particulate mercury (PHg) to total mercury (THg) (PHg/THg) in inflow and outflow rivers of six reservoirs in Wujiang River Basin, Guizhou Province, China (PD-I, inflow of PD; PD-O, outflow of PD; YZD-I, inflow of YZD; YZD-O, outflow of YZD; HJD-I, inflow of HJD; HJD-O, outflow of HJD; DF-I, inflow of DF; DF-O, outflow of DF; SFY-I, inflow of SFY; SFY-O, outflow of SFY; WJD-I, inflow of WJD; WJD-O, outflow of WJD) (reprinted from Zhao et al. 2017, with permission from Elsevier Ltd.)

during the sampling periods (T-test, $p < 0.05$), which can be explained by the interception effect of reservoirs to particulate materials but not the dissolution of PHg in reservoir. However, the ratios of PHg/THg in inflow rivers of Wujiang River Basin were much lower than the observations in inflow rivers (Fox river) of Lake Michigan in North American (93.6%) (Hurley et al. 1998). In this study, the average ratio of PHg/THg in the inflow rivers of reservoirs during the rainy season (61%) was slightly higher than that during the dry season (57%) (T-test, $p > 0.05$). Although the temporal/seasonal trend of PHg/THg was not clear, the spatial distribution of PHg/THg in inflow rivers was clearly observed in this study (Fig. 5.4). In detail, the average ratios of PHg/THg in inflow rivers of PD (63%) and HJD (66%), which are located in the upstream section of Wujiang River Basin, were significantly higher than those in inflow rivers of downstream reservoirs (e.g., YZD, 44%; DF, 55%; SFY, 53%) (T-test, $p < 0.05$). The clear spatial distribution patterns of PHg/THg in inflow rivers of Wujiang River Basin further confirmed that particulate matter from soil erosion was the potential sources of PHg in the inflow rivers of the upstream reservoirs in Wujiang River Basin (Zhang et al. 2009; Meng et al. 2010; Yao et al. 2011).

5.3.3 Dissolved Mercury

The annual mean concentration of DHg in river water (inflow and outflow rivers) in Wujiang River Basin was 1.36 ± 0.44 ng L^{-1} (ranging from 0.57 ng L^{-1} to 3.1 ng L^{-1}) throughout all the sampling sites. The annual mean concentration of DHg in inflow rivers (1.44 ± 0.48 ng L^{-1}) was slightly higher than that in outflow

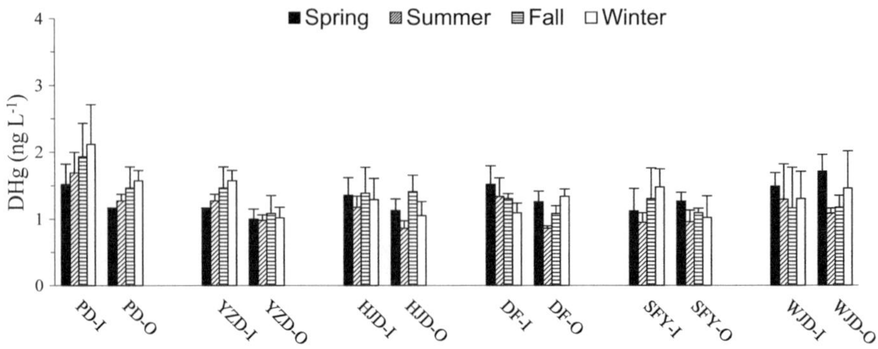

Fig. 5.5 Annual mean concentrations of dissolved mercury (DHg) in inflow and outflow rivers of six reservoirs in Wujiang River Basin, Guizhou Province, China (PD-I, inflow of PD; PD-O, outflow of PD; YZD-I, inflow of YZD; YZD-O, outflow of YZD; HJD-I, inflow of HJD; HJD-O, outflow of HJD; DF-I, inflow of DF; DF-O, outflow of DF; SFY-I, inflow of SFY; SFY-O, outflow of SFY; WJD-I, inflow of WJD; WJD-O, outflow of WJD) (reprinted from Zhao et al. 2017, with permission from Elsevier Ltd.)

rivers (1.18 ± 0.28 ng L^{-1}). However, no discernable spatial and seasonal trends of DHg in the rivers were observed throughout the six reservoirs during the sampling periods (Fig. 5.5), which was different from that for THg and PHg.

The Partition Coefficient (K_d) between Hg in the particulate phase (solid) and the dissolved phase (liquid) was calculated based on the measured concentrations of PHg, DHg, and TSS, as described in the following equation (Hurley et al. 1998; Covelli et al. 2006):

$$K_d = {S_{Hg}}/{C_{DHg}} \tag{5.1}$$

where K_d is the Partition Coefficient between Hg in the particulate phase and the dissolved phase (L kg^{-1}); S_{Hg} is the concentration of Hg species absorbed in particulate materials (ng kg^{-1}); and C_{DHg} is the concentration of dissolved Hg species in water sample (ng L^{-1}).

S_{Hg} is calculated using the following equation:

$$S_{Hg} = {C_{PHg}}/{C_{TSS}} \tag{5.2}$$

where C_{PHg} is the concentration of particulate Hg species in a water sample (ng L^{-1}); and C_{TSS} is the content of TSS in the water sample (kg L^{-1}).

Our calculated data showed that the average $\log K_d$ in rivers of the Wujiang River Basin was 5.8 ± 0.20 (ranging from 5.1 to 6.6), with values in inflow rivers and outflow rivers of 5.8 ± 0.2 (ranging from 5.1 to 6.3) and 5.9 ± 0.3 (ranging from 5.2 to 6.6), respectively. The $\log K_\mathrm{d}$ observed in this study was slightly higher than that in other rivers globally, such as the inflow rivers of the Chesapeake Bay (5.06–5.52) (Lawson et al. 2001) and the Fox River (5.4–6.0, Hurley et al. 1998). However, our results showed similar levels of $\log K_d$ when compared to the Isonzo River (5.5–6.2, Covelli et al. 2006) and the Lot-Garonne River (5.2–6.2, Schaefer et al. 2006).

5.3.4 Reactive Mercury

The annual mean concentration of RHg in river water (inflow and outflow rivers) in Wujiang River Basin was 0.24 ± 0.11 ng L^{-1} (ranging from 0.09 ng L^{-1} to 1.01 ng L^{-1}) throughout all the sampling sites. The annual mean concentrations of RHg in inflow rivers and outflow rivers were 0.25 ± 0.12 ng L^{-1} and 0.22 ± 0.08 ng L^{-1}, respectively. There was no significant difference in RHg concentrations between inflow rivers and outflow rivers because RHg in water primary exists in a dissolved phase. (T-test, $p > 0.05$) (Fig. 5.6). However, the ratio of RHg/THg in inflow rivers (7.4%) was significantly lower than that in outflow rivers (9.7%) (T-test, $p < 0.05$). Compared with inflow rivers, concentrations of THg in outflow rivers were significantly decreased due to the interception effect of the reservoir. Therefore, the relatively higher RHg/THg in outflow rivers (compared with inflow rivers) can be explained by the decreased THg in outflow rivers (compared with inflow rivers) together with the stable RHg concentrations between inflow rivers and outflow rivers.

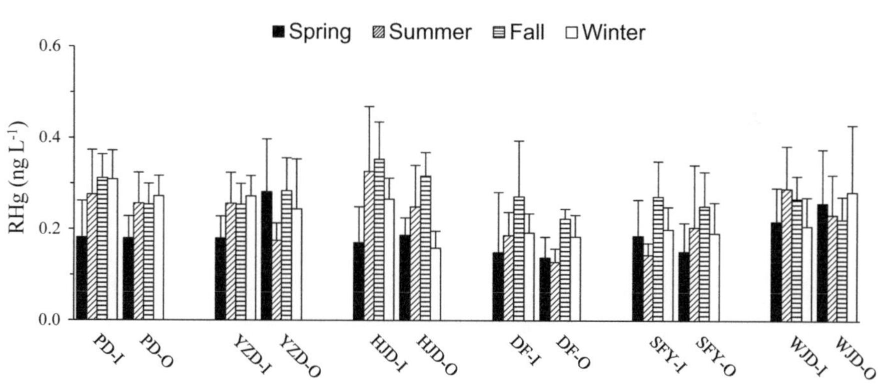

Fig. 5.6 Annual mean concentrations of reactive mercury (RHg) in inflow and outflow rivers of six reservoirs in Wujiang River Basin, Guizhou Province, China (PD-I, inflow of PD; PD-O, outflow of PD; YZD-I, inflow of YZD; YZD-O, outflow of YZD; HJD-I, inflow of HJD; HJD-O, outflow of HJD; DF-I, inflow of DF; DF-O, outflow of DF; SFY-I, inflow of SFY; SFY-O, outflow of SFY; WJD-I, inflow of WJD; WJD-O, outflow of WJD) (reprinted from Zhao et al. 2017, with permission from Elsevier Ltd.)

5.3.5 Methyl Mercury

The annual mean concentration of TMeHg in river water (inflow and outflow rivers) in Wujiang River Basin was 0.15 ± 0.06 ng L^{-1} (ranging from 0.07 ng L^{-1} to 0.70 ng L^{-1}) throughout all the sampling sites. The annual mean concentrations of TMeHg in inflow rivers and outflow rivers were 0.14 ± 0.06 ng L^{-1} and 0.17 ± 0.06 ng L^{-1}, respectively. Seasonal distribution patterns and concentrations of TMeHg in inflow rivers and outflow rivers are illustrated in Fig. 5.7. The statistical analysis showed that TMeHg concentrations in inflow rivers in the summer were significantly higher than those in the three other seasons (winter, spring, and fall) (T-test, $p < 0.05$). The elevated TMeHg concentrations in inflow rivers during the summer can be explained by the increased surface runoff due to the sufficient precipitation (Zhang et al. 2009; Meng et al. 2010), which subsequently resulted in the increase of MeHg entering into river water through surface runoff and sediment re-suspension pathways. Furthermore, another reason could be the higher water temperature together with the increasing amount of algal-derived NOM during summer, which can stimulate bacterial activity and Hg methylation (Bravo et al. 2017).

A significantly positive correlation between TMeHg and TSS in inflow rivers was observed in this study, with a correlation coefficient of 0.47 ($p < 0.001$, $n = 168$) (Table 5.7). Moreover, TMeHg concentrations in inflow rivers were positively correlated with PMeHg ($r = 0.93$, $P < 0.001$, $n = 168$). The significant correlations between TMeHg and TSS as well as PMeHg in inflow rivers implied that river erosion and surface runoff could be the main sources of MeHg in the inflow rivers of the six reservoirs in Wujiang River Basin. The outflow river water

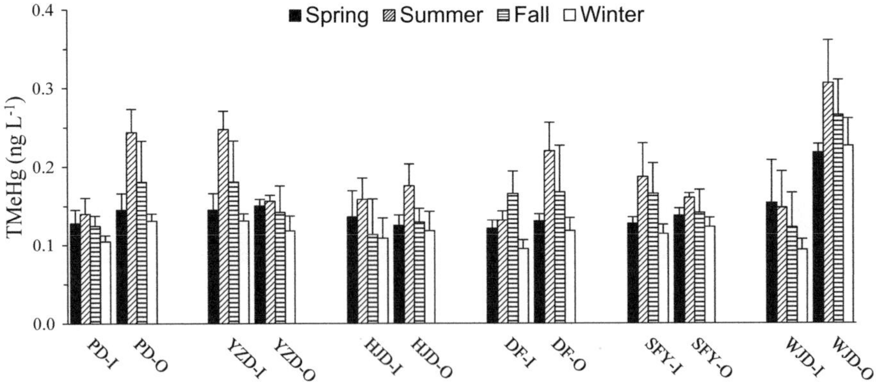

Fig. 5.7 Annual mean concentrations of total methylmercury (TMeHg) in inflow and outflow rivers of six reservoirs in Wujiang River Basin, Guizhou Province, China (PD-I, inflow of PD; PD-O, outflow of PD; YZD-I, inflow of YZD; YZD-O, outflow of YZD; HJD-I, inflow of HJD; HJD-O, outflow of HJD; DF-I, inflow of DF; DF-O, outflow of DF; SFY-I, inflow of SFY; SFY-O, outflow of SFY; WJD-I, inflow of WJD; WJD-O, outflow of WJD) (reprinted from Zhao et al. 2017, with permission from Elsevier Ltd.)

was mainly from the discharged water of the reservoirs. In this study, concentrations of TMeHg in outflow rivers in warm seasons (especially during the period from July to September) were significantly higher than those in the dry seasons (T-test, $P < 0.01$). Furthermore, TMeHg concentrations in outflow rivers were also significantly higher than that in inflow rivers especially in rainy seasons (T-test, $P < 0.05$). These observations implied the active Hg methylation in reservoirs during warm seasons. Canavan et al. (2000) observed that concentrations of MeHg in the discharged water of a reservoir in New Mexico were approximately six times higher from late summer to early fall. This observation shown above was consistent with the results in this study.

The average ratio of TMeHg to THg (TMeHg/THg) in inflow–outflow rivers throughout the six reservoirs was 5.3 ± 2.9%. The ratios of TMeHg/THg in inflow rivers throughout the six reservoirs exhibited the following distribution patterns (Fig. 5.8): summer (5%) > spring, fall (4%) > winter (3%), with the corresponding trend of summer (11%) > fall (7%) > spring, winter (6%) in outflow rivers. Furthermore, the ratio of TMeHg/THg in outflow rivers (7.4%) was significantly higher than that in inflow rivers (4.3%) (T-test, $p < 0.05$), which further confirmed the net Hg methylation in the reservoirs.

The annual mean concentration of DMeHg in river water (inflow and outflow rivers) in Wujiang River Basin was 0.08 ± 0.03 ng L^{-1} (ranging from 0.03 ng L^{-1} to 0.26 ng L^{-1}) throughout all the sampling sites. The annual mean concentrations of DMeHg in inflow rivers and outflow rivers were 0.07 ± 0.03 ng L^{-1} and 0.10 ± 0.04 ng L^{-1}, respectively. Concentrations of DMeHg in outflow rivers were significantly higher than those in inflow rivers (T-test, $p < 0.05$). The higher DMeHg concentrations in outflow rivers can be attributed to the discharge of water from the hypolimnion with high levels of MeHg. Muresan et al. (2008) observed

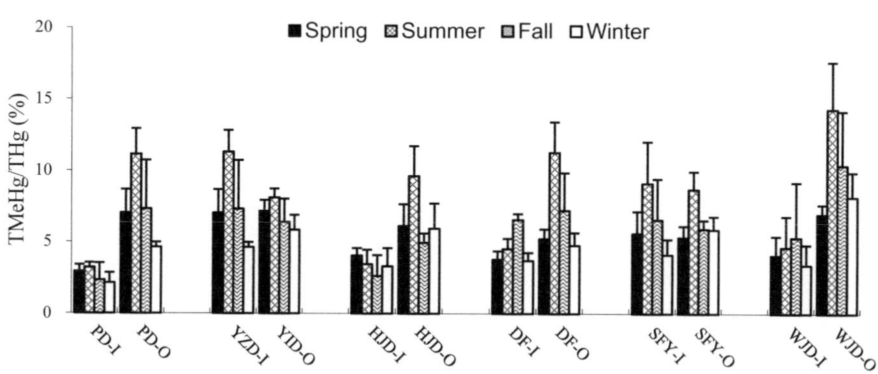

Fig. 5.8 Annual ratios of total methylmercury (TMeHg) to total mercury (THg) (TMeHg/THg) in inflow and outflow rivers of six reservoirs in Wujiang River Basin, Guizhou Province, China (PD-I, inflow of PD; PD-O, outflow of PD; YZD-I, inflow of YZD; YZD-O, outflow of YZD; HJD-I, inflow of HJD; HJD-O, outflow of HJD; DF-I, inflow of DF; DF-O, outflow of DF; SFY-I, inflow of SFY; SFY-O, outflow of SFY; WJD-I, inflow of WJD; WJD-O, outflow of WJD) (reprinted from Zhao et al. 2017, with permission from Elsevier Ltd.)

that MeHg concentrations in the hypolimnion of reservoirs were approximately 3 times higher than those in the upper water column, which support our results.

The seasonal distribution patterns of DMeHg in the inflow–outflow rivers of six reservoirs are illustrated in Fig. 5.9. There were no discernable differences in the DMeHg of inflow rivers among spring, summer, fall, and winter (K–W test, $p > 0.05$). However, DMeHg concentrations in outflow rivers were significantly higher than that in inflow rivers especially in rainy seasons (T-test, $P < 0.01$). Furthermore, DMeHg concentrations in outflow rivers of the six reservoirs in rainy seasons (summer and fall) were significantly higher than those in dry seasons (spring and winter) (T-test, $p < 0.05$), which was consistent with the seasonal trend of TMeHg in outflow rivers. Recently, we observed that both the newly constructed reservoir (e.g., YZD) and old reservoir (e.g., WJD) in Wujiang River Basin were completely stratified during rainy seasons, especially at the downstream section of the reservoirs (Meng et al. 2010). Furthermore, MeHg concentrations and the ratios of MeHg/THg in the hypolimnion of reservoirs were highly elevated compared with overlying water, indicating the active net Hg methylation in the stratification zone of water column (Meng et al. 2010). A previous study confirmed that the elevated DMeHg in the hypolimnion of reservoirs during summer and fall can result in the increase of DMeHg concentrations in outflow rivers (Watras et al. 1995). Choe and Gill (2003) reported that the DMeHg bound to organic colloids can pass through the 0.45 μm size of the filter membrane and exist in the liquid phase. These observations shown above support our results.

The average ratio of DMeHg to TMeHg (DMeHg/TMeHg) in inflow–outflow rivers throughout the six reservoirs was 50.7 ± 7.7%. Specially, the ratio of DMeHg/TMeHg in outflow rivers (55%) was slightly higher than that in inflow rivers (48%), but was not significantly different (T-test, $p > 0{,}05$), which could be

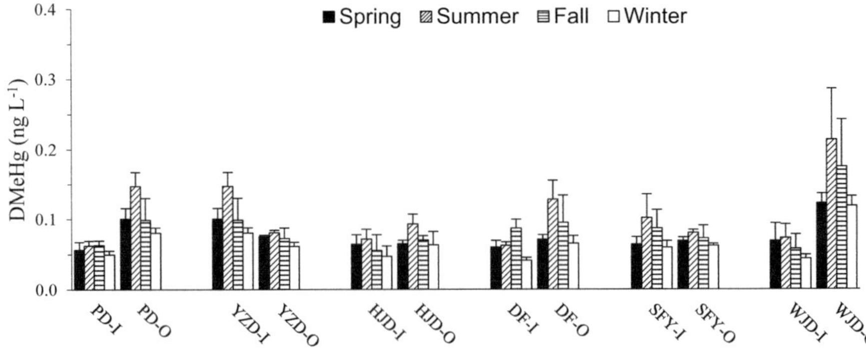

Fig. 5.9 Annual mean concentrations of dissolved methylmercury (DMeHg) in inflow and outflow rivers of six reservoirs in Wujiang River Basin, Guizhou Province, China (PD-I, inflow of PD; PD-O, outflow of PD; YZD-I, inflow of YZD; YZD-O, outflow of YZD; HJD-I, inflow of HJD; HJD-O, outflow of HJD; DF-I, inflow of DF; DF-O, outflow of DF; SFY-I, inflow of SFY; SFY-O, outflow of SFY; WJD-I, inflow of WJD; WJD-O, outflow of WJD) (reprinted from Zhao et al. 2017, with permission from Elsevier Ltd.)

attributed to the dissolution of PMeHg in water body along the reservoirs. We further calculated the ratios of DMeHg to DHg (DMeHg/DHg) in the river water of the six reservoirs. Our results showed that the average ratios of DMeHg/DHg in inflow rivers (5.2 ± 2.7%) were significantly lower than those in outflow rivers (8.4 ± 4.3%) (T-test, $p < 0.05$), which could be explained by the net Hg methylation in reservoirs. Coquery et al. (2003) further observed that the levels of DMeHg/DHg in surface water of reservoirs were similar to inflow rivers; however, the ratios of DMeHg/DHg in the hypolimnion of reservoirs were approximately 10 times higher than those in surface water, indicating active Hg methylation in the hypolimnion.

5.4 Comparisons with Other Reservoirs

The magnitude of THg, DHg, RHg, TMeHg, and DMeHg concentrations in river water samples from the six reservoirs was compared with data from the literature globally (listed in Table 5.8).

Generally, concentrations of THg, DHg, and RHg in the river water of the six reservoirs were significantly lower compared with Hg-contaminated lakes/reservoirs worldwide, including Anacostia River, San Carlos Creek, Fox River, and Carson River in the USA, Riou Mort River in France, and Pra River in Ghana. However, concentrations of Hg species in the river water of the six reservoirs appear to be higher than those reported for other uncontaminated rivers/lakes, including Blacklick Run River, Patuxent River, and Mobile Alabama River in the USA, but slightly lower than the observations in Lot River and Garonne River in France. Similarly, THg, DHg, RHg, TMeHg, and DMeHg concentrations in river water samples from the six reservoirs appear to be lower than those reported for the other lakes in China, including the Hongfeng Reservoir, Aha Reservoir, and Baihua Reservoir, which are located in the same region.

5.5 Influence of Cascade Reservoirs on the Distributions of Mercury Species in River Water

Rivers are major pathways for a variety of materials to flow in and out of reservoirs. Impounding rivers changes the characteristics of a water body from "rivers" to "reservoirs", affecting not only their hydrology but also their physical, chemical, and biological characteristics. In particular, the transportation and transformation of Hg in the reservoir ecosystem will be significantly changed with the change in hydrodynamic conditions, synchronously. After the river enters the reservoir, the capacity of the river to carry particulate matters is weakened. Consequently, the "river's transportation function" under strong hydrodynamic conditions will be

Table 5.8 Comparison of levels of Hg species in inflow and outflow rivers of reservoirs in Wujiang River Basin, Guizhou province, China, with literature data globally (ng L^{-1}, mean (range))

Rivers, country		THg	DHg	RHg	TMeHg	DMeHg	Pollution source of Hg	References
Minnesota, USA		6.6–18			0.20			Balogh et al. (2003)
Mississippi, USA		10			0.35			Balogh et al. (2003)
St. Lawrence, Canada			0.08–2.05				No	Quemerais et al. (1999)
Loire estuary, France			0.40–1.20				No	Coquery et al. (1997)
Seine estuary, France			0.20–0.60				No	Coquery et al. (1997)
Blacklick Run, USA		1.7 ± 1.06			0.014 ± 0.012		No	Lawson et al. (2001)
Herrington Creek, USA		2.1 ± 0.94			0.06 ± 0.06		No	Lawson et al. (2001)
Patuxent, USA		0.5–6	0.2–1.5		0.025	0.02	No	Benoit et al. (1998)
Anacostia, USA	Dry season	6.59		0.91	0.10	0.07	Flow across city	Mason and Sullivan (1998)
	Wet season	59.9		1.64	0.58	0.10		
San Carlos Creek, USA	Control site	12 (4–13)	2.4		0.02		No	Ganguli et al. (2000)
	Hg mine waste water	5.2–41			0.3		Historical Hg mining area	
	7.5 km distance to mining site	700			1.1			

(continued)

Table 5.8 (continued)

Rivers, country		THg	DHg	RHg	TMeHg	DMeHg	Pollution source of Hg	References
Mobile Alabama, USA		1.1 (0.2–3.8)			0.2 (0.01–1.5)		No	Warner et al. (2005)
Lot-Garonne Rive, France	Lot river	2.2–336	0.4–3.2	0.1–1.7			Historical PE factory	Schäfer et al. (2006)
	Garonne rive	0.3–120	0.3–2.1	0.2–1.2				
	Riou Mort	~880	0.8–4.6	0.1–2.4				
Pra, Ghana		28.7–462.1			0.03–19.64		Gold mining	Donkor et al. (2006)
Michigan lake	Fox river	27.9	1.0–1.6		0.17	0.02	Paper mill	Hurley et al. (1998)
	Other Tributaries	1.05–10.3						
Tributary of Sepetiba Bay, Brazil			0.1–66.6	0.1–18.1				Paraquetti et al. (2004)
Carson River, USA	Upstream of mining site	4–28	2–7	0.4	0.3–0.5	0.1–0.4	Gold mining	Bonzongo et al. (1996)
	Downstream of mining site	645–2107	9–46	2.72–3.8	1.8–7.2	0.8–1.2		
Aha lake, China	Inflow river	19.6 (7.1–47.9)	4.6 (2.3–10.1)	1.3 (0.3–2.4)	1.03 (0.16–2.50)	0.36 (0.04–1.22)	domestic sewage	Bai (2006)
	Outflow river	5.45	4.65	2.04	0.63	0.23	No	
Hongfeng Lake, China	Inflow river	25.2 (1.2–150)	4.2 (0.5–21)		0.36 (0.09–0.86)	0.16 (0.04–0.27)	Chemical plant	He (2007)
	Outflow river	3.7	1.55		0.93	0.45	No	

(continued)

Table 5.8 (continued)

Rivers, country		THg	DHg	RHg	TMeHg	DMeHg	Pollution source of Hg	References
Baihua reservoir, China	Inflow river	34.5 (18.7–67.6)	11.9 (6.8–18.3)	3.48 (0.93–8.71)	0.74 (0.25–1.43)	0.24 (0.14–0.43)	Chemical plant	Yan (2005)
	Outflow river	14.6	7.41	1.4	0.36	0.23	No	
Wujiang river, china	Inflow river	3.90 ± 2.21 (3.17[a])	1.44 ± 0.48 (1.22[a])	0.25 ± 0.12 (0.24[a])	0.14 ± 0.06 (0.14[a])	0.07 ± 0.03 (0.07[a])	No	This study
	Outflow river	2.33 ± 0.45 (2.34[a])	1.18 ± 0.28 (1.17[a])	0.22 ± 0.08 (0.22[a])	0.17 ± 0.06 (0.18[a])	0.10 ± 0.04 (0.11[a])	No	

[a]Weighted average concentration

gradually replaced by the "reservoir's deposition function" under relative weakened hydrodynamic conditions, which will subsequently impact the transportation processes of particulate matter and related materials within the river-reservoir ecosystems. Therefore, it is logically accepted that Hg that is bound to particulate matters will be deposited into sediment along with the sedimentation of the particulates.

As concluded above, PHg was the main Hg fraction present in the inflow rivers of the Wujiang River Basin. Furthermore, surface runoff and sediment re-suspension were the primary sources of PHg in the inflow rivers of reservoirs. It is reasonable to believe that a considerable amount of Hg entered into the sediment by way of sedimentation. With the change in hydrodynamic conditions, dam construction will significantly impact the transportation feature of rivers on particulate matters, subsequently impacting the transportation and transformation of Hg in river-reservoir ecosystems in the Wujiang River Basin.

Current study observed the clear sedimentation of Hg in reservoirs within Wujiang River Basin (Fig. 5.10). Similar to the spatial distribution patterns of TSS in inflow–outflow rivers of reservoirs, concentrations of THg also gradually decreased from upstream to downstream of Wujiang River Basin (Fig. 5.10). For example, the annual mean THg concentrations in the inflow rivers of PD (SC) and HJD (LC) were 4.65 ± 1.28 ng L^{-1} and 4.50 ± 2.13 ng L^{-1}, respectively. Under the sedimentation processes of Hg in PD and HJD, concentrations of THg (annual mean concentration) in the outflow rivers (reservoir discharge) of PD and HJD significantly decreased (approximately 2 times), to 2.43 ± 0.40 ng L^{-1} and 2.18 ± 0.53 ng L^{-1}, respectively. These results indicated that a great quantity of Hg was intercepted by the reservoirs and stored probably in the sediment. On the other hand, reservoir's scavenging effect on Hg was clearly weakened in the four other reservoirs, which were located in the downstream portions of Wujiang River Basin. It is generally accepted that sediment and/or hypolimnion are very important site for Hg methylation. Therefore, the dam intercepted Hg can be transformed into MeHg in reservoir, posing a potential threat to river-reservoir ecosystems.

Previous studies showed that dam construction significantly changed the concentrations and spatial distribution patterns of MeHg in the river (Canavan et al. 2000) and subsequently resulted in the increase of MeHg levels in the downstream river. Canavan et al. (2000) further observed that the concentrations of MeHg were still very high in the outflow river of the reservoir, approximately 22 km away from the dam. To better understand the influence of dam construction on the distribution of MeHg in the Wujiang River, the six selected reservoirs in Wujiang River Basin are classified as the oligotrophic–mesotrophic stage (YZD, SFY, and HJD), the mesotrophic–eutrophic stage (PD and DF), and the hyper-eutrophic stage (WJD) based on the trophic state of the specific reservoirs (Table 5.2). We further observed the distributions of TMeHg and DMeHg in inflow–outflow rivers of different trophic stages of reservoirs (Figs. 5.11 and 5.12). As shown in Fig. 5.11, the TMeHg concentrations in outflow rivers of PD, DF, and WJD significantly increased by the factors of 46, 53, and 92% compared with the data in the inflow rivers of these three reservoirs, respectively.

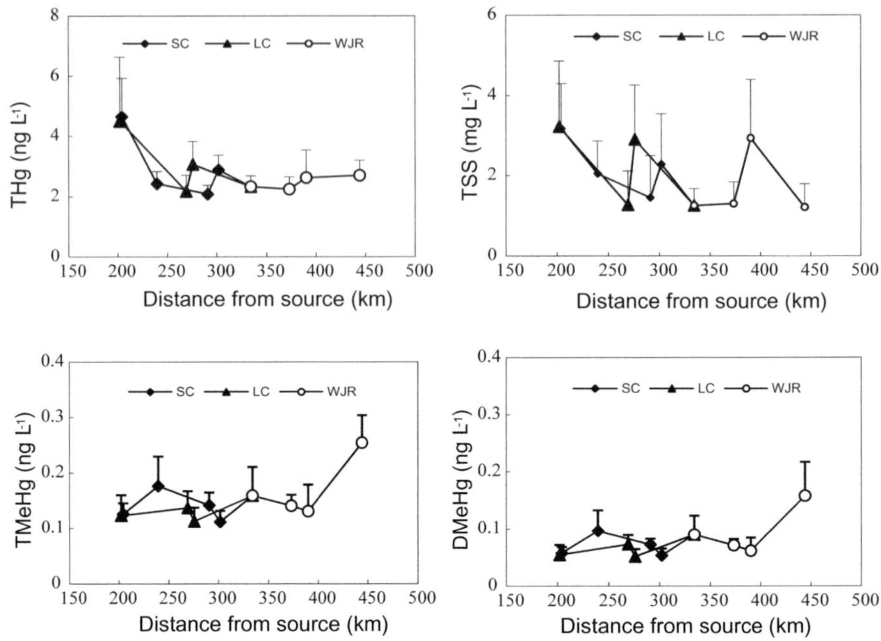

Fig. 5.10 Spatial distributions of total Hg (THg), total suspended solid (TSS), total methylmercury (TMeHg), and dissolved methylmercury (DMeHg) in different sections of Wujiang River, Guizhou Province, China (SC, Sancha River; LC, Liuchong River; WJR, Mainstream of Wujiang River; the source of Wujiang River is Xianglu mountain in Weining county, Guizhou Province, China) (redrawed from Guo et al. 2009, with permission from Resources and Environment in The Yangtze Basin; Reprinted from Zhao et al. 2017, with permission from Elsevier Ltd.)

Previous studies showed that apart from the age of the reservoir, many other factors including flooded soil type (e.g., organic matter content and Hg concentration), water volume, water residence time, water temperature, water chemistry, water depth in reservoir, may also exert an influence on net MeHg methylation in reservoirs (St. Louis et al. 2004; Ullrich et al. 2001). As listed in Table 5.2, the average water residence time in YZD (44 days) was 6.3 times higher than that in SFY (7 days), but was an order of magnitude lower than that in HJD (380 days). Furthermore, the total water volume in YZD was ~2 times higher than that in SFY, but was ~11 times lower than that in HJD. However, the distribution patterns of MeHg within inflow–outflow rivers (elevated from inflow rivers to outflow rivers of reservoirs) were less pronounced in the newly constructed reservoirs in Wujiang River Basin (e.g., YZD, SFY, and HJD). In contrast, both the water residence time and total water volume in YZD, PD, DF, and WJD was similar. However, the statistical analysis showed that concentrations of TMeHg and DMeHg in the outflow rivers in the hyper-eutrophic stage reservoirs (WJD) were significantly higher than those in oligotrophic–mesotrophic (YZD, SFY, and HJD) and mesotrophic–eutrophic (PD and DF) stage reservoirs (K–S test, $p < 0.01$, $n = 18$), indicating that

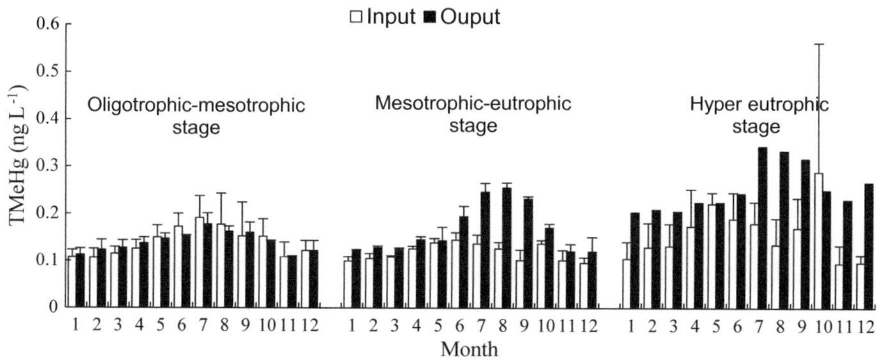

Fig. 5.11 Monthly mean concentrations of total methylmercury (TMeHg) in inflow–outflow rivers in different trophic stages of reservoirs in Wujiang River Basin, Guizhou Province, China (YZD, SFY, and HJD with oligotrophic–mesotrophic stage; PD and DF with mesotrophic–eutrophic stage; WJD with hyper-eutrophic stage) (redrawn from Guo et al. 2009, with permission from Resources and Environment in The Yangtze Basin; Reprinted from Zhao et al. 2017, with permission from Elsevier Ltd.)

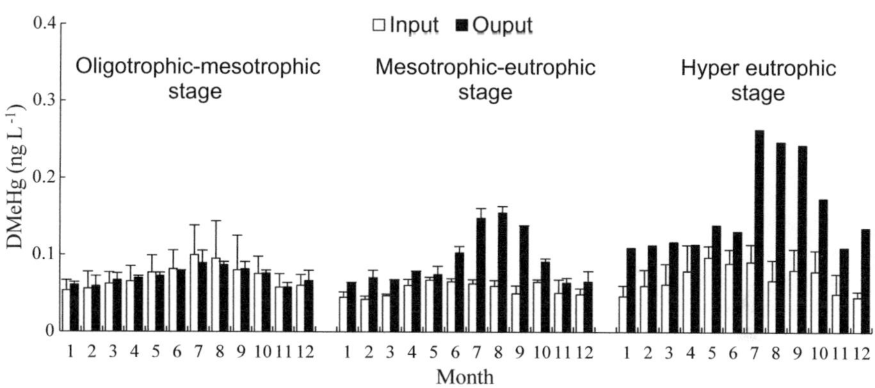

Fig. 5.12 Monthly mean concentrations of dissolved methylmercury (DMeHg) in inflow–outflow rivers in different trophic stages of reservoirs in Wujiang River Basin, Guizhou Province, China (YZD, SFY, and HJD with oligotrophic–mesotrophic stage; PD and DF with mesotrophic–eutrophic stage; WJD with hyper-eutrophic stage) (redrawed from Guo et al. 2009, with permission from Resources and Environment in The Yangtze Basin; Reprinted from Zhao et al. 2017, with permission from Elsevier Ltd.)

the net Hg methylation and the output of MeHg from reservoirs increased with the increase of trophic stage in reservoir ecosystem.

The six cascade reservoirs which are located in the karstic environment of Wujiang River Basin are typical deep-valley and high-mountain gorges. Hence, it was reasonable that the selected reservoirs represented similar geological background (e.g., water depth of reservoirs, organic matter content and Hg concentrations in submerged soil) (Meng et al. 2010, 2016; Yao et al. 2011; Feng et al. 2009a, b). Therefore, a current study jointly with the previous observations implied that the primary indicator for Hg methylation could not be the water residence time, total water volume, and basis geological background in reservoirs. In comparison, our observations illustrated that primary productivity (trophic state) could be one of the most important factors controlling the net Hg methylation in reservoirs in Wujiang River Basin. The increase in planktonic derived NOM could be the possible reason to explain the enhanced MeHg production in the reservoirs in Wujiang River Basin (Bravo et al. 2017). Furthermore, the construction of the cascade reservoirs resulted in the elevation of MeHg in several sections of the Wujiang River (Fig. 5.10). For example, the MeHg concentration exhibited the highest level in the section that was approximately 445 km distant from the source of the Wujiang River (Fig. 5.10). The elevated MeHg levels in specific sections of Wujiang River can be attributed to the net Hg methylation in reservoirs and discharge of MeHg into the downstream areas of the Wujiang River. MeHg-enriched water in outflow rivers could be transported to downstream areas, posing a potential threat to the aquatic food web and human health.

References

Bai WY (2006) The primary study on the distributions and transformations of the different species of mercury in Aha reservoir. Master dissertation, Graduate School of the Chinese Academy of Sciences, Beijing. (In Chinese, with English abstract)

Balogh SJ, Huang Y, Offerman HJ (2003) Methylmercury in rivers draining cultivated watersheds. Sci Total Environ 304:305–313

Benoit JM, Gilmour CC, Mason RP, Riedel GS, Riedel GF (1998) Behavior of mercury in the Patuxent River estuary. Biogeochemistry 40:249–265

Bonzongo JC, Heim KJ, Warwick JJ, Lyons WB (1996) Mercury levels in surface waters of the Carson River-Lahontan reservoir system, Nevada: influence of historic mining activity. Environ Pollut 92:193–201

Bravo AG, Bouchet S, Tolu J, Bjorn E, Mateos-Rivera A, Bertilsson S (2017) Molecular composition of organic matter controls methylmercury formation in boreal lakes. Nature Commun 8:14255

Canavan CM, Caldwell CA, Bloom NS (2000) Discharge of methylmercury-enriched hypolimnetic water from a stratified reservoir. Sci Total Environ 260:159–170

Choe KY, Gill GA (2003) Distribution of particulate, colloidal, and dissolved mercury in San Francisco Bay estuary: 2. Monomethyl mercury. Limnol Oceanogr 48:1547–1556

Coquery M, Cossa D, Azemard S, Peretyazhko T, Charlet L (2003) Methylmercury formation in the anoxic waters of the Petit-Saut reservoir (French Guiana) and its spreading in the adjacent Sinnamary river. J Phys 107:327–331

Coquery M, Cossa D, Sanjuan J (1997) Speciation and sorption of mercury in two macro-tidal estuaries. Mar Chem 58:213–217

Covelli S, Piani R, Kotnik J, Horvat M, Faganeli J, Brambati A (2006) Behaviour of Hg species in a microtidal deltaic system: the Isonzo River mouth (north Adriatic Sea). Sci Total Environ 368:210–223

Donkor AK, Bonzongo JC, Nartey VK, Adotey DK (2006) Mercury in different environmental compartments of the Pra River Basin, Ghana. Sci Total Environ 368:164–176

Duan SW (2000) Study of delivery and sources of nutrient of Yangtze River. Ph.D. dissertation, Graduate School of the Chinese Academy of Science, Beijing. (In Chinese, with English abstract)

Feng XB, Jiang HM, Qiu GL, Yan HY, Li GH, Li ZG (2009a) Geochemical processes of mercury in Wujiangdu and Dongfeng reservoirs, Guizhou, China. Environ Pollut 157:2970–2984

Feng XB, Jiang HM, Qiu GL, Yan HY, Li GH, Li ZG (2009b) Mercury mass balance study in Wujiangdu and Dongfeng Reservoirs, Guizhou, China. Environ Pollut 157:2594–2603

Ganguli PM, Marson RP, AbuSaba KE, Anderson RS, Flegal AR (2000) Mercury speciation in drainage from the New Idria mercury mine, California. Environ Sci Technol 34:4773–4779

Guentzel JL, Portilla E, Keith KM, Keith EO (2007) Mercury transport and bioaccumulation in riverbank communities of the Alvarado Lagoon system, Veracruz State, Mexico. Sci Total Environ 388:316–324

Guo YN, Feng XB, Yan HY, Qian XL, Meng B, Yao H (2009) Temporal and spatial distribution of total and methyl mercury in inflow and outflows of cascade reservoirs in Wujiang River. Resour Environ Yangtze Basin 18:356–360. (In Chinese, with English abstract)

Guo YN (2008) Input and output fluxes of mercury in different evolutive reservoirs in Wujiang River Basin. Ph.D. dissertation, Graduate School of the Chinese Academy of Sciences, Beijing. (In Chinese, with English abstract)

He TR (2007) Biogeochemical cycling of mercury in Hongfeng reservoir, Guizhou, China. Ph.D. dissertation, Graduate School of the Chinese Academy of Sciences, Beijing. (In Chinese, with English abstract)

Hurley JP, Cowell SE, Shafer MM, Hughes PE (1998) Partitioning and transport of total and methyl mercury in the lower Fox River, Wisconsin. Environ Sci Technol 32:1424–1432

Lawson NM, Mason PR, Laporte JM (2001) The fate and transport of mercury, methylmercury, and other trace metals in Chesapeake Bay tributaries. Water Res 35:501–515

Mason RP, Sullivan KA (1998) Mercury and methylmercury transport through an urban watershed. Water Res 32:321–330

Meng B, Feng XB, Chen CX, Qiu GL, Sommar J, Guo YN, Wan Q (2010) Influence of eutrophication on the distribution of total mercury and methylmercury in hydroelectric reservoirs. J Environ Qual 39:1624–1635

Meng B, Feng XB, Qiu GL, Li ZG, Yao H, Shang LH, Yan HY (2016) The impacts of organic matter on the distribution and methylation of mercury in a hydroelectric reservoir in Wujiang River, Southwest China. Environ Toxicol Chem 35:191–199

Minstry of Environmental Protection (2002) Environmental standard for surface water (GB3838-2002). China Environmental Science Press, Beijing

Muresan B, Cossa D, Richard S, Dominique Y (2008) Monometylmercury sources in a tropical artificial reservoir. Appl Geochem 23:1101–1126

Paraquetti HH, Ayres GA, Dominguez de Almeida M, Molisani MM, de Lacerda LD (2004) Mercury distribution, speciation and flux in the Sepetiba Bay tributaries, SE Brazil. Water Res 38:1439–1448

Quemerais B, Cossa D, Rondeau B, Pham TT, Gagnon P, Fortin B (1999) Sources and fluxes of mercury in the St. Lawrence river. Environ Sci Technol 33:840–849

Schaefer J, Blanc G, Audry S, Cossa D, Bossy C (2006) Mercury in the Lot-Garonne river system (France); sources, fluxes and anthropogenic component. Appl Geochem 21:515–527

St. Louis VL, Rudd JWM, Kelly CA, Bodaly RAD, Paterson MJ, Beaty KG, Hesslein RH, Heyes A, Majewsk AR (2004) The rise and fall of mercury methylation in an experimental reservoir. Environ Sci Technol 38:1348–1358

Ullrich SM, Tanton TW, Abdrashitova SA (2001) Mercury in the aquatic environment: a review of factors affecting methylation. Crit Rev Environ Sci Technol 31:241–293

USEPA (1992) Water quality standards; establishment of numeric criteria for priority toxic pollutants; states' compliance; final rule. Fed Reg USEPA, 40 CFR Part 131, 57/246, 60847–60916

Warner KA, Bonzongo JC, Roden EE, Ward GM, Green AC, Chaubey I, Lyons B, Arrington AD (2005) Effect of watershed parameters on mercury distribution in different environmental compartments in the Mobile Alabama River Basin, USA. Sci Total Environ 347:187–207

Watras CJ, Bloom NS, Claas SA, Morrison KA, Gilmour CC, Craig SR (1995) Methylmercury production in the anoxic hypolimnion of a dimictic seepage lake. Water Air Soil Pollut 80:735–745

Yan H (2005) The methodological development of mercury species in environmental samples and the mass balance of mercury in Baihua Reservoir, Guizhou, China. Ph.D. dissertation, Graduate School of the Chinese Academy of Sciences, Beijing. (In Chinese, with English abstract)

Yao H, Feng XB, Guo YN, Yan HY, Fu XW, Li ZG, Meng B (2011) Mercury and methylmercury concentrations in 2 newly constructed reservoirs in the Wujiang River, Guizhou, China. Environ Toxicol Chem 30:530–537

Yu YX (2008) The effects of cascaded damming on biogeochemical cycling of carbon in the middle-upper reaches of Wujiang River, SW China. Ph.D. dissertation, Graduate School of the Chinese Academy of Sciences, Beijing. (In Chinese, with English abstract)

Zhang JF, Feng XB, Yan HY, Guo YN, Yao H, Meng B, Liu K (2009) Seasonal distributions of mercury species and their relationship to some physicochemical factors in Puding Reservoir, Guizhou, China. Sci Total Environ 408:122–129

Zhang JF (2009) Mercury distribution and mass balance study on Puding and Dongfeng reservoirs in Wujiang Watershed. Ph.D. dissertation, Graduate School of the Chinese Academy of Science, Beijing. (In Chinese, with English abstract)

Zhao L, Guo YN, Meng B, Yao H, Feng XB (2017) Effects of damming on the distribution and methylation of mercury in Wujiang River, Southwest China. Chemosphere 185:780–788

Zhu J (2005) Effects of dams on the biogeochemical cycles of nutrients in the Wujiang River. Ph.D. dissertation, Graduate School of the Chinese Academy of Sciences, Beijing. (In Chinese, with English abstract)

Chapter 6
Biogeochemical Process of Mercury in Reservoirs in the Main Stream of the Wujiang River

Abstract To understand the biogeochemical process of mercury (Hg) in reservoir in Wujiang River Basin, Southwest China, six cascade reservoirs including Yinzidu Reservoir (YZD), Suofengying Reservoir (SFY), Hongjiadu Reservoir (HJD), Puding Reservoir (PD), Dongfeng Reservoir (DF), and Wujiangdu Reservoir (WJD) in the mainstream of Wujiang River were selected in this study. The primary objectives of this chapter were: (1) to investigate spatial and seasonal variations of Hg species in different sectors of reservoirs (e.g., water column, sediment, sediment pore water); (2) to reveal the processes of Hg methylation and their possible controlling factors in the reservoirs in the Wujiang River Basin, Southwest China.

Keywords Biogeochemical process · Mercury · Reservoir · Wujiang river

6.1 Biogeochemical Cycling of Mercury in the Hongjiadu, Suofengying, and Yinzidu Reservoir

HJD is the largest reservoir of the 11 cascade reservoirs created in the Wujiang River, whereas SFY is the smallest. The Liuchong and Aoshui Rivers provide the predominant inflows into the HJD. The inflows to the SFY include a major one from the DF and a relatively minor one from the Maotiao River. In addition to river inputs, upland runoff contributes seasonally to the reservoirs. The YZD is located in the lower portion of the Wujiang river basin. The average water residence time in the YZD is 43.7 d. At present, there are no significant point source discharges in the drainage basins of these reservoirs.

The reservoirs experience a typical subtropical humid monsoon climate. The rainy season occurs in the spring and summer seasons and accounts for more than 70% of annual precipitation. The reservoirs were created for a variety of purposes, including the production of hydroelectricity, irrigation, flood control, fishery production, and recreation. These reservoirs are located in rural areas, remote from industrial activities and densely populated urban centers. Seventy-two percent of

X. Feng et al., *Biogeochemical Cycle of Mercury in Reservoir Systems in Wujiang River Basin, Southwest China*, https://doi.org/10.1007/978-981-10-6719-8_6

the flooded area is composed of former agricultural lands with the underlying bedrock of limestone and dolomite.

6.1.1 Sampling Location, Sample Collection, and QA/QC

Water samples were collected on a seasonal basis during 2007 at stations representing the upstream, midstream, and proximal to dam sections of the reservoirs, as well as within the tributaries. In all three of the reservoirs, sampling sites were oriented in a west-to-east direction, following the water flow (Fig. 6.1). Seasonal vertical column profiles were established at each site, with sampling at 6–8 depths. Filtered water column samples were analyzed for dissolved Hg (DHg) and dissolved methylmercury (DMeHg) levels. Total Hg (THg), reactive Hg (RHg), total methylmercury (TMeHg), and total suspended solids (TSS) levels were analyzed in each of the unfiltered samples.

Surface soil samples were collected from all three reservoirs along the water edge to represent flooded soil. Figure 6.1 shows pairs of soil and water sampling sites from the water column. All samples were stored in acid-cleaned high-density polyethylene centrifuge tubes, transported to the laboratory and stored frozen. Soil samples were freeze-dried and homogenized with a mortar and pestle and then sieved through an 80-mm sieve to remove coarse particles and biologic debris before THg analysis.

Quality control was exercised using duplicates, method blanks, blank spikes and matrix spikes, and standard reference material. Blank spikes and duplicates were taken regularly (>10% of samples) throughout each sampling campaign. The

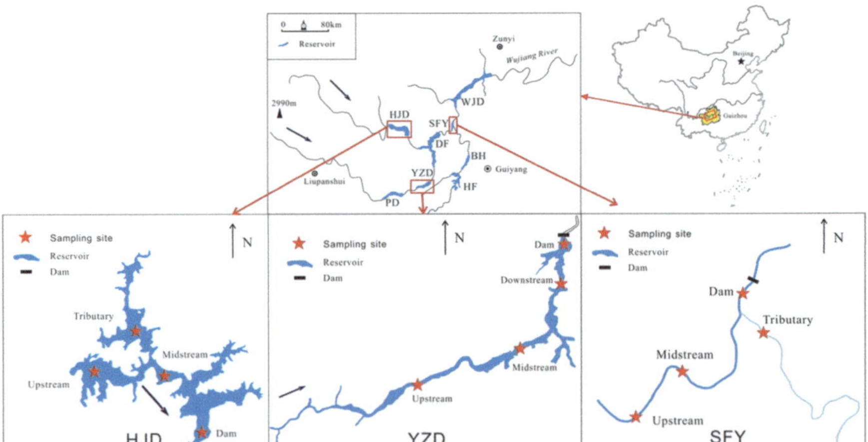

Fig. 6.1 Location of the sampling sites near the Hongjiadu (HJD), Suofengying (SFY), and Yinzidu Reservoir (YZD) reservoirs within the Wujiang River, Guizhou Province, China

method detection limit (MDL), based on three times the standard deviation of replicate measurements of a blank solution in the water sample, was as follows: 0.20 ng L^{-1} for THg, 0.032 ng L^{-1} for MeHg, respectively, and the method blank was found to be less than the detection limits in all cases. The average relative standard deviations for the duplicate analysis of THg and MeHg for water samples were from 2.5 to 12.6% and from 4.4 to 7.5%, respectively. Spike recoveries for THg and TMeHg in water samples were between 90 and 110%, and between 84.1 and 113.1%, respectively. The mean THg concentration of standard materials of GBW07305 and GBW07405 was 0.10 ± 0.10 and 0.30 ± 0.08 mg kg^{-1} in soil, which is comparable with the certified value of 0.10 ± 0.02 and 0.29 ± 0.04 mg kg^{-1}.

The assumption of the parametric procedure was examined using one-sample K–S test, and the result showed that the data sets followed a normal distribution; then, we used one-way analysis of variance (ANOVA) and multiple comparisons to analyze the significance of differences in concentrations (e.g., TMeHg levels) among seasons and sites. Pearson's values were used as a guide to determine significance at the 5% level. Linear regression and ANOVA were performed with R. All statistical analyses were performed in the software Origin 8.0 (OriginLab) and SPSS 16.0 (SPSS).

6.1.2 General Water Quality Characteristics

The temporal and spatial characteristics for temperature (*T*) and dissolved oxygen (DO) from the HJD, SFY, and YZD are displayed in Figs. 6.2, 6.3, 6.4, 6.5, 6.6 and

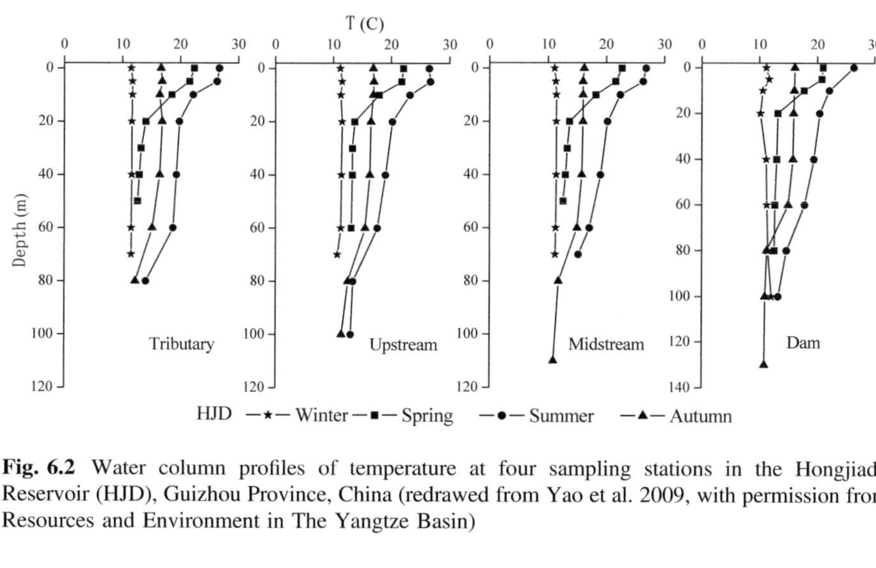

Fig. 6.2 Water column profiles of temperature at four sampling stations in the Hongjiadu Reservoir (HJD), Guizhou Province, China (redrawed from Yao et al. 2009, with permission from Resources and Environment in The Yangtze Basin)

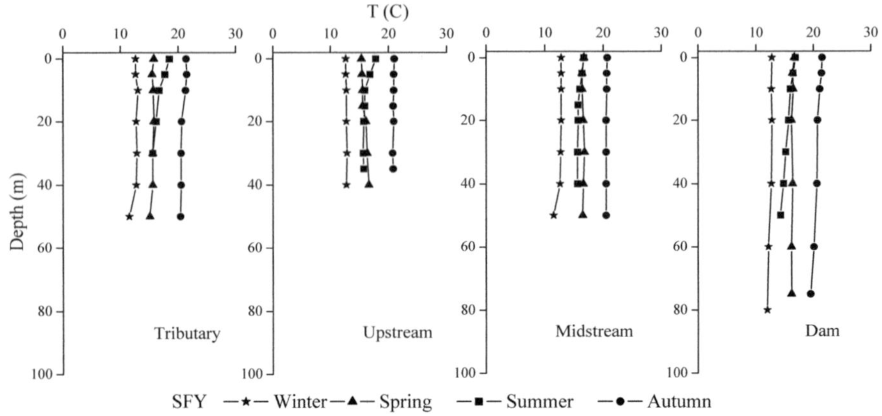

Fig. 6.3 Water column profiles of temperature at four sampling stations in the Suofengying Reservoir (SFY), Guizhou Province, China (redrawed from Yao et al. 2009, with permission from Resources and Environment in The Yangtze Basin)

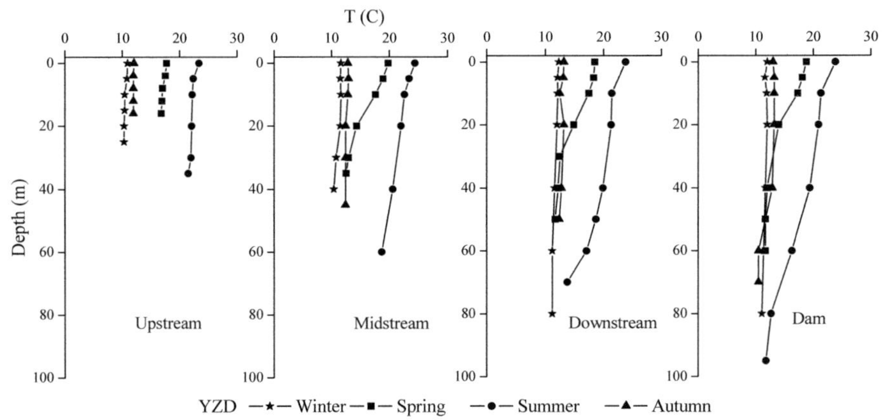

Fig. 6.4 Water column profiles of temperature at four sampling stations in the Yinzidu Reservoir (YZD), Guizhou Province, China (redrawed from Meng et al. 2010, with permission from The Alliance of Crop, Soil, and Environmental Science Societies; redrawed from Meng et al. 2011, with permission from Chinese Journal of Ecology)

6.7. Figures 6.2, 6.3 and 6.4 show that the surface water *T* is higher in the summer than in the spring and autumn in the three newly built reservoirs and lowest in the winter.

In the vertical direction, *T* and DO are obviously different, except in the winter when they show a gradually decreasing trend from the surface to the bottom. The water column showed obvious stratification, which can affect water environment conditions in these reservoirs (Wang 2005). *T* and DO had maximum values from 0

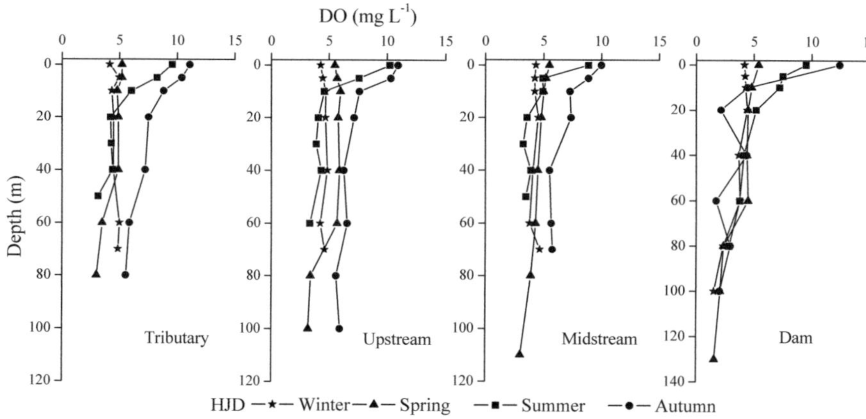

Fig. 6.5 Water column profiles of dissolved oxygen at four sampling stations in the Hongjiadu Reservoir (HJD), Guizhou Province, China (redrawed from Yao et al. 2009, with permission from Resources and Environment in The Yangtze Basin)

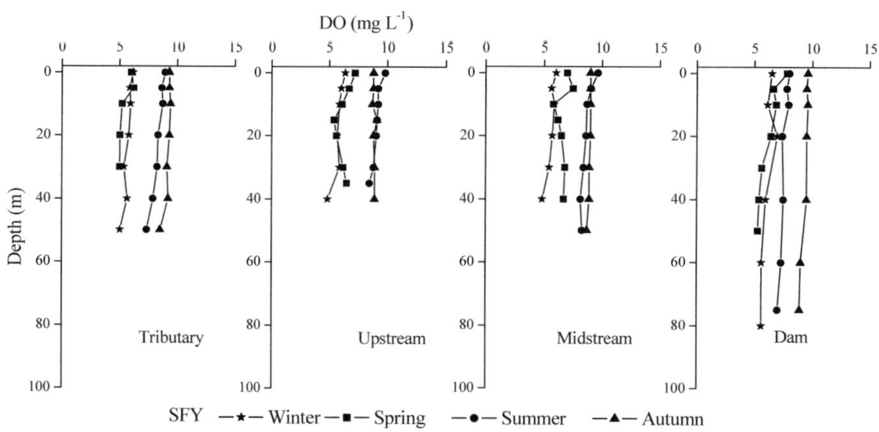

Fig. 6.6 Water column profiles of dissolved oxygen at four sampling stations in the Suofengying Reservoir (SFY), Guizhou Province, China (redrawed from Yao et al. 2009, with permission from Resources and Environment in The Yangtze Basin)

to 10 m down the water column. They decreased gradually with increasing depth, and at the bottom of the water column, they reached the minimum value. Water stratification leads to a low temperature, anaerobic environment, but photosynthesis of algae (algae absorb carbon dioxide and release oxygen) and an enriched oxygen environment in the bottom can result in higher DO. However, algae reduction and weakening of illumination intensity can weaken or even stop photosynthesis. Moreover, the endogenous organic matter will decompose in reservoirs, producing carbon dioxide. Water temperature stratification forces free carbon dioxide to stay

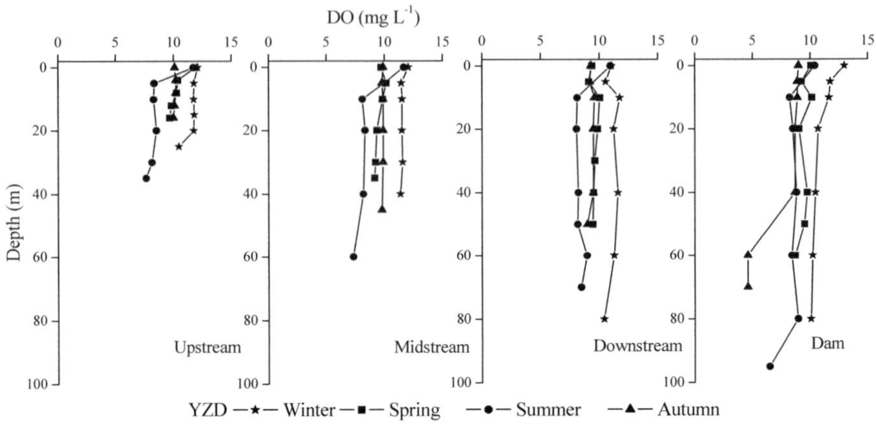

Fig. 6.7 Water column profiles of dissolved oxygen at four sampling stations in the Yinzidu Reservoir (YZD), Guizhou Province, China (redrawed from Meng et al. 2010, with permission from The Alliance of Crop, Soil, and Environmental Science Societies; redrawed from Meng et al. 2011, with permission from Chinese Journal of Ecology)

in the bottom water, resulting in DO gradually decreasing with increasing water depth and bottom water having a very low value. The reservoirs were well mixed in winter, with no significant changes in the water column, but DO had low values at the bottom of the dam.

SFY showed differences from HJD. Generally, *T* and DO were invariant from the surface to bottom ($p > 0.05$), although they did show slight changes in the 10 m or 20 m sampling depths at the upstream site and dam in the spring.

YZD was slightly temperature stratified in the spring and summer, with strong trends at the downstream and dam stations of the reservoirs. However, the vertical profiles of DO showed no changes from the surface to the bottom water.

Clearly, HJD has the strongest stratification in the spring, summer, and autumn; YZD had a slight stratification, and SFY showed little variation within the water column. These differences may be depending on water residence time and reservoir volume (see Table 6.1). When the reservoirs were well mixed, *T* and DO showed no obvious change from the surface to the bottom in the water column.

Table 6.1 shows the basic physical and chemical parameters in different seasons at each sampling site at the HJD, SFY, and YZD in 2007. Sulfate concentrations ranged from 49.1 to 81.6 mg L^{-1} in HJD, from 55.0 to 121.9 mg L^{-1} in SFY, and from 71.6 to 136.8 mg L^{-1} in YZD. Sulfate concentrations were very high, which is primarily related to atmospheric rainfall, due to the release of SO_2 from coal combustion (Guo et al. 2008a, b), and dissolution of sulfate from the carbonate bedrock (Jiang 2007; Jiang et al. 2006).

Due to the karstic geology of the Wujiang River basin, the water was slightly alkaline in most samples (HJD: 6.8–8.7; SFY: 7.2–8.2; YZD: 6.6–8.4), and no

Table 6.1 Water column spatial distribution of main physical characteristics in the Hongjiadu Reservoir (HJD), Suofengying Reservoir (SFY) and Yinzidu Reservoir (YZD), Guizhou Province, China

Sampling	Sampling	Temperature (°C)				pH					Dissolved oxygen (mg L^{-1})					
Sites	Time	Tributary	Upstream	Midstream	Downstream	Dam	Tributary	Upstream	Midstream	Downstream	Dam	Tributary	Upstream	Midstream	Downstream	Dam
HJD	2007.01	11.6	11.3	11.3	–	11.2	7.4	7.4	7.3	–	7.1	4.6	4.5	4.3	–	3.6
	2007.04	16.5	16.5	16.4	–	15.9	8	8	8	–	7.9	5.7	5.4	4.7	–	5.7
	2007.08	21	20	21	–	19.2	7.8	7.6	7.9	–	7.7	8.1	7.6	7.2	–	6.9
	2007.11	15.8	15.4	14.7	–	14	7.6	7.6	7.6	–	7.6	4.5	5.2	4.5	–	3.7
SFY	2007.01	12.6	12.7	12.8	–	12.5	8	8	8.1	–	8.1	5.7	5.8	5.6	–	6.3
	2007.04	16.9	16.2	16	–	15.6	7.5	7.5	7.4	–	7.3	5.5	6.2	6.7	–	6.4
	2007.08	20.9	20.9	20.6	–	20.7	7.8	7.9	7.8	–	7.8	8	9.7	9	–	7.4
	2007.11	15.6	15.8	16.6	–	16.4	8	8	7.9	–	8	9.2	8.9	8.9	–	9.4
YZD	2007.01	–	11	11	12	12	–	7.8	7.6	7.5	7.5	–	12	12	11	11
	2007.05	–	17	16	15	15	–	8.1	8	7.9	8	–	10	9.6	9.6	9.5
	2007.08	–	22	22	19	18	–	7.2	7.5	7.2	7.3	–	8.8	8.9	8.7	8.6
	2007.11	–	12	13	13	12	–	8.3	8.3	8.3	8	–	10	9.9	9.4	7.6
Sampling	Sampling	Total suspended solids (mg L^{-1})				DOC (mg L^{-1})					SO_4^{2-} (mg L^{-1})					
Sites	Time	Tributary	Upstream	Midstream	Downstream	Dam	Tributary	Upstream	Midstream	Downstream	Dam	Tributary	Upstream	Midstream	Downstream	Dam
HJD	2007.01	1.2	1.3	1	–	1.1	0.7	0.7	0.8	–	0.7	76.1	57.2	57.2	–	74.9
	2007.04	1.0	1.3	1.3	–	1.1	2.2	3.4	3	–	4.6	61.4	60.5	60.2	–	63.7
	2007.08	2.8	2.1	1.9	–	1.4	1.6	1.5	1.8	–	1.2	58.6	59.4	61.2	–	56
	2007.11	1.9	1.3	1.5	–	1.3	1.3	1.1	1.2	–	1.2	70.2	69.1	71.1	–	66.3
SFY	2007.01	1.3	1.2	0.8	–	1.1	0.8	0.5	0.6	–	0.8	64.3	62.3	62.7	–	69.7
	2007.04	1.9	0.9	0.8	–	0.9	3.2	3.5	3.2	–	3.8	99.9	71.6	66.2	–	87.8
	2007.08	1.7	0.7	0.7	–	1.4	2.7	2.3	2.6	–	1.8	89.8	87.3	87.6	–	87.9
	2007.11	2.2	1.2	1.1	–	1.7	1	0.9	1	–	1.2	82.7	77.2	77.9	–	81.6
YZD	2007.01	–	3.2	2.9	3.3	3.1	–	0.6	0.5	0.5	0.6	–	94.7	93.6	91.4	92.7
	2007.05	–	4.9	2.2	1.2	1.4	–	0.8	0.9	0.8	0.7	–	91.9	93.6	93.3	91.6
	2007.08	–	3.8	2.6	1.7	1.1	–	5	3.7	4.9	3.4	–	80.7	86.6	84.2	90.6
	2007.11	–	1.8	1.6	1.2	0.63	–	0.2	0.2	0.4	0.1	–	121.4	108.4	111.5	112.9

The data listed in the table are the mean values of water column significant seasonal pH variations were observed at these three reservoirs during this study ($p > 0.05$).

Total suspended solids (TSS) concentrations ranged from 0.06 to 8.9 mg L^{-1}, with a mean concentration of 1.7 mg L^{-1}. TSS in the HJD showed a seasonal pattern, with the highest average TSS values in wet seasons and the lowest values in dry seasons. The TSS values in the YZD in the spring and summer were significantly higher than those in the winter ($p < 0.01$), but the TSS values of the SFY showed no seasonality. Because HJD is located upstream of the Wujiang River in these reservoirs, the river channel erosion and surface runoff were the main sources of TSS to the HJD during our sampling.

The mean concentration of DOC was 1.7 mg L^{-1} in these reservoirs (range: 0.1–7.0 mg L^{-1}). DOC concentrations in the spring and summer were higher than those in the autumn and winter, which may be due to the transportation of DOC from the soil by precipitation-induced soil erosion.

Overall, these newly constructed reservoirs were located in a karst environment, with the same geographical environment, so the water was neutral to partially alkaline, with low DOC, low TSS, and high sulfate.

6.1.3 Distribution of Mercury Species in Water Column

1. The spatial distribution characteristics of total mercury, dissolved mercury, and particulate mercury

We collected water samples from each season during 2007 at different sites in these three newly constructed reservoirs (HJD, SFY, and YZD) and analyzed the mercury speciation. We compared the seasonal and spatial distribution of mercury species in the water column in the HJD, SFY, and YZD. The mean annual concentrations and variation ranges of different mercury species in different seasons are shown in Tables 6.2, 6.3 and 6.4. As shown in the table, THg, DHg, and PHg concentrations decreased in the following order in the HJD: summer > autumn > spring > winter ($p < 0.01$); the concentrations clearly changed seasonally. The SFY and YZD showed different results than the HJD. There were no obvious changes in the spring, summer, and winter at the SFY and YZD ($p > 0.05$), except the DHg levels in summer were lower than those in winter ($0.01 < p < 0.05$). However, the concentration of THg indicated no discernible trend during all seasons ($p > 0.05$) in the water column of the YZD. The mean annual concentration of DHg in spring was lower than those of other seasons ($p > 0.05$), possibly as a result of more extensive settling of particles in the DF, which is located upstream of the SFY. The main input for the YZD comes from the outlet of the upstream PD.

We found that the mean ratios of PHg/THg in summer and autumn were 69.5 and 57.9% in the HJD and decreased to 36.0 and 40.4% in winter and spring, respectively. The mean ratios of PHg/THg in summer and autumn were 57.1 and 57.9% in the SFY and decreased to 21.6 and 41.8% in winter and spring,

Table 6.2 The range and mean values of different mercury species concentrations in the water column in Hongjiadu Reservoir (HJD) (ng L^{-1})

Sampling time	RHg	DHg	PHg	THg	DMeHg	PMeHg	TMeHg
2007.01	0.09 ± 0.03	0.37 ± 0.06	0.22 ± 0.11	0.59 ± 0.11	0.06 ± 0.01	0.04 ± 0.02	0.09 ± 0.01
	(0.04–0.15)	(0.23–0.47)	(0.05–0.45)	(0.39–0.76)	(0.04–0.08)	(0.01–0.07)	(0.06–0.13)
2007.04	0.17 ± 0.08	0.38 ± 0.13	0.26 ± 0.18	0.64 ± 0.26	0.05 ± 0.01	0.09 ± 0.02	0.14 ± 0.02
	(0.08–0.43)	(0.23–0.59)	(0.04–0.94)	(0.32–1.52)	(0.03–0.08)	(0.05–0.13)	(0.10–0.18)
2007.08	0.18 ± 0.09	1.65 ± 0.35	1.01 ± 1.13	2.66 ± 1.26	0.04 ± 0.01	0.07 ± 0.02	0.11 ± 0.03
	(0.09–0.49)	(1.04–2.27)	(0.03–4.51)	(1.49–6.57)	(0.02–0.07)	(0.03–0.12)	(0.06–0.16)
2007.11	0.09 ± 0.04	0.75 ± 0.14	0.69 ± 0.34	1.44 ± 0.37	0.05 ± 0.01	0.02 ± 0.01	0.08 ± 0.02
	(0.04–0.17)	(0.48–0.98)	(0.20–1.65)	(1.00–2.52)	(0.03–0.08)	(0.01–0.05)	(0.04–0.13)

Table 6.3 The range and mean values of different mercury species concentrations in the water column in Suofengying (SFY) (ng L^{-1})

Sampling time	RHg	DHg	PHg	THg	DMeHg	PMeHg	TMeHg
2007.01	0.11 ± 0.04	0.79 ± 0.07	0.23 ± 0.19	1.02 ± 0.20	0.07 ± 0.01	0.02 ± 0.01	0.09 ± 0.01
	(0.04–0.20)	(0.70–1.03)	(0.01–0.67)	(0.75–1.44)	(0.05–0.09)	(0.01–0.05)	(0.07–0.12)
2007.04	0.12 ± 0.05	0.53 ± 0.13	0.38 ± 0.33	0.91 ± 0.43	0.05 ± 0.02	0.10 ± 0.03	0.15 ± 0.03
	(0.06–0.29)	(0.36–0.82)	(0.01–1.30)	(0.44–2.04)	(0.01–0.10)	(0.04–0.15)	(0.10–0.20)
2007.08	0.09 ± 0.03	0.34 ± 0.12	0.57 ± 0.36	0.91 ± 0.40	0.04 ± 0.02	0.07 ± 0.03	0.12 ± 0.04
	(0.04–0.20)	(0.21–0.63)	(0.06–1.36)	(0.42–1.78)	(0.01–0.11)	(0.03–0.15)	(0.06–0.18)
2007.11	0.10 ± 0.02	0.64 ± 0.26	0.88 ± 0.68	1.52 ± 0.81	0.04 ± 0.01	0.02 ± 0.02	0.06 ± 0.02
	(0.06–0.14)	(0.24–1.15)	(0.29–3.74)	(0.75–4.49)	(0.02–0.07)	(0.01–0.09)	(0.03–0.12)

Table 6.4 The range and mean values of different mercury species concentrations in the water column in Yinzidu Reservoir (YZD) (ng L^{-1})

Sampling time	RHg	DHg	PHg	THg	DMeHg	PMeHg	TMeHg
2007.01	0.11 ± 0.06	0.54 ± 0.09	0.47 ± 0.34	1.01 ± 0.33	0.05 ± 0.01	0.03 ± 0.01	0.07 ± 0.01
	(0.04–0.28)	(0.37–0.71)	(0.12–1.39)	(0.66–1.84)	(0.03–0.07)	(0.00–0.05)	(0.05–0.11)
2007.05	0.16 ± 0.09	0.42 ± 0.13	0.68 ± 0.31	1.10 ± 0.30	0.11 ± 0.04	0.05 ± 0.04	0.16 ± 0.05
	(0.04–0.37)	(0.22–0.72)	(0.09–1.36)	(0.66–1.9)	(0.03–0.20)	(0.004–0.12)	(0.07–0.25)
2007.08	0.37 ± 0.17	0.53 ± 0.18	0.43 ± 0.28	0.96 ± 0.31	0.06 ± 0.04	0.12 ± 0.08	0.18 ± 0.10
	(0.12–0.78)	(0.22–1.04)	(0.10–1.07)	(0.40–1.89)	(0.03–0.22)	(0.01–0.32)	(0.08–0.44)
2007.11	0.12 ± 0.03	0.51 ± 0.14	0.45 ± 0.21	0.96 ± 0.19	0.04 ± 0.02	0.02 ± 0.02	0.06 ± 0.03
	(0.08–0.17)	(0.25–0.85)	(0.06–0.89)	(0.62–1.37)	(0.02–0.07)	(0.00–0.09)	(0.03–0.14)

These data listed in the table with a mean (±SD) values, and range values in the brackets respectively. The mean ratios of PHg/THg in the spring were 60.0% in the YZD and decreased to 42.0, 45.0 and 41.9% in spring, autumn, and winter. Statistical analyses yielded significant positive correlations between PHg and THg in the HJD ($p < 0.001$, $r = 0.88^{**}$, $n = 117$), SFY ($p < 0.0001$, $r = 0.94^{**}$, $n = 107$) and YZD ($p < 0.0001$, $r = 0.88^{**}$, $n = 100$). Corresponding to an increase in rainfall during spring and summer, abundant surface water carried more particulate matter with higher THg concentrations to the HJD, causing higher PHg in summer than in winter. At the same time, THg concentrations in HJD were significantly higher than those in other reservoirs. This result further illustrates the impact of the HJD as the largest reservoir of the eleven cascading reservoirs creating the Wujiang River. Due to reservoir construction, TSS levels downstream decreased, demonstrating that reservoirs have potential to reduce THg in water of the reservoir-river system. TSS concentrations and inorganic mercury species in HJD Reservoir were significantly higher than those in YZD and SFY.

Vertical profiles for THg, PHg, and DHg from these three reservoirs are displayed in Figs. 6.8, 6.9, 6.10, 6.11, 6.12, 6.13, 6.14, 6.15 and 6.16. THg concentrations (range: 5.8–6.6 ng L^{-1}) and PHg (range: 3.5–4.5 ng L^{-1}) significantly increased in the summer from 0 to 10 m in the water column. However, in the SFY, THg, and PHg concentrations clearly increased in the autumn at the dam and tributaries potentially related to a seasonal input of PHg. The distribution trends of THg and PHg in the YZD were highly consistent and showed higher concentrations in the middle and bottom than other sites. This may be related to the higher frequency of water exchange and the greater disturbance of sediments, especially from the middle to the dam.

The spatial distributions of inorganic mercury species showed no obvious changes between the three reservoirs. THg concentrations decreased gradually from

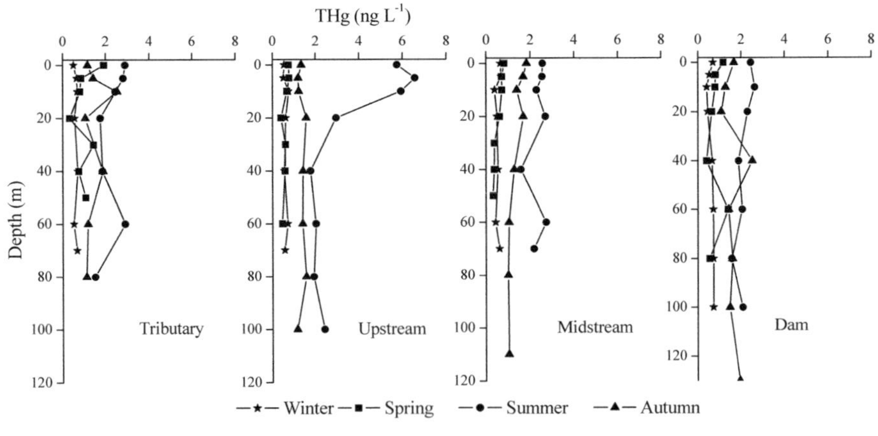

Fig. 6.8 Water column profiles of the concentrations of total mercury (THg) in Hongjiadu Reservoir (HJD) (redrawed from Yao et al. 2011, with permission from Chinese Journal of Ecology)

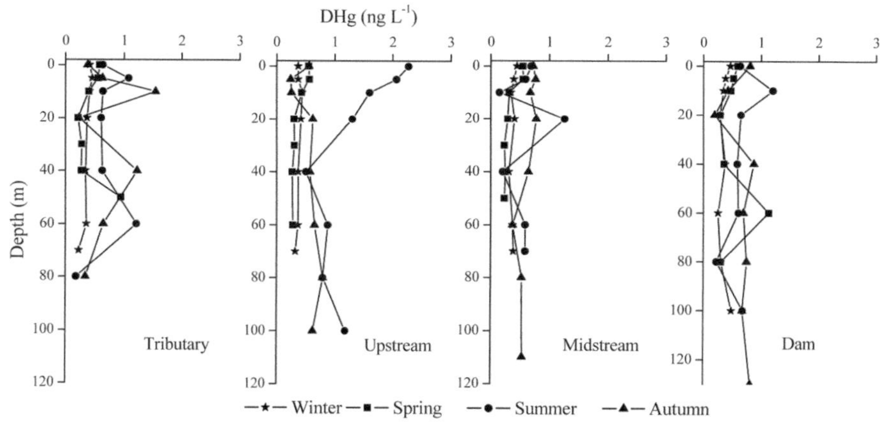

Fig. 6.9 Water column profiles of the concentrations of dissolved mercury (DHg) in Hongjiadu Reservoir (HJD) (redrawed from Yao et al. 2011, with permission from Chinese Journal of Ecology)

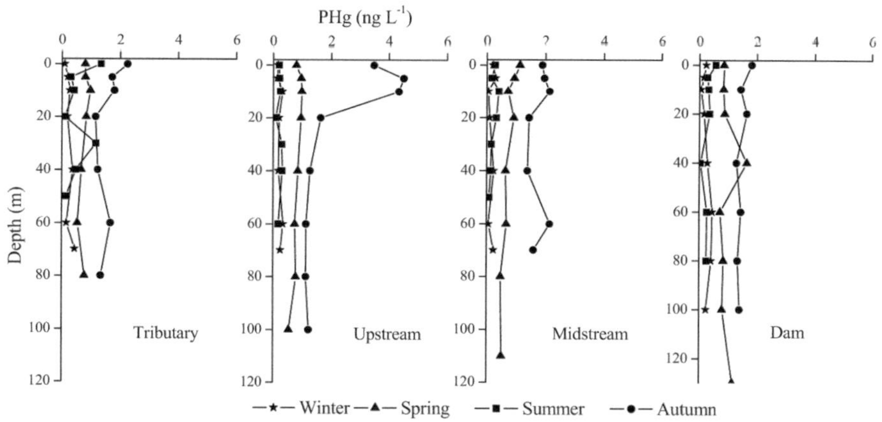

Fig. 6.10 Water column profiles of the concentrations of particulate mercury (PHg) in Hongjiadu (HJD)

upstream to the dam in the HJD in the water column. As Fig. 6.8 shows, the mean annual THg concentration was 3.7 ± 2.1 ng L^{-1} in the upstream region and 2.1 ± 0.4 ng L^{-1} in the dam. PHg concentrations decreased gradually from 2.3 ± 1.5 ng L^{-1} in the upstream region to 1.5 ± 0.2 ng L^{-1} in the dam. We found that the THg concentration decreased by 43.2% in the dam compared with the upstream site, and the PHg concentrations decreased by 34.8% in the dam compared to the upstream site. This similarity in trend suggests that mercury transported by rivers is partially settling out in the reservoirs, especially particulate matter. The hydrodynamic condition weakens when rivers enter reservoirs, and

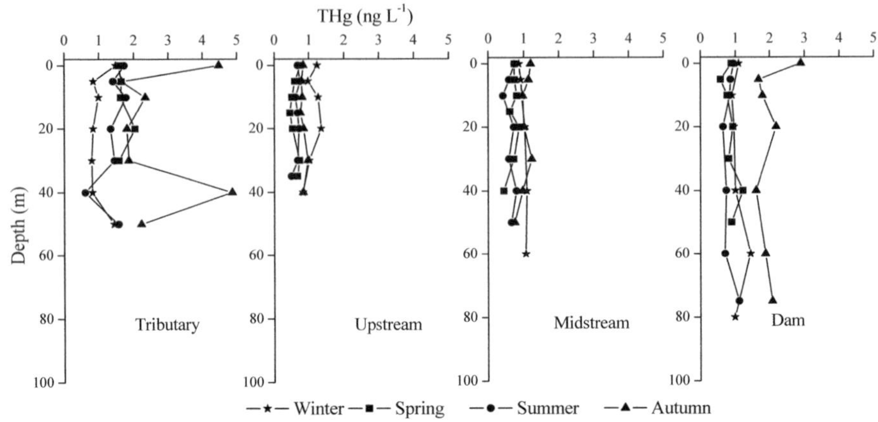

Fig. 6.11 Water column profiles of the concentrations of total mercury (THg) in Suofengying Reservoir (SFY)

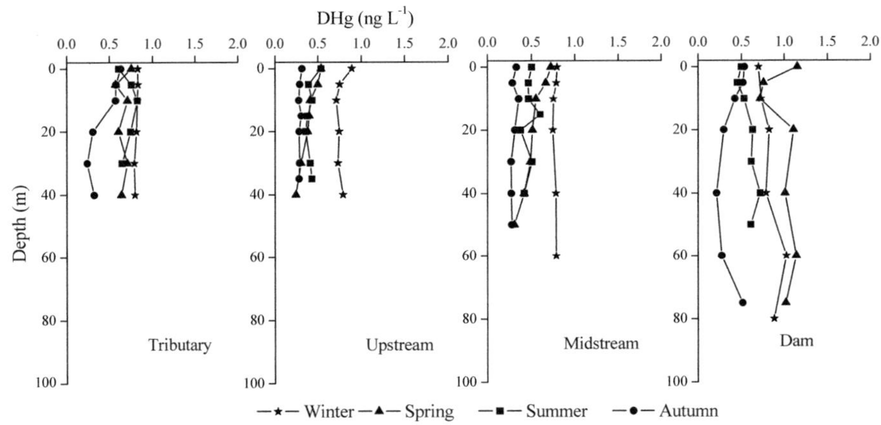

Fig. 6.12 Water column profiles of the concentrations of dissolved mercury (DHg) in Suofengying Reservoir (SFY)

most particles from rivers settled into sediments in reservoirs. This leads to the observation that particles decreased downstream, suggesting that the HJD scavenged the most THg. When there are reservoirs upstream, the scavenging ability gradually decreases, such as in the SFY and YZD; THg concentrations at the dam sites had reduced 21 and 32% compared with those at the upstream sites. The three newly built reservoirs have low primary productivity, low human impact and lack of local point sources, implying that no discernible vertical and spatial seasonal trends in the different mercury species in the water column occurred during our sampling.

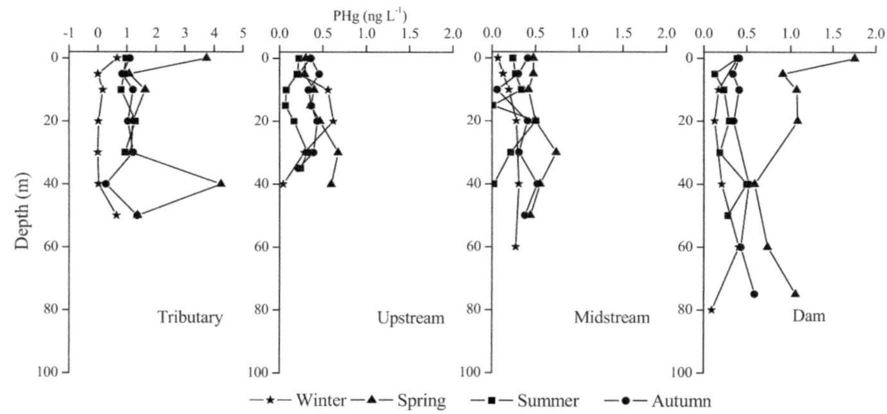

Fig. 6.13 Water column profiles of the concentrations of particulate mercury (PHg) in Suofengying Reservoir (SFY)

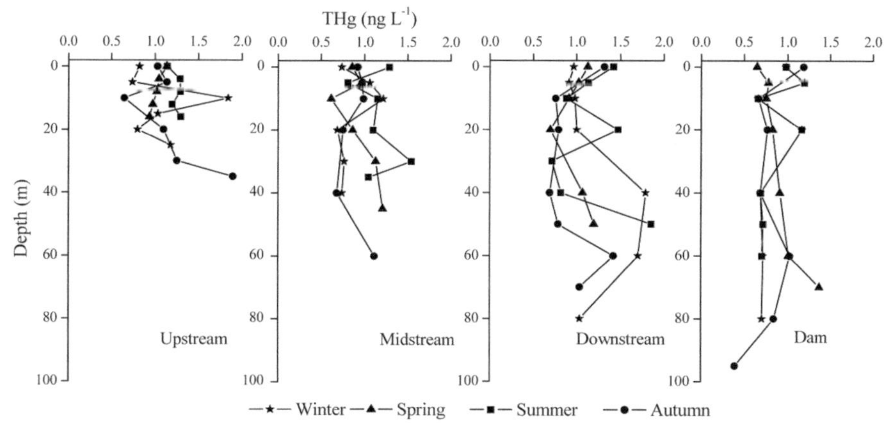

Fig. 6.14 Water column profiles of the concentrations of total mercury (THg) in Yinzidu Reservoir (YZD) (redrawed from Meng et al. 2010, with permission from The Alliance of Crop, Soil, and Environmental Science Societies; redrawed from Meng et al. 2011, with permission from Chinese Journal of Ecology)

2. Reactive mercury

Reactive mercury (RHg) is easily absorbed by organisms in the water column and is easily reducible to unlabelled mercury. RHg can convert into elemental mercury (Hg^0), and into methylmercury (MeHg) (Mason and Fitzgerald 1990). Therefore, RHg concentrations can reflect the activity and methylation ability of mercury in the water column.

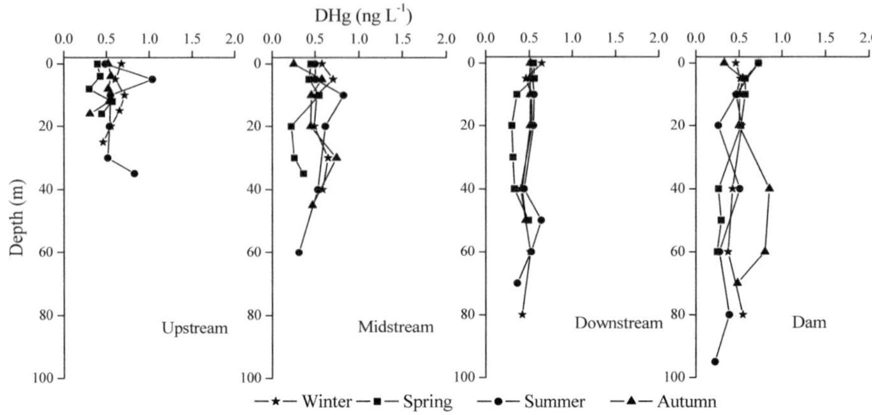

Fig. 6.15 Water column profiles of the concentrations of dissolved mercury (DHg) in Yinzidu Reservoir (YZD) (redrawed from Meng et al. 2010, with permission from The Alliance of Crop, Soil, and Environmental Science Societies; redrawed from Meng et al. 2011, with permission from Chinese Journal of Ecology)

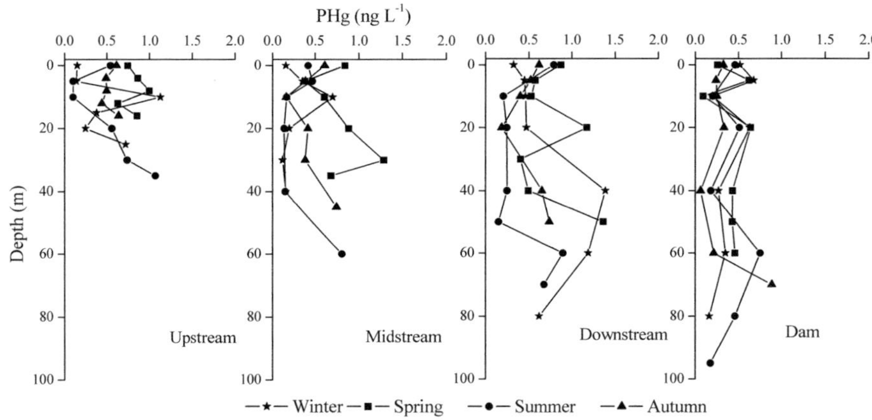

Fig. 6.16 Water column profiles of the concentrations of particulate mercury (PHg) in Yinzidu Reservoir (YZD) (redrawed from Meng et al. 2010, with permission from The Alliance of Crop, Soil, and Environmental Science Societies; redrawed from Meng et al. 2011, with permission from Chinese Journal of Ecology)

We found that RHg concentrations were relatively low in these three reservoirs, as shown in Tables 6.2, 6.3 and 6.4 (mean = 0.14 ± 0.07 ng L^{-1} in HJD, mean = 0.11 ± 0.03 ng L^{-1} in SFY, mean = 0.19 ± 0.15 ng L^{-1} in YZD). RHg concentrations in the present study were lower than other unpolluted reservoirs (1 ng L^{-1}) (Meuleman et al. 1995; Lucotte et al. 1999; Kotnik et al. 2007). There were no discernible differences in RHg concentrations between the HJD and SFY during the sampling seasons ($p > 0.05$), while there was a significant difference in the YZD ($0.01 < p < 0.05$). Furthermore, we found a seasonal variation in RHg concentrations for both the HJD and YZD. RHg concentrations were obviously higher in summer than in autumn or winter in the HJD and YZD ($p < 0.01$) (see Figs. 6.17 and 6.19). These elevated levels of RHg are potentially due to surface runoff and other exogenous inputs from rainfall in summer. However, RHg concentrations in the SFY had no obvious seasonal variations (see Fig. 6.18) ($p > 0.05$), possibly related to storage time in the water column or the absence of a sediment layer.

Correlation analysis showed that there was no obvious correlation between RHg and THg in the three reservoirs (HJD: $p > 0.05$, SFY: $p > 0.05$, YZD: $p > 0.05$). RHg was positively correlated with TMeHg in the water column (HJD: $p < 0.01$, SFY: $p < 0.05$, YZD: $0.01 < p < 0.05$), indicating that RHg and TMeHg have similar origins in the three reservoirs. The high RHg concentrations would increase mercury methylation, which would cause MeHg concentrations to increase. Therefore, this result demonstrated that low MeHg concentrations in both reservoirs were associated with low RHg in the water column.

3. Methylmercury

The spatial distribution patterns and concentrations of total methylmercury (TMeHg), dissolved methylmercury (DMeHg), and particulate methylmercury (PMeHg) in these reservoirs are illustrated from Figs. 6.20, 6.21, 6.22, 6.23, 6.24, 6.25, 6.26, 6.27, and 6.28, and supplementary from Tables 6.2 to 6.4. The concentration of TMeHg, DMeHg, and PMeHg in spring and summer were slightly higher than those in winter and autumn ($p < 0.001$). It is clear that MeHg species did not significantly increase with depth based on the vertical concentration-depth profiles from the water column ($p > 0.05$). In addition, MeHg concentrations slightly increased in summer and autumn at a depth of 40 m between the dam and

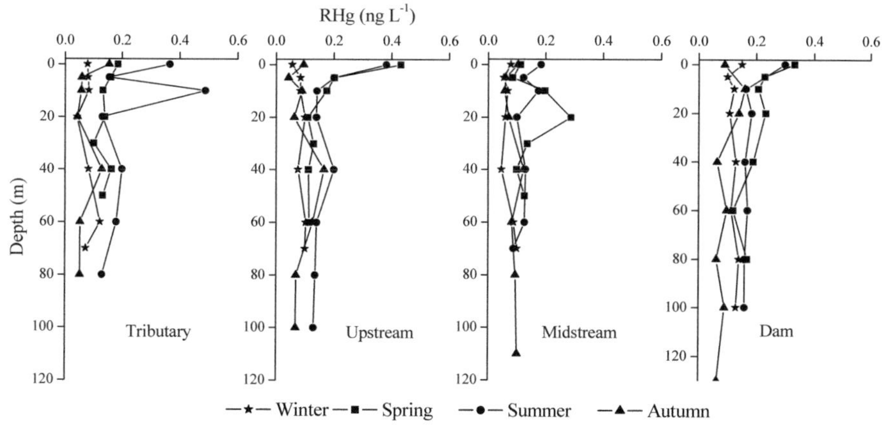

Fig. 6.17 Water column profiles of the concentrations of reactive mercury (RHg) in Hongjiadu Reservoir (HJD)

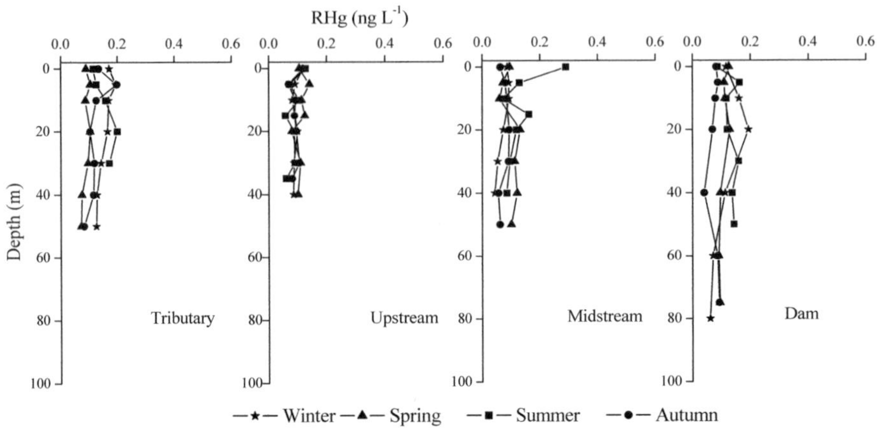

Fig. 6.18 Water column profiles of the concentrations of reactive mercury (RHg) in Suofengying Reservoir (SFY)

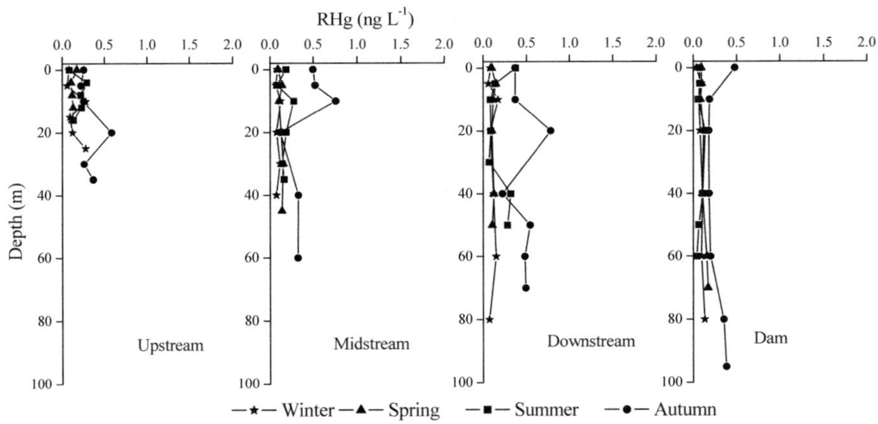

Fig. 6.19 Water column profiles of the concentrations of reactive mercury (RHg) in Yinzidu Reservoir (YZD) (redrawed from Meng et al. 2010, with permission from The Alliance of Crop, Soil, and Environmental Science Societies; redrawed from Meng et al. 2011, with permission from Chinese Journal of Ecology)

midstream sites, and elevated MeHg primarily existed in particulate form (PMeHg/THg ratios were 91% in the dam and 83% in the midstream). Because MeHg concentration in summer atmospheric precipitation is relatively low (see Chap. 3), the elevated MeHg concentration cannot be from atmospheric precipitation. At the same time, it cannot be plausibly attributed to in situ methylation due to high DO and pH in this layer, which is an unsuitable environment for active Hg methylation. This result implies that the elevated MeHg in the 40-m-deep dam and midstream bottom water in the spring and summer is not only due to the

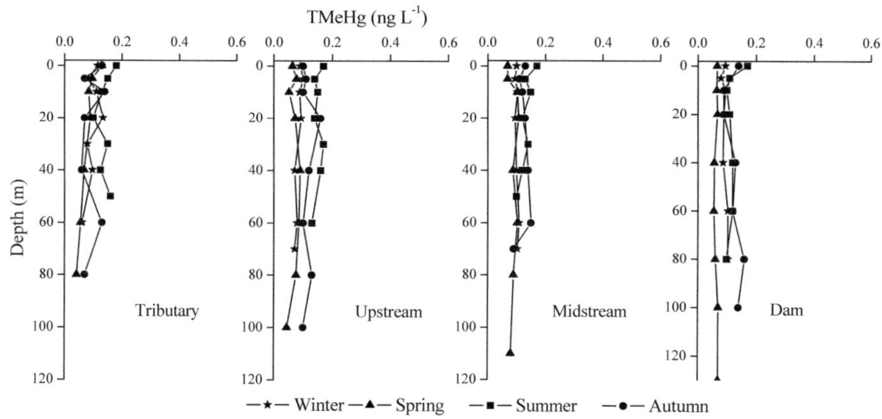

Fig. 6.20 Water column profiles of the concentrations of total methylmercury (TMeHg) in Hongjiadu Reservoir (HJD) (redrawed from Yao et al. 2011, with permission from John Wiley and Sons, Inc.; redrawed from Yao et al. 2011, with permission from Chinese Journal of Ecology; redrawed from Yao et al. 2009, with permission from Resources and Environment in The Yangtze Basin)

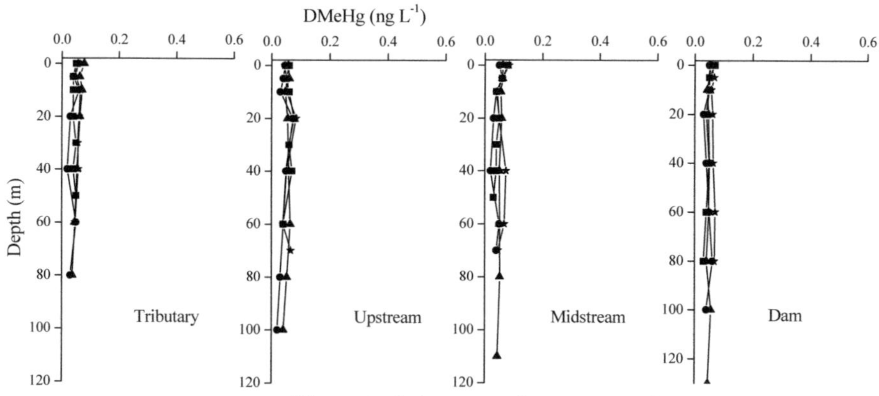

Fig. 6.21 Water column profiles of the concentrations of dissolved methylmercury (DMeHg) in Hongjiadu Reservoir (HJD) (redrawed from Yao et al. 2011, with permission from Chinese Journal of Ecology)

accumulation of settling particulate matter from overlying water but also from river erosion and surface runoff bringing PMeHg into the reservoir. The maximum MeHg concentration was not found in the bottom water, suggesting that submerged soil did not release MeHg into the water column in these three newly constructed reservoirs, and these reservoirs are not a source of MeHg.

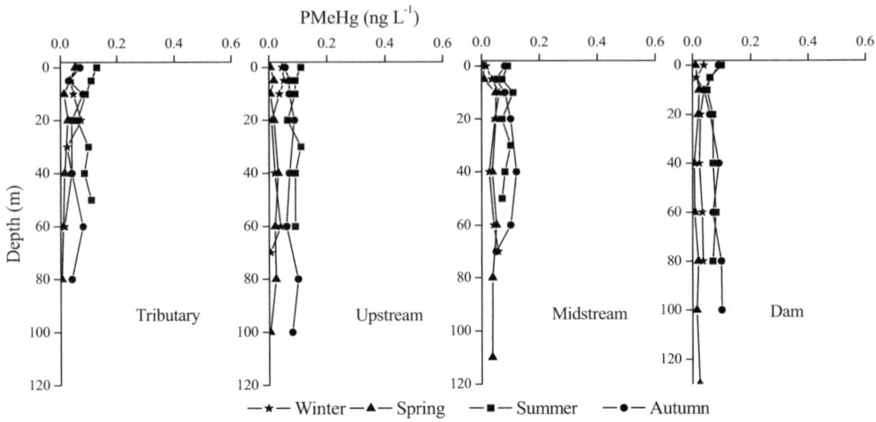

Fig. 6.22 Water column profiles of the concentrations of particulate methylmercury (PMeHg) in Hongjiadu Reservoir (HJD) (redrawed from Yao et al. 2011, with permission from Chinese Journal of Ecology)

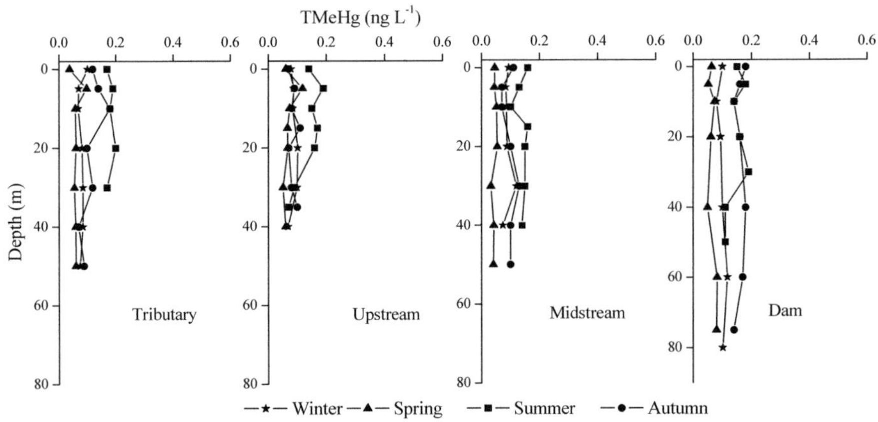

Fig. 6.23 Water column profiles of the concentrations of total methylmercury (TMeHg) in Suofengying Reservoir (SFY) (redrawed from Yao et al. 2011, with permission from John Wiley and Sons, Inc.; redrawed from Yao et al. 2011, with permission from Chinese Journal of Ecology; redrawed from Yao et al. 2009, with permission from Resources and Environment in The Yangtze Basin)

The mean PMeHg/TMeHg ratios are different seasonally in the HJD (63.0% in spring, 62.4% in summer, 37.6% in winter and 28.0% in autumn) during the four sampling periods in 2007. The same trend also appeared in the SFY (67.3% in the spring, 64.3% in the summer, 24% in the winter and 35.7% in the autumn) and YZD (37.2% in the spring, 60.9% in the summer, 33.2% in the winter and 29.2% in the autumn). PMeHg was significantly correlated with TMeHg in the spring and

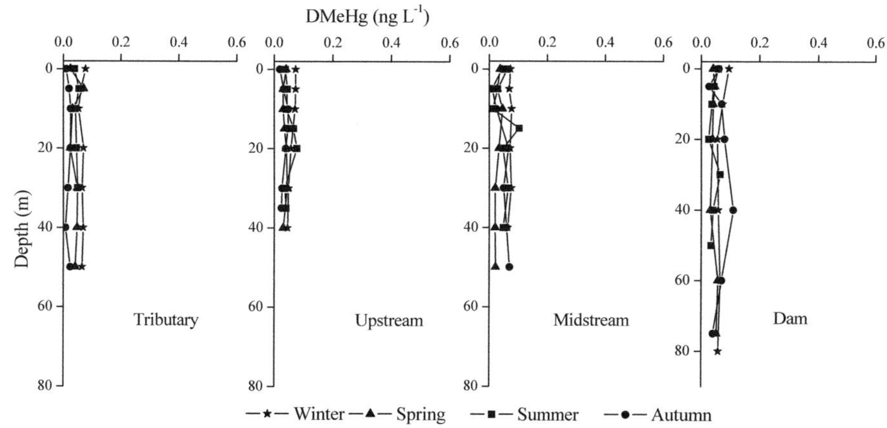

Fig. 6.24 Water column profiles of the concentrations of dissolved methylmercury (DMeHg) in Suofengying Reservoir (SFY) (redrawed from Yao et al. 2011, with permission from Chinese Journal of Ecology)

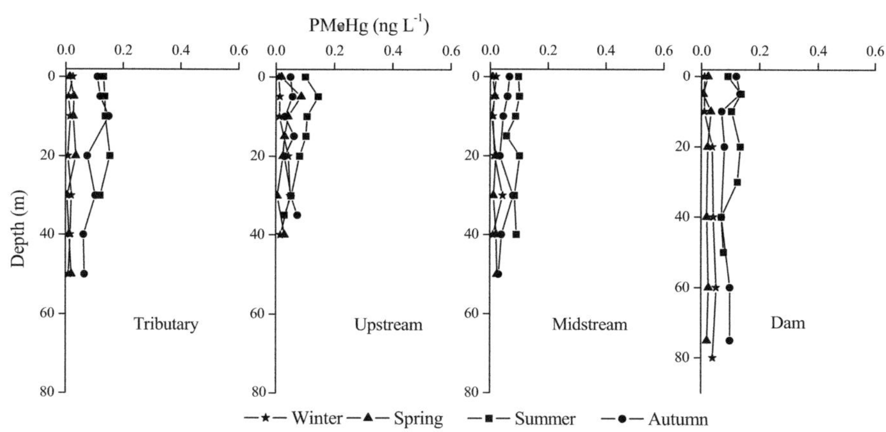

Fig. 6.25 Water column profiles of the concentrations of particulate methylmercury (PMeHg) in Suofengying Reservoir (SFY) (redrawed from Yao et al. 2011, with permission from Chinese Journal of Ecology)

summer ($p < 0.001$, $r = 0.87$ in HJD, $p < 0.0001$, $r = 0.83$ in SFY, $p < 0.0001$, $r = 0.79$ in YZD). DMeHg was significantly correlated with TMeHg in the autumn and winter ($p < 0.0001$, $r = 0.79$ in HJD, $p < 0.0001$, $r = 0.72$ in SFY, $p < 0.0001$, $r = 0.65$ in YZD), indicating that PMeHg and DMeHg are the main factors controlling elevated MeHg in both reservoirs during 2007.

In spatial distribution, no discernible change in TMeHg could be detected from upstream to the dam moving along the sampling section of the reservoirs. All mean

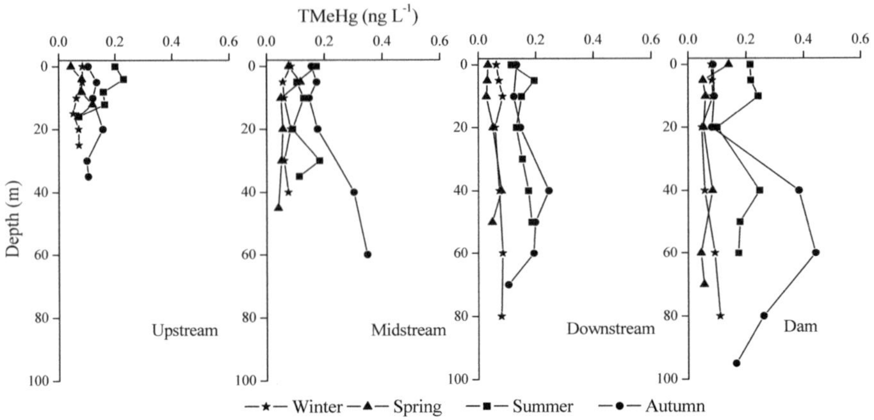

Fig. 6.26 Water column profiles of the concentrations of total methylmercury (TMeHg) in Yinzidu Reservoir (YZD) (redrawed from Meng et al. 2010, with permission from The Alliance of Crop, Soil, and Environmental Science Societies; redrawed from Meng et al. 2011, with permission from Chinese Journal of Ecology)

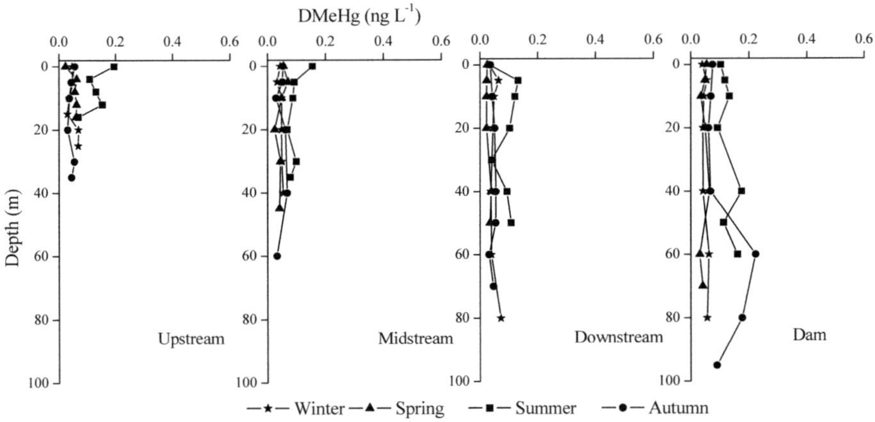

Fig. 6.27 Water column profiles of the concentrations of dissolved methylmercury (DMeHg) in Yinzidu Reservoir (YZD) (redrawed from Meng et al. 2010, with permission from The Alliance of Crop, Soil, and Environmental Science Societies; redrawed from Meng et al. 2011, with permission from Chinese Journal of Ecology)

values were tested by significant differences, and there was no obvious difference ($p > 0.05$) between their concentrations. The mean MeHg concentrations in the HJD were consistent with values observed in the SFY and YZD, and there was no significant difference ($p > 0.05$). The percentage of THg that occurs as MeHg (% MeHg) is a good indicator of MeHg production in ecosystems (St. Louis et al. 1994; Rudd 1995; Gilmour et al. 1998). Therefore, we calculated the percentage of MeHg in the HJD, SFY, and YZD. The TMeHg/THg percentages were

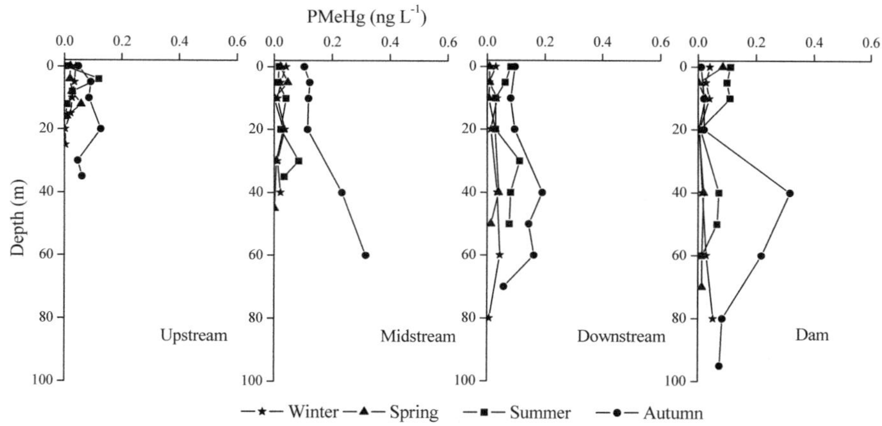

Fig. 6.28 Water column profiles of the concentrations of particulate methylmercury (PMeHg) in Yinzidu Reservoir (YZD) (redrawed from Meng et al. 2010, with permission from The Alliance of Crop, Soil, and Environmental Science Societies; redrawed from Meng et al. 2011, with permission from Chinese Journal of Ecology)

11.8% ± 9.3%, 13.7% ± 8.8% and 13.0% ± 10.0% in the HJD, SFY, and YZD, respectively. The percentage MeHg at sites close to the dam were somewhat lower than those at the upstream sites in both reservoirs. Furthermore, the percentage MeHg in all three reservoirs were much lower than the percentages in newly built reservoirs reported in North America and Europe (50–80%) (Tremblay et al. 1996; Kannan et al. 1998; St. Louis et al. 2004; Hall et al. 2005). Moreover, Guo (2008) conducted a mass balance study of both THg and MeHg based on one year of continuous measurements of monthly average THg and MeHg concentrations in inflows and outflows in these three reservoirs. They found that both reservoirs were net sinks for THg and TMeHg, as shown in Chap. 5, Sect. 5.4. Therefore, our data confirm that these three new reservoirs were not active sites for net Hg methylation.

A comparison of Hg concentrations between these three newly built reservoirs and reservoirs in other areas is shown in Table 6.5. THg concentrations in the present study were somewhat lower than those in reservoirs in North America and Europe. This indicated that the newly built reservoirs were not obviously polluted by THg in the Wujiang River basin. However, the TMeHg concentrations in the water column of both reservoirs were significantly lower than the concentrations in newly built reservoirs in North America and Europe but were similar to the concentrations in old reservoirs in North America and Europe. Compared with other Chinese lakes/reservoirs located in the same region, including Hongfeng Reservoir (0.05–0.92 ng L^{-1}), Aha (0.03–2.05 ng L^{-1}), Baihua Reservoir (1.29 ng L^{-1}), and Wujiang River (0.07–0.7 ng L^{-1}) (Bai 2006; He 2007; Yan 2005; Guo 2008), the concentrations of MeHg were significantly lower. This comparison confirms that no net MeHg production occurred in either of these reservoirs, contrary to the findings in newly built reservoirs in North America and Europe.

Table 6.5 Comparison of total mercury (THg) and total methylmercury (TMeHg) concentrations in the water columns in Hongjiadu, Suofengying and Yinzidu, China, with reservoirs in North America and Europe

Sampling location	Periods (a)	THg (ng L^{-1})	TMeHg (ng L^{-1})	Data source
Flood-control imploundments, USA	2–4	0.74–6.97	0.06–6.6	Brigham et al. (2002)
Experimental reservoir, Canada	0.04	0.98–6.95	0.05–3.2	Kelly et al. (1997)
Experimental reservoir, Canada	3	1.1–6.0	0.1–2.1	Hall et al. (2005)
Quebec reservoir, Canada	3	<5	0.01–2	Lucotte et al. (1999)
Maryland reservoir, USA	12–133	0.4–6.8	0.048–0.38	Mason and Sveinsdottir (2003)
Narraguinnep reservoir, USA	16	0.47–1.06	0.010–0.043	Gray et al. (2005)
Caniapiscau reservoir, Canada	17	1.19–1.69	0.06–0.09	Schetagne et al. (2000)
SFY Resrvoir, Wujiang River	3	0.42–4.9	0.030–0.22	Present study
HJD Resrvoir, Wujiang River	4	0.32–6.6	0.05–0.17	Present study
YZD Resrvoir, Wujiang River	6	0.40–1.9	0.0028–0.44	Present study

6.1.4 Distribution of Mercury Species in Flooded Soil

1. Organic matter

We did not collect samples of flooded soil; however, as proxies, we collected numerous soil samples along the banks of the reservoirs to represent flooded soil. We analyzed the concentration of organic matter (OM), and the results are shown in Table 6.6. OM concentrations were relatively low in the submerged soil of the HJD and SFY, varying from 0.4 to 6.9%. These reservoirs remained in an oligo-mesotrophic state in 2007 (see the Chap. 10, Sect. 10.2), suggesting that these two reservoirs have low primary productivity, lack of phytoplankton, and low organic carbon content. Therefore, OM in submerged soil was sourced primarily from surface runoff and river input.

2. Total mercury and methylmercury

As described in Table 6.6, during the sampling period, the mean concentrations of THg and TMeHg in submerged soil were 68.8 ± 34.5 ng g^{-1} and 1.2 ± 0.5 ng g^{-1} in the HJD (n = 25) and 268.3 ± 72.9 ng g^{-1} and 1.3 ± 0.4 ng g^{-1} in SFY (n = 19).

Table 6.6 Total mercury (THg), total methylmercury (TMeHg) and organic matter (OM) concentrations in flooded soil in Hongjiadu (HJD) and Suofengying (SFY) reservoirs

		HJD	SFY
Number	*n*	25	19
OM (%)	Range	0.4–3.9	2.6–6.9
	Mean ± SD	1.9 ± 1.1	4.1 ± 1.3
THg (ng g^{-1})	Range	20.1–137.2	141.3–372.4
	Mean ± SD	68.8 ± 34.5	268.3 ± 72.9
TMeHg (ng g^{-1})	Range	0.4–1.9	0.7–2.
	Mean ± SD	1.2 ± 0.5	1.3 ± 0.4

As a result, THg concentrations in the soil along the bank of the SFY were significantly higher ($p < 0.05$) than those in the soil of the HJD; however, the SFY is located downstream of an Hg mining area, as shown in Fig. 6.1. Therefore, this is not a typical background site but one directly influenced by nearby Hg mining activity. Overall, the background THg concentrations of soil samples in these two reservoirs (especially SFY) are much higher than the background Hg concentration of soil in China, which is 38.0 ng g^{-1} (China National Environmental Monitoring Centre 1992; Feng et al. 2006). However, the soil TMeHg concentrations in these two reservoirs were obviously lower than the TMeHg concentrations of soil in North America, which is 2–12 ng g^{-1} (Tremblay et al. 1996). TMeHg/THg ratio differences were due to varying THg; the TMeHg/THg ratios were 2.11 ± 1.0% in HJD soil, and the TMeHg/THg ratios were 0.55 ± 0.23% in SFY soil. Moreover, we found that THg concentrations in precipitation (HJD: 41.2 ± 24.1 ng L^{-1}; SFY: 51.6 ± 38.4 ng L^{-1}) were much higher than those reported in relatively pristine areas in North America and Europe, which are generally lower than 10.0 ng L^{-1} (Lindqvist et al. 1991; St. Louis et al. 2004; Hall et al. 2005). The enhancement is attributed to regional Hg pollution by coal combustion. In contrast, the volume-weighted mean concentrations of MeHg (±SD) in precipitation collected at the HJD and SFY were merely 0.10 ± 0.05 and 0.12 ± 0.06 ng L^{-1}, respectively, which are low levels in the same reference framework. Therefore, the contribution of MeHg from precipitation to the studied reservoirs would be considered trivial in comparison to the impact from other sources. Our data have shown that the THg concentrations in precipitation and submerged soil at the HJD and SFY were significantly higher than the concentrations at newly built reservoirs in North America and Europe.

However, the reservoirs of the present study were deduced to be net sinks for THg and TMeHg because they lacked active sites for Hg methylation. This result is in spite of the fact that THg concentration in precipitation and regional soil was elevated from an international perspective. It is generally believed that flooded soil is propitious for MeHg production. The submerged soil in the relevant investigations from North America and Europe was normally from boreal forest or wetland, where OM concentrations can vary from 30 to 50% (Verdon et al. 1991; Tremblay et al. 1996; Lucotte et al. 1999). As shown in Fig. 6.29, St. Louis experimentally

Fig. 6.29 St. Louis experimentally flooded a wetland complex (peatland surrounding an open water pond) at the Experimental Lakes Area (ELA), northwestern Ontario, Canada (St. Louis et al. 2004)

flooded a wetland complex (peatland surrounding an open water pond) at the Experimental Lakes Area (ELA), northwestern Ontario, Canada, to examine the biogeochemical cycling of MeHg in reservoirs. Figure 6.29 shows aerial photographs of the wetland complex (peatland surrounding an open water pond) prior to flooding (1992; A) and the reservoir after the first (1993; B), fifth (1997; C), and ninth (2001; D) years of flooding. Most of the inundated surface peat floated in the reservoir by 1996. By 2001, vegetation had regrown on most of the floating peat surfaces. They studied the changes in MeHg production rate before and after flooding. Prior to the flooding, the wetland was a net source of 1.7 mg ha^{-1} $year^{-1}$ MeHg to downstream ecosystems, yielding 130% of the MeHg amount that entered the wetland. In the first year of flooding, the net export of MeHg from the reservoir increased 40-fold to approximately 70 mg ha^{-1} $year^{-1}$ MeHg, yielding over 860% of MeHg inputs (St. Louis et al. 2004). The MeHg concentration is elevated in environments that favor Hg methylation processes such as the degradation of OM in the soil. These environments also produce many nutrients through microbiological activity that promote Hg II conversion into MeHg, which results in increased MeHg concentrations in water columns and elevated MeHg concentrations in fish by bioaccumulation and biomagnification through the food chain (Furutani and Rudd 1980; Bodaly et al. 1984; Ramlal et al. 1987; Ramsey 1990; Hecky et al. 1991;

Lucotte et al. 1999). However, in the following years, with the degradation of OM in flooded soil, it was exhausted of methylation sources, which significantly reduced the MeHg production rate in submerged soil.

Guizhou is located in the southwestern area of China where agricultural land accounts for 86.78% of the total land area. The thickness of soil layer is thin in most of the region, soil fertility is poor, the slope is steep, and approximately 80% of cultivated land is subject to soil erosion (Guizhou Statistical Yearbook 2008). In our study, these three newly built reservoirs are located in rocky, plateau-like mountains with few farmlands; 72% of the land area is dry land, and 4.5% of the farmland is located on steep slopes. Loping arable land and Shi Kala coverage account for 60% of cultivated land, resulting in very poor land quality in this region (see Fig. 6.30).

Due to the limited land resources in the Southwest area, river valleys have become farmland, and the farmers have overstressed the farmland, resulting in reduced OM concentrations in flooded soil. In addition, prior to flooding, the people exploited forests and scrubland around the reservoirs (forest cut away, leaving the root flat ground) and cleared residential areas, and other obstacles have been cleared (Hydropower Consulting Group Guiyang Survey Design and Research Institute 1987). Therefore, extensive agricultural activity in this region is the main cause of low OM concentrations in soil (see Table 6.7). This is consistent with the results from the China National Environmental Monitoring Center, and Liu (2009), who reported that OM contents in soil ranged from 1.2 to 7.8% in southwestern China. It has been demonstrated that low MeHg concentrations in these reservoirs were associated with low OM in submerged soil. These results suggest that OM in soil is a key factor governing Hg methylation rates in reservoirs. It is clear that further study is needed to elucidate the mechanisms for soil OM governing MeHg distribution in the water columns of these reservoirs.

Apart from OM levels in the soil, MeHg production may be dependent on other physiochemical factors, such as pH, DO, DOC, trophic status, and watershed characteristics. In the primary evolutionary phases, we found that DOC was

Fig. 6.30 Surrounding environment map in the Hongjiadu Reservoir (HJD) and Suofengying Reservoir (SFY) (the photo was taken on December, 2009)

Table 6.7 Comparison of organic matter levels in soil in the catchment of Hongjiadu (HJD) and Suofengying reservoirs (SFY) in Southwestern China

Sampling location	OM (%)	Data source
HJD, Guizhou	1.9 ± 1.1	Present study
SFY, Guizhou	4.1 ± 1.3	Present study
Guizhou	4.3 ± 2.7	China National Environmental Monitoring Center (1990)
Yunnan	3.9 ± 2.4	
Sichuan (including Chongqing area)	3.3 ± 4.5	
Tibet	4.6 ± 4.7	
Southwestern China	1.2–7.8%	Liu et al. (2009)

Reprinted from Yao et al. (2011), with permission from John Wiley and Sons, Inc

positively correlated with TMeHg in the water column (HJD, $R^2 = 0.14$, $p < 0.0001$, $n = 116$; SFY, $R^2 = 0.31$, $p < 0.0001$, $n = 106$; YZD, $R^2 = 0.22$, $p < 0.0001$, $n = 101$) (see Fig. 6.31), which indicated that RHg and TMeHg have similar origins in the three reservoirs. DOC levels (ranged from 0.5 to 4.9 mg L^{-1}) were not as high as those observed in peat bog lakes in North America, where DOC concentrations reached up to 11 mg L^{-1} (Lucotte et al. 1999; Porvari and Verta 2003; Hines et al. 2004; Liu et al. 2008). We have shown that the concentration of MeHg increased with the concentration of DOC when DOC was greater than 5 mg L^{-1} (McMurtry et al. 1989; Barkay et al. 1997). Low pH is conducive to the release of heavy metals from sediments and particulate matter, and water acidification will affect the distribution of MeHg. Many studies have found that acidic conditions are favorable to the formation of MeHg in the lake, and low pH in lakes can improve the rate of Hg methylation (Bloom 1992; Miskimmin et al. 1992; Hudson et al. 1994). In the HJD, SFY, and YZD, we found that pH was positively correlated with TMeHg in the water column (HJD, $R^2 = 0.16$, $p = 0.002$, $n = 116$; SFY, $R^2 = 0.36$, $p < 0.0001$, $n = 106$; YZD, $R^2 = 0.36$, $p < 0.0001$, $n = 101$). The pH levels were slightly alkaline in both reservoirs and slightly higher than those reported in North America and Europe, which were generally lower than 7.0 (Lucotte et al. 1999; Porvari and Verta 2003). Therefore, low TMeHg may be

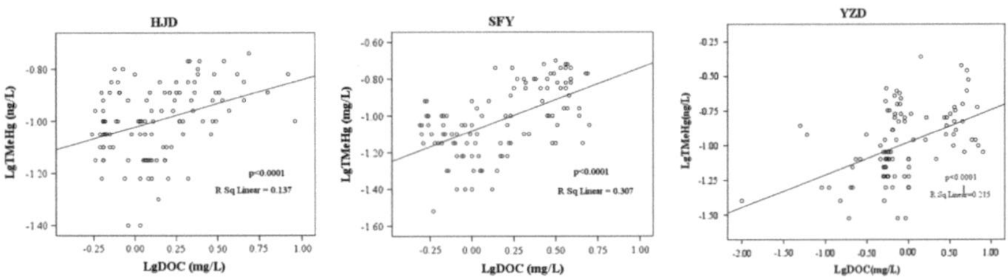

Fig. 6.31 Scatter diagram of LgTMeHg (ng L^{-1}) and LgDOC (mg L^{-1}) in the water column with these three newly built reservoirs (HJD, SFY and YZD)

related to pH in the water column in these three reservoirs, although it is not a direct controlling factor. Because the variation in pH does not directly affect the rate of Hg methylation, pH may affect the solubility and mobility of MeHg and other mercury species in the aquatic environment. The mercury concentration is elevated in the water environment from the watershed, and it affects MeHg concentrations in the water column (Lee and Hultberg 1990).

These three newly built reservoirs have different total water volumes and water residence times. HJD has the largest water volume (490 million m^3) of the three cascading reservoirs created in the Wujiang River, whereas SFY has the smallest (20 million m^3). In the present study, no obvious anoxic conditions were found in any of the three reservoirs; pH levels were slightly alkaline, DOC concentrations were very low, and primary productivity was low. It is well documented that the high pH and low DOC do not favor Hg methylation processes in aquatic systems, although these three reservoirs have different watershed areas, total water volumes, and water residence times. Therefore, the environments of all reservoirs were not conducive to Hg methylation.

6.2 Biogeochemical Cycling of Mercury in the Dongfeng Reservoir and Puding Reservoir

6.2.1 Sampling Site Descriptions, Sample Collection, Analytical Methods, and QA/QC

As outlined in Fig. 6.13, four sampling stations from PD reservoir and two sampling stations from DF were chosen in this study. These sampling stations were spatially distributed from upstream to downstream of PD and DF. In detail, DF-1 and PD-1 are located at the upper end of the reservoirs. PD-2 is located in the middle part of the reservoir. PD-3 is situated in the downstream part of the reservoir. DF-2 and PD-4 are located adjacent to the dams (within approximately 500 m). Water samples (water column profiles with 5–8 different depths), sediment cores (liquid phase), sediment pore water (solid phase), and water samples from the water–sediment interface were collected at each sampling station during summer (July, 2006) and winter (January, 2007) (Fig. 6.32).

Filtered water column samples were analyzed for DHg and DMeHg levels. THg, RHg, TMeHg, and TSS levels were analyzed in each of the unfiltered samples. Water samples from the water–sediment interface and sediment pore water samples were divided for the DHg and DMeHg analyses. The freeze-dried sediment samples were ground and homogenized for analyses of the solid-phase THg, MeHg, and organic matter concentrations. The TSS content in water samples was determined gravimetrically by filtering an aliquot of water (typically 1500 mL) through a pre-weighed 0.45 μm pore-size, 47 mm (diameter) polycarbonate membrane filter. Water quality parameters, such as pH, *T*, and DO, were measured in situ using a

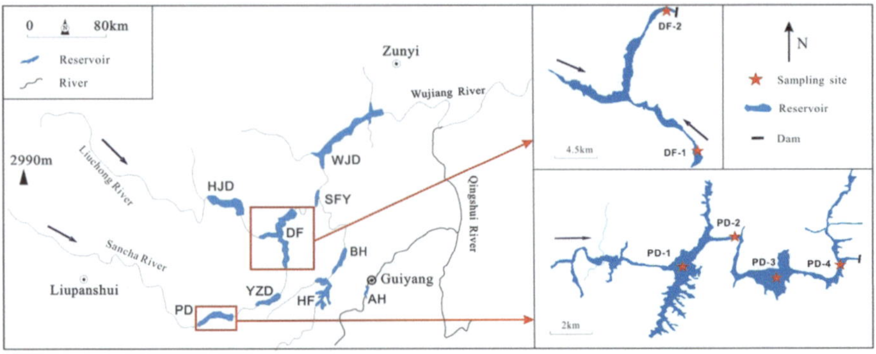

Fig. 6.32 Map of the study area and location of sampling stations at DF and PD

portable analyzer (Pioneer 65 Portable Multi-parameter Instrument, France). The DOC was qualified in the laboratory using a high-temperature combustion technique with a TOC analyzer (Elementar High TOC II). Water anions such as chloride (Cl^-), nitrate (NO_3^-) were detected by ion chromatography (Dionex). Detailed information about sample collection, preparation, and analyses are available in Chap. 2.

Quality assurance and quality control of the analytical processes were performed using field blanks, system blanks, spike recoveries, and sample duplicates. Field blanks and duplicates were collected regularly (>10%) throughout each sampling campaign. Detection limits were estimated as three times the standard deviation of the blank measurement and were 0.10 and 0.029 ng/L for THg and MeHg in water samples, respectively. The limits of determination were 0.01 ng g^{-1} for THg and 0.002 ng g^{-1} for MeHg in sediment samples. The reproducibility of the THg measurements had an analytical precision (coefficient of variation) of 2–6% and was <11% for MeHg in water samples. Recoveries for the matrix spikes ranged from 88–111% to 89–111% for the THg and MeHg analysis in water samples, respectively. Recoveries for the IAEA-405 geological standard ranged from 80 to 110% for the THg and MeHg analyses, with an analytical precision (coefficient of variation) <10%.

Statistical evaluations were performed using SPSS 11.5 software. Relationships between covariant sets of data were analyzed with a regression analysis. The correlation coefficient (r) and test of significance (p) were computed for the correlation analysis. Correlations were significant at 0.05 (1-tailed). Pearson's correlation matrix, which yielded the linear correlation coefficients (r) between Hg species and ancillary water quality parameters, were calculated in this study to investigate the possible factors controlling the distribution of Hg species in the water column. In addition, Kolmogorov–Smirnov (K–S) and Kruskal–Wallis (K–W) tests were performed to compare significant differences between two or more independent datasets. Differences were significant at $p < 0.05$.

6.2.2 Mercury Species in the Water Column

1. General water quality characteristics

Vertical profiles for *T*, pH, and DO from the PD and DF are displayed in Figs. 6.33, 6.34, 6.35, 6.36 and 6.37, and summary data are shown in Table 6.8.

The water temperature profiles exhibited the expected seasonal pattern, with temperatures in the PD and DF ranging from 7.9 to 29 °C and ranging from 11 to 27 °C, respectively. The water bodies of PD and DF were well mixed in winter, and stratified in summer. Due to the karstic geology of the Wujiang River Basin, the water samples from both PD and DF were slightly alkaline (DF, pH = 7.80 ± 0.35; PD, pH = 7.72 ± 0.27) but were slightly acidic in the surface layer at PD-4 in winter (pH = 6.53), possibly due to the input of sewage and/or living pollutants through human activities near the reservoir dam. As shown in Fig. 6.33, the maximum pH values in each of the sampling stations at PD was observed in the surface layer in summer, which then decreased with water depth. In winter, the minimum pH values at PD occurred in the surface water layer, with the exception of the PD-2 sampling station. The distribution of pH values in the water columns at DF was different from the distribution at PD (Fig. 6.37), as the peak values were observed in the subsurface water layer (depth of 10 m) and decreased slightly with water depth.

As shown in Figs. 6.35 and 6.36, no clear vertical distributions of DO were observed in winter at PD and DF, with the exception of slight differences between the surface water layer and subsurface water layer, which was similar to the thermal stratification (well mixed during winter). However, the maximum DO values at PD and DF occurred in the surface water layer during summer for each of the sampling

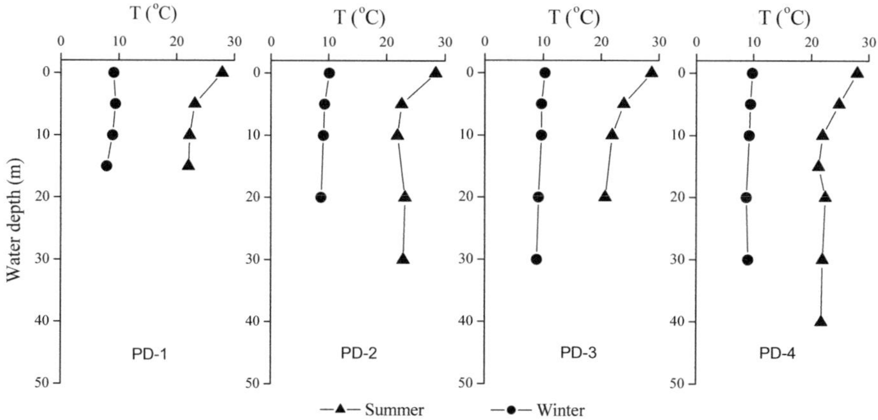

Fig. 6.33 Water column profiles of the temperature (*T*) at four sampling stations (PD-1, PD-2, PD-3, PD-4) in PD (redrawed from Zhang et al. 2009a, b, with permission from Elsevier; redrawed from Zhang et al. 2009a, b, with permission from Earth and Environment)

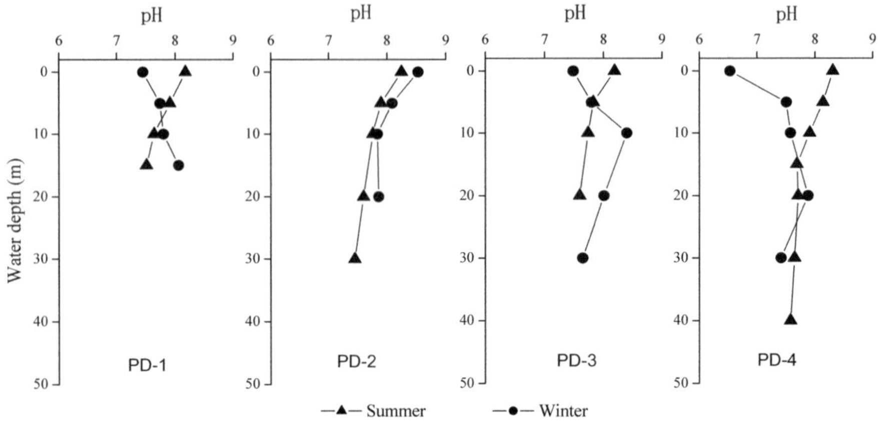

Fig. 6.34 Water column profiles of the pH at four sampling stations (PD-1, PD-2, PD-3, PD-4) in PD (redrawed from Zhang et al. 2009a, b, with permission from Elsevier; redrawed from Zhang et al. 2009a, b, with permission from Earth and Environment)

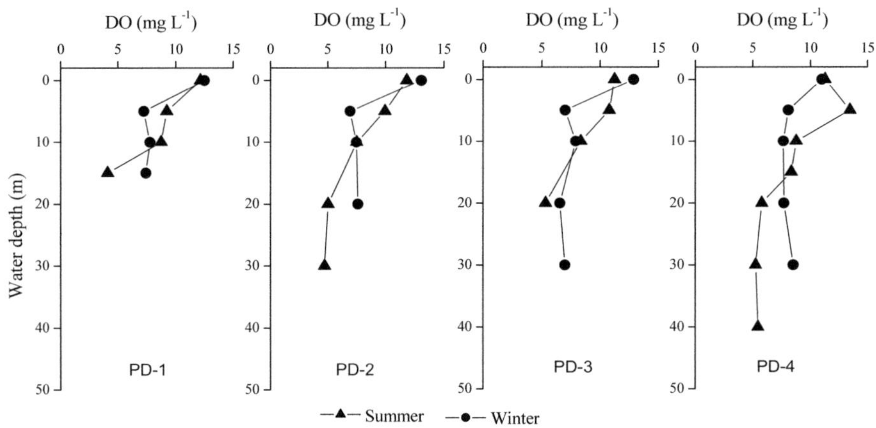

Fig. 6.35 Water column profiles of the dissolved oxygen (DO) at four sampling stations (PD-1, PD-2, PD-3, PD-4) in PD (redrawed from Zhang et al. 2009a, b, with permission from Elsevier; redrawed from Zhang et al. 2009a, b, with permission from Earth and Environment)

stations, with the exception of sampling stations PD-4 and DF-2, and decreased with water depth. During summer, the lowest DO values were observed in the bottom water at PD and DF. Notably, the peaks of DO in DF-2 and PD-4 in summer did not occur in the surface water layer, but were observed in the subsurface water layer (depth of 5–10 m), consistent with the peak total suspended solid levels. The peak algae density is generally observed at approximately a 10 m depth in the reservoir water column (Lu et al. 1999). Therefore, the summer maximum DO values in the subsurface water layers at PD-4 and DF-2 may be explained by algal

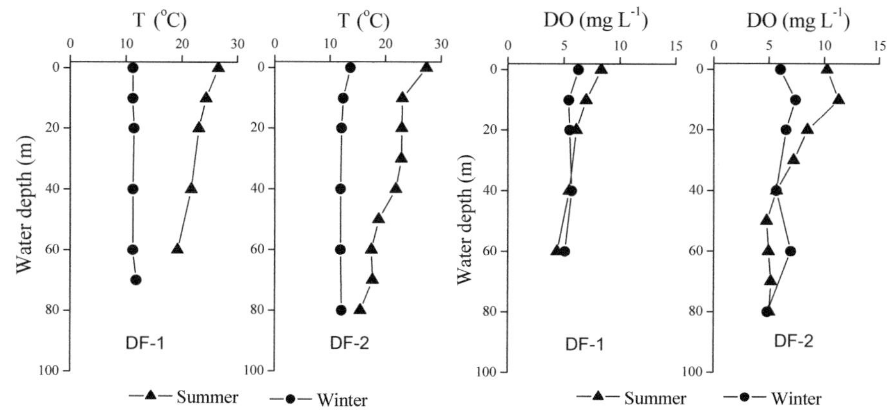

Fig. 6.36 Water column profiles of the temperature (*T*) and dissolved oxygen (DO) at two sampling stations (DF-1, DF-2) in DF

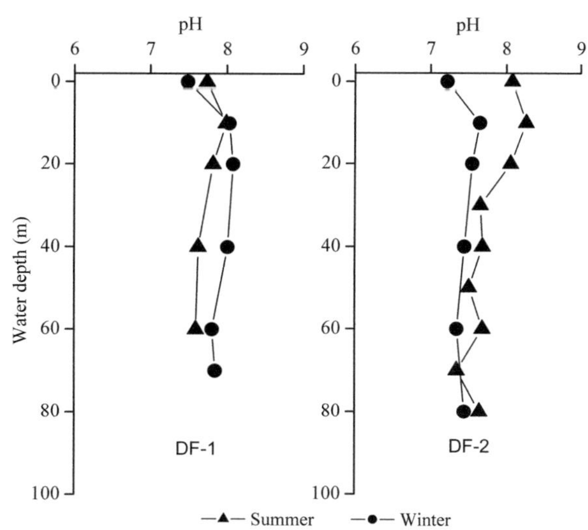

Fig. 6.37 Water column profiles of the pH at two sampling stations (DF-1, DF-2) in DF

blooms, as previously reported (He et al. 2008a). However, the peak DO value was not observed in the subsurface water layer at PD-1, PD-2, PD-3, DF-1, and DF-2 during summer, possibly indicating the absence of an algal bloom at these locations.

Overall, the water bodies at PD and DF were well mixed during winter. The general water parameters of the surface and subsurface water layers were primarily impacted by external factors such as human activities, wind fluctuation, and sailing activities. Therefore, the water *T*, pH, and DO were slightly different between the surface and subsurface water layers. In contrast, a clear vertical distribution from subsurface water layer to the bottom water was not observed at either PD or DF.

Table 6.8 Distribution of routine sampling water parameters collected from the PD and DF (mean ± SD, range)

Reservoir	Season	TSS (mg L^{-1})	DOC (mg L^{-1})	T (°C)	DO (mg L^{-1})	pH	NO_3^- (mg L^{-1})	Cl^- (mg L^{-1})
PD	Summer	5.45 ± 1.84, 1.43–8.75	1.86 ± 1.83, 0.38–8.15	23.7 ± 2.5, 20.7–28.7	8.38 ± 2.86, 4.14–13.5	7.83 ± 0.26, 7.45–8.31	13.8 ± 1.17, 12.3–17.6	3.12 ± 0.53, 2.58–4.26
	Winter	3.74 ± 1.15, 1.23–5.90	1.12 ± 0.41, 0.65–2.05	9.3 ± 0.6, 7.90–10.30	8.58 ± 2.17, 6.57–13.06	7.76 ± 0.43, 6.53–8.53	11.1 ± 0.58, 9.43–12.2	4.58 ± 0.88, 3.26–6.25
DF	Summer	3.93 ± 1.08, 1.89–5.29	1.86 ± 2.20, 0.63–8.50	21.6 ± 3.5, 15.5–27.4	6.72 ± 2.15, 4.38–11.31	7.77 ± 0.25, 7.35–8.27	14.50 ± 3.66, 7.74–19.33	3.88 ± 1.35, 2.65–6.42
	Winter	2.81 ± 1.45, 0.78–6.00	0.79 ± 0.22, 0.47–1.22	11.8 ± 0.7, 11.2–13.7	6.06 ± 0.85, 4.87–7.4	7.66 ± 0.29, 7.22–8.08	12.29 ± 0.56, 11.16–12.86	2.86 ± 0.17, 2.65–3.19

TSS total suspended solids; *DOC* dissolved organic carbon; *T* temperature; *DO* dissolved oxygen

However, PD and DF were completely stratified in summer, particularly at the downstream stations. Clear vertical distributions of water *T*, pH, and DO were observed at PD and DF, and the peak *T*, DO, and pH values occurred in the surface water layer or subsurface water layer and decreased with depth. Therefore, the stratification of the water body resulting from the temperature during different seasons was the primary factor controlling the water column characteristics in the stratified PD and DF reservoirs.

The mean concentrations of total suspended solids at PD in summer and winter were 5.45 ± 1.84 and 3.74 ± 1.15 mg L^{-1}, respectively. The corresponding concentrations at DF were 3.93 ± 1.08 and 2.81 ± 1.45 mg L^{-1}, respectively. According to the statistical analysis, the TSS in summer was significantly higher than the TSS in winter both for PD and DF ($p < 0.01$), possibly due to the input of particulate material through surface runoff during the summer wet season. Moreover, the abundant algae observed in summer are an important source of TSS in the water bodies of the reservoirs. Moreover, the TSS in the water column of PD was significantly higher than the TSS at DF during the sampling campaigns in 2007 ($p < 0.01$). As described in Chap. 10, the reservoir eutrophication evaluation showed that a state of hypereutrophication existed in the PD, whereas the DF was oligotrophic–mesotrophic during our sampling periods. Therefore, compared with DF, the relatively higher primary productivity at PD is one of the most important factors contributing to the higher levels of TSS in the water column. PD is located upstream of the Wujiang River, whereas DF is located downstream of PD (Fig. 6.32).

The hydrodynamic conditions within the river are weakened after the reservoir starts to impound. Thus, most of the particulate matter in river water will be deposited upstream of the reservoir. In contrast, only a limited quantity of particulate matter is transported to sites downstream of the Wujiang River. Therefore, DF presented a relatively higher TSS in water than PD during our sampling periods, indicating a sedimentation effect of the reservoir dam on the particulate matter in river water.

The nitrate (NO_3^-) concentrations in water samples collected from PD and DF showed significant differences between the two sampling seasons (winter and summer) (K–S test, $p < 0.001$). In addition, the water NO_3^- concentrations in summer at both PD and DF were significantly higher than the concentrations in winter (K–S test, $p < 0.001$), similar to the seasonal distribution patterns of rainfall. For example, the rainfall within the PD catchment during July, 2007 was 2.1×10^6 m^3, which was approximately three times higher than the rainfall during January, 2007 (0.71×10^6 m^3). In the summer season, agricultural activities were associated with increased usage of chemical fertilizers, including nitrate. Higher runoff from precipitation transported nitrate (from fertilizer) into the reservoirs, resulting in the elevated water NO_3^- concentrations in summer at both PD and DF. In contrast, the NO_3^- concentrations were comparatively lower in the winter, consistent with the decreased precipitation and agricultural activities.

The highest Cl^- concentrations (6.25 mg L^{-1}) in water samples from PD were observed in winter. According to the statistical analysis, the water Cl^-

concentrations at PD in winter were significantly higher than the concentrations in summer (K–S test, $p < 0.01$), probably due to the dilution of water through river runoff during the wet season. However, dilution from river runoff was less pronounced during the dry season, resulting in the comparatively higher Cl^- concentrations in the water samples from PD. The seasonal distribution patterns of Cl^- concentrations in the water samples from DF were completely different from PD, with higher Cl^- concentrations observed during summer than in winter. A possible reason for the discrepancy is that human activities (particularly at sampling station DF-2), but not the seasonal surface runoff variations, were the primary factor controlling the Cl^- concentrations in water samples from DF.

2. Distribution of mercury species in the water column

The spatial and seasonal distributions of the reactive mercury (RHg), THg, DHg, RHg, and PHg concentrations in water samples from PD and DF are shown in Figs. 6.38, 6.39, 6.40, 6.41, 6.42, 6.43, 6.44 and 6.45, and summary data are shown in Table 6.9.

(1) Reactive mercury

The seasonal distribution patterns of RHg in water samples from PD and DF are presented in Figs. 6.38 and 6.39. The mean RHg concentrations in water samples from PD were 0.48 ± 0.28 ng L^{-1} (0.25–1.48 ng L^{-1}) in summer and 0.08 ± 0.06 ng L^{-1} (0.04–0.25 ng L^{-1}) in winter. The corresponding data from DF were 0.28 ± 0.18 ng L^{-1} (0.12–0.85 ng L^{-1}) in summer and 0.25 ± 0.15 ng L^{-1} (0.12–0.63 ng L^{-1}) in winter. No discernable difference in the concentrations of RHg was observed between PD and DF (K–S test, $p > 0.05$) during the sampling periods.

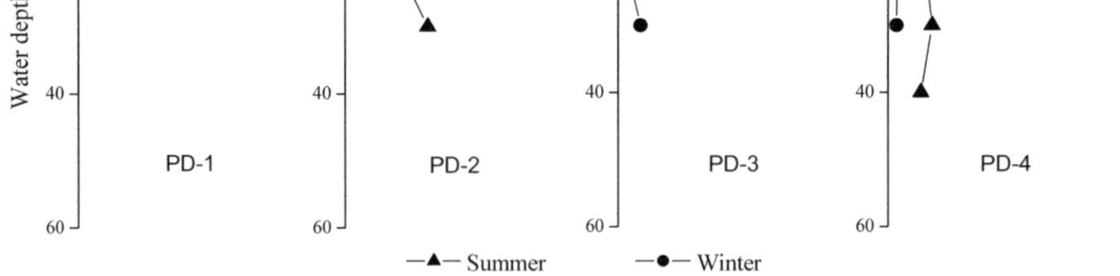

Fig. 6.38 Distributions of reactive mercury (RHg) in water column profiles of the four sampling stations (PD-1, PD-2, PD-3, PD-4) in PD

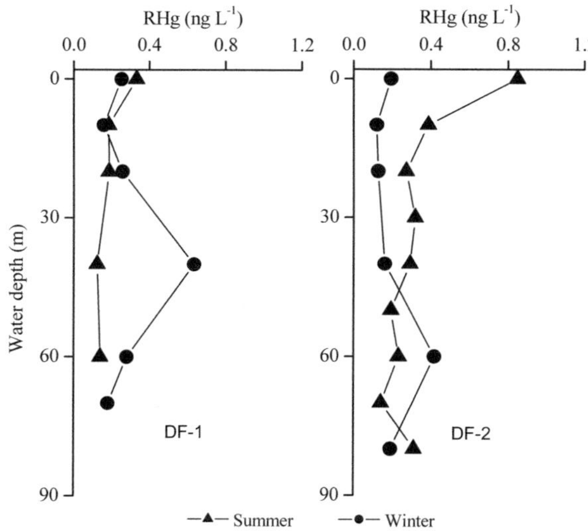

Fig. 6.39 Distributions of reactive mercury (RHg) in water column profiles of the two sampling stations (Df-1 and DF-2) in DF

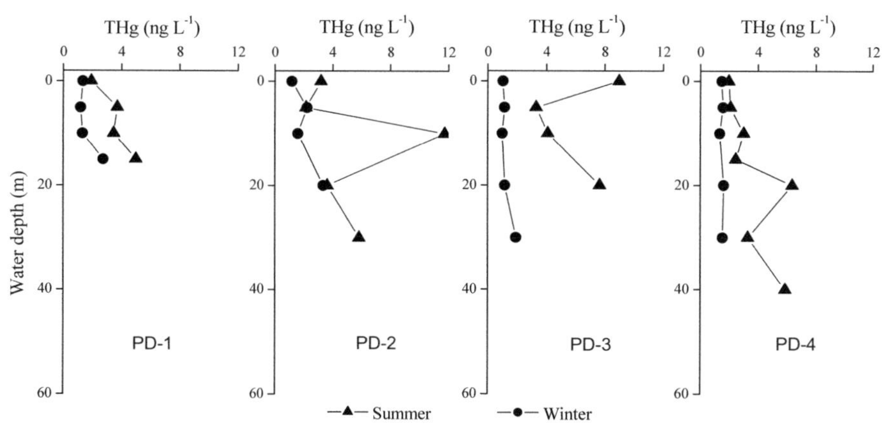

Fig. 6.40 Distributions of total mercury (THg) in water column profiles of the four sampling stations (PD-1, PD-2, PD-3, PD-4) in PD (redrawed from Zhang et al. 2009a, b, with permission from Elsevier)

RHg concentrations in water samples from uncontaminated rivers/reservoirs/lakes are usually less than 1 ng L^{-1} (Meuleman et al. 1995). In this study, the mean RHg concentrations in water samples collected from PD and DF were significantly less than 1 ng L^{-1}, indicating that DF and PF were not obviously impacted by nearby Hg sources. Statistically significant differences were observed in the RHg

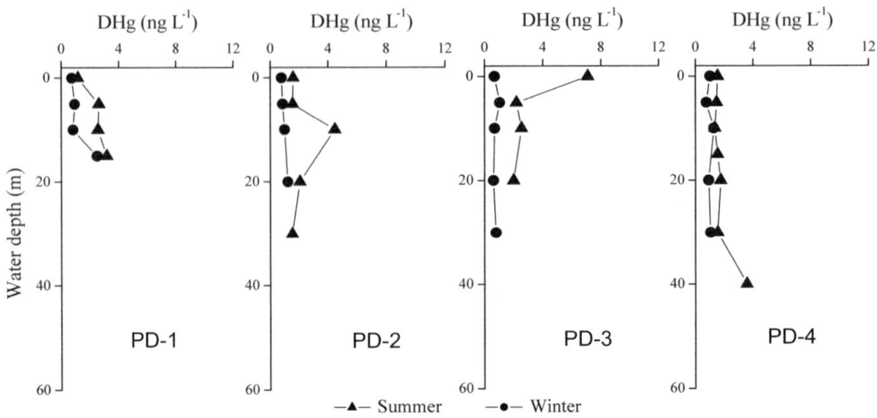

Fig. 6.41 Distributions of dissolved mercury (DHg) in water column profiles of the four sampling stations (PD-1, PD-2, PD-3, PD-4) in PD (redrawed from Zhang et al. 2009a, b, with permission from Elsevier)

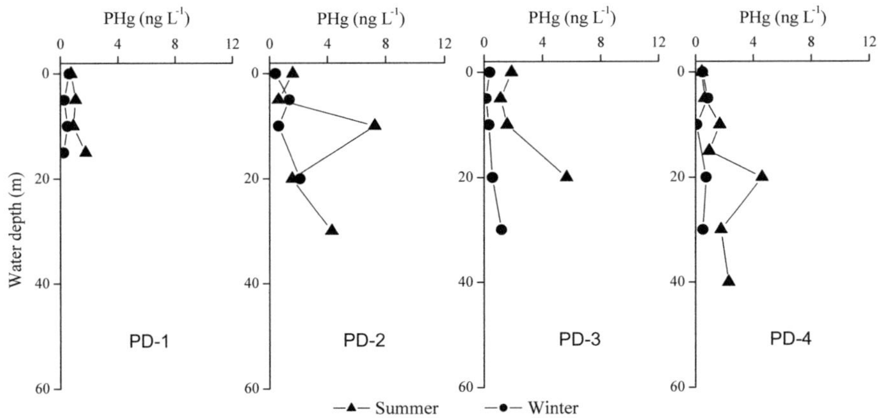

Fig. 6.42 Distributions of particulate mercury (PHg) in water column profiles of the four sampling stations (PD-1, PD-2, PD-3, PD-4) in PD (redrawed from Zhang et al. 2009a, b, with permission from Elsevier)

concentrations at PD between summer and winter (K–S test, $p < 0.001$). In addition, the RHg concentrations at PD in summer were significantly higher than the concentrations in winter (K–S test, $p < 0.05$). However, no discernable differences were observed in the RHg concentrations at DF between winter and summer during the sampling periods (K–S test, $p > 0.05$).

Although the RHg concentrations in river/reservoir/lake water are very low (Meuleman et al. 1995), the transportation/transformation processes for RHg play very important roles in the biogeochemical cycle of Hg in the aquatic ecosystem. In

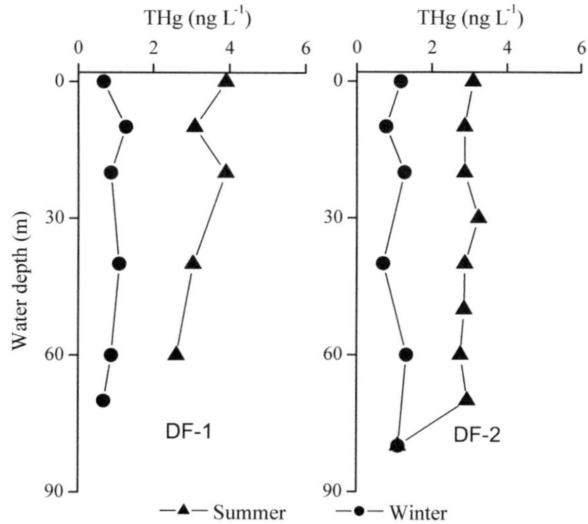

Fig. 6.43 Distributions of total mercury (THg) in water column profiles of the two sampling stations (Df-1 and DF-2) in DF

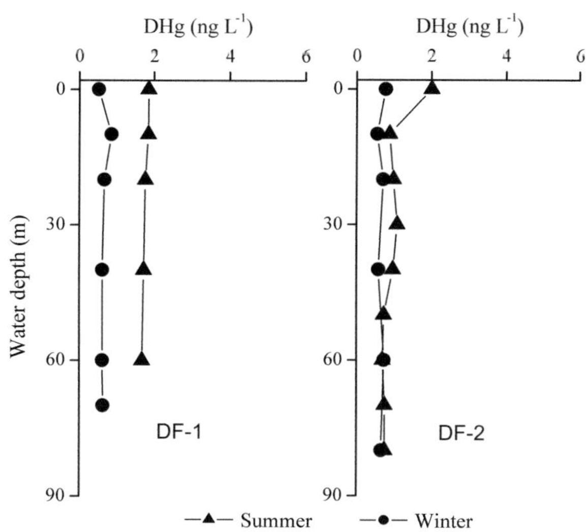

Fig. 6.44 Distributions of dissolved mercury (DHg) in water column profiles of the two sampling stations (Df-1 and DF-2) in DF

natural water, RHg is transformed into Hg^0 or MeHg. Methylation of RHg will greatly increase the toxicity of Hg to ecosystems; in contrast, the transformation of RHg into Hg^0 in response to light and biological conditions reduces the Hg burden in the water column, and then reduces the potential threat to ecosystems. The transformation of RHg into Hg^0 is affected by various factors, such as pH, biological activity, and light (Amyot et al. 1997; Lanzillotta et al. 2002; Tseng et al. 2004; Grace et al. 1991; Steven et al. 2002).

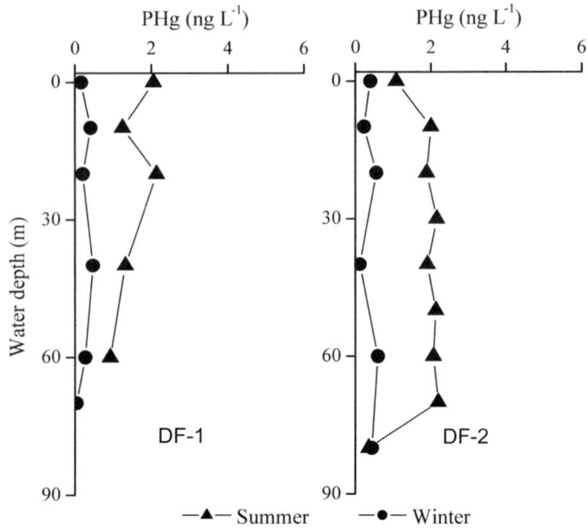

Fig. 6.45 Distributions of particulate mercury (PHg) in water column profiles of the two sampling stations (Df-1 and DF-2) in DF

The RHg concentrations at PD represented 12.5 ± 5.8% of THg in summer and 2.5 ± 5.8% of THg in winter. The regression analysis revealed a significantly positive correlation between RHg and THg in PD ($r = 0.68$, $p < 0.001$, $n = 56$), suggesting that RHg and THg have a similar source in the water column at PD. During the sampling periods, the ratios of RHg/THg at DF were 10.5 ± 5.87.8% and 25.8 ± 13.3% in summer and winter, respectively. No correlation was observed between RHg and THg in the water samples from DF during the sampling seasons ($r = 0.12$, $p = 0.55$, $n = 26$).

(2) Total mercury, dissolved mercury, and particulate mercury

According to the statistical analysis, the concentrations THg, DHg, and PHg at PD and DF in summer were significantly higher than the concentrations in winter ($p < 0.001$). During our sampling season, the THg concentrations in the water samples ranged from 1.00 to 11.7 ng L^{-1} at PD and from 0.68 to 3.92 ng L^{-1} at DF, respectively, which were comparable to the levels reported for reservoirs in North America (e.g., Brigham et al. 2002).

Generally, the THg and PHg concentrations at PD and DF were relatively higher in the summer than in the winter (Figs. 6.40, 6.42, 6.43 and 6.45) Furthermore, the mean THg concentrations at PD in summer (4.48 ± 2.62 ng L^{-1}) were approximately 2.8 times higher than the concentrations in winter (1.60 ± 0.62 ng L^{-1}). Similarly, the mean THg concentrations at DF in summer (2.94 ± 0.65 ng L^{-1}) were approximately 2.8 times higher than the concentrations in winter (0.99 ± 0.25 ng L^{-1}). The highest average THg and PHg concentrations were observed in summer at the PD-3 (6.01 ng L^{-1}) and PD-2 (3.06 ng L^{-1}) sampling stations, respectively. Similarly, highly elevated TSS concentrations were also observed at PD-3 (mean TSS = 6.26 mg L^{-1}) and PD-2 (5.78 mg L^{-1}) in winter.

Table 6.9 Seasonal and spatial distribution of mercury species collected from the PD and DF (ng L^{-1}, mean ± SD, range)

Reservoir	Season	RHg	DHg	PHg	THg	DMeHg	PMeHg	TMeHg
PD	Summer	0.48 ± 0.28, 0.25–1.48	2.37 ± 1.40, 1.17–7.11	2.11 ± 1.86, 0.43–7.27	4.48 ± 2.59, 1.91–11.74	0.19 ± 0.05, 0.12–0.30	0.13 ± 0.10, 0.02–0.33	0.32 ± 0.10, 0.19–0.51
	Winter	0.08 ± 0.06, 0.04–0.25	0.97 ± 0.43, 0.61–2.52	0.62 ± 0.50, 0.09–2.12	1.60 ± 0.62, 1.00–3.34	0.06 ± 0.02, 0.03–0.10	0.06 ± 0.03, 0.01–0.13	0.12 ± 0.03, 0.09–0.19
DF	Summer	0.28 ± 0.18, 0.12–0.85	1.26 ± 0.51, 0.69–2.01	1.68 ± 0.58, 0.36–2.20	2.94 ± 0.65, 1.10–3.92	0.19 ± 0.11, 0.03–0.40	0.13 ± 0.08, 0.03–0.30	0.33 ± 0.11, 0.18–0.50
	Winter	0.25 ± 0.15, 0.12–0.63	0.66 ± 0.10, 0.52–0.86	0.33 ± 0.18, 0.12–0.60	0.99 ± 0.25, 0.68–1.32	0.08 ± 0.01, 0.05–0.09	0.03 ± 0.02, 0.01–0.08	0.11 ± 0.02, 0.07–0.16

RHg reactive mercury; *DHg* dissolved mercury; *PHg* particulate mercury; *THg* total mercury; *DMeHg* dissolved methylmercury; *PMeHg* particulate methylmercury; *TMeHg* total methylmercury

In addition, the peak values for both THg and PHg matched the highest TSS concentrations.

This seasonal distribution of Hg may be due to agricultural runoff. In summer, the high runoff volume from the abundant precipitation carried Hg-containing particulate matter into the reservoir, whereas less precipitation occurred in the winter. This observation was verified by the positive relationship between THg and TSS (PD: $r = 0.58$, $p < 0.001$; DF: $r = 0.55$, $p < 0.001$) and PHg and SPM (PD: $r = 0.51$, $p < 0.001$; DF: $r = 0.47$, $p < 0.01$), suggesting that TSS may play an important role in the distribution of THg and PHg in the water columns of the reservoirs. The highest THg and TSS concentrations were observed in the middle section of PD (instead of the upstream section) and may be explained by the specific catchment topography. In the middle section of the reservoir, the catchment topography is generally steeper than the upstream and downstream sections. In addition, agricultural activities in the steep slopes accelerated soil erosion to the middle section, where the THg and TSS concentrations were higher.

The DHg concentrations in the water samples ranged from 0.61 to 7.11 ng L^{-1} at PD and from 0.52 to 2.01 ng L^{-1} at DF, respectively. Significantly positive correlations between DHg and THg were observed at both PD ($r = 0.81, p < 0.001$, $n = 38$) and DF ($r = 0.75, p < 0.001$, $n = 26$). DHg accounts for a major fraction of THg during the sampling periods. The mean DHg/THg ratios at DF were $57 \pm 17\%$ in summer, whereas the ratios were elevated to $63 \pm 17\%$ in winter. Similarly, the DHg/THg ratios at DF were higher in winter ($69 \pm 12\%$) than in summer ($44 \pm 16\%$). These ratios are similar to previous studies of the Hongfeng Reservoir (He et al. 2008a, b), one of the reservoirs in the Wujiang River Basin.

The proportion of DHg increased in winter, probably because of the reduced precipitation, which resulted in reduced numbers of particles (or TSS). Furthermore, the seasonal distributions of the DHg/THg ratios at PD and DF suggested that (1) PHg was the primary Hg fraction present in the water column in summer and (2) Hg was mainly present as DHg in the water column in winter. Several possible reasons may explain these findings. (1) The particulate matter contained in the water column during the summer (wet season) was higher than in winter (dry season) due to sufficient water input through surface runoff. Therefore, the ratio of PHg to THg in the water column in summer was relatively higher than the ratio in winter. (2) The PHg concentration and TSS in the water column of PD were higher than the values at DF, suggesting that the erosion caused by surface runoff within the catchment of PD was much more active than in the catchment of DF. In addition, TSS and PHg were higher at PD than at DF due to surface runoff input. (3) As shown above, PHg was the primary Hg fraction present in the water columns at PD and DF in winter. Furthermore, no discernable differences in the DHg/THg ratios were observed between PD and DF in winter (K–S test, $p > 0.05$), implying that the TSS and PHg obtained through surface runoff were not the primary factors controlling the Hg levels/fractions in the water columns of the reservoirs. In contrast, the distributions and concentrations of the Hg fractions in the water columns at PD and DF were mainly impacted by other factors, such as wet/dry deposition and human activities.

(3) Methylmercury

The distribution patterns and concentrations of TMeHg, DMeHg, and PMeHg at PD and DF are illustrated in Figs. 6.46, 6.47, 6.48, 6.49, 6.50 and 6.51, and the summary data are shown in Table 6.9.

The annual mean concentrations of TMeHg, DMeHg, and MeHg in the water column of PD were 0.23 ± 0.12 ng L^{-1} (0.09–0.51 ng L^{-1}), 0.13 ± 0.07 ng L^{-1} (0.03–0.30 ng L^{-1}), and 0.10 ± 0.08 ng L^{-1} (0.01–0.33 ng L^{-1}), respectively. The corresponding data in the water column of DF were 0.23 ± 0.14 ng L^{-1} (0.08–0.50 ng L^{-1}), 0.14 ± 0.11 ng L^{-1} (0.03–0.40 ng L^{-1}), and 0.09 ± 0.08 ng L^{-1} (0.01–0.30 ng L^{-1}), respectively. During our sampling periods, the annual mean DMeHg/TMeHg (PMeHg/TMeHg) ratios were 57 ± 18% (43% ± 19%) and 65 ± 21% (35% ± 21%) at PD and DF, respectively. Statistically significant differences in the TMeHg, DMeHg, and PMeHg concentrations in the water columns at the reservoirs (PD and DF) were observed between summer and winter ($p < 0.01$). In addition, the concentrations of MeHg species were significantly higher in summer than in winter (K–S test, $p < 0.001$ for both PD and DF). For example, the mean concentrations of TMeHg in the water columns at PD (0.32 ± 0.10 ng L^{-1}) and DF (0.33 ± 0.11 ng L^{-1}) in summer were approximately three times higher than the concentrations observed in winter (PD, 0.12 ± 0.03 ng L^{-1}; DF, 0.11 ± 0.03 ng L^{-1}).

At PD, the highest TMeHg concentration (0.51 ng L^{-1}) was observed in summer at sampling station PD-4, which was much higher than the highest value observed in the winter (0.19 ng L^{-1}). The spatial and temporal distributions of the TMeHg concentrations (Fig. 6.46) showed that the TMeHg concentrations were significantly increased in the hypolimnion in summer, particularly at sampling

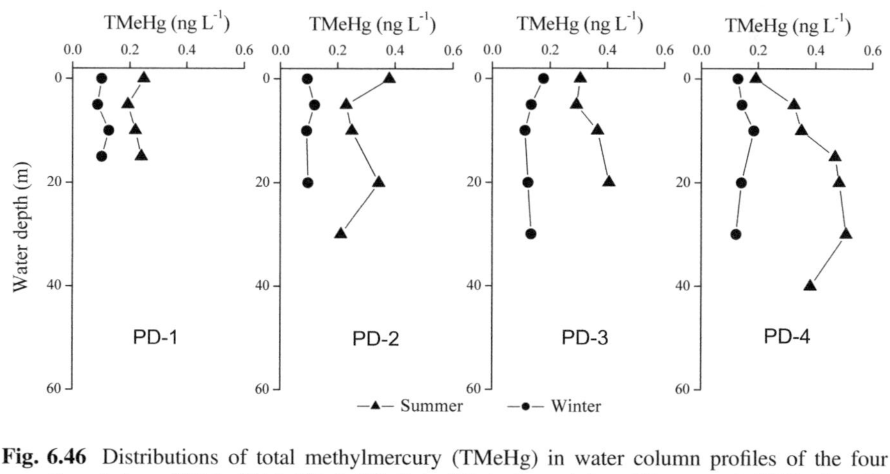

Fig. 6.46 Distributions of total methylmercury (TMeHg) in water column profiles of the four sampling stations (PD-1, PD-2, PD-3, PD-4) in PD (redrawed from Zhang et al. 2009a, b, with permission from Elsevier; redrawed from Zhang et al. 2009a, b, with permission from Earth and Environment)

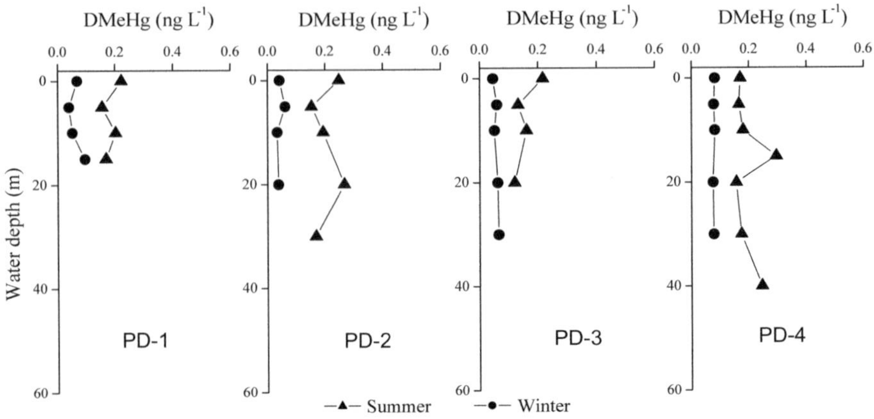

Fig. 6.47 Distributions of dissolved methylmercury (DMeHg) in water column profiles of the four sampling stations (PD-1, PD-2, PD-3, PD-4) in PD (redrawed from Zhang et al. 2009a, b, with permission from Elsevier; redrawed from Zhang et al. 2009a, b, with permission from Earth and Environment)

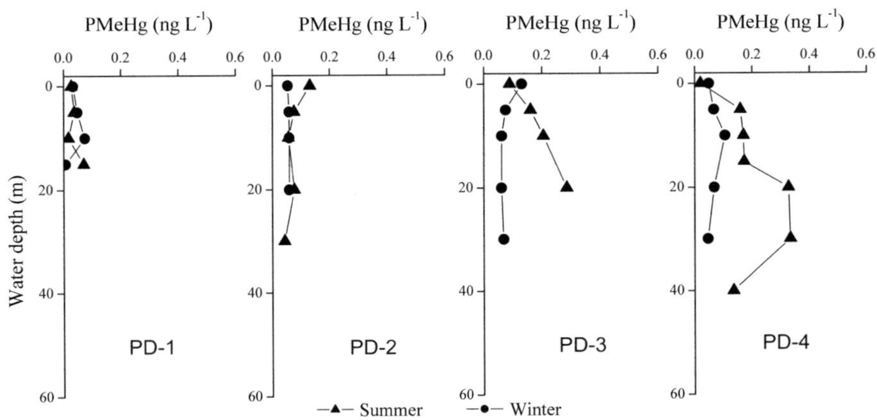

Fig. 6.48 Distributions of particulate methylmercury (PMeHg) in water column profiles of the four sampling stations (PD-1, PD-2, PD-3, PD-4) in PD (redrawed from Zhang et al. 2009a, b, with permission from Elsevier; redrawed from Zhang et al. 2009a, b, with permission from Earth and Environment)

station PD-4, where the TMeHg concentration increased from 0.19 ng L^{-1} in the surface water layer to 0.51 ng L^{-1} in the hypolimnion. Similarly, obvious peaks of TMeHg concentrations occurred at DF-2 during the summer (Fig. 6.49).

According to previous studies, an increase in methylation in the water column would be expected under anoxic conditions, because sulfur-reducing bacteria (SBR) are active in an oxygen-depleted environment (e.g., Ullrich et al. 2001; Eckley and Hintelmann 2006). However, in this study, no obvious anoxic

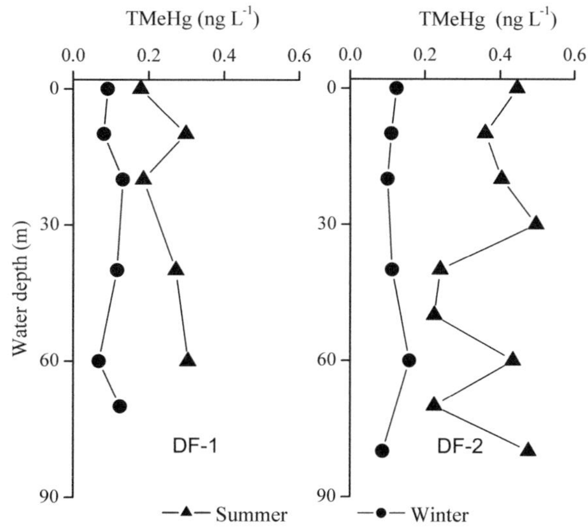

Fig. 6.49 Distributions of total methylmercury (TMeHg) in water column profiles of the two sampling stations (DF-1 and DF-2) in DF

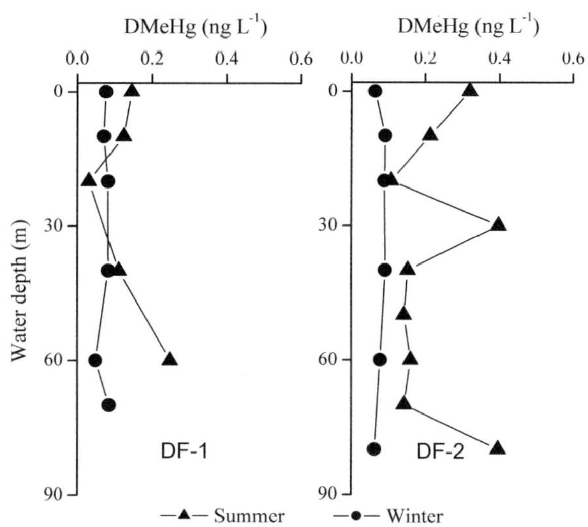

Fig. 6.50 Distributions of dissolved methylmercury (DMeHg) in water column profiles of the two sampling stations (DF-1 and DF-2) in DF

conditions were observed in the hypolimnion of PD and DF during the summer (Figs. 6.35 and 6.36). Furthermore, DMeHg, which may well reflect methylation, did not show a visible increasing trend in the hypolimnion of the reservoirs (Fig. 6.47), with the exception of slight DMeHg peaks at depths of 30 m and 80 m at DF (Fig. 6.31). Although TMeHg showed an increasing trend in summer, we could not conclusively state that hypolimnion methylation commonly occurred. In summer, the linear correlation coefficient between TMeHg and PMeHg at PD was higher ($r = 0.89$, $p < 0.001$) than TMeHg and DMeHg ($r = 0.23$, $p < 0.001$),

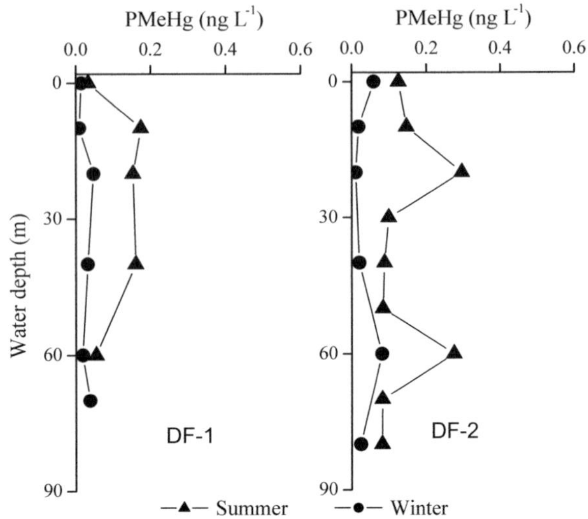

Fig. 6.51 Distributions of particulate methylmercury (PMeHg) in water column profiles of the two sampling stations (DF-1 and DF-2) in DF

suggesting that the levels of PMeHg in summer likely determined the distribution of MeHg. Phytoplankton blooms in July absorbed MeHg from the water, which may explain the higher PMeHg concentrations (Hurley et al. 1994).

The DMeHg concentrations in the water column of PD ranged from 0.12 to 0.30 ng L^{-1}, with an average concentration (±SD) of 0.19 ± 0.05 ng L^{-1} in the summer, which was significantly higher than the concentrations in the winter (Table 6.9). Similar seasonal distribution patterns of DMeHg in the water column were also observed at DF during our sampling campaigns. Statistical analyses yielded significant positive correlations between DMeHg and TMeHg at PD ($r = 0.66$, $p < 0.0011$) and DF ($r = 0.84$, $p < 0.001$). The mean DMeHg/TMeHg ratio in the water column of PD was 64 ± 20% in summer, whereas the ratio decreased to 50 ± 17% in winter. In contrast, the mean DMeHg/TMeHg ratio at DF in summer (58 ± 22%) was lower than the ratio in winter (73 ± 14%). DMeHg accounted for a major fraction of TMeHg at both PD and DF during each of the sampling periods. As shown above, the majority of THg in the water column of PD was present in the dissolved phase during the dry season (winter); however, a higher proportion of THg was bound to particles during the rainy season (summer), suggesting that the source of Hg was erosion within the watershed. The increase in the MeHg concentrations in the dissolved phase during the summer suggested that MeHg was produced within the reservoir in situ as opposed to being transported into the reservoir from the watershed.

We calculated the MeHg/THg ratio (the percentage of MeHg to THg), which has been used by several studies as a surrogate measurement for methylation activity (e.g., Gilmour et al. 1998; Mitchell et al. 2008), to further investigate MeHg production in PD. In the dry season (winter), the TMeHg/THg and DMeHg/THg ratios in the water column of PD were 8.8 ± 4.3% and 4.9 ± 2.6%, respectively, which

Table 6.10 Pearson's correlation matrix, giving the linear correlation coefficients (*r*) between the variables in PD

	THg	DHg	PHg	RHg	TMeHg	DMeHg	PMeHg	DOC	DO	*T*	pH	TSS	Cl^-	NO_3^-
THg	1													
DHg	0.81***	1												
PHg	0.88***	0.43**	1											
RHg	0.66***	0.79***	0.38*	1										
TMeHg	0.53***	0.39*	0.49**	0.51***	1									
DMeHg	0.43**	0.45**	0.28	0.56***	0.66***	1								
PMeHg	0.28	0.052	0.39*	0.22	0.70***	0.19	1							
DOC	−0.02	0.11	−0.12	0.071	0.11	0.22	−0.07	1						
DO	−0.30	−0.07	−0.41*	−0.01	−0.12	−0.14	−0.23	0.19	1					
T	0.55***	0.57***	0.38*	0.73***	0.72***	0.79***	0.25	0.24	0.12					
pH	−0.03	0.09	−0.11	0.12	0.01	0.01	0.06	0.17	0.26	0.20	1			
TSS	0.58***	0.45**	0.52***	0.50***	0.56***	0.35*	0.44**	0.16	−0.05	0.41*	0.06	1		
Cl^{-1}	−0.44**	−0.30	−0.42**	−0.45**	−0.66***	−0.55***	−0.36*	−0.27	−0.07	−0.66***	0.073	−0.55***	1	
NO_3^{-1}	0.53***	0.51***	0.39*	0.56***	0.67***	0.72***	0.30	0.01	−0.28	0.78***	−0.03	0.29	−0.39*	1

*$p < 0.05$, **$p < 0.01$, ***$p < 0.001$

increased to 9.0 ± 4.9%, 5.5 ± 3.3% in the summer, indicating that MeHg was produced within PD.

A previous study confirmed that the seasonal changes in the MeHg concentrations in the water column of a reservoir have been most commonly attributed to increased methylation in the water column and sediment (e.g., Gilmour and Henry 1991).

Since the water column was not anoxic, the seasonal increase in MeHg production in the reservoir probably occurred in anoxic sediments. Both the resuspension of the sediment and diffusion of MeHg from sediment pore water can carry MeHg into the water column (Feng et al. 2009a, b). In addition, increased temperatures in summer possibly contributed to increased methylation in anoxic bottom sediments, which produced higher MeHg/THg ratios in the water column. As shown in Chap. 5, we measured the monthly inflows and outflows of MeHg in PD in 2006 and showed that approximately two times more MeHg was exported in the summer that was imported an approximately two times higher level than was observed in the other seasons. In addition, according to our mass balance calculation, the net MeHg flux in PD in 2006 was 69.4 g $year^{-1}$. Based on the results from the present study, MeHg was produced within PD, particularly in summer.

3. Physicochemical factors controlling the distribution of mercury species in the water column

The distribution and occurrence of Hg species in aquatic environments are regulated by many chemical and biological parameters (Ullrich et al. 2001; Gill and Bruland 1990; Bloom and Effler 1990). MeHg contents in water are influenced by a wide variety of environmental factors, such as the total and reactive Hg contents, water temperature, redox potential, pH, the inorganic and organic solutes, and microbial activity (Ullrich et al. 2001). For the purpose of investigating the possible controlling factors, the correlations between Hg species and seven ancillary water quality parameters (TSS, DOC, DO, *T*, pH, Cl^-, and NO_3^-) were examined for all data sets obtained during the sampling campaigns. The correlation matrixes are shown in Tables 6.10 and 6.11. In general, TSS, *T*, and NO_3^- had a greater correlation with the distribution of Hg species in the reservoirs, particularly in PD, whereas DOC, DO, and pH were less important. Cl^- was also significantly negatively correlated with a few Hg species. The correlations between the physicochemical factors and Hg species implied that the water quality parameters played important roles in the distribution of the different Hg species in the water columns of the reservoirs. A detailed discussion is provided below.

(1) Nitrate (NO_3^-)

As shown in Table 6.10, nitrate concentrations were significantly positively correlated to all the Hg species at PD. In addition, significant positive correlations between nitrate and particulate Hg species (PHg and PMeHg) were also observed at DF (Table 6.11). The clear correlations shown above may be explained by agricultural activities. In the summer season, agricultural activities were associated with increased usage of chemical fertilizers, including nitrate. Higher runoff from precipitation transported nitrate (from fertilizer) and Hg-containing particulate matter (from erosion) to the reservoir. The NO_3^- and Hg concentrations were lower in the winter, when both precipitation and agricultural activities decreased. Furthermore, the concentrations of Hg species varied exponentially with the NO_3^- concentrations, which roughly reflected the frequency of agricultural activities (Zhang et al. 2009a, b). Thus, agricultural activity in the summer increased the Hg levels in the reservoirs of the Wujiang River Basin.

(2) Total suspended solids

Total suspended solids (TSS) levels were significantly related to the Hg species in PD (Table 6.10), with a significance level (p) of less than 0.001. Furthermore, significant correlations between TSS and THg, and PHg and DHg in DF were also observed during the sampling campaigns (Table 6.11). As summarized above, the levels of TSS and the corresponding Hg species were significantly higher in summer than in winter at both PD and DF ($p < 0.01$), possibly due to the input of particulate material through surface runoff during the wet season in summer. According to the statistical analysis, TSS from surface runoff appeared to play an important role in the distributions of Hg species (particularly PHg and PMeHg) in water columns of PD and DF.

(3) Temperature, Dissolved oxygen, and pH

Warmer temperatures (T) may increase microbial Hg methylation (Hecky et al. 1991). In our investigations, the water temperatures were significantly positively correlated with different Hg species in PD and DF, as shown in Tables 6.10 and 6.11. Although the seasonal differences observed for MeHg are related to TSS, they are also likely influenced by water temperature. With seasonal mean temperature differences of up to 14.4 °C between summer and winter, the significantly lower DMeHg and PMeHg levels in the winter campaigns (Figs. 6.28, 6.29, 6.30 and 6.31) may be attributed to reduced microbial methylation (Winfrey and Rudd 1990; Bodaly et al. 1993; Ramlal et al. 1993).

Anaerobic and low pH conditions are favorable for net Hg methylation (Ullrich et al. 2001). However, no correlations were observed between MeHg and DO or pH at PD and DF during any of the sampling seasons, implying that DO and pH were not the key factors controlling Hg methylation in PD and DF. DO concentrations were poorly related to the dissolved fractions of Hg in PD and DF, indicating that only a small fraction of DHg was complexed to dissolved organic ligands (Bonzongo et al. 1996). Therefore, we hypothesized that a large fraction of DHg in the studied area is available for conversion processes and uptake by aquatic organisms.

(4) Chloride

Chloride (Cl^-) concentrations are higher in the more urbanized portions of water bodies due to domestic wastewater discharges. For example, in the study by Lyons

Table 6.11 Pearson's correlation matrix, giving the linear correlation coefficients (*r*) between the variables in DF

	THg	DHg	PHg	RHg	TMeHg	DMeHg	PMeHg	DOC	DO	*T*	pH	TSS	Cl^{-1}	NO_3^{-1}
THg	1													
DHg	0.75***	1												
PHg	0.92***	0.44*	1											
RHg	0.06	0.16	−0.01	1										
TMeHg	0.58**	0.40*	0.57**	0.24	1									
DMeHg	0.19	0.12	0.20	0.32	0.77***	1								
PMeHg	0.63***	0.38	0.63***	−0.07	0.66***	0.16	1							
DOC	0.33	0.04	0.44*	−0.28	0.05	−0.12	0.21	1						
DO	0.30	0.30	0.24	0.52**	0.33	0.17	0.08	−0.14	1					
T	0.935***	0.802***	0.801***	0.208	0.688***	0.346	0.595***	0.176	0.466*	1				
pH	0.177	0.232	0.096	0.471*	0.286	0.116	0.104	−0.234	0.500**	0.276	1			
TSS	0.54**	0.40*	0.52**	−0.04	0.28	0.01	0.34	0.347	0.20	0.45*	0.04	1		
Cl^{-1}	0.46*	0.25	0.49*	−0.13	0.27	0.15	0.27	−0.20	−0.17	0.40	−0.22	0.08	1	
NO_3^{-1}	0.40	0.21	0.41*	−0.33	0.26	0.19	0.45*	0.22	−0.35	0.31	−0.33	0.21	0.57**	1

*$p < 0.05$, **$p < 0.01$, ***$p < 0.001$

et al. (2006), a peak Cl^- concentration of 52 mg L^{-1} was observed downstream of sewage treatment facilities. The catchment area of PD is dominated by agricultural farming. For this study, the peak Cl^- concentration in PD was observed in the winter campaign, and, in general, the average Cl^- concentrations were higher in the winter and lower in the summer. This result may be due to dilution from precipitation (since most runoff is from agricultural activities, not urban runoff). High precipitation or runoff in summer diluted the Cl^- concentrations in PD, but the Hg levels were not diluted due to high TSS levels in the reservoir. Hence, significantly negative correlations between Cl^- and some Hg species were observed, as shown in Table 6.10. However, the correlations between the Cl^- concentrations and Hg species in the water column of DF were less pronounced in this study (Table 6.11), indicating that the influence of Cl^- on the distributions of Hg species in PD and DF is different.

6.2.3 *Mercury Species in Sediment Cores*

1. General physical properties of sediment samples

(1) Water content

In general, the vertical distributions of water content in sediments do not display a significant difference from upstream to downstream regions (Fig. 6.52) in PD and DF. Thus, the sources of sediments are similar at upstream and downstream regions in these reservoirs (PF and PD).

When the river runs into the catchment area of the reservoir, its hydrodynamic conditions decline; therefore, particulate matter is easily deposited at the entrance of

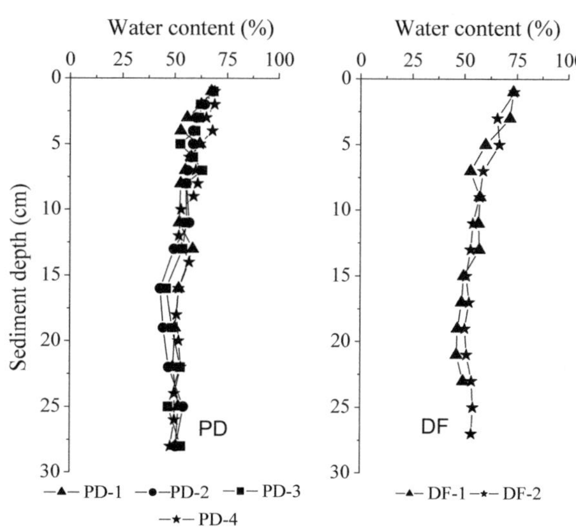

Fig. 6.52 Distributions of water content in sediment profiles in PD and DF

the reservoir, i.e., reservoir construction promotes a "blocking effect" for a particular matter. Meanwhile, only a small fraction of the suspended particles are transported downstream. If the upstream river input is the main source of reservoir sediments, then the water content in the sediments will be different in upstream and downstream (dam) regions. However, no significant spatial differences in water content were observed in PD and DF. In addition to the river, other sediment sources existing in these reservoirs, e.g., surface runoff and atmospheric deposition, probably explain this phenomenon. The vertical distribution of water content in sediments varied significantly in PD and DF, where the peak value was obtained in surface sediments (70%) and decreased with increasing depths, but was stable (40–50%) at depths of 10 cm, showing the mechanical compression of the sediments.

(2) Organic matter

The vertical profiles of organic matter content (OM) in sediment cores from the PD and DF are shown in Fig. 6.53. OM was highest in sediment samples collected from the surface layer of PD-4 (8.1%), and then gradually decreased to 4% at depth of 12 cm. The vertical distribution of OM in upstream (PD-1) sediments largely changed in a jagged distribution, and the maximum value (9.02%) appeared at 25 cm. Strong hydrodynamic conditions exist upstream, where sediment disturbance is susceptible to water, resulting in the irregular vertical distribution of OM in upstream sediments. OM of sediments shows a vertically homogenized profile in PD-2 and PD-3, with a small range of changes (range: 6.3–7.7%).

Overall, the OM distribution varied largely in different sampling sites in PD. The OM changed from 3.0 to 9.0%, with a mean value of 6.2% in the whole reservoir; the OM gradually increased from upstream to downstream (dam): 6.3% in PD-1, 6.9% in PD-2, 7.4% in PD-3 and 8.1% in PD-4, respectively. The OM in the surface sediments in the PD reservoir was higher than the OM in surface soil

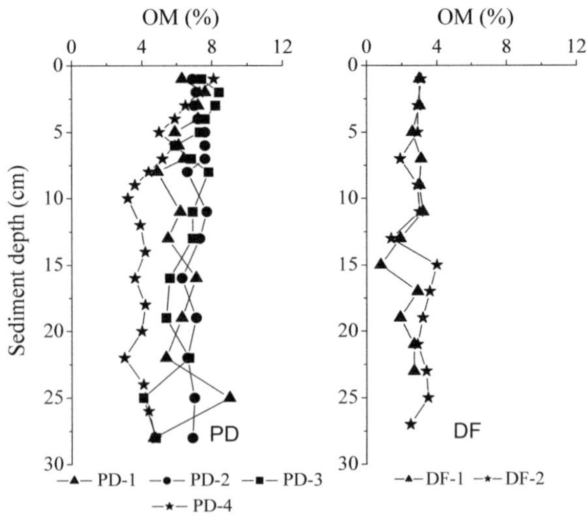

Fig. 6.53 Distributions of organic matter (OM) in sediment profiles in PD and DF

from the Wujiang River Basin. Thus, runoff has less of a contribution to OM in PD than internal sources.

No obvious variations in the sedimentary OM distribution in the DF Reservoir were observed (ranging from 1.40 to 3.98%, with a mean value of 2.77%); no significant trend was observed in the vertical distribution of OM in DF. The range of OM in DF sediments was similar to OM in surface soil in Wujiang River Basin, implying that river runoff and external input are the main sources of OM in DF, but internal input has a small contribution.

The evolutionary histories of PD and DF are relatively similar. However, in the former, the OM in surface sediments is significantly higher than in the latter, suggesting that the OM input in PD is higher than in DF. In addition, the OM in surface sediments gradually increased from upstream (PD-1) to downstream/dam (PD-3/PD-4) in PD. If OM is principally obtained from exogenous input (river and runoff) in PD, then the OM will decrease in response to sedimentation; thus, the OM in surface sediments gradually decreases from upstream to downstream (dam). However, the opposite situation was observed, indicating that OM in PD is not principally controlled by the external input. During our sampling periods, PD exhibited a eutrophic state, whereas DF exhibited an oligotrophic–mesotrophic status, as shown in Chap. 10. Therefore, the abundant algae in water may be the main source of OM in sediments from PD. Thus, OM is mainly obtained from internal input in this reservoir. In general, the eutrophic level of water gradually strengthens from upstream to downstream, as shown in the study of WJD, where the water eutrophic level changes from mesotrophication to eutrophication from upstream to downstream (dam). Consequently, the increase in the OM of surface sediments from upstream to downstream (dam) is most likely related to the extent of eutrophication, which gradually increases from upstream to downstream (dam).

The OM in sediments from DF and PD exhibits the following characteristics compared with WJD: on the one hand, the OM in PD surface sediments is higher than the OM in the upstream regions of WJD; on the other hand, the OM in surface sediments of PD is significantly lower than the OM in downstream and dam areas of WJD. The main reason may be the abundant internal OM input (artificial bait and fish feces) in WJD. In addition, the OM in surface sediments of DF is similar to the upstream WJD, but much lower than the downstream and dam areas of WJD.

2. Distribution of mercury species in sediment cores

(1) Total Mercury

The vertical distributions of THg in sediments collected from sampling stations in PD and DF are illustrated in Figs. 6.54 and 6.55. The THg concentrations in sediments collected from PD-1, PD-2, PD-3 and PD-4 are 184 ± 22 ng g^{-1} (range: 152–245 ng g^{-1}), 191 ± 12 ng g^{-1} (range: 156–220 ng g^{-1}), 189 ± 24 ng g^{-1} (range: 148–261 ng g^{-1}), and 259 ± 18 ng g^{-1} (range: 219–298 ng g^{-1}), respectively. The THg concentrations in sediments collected from DF-1 and DF-2 in DFr are 174 ± 32 ng g^{-1} (range: 130–245 ng g^{-1}) and 234 ± 33 ng g^{-1} (range: 174–330 ng g^{-1}), respectively.

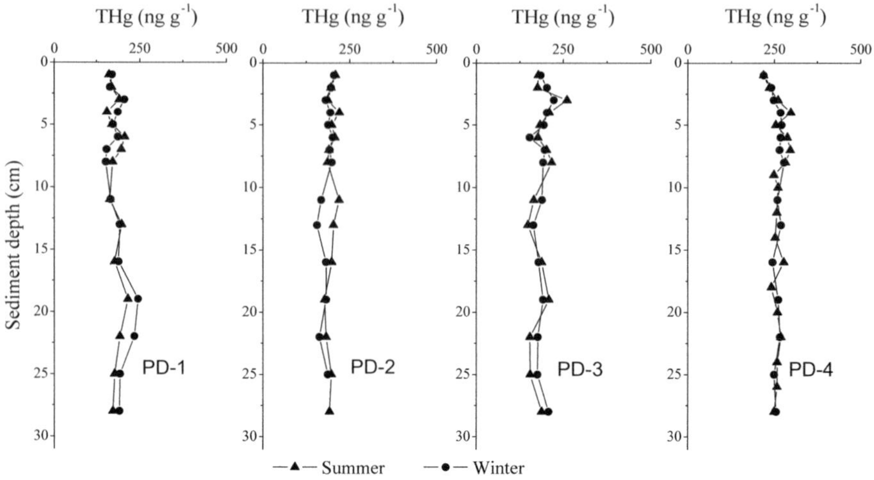

Fig. 6.54 Distributions of total mercury (THg) in sediment profiles of the four sampling stations (PD-1, PD-2, PD-3, PD-4) in PD

Fig. 6.55 Distributions of total mercury (THg) in sediment profiles of the two sampling stations (DF-1, DF-2) in DF

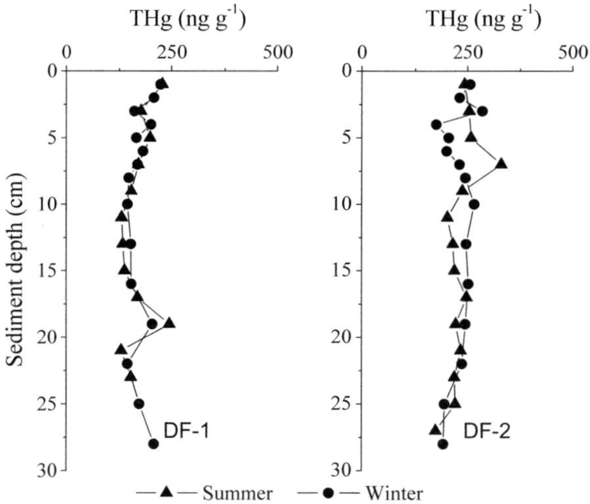

No significant seasonal trend in THg concentrations in sediments from PD and DF was observed. However, the higher THg concentrations ($p < 0.01$) measured in sediments collected from dam area (PD-4 and DF-2) than in other sampling stations at each reservoir suggests potential influences of emission sources adjacent to the dam, which may be subject to anthropogenic activities. The insignificant correlation ($r = -0.16$, $p = 0.12$, $n = 90$) between the THg concentrations and OM in sediments collected from the PD and DF reservoirs suggests that the OM does not have an influence on the THg distribution in sediments of PD and DF in this study.

Compared to the large THg concentration range measured in surface soils distributed in the Wujiang watershed (THg concentration range: 20–372 ng g^{-1}), relatively stable seasonal THg vertical distributions were observed in sediments collected from PD and DF (THg range: 148–298 ng g^{-1} in PD and 130–330 ng g^{-1} in DF), suggesting limited contamination from anthropogenic emissions and stable Hg sources for Hg in sediments from the two reservoirs. A highly likely source of Hg in the sediments of the two reservoirs is surface runoff, by which the soil particles are deposited over the water bed.

Compared to reported data observed in sediments obtained from other uncontaminated reservoirs, e.g., reservoirs from Newfoundland, Canada (mean THg concentration: 39 ng/g) reported by French et al. (1999), the elevated THg concentrations observed in sediments from this study may have been produced by several factors. Subject to the impact of the global Hg mineralization belt passing through the WJ watershed, elevated background Hg concentrations in surface soil from this region increase the Hg concentration in surface runoff, leading to a high concentration in sediments in the two reservoirs. In addition, regional coal combustion leads to an elevated atmospheric Hg concentration, which may increase the Hg concentrations in sediments. Other possible factors, e.g., contamination from released industrial water, may contribute to the elevated Hg concentrations in regional sediments.

(2) Methyl mercury

Seasonal variations in the MeHg concentrations in the sediment profiles of PD and DF are illustrated in Figs. 6.56 and 6.57. The distribution patterns of MeHg in sediment cores of PD are characterized below.

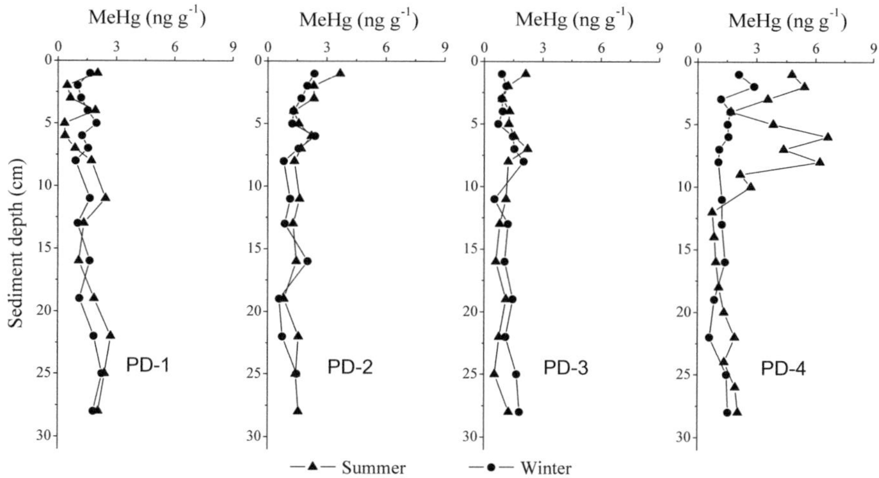

Fig. 6.56 Distributions of methlmercury (MeHg) in sediment profiles of the four sampling stations (PD-1, PD-2, PD-3, PD-4) in PD

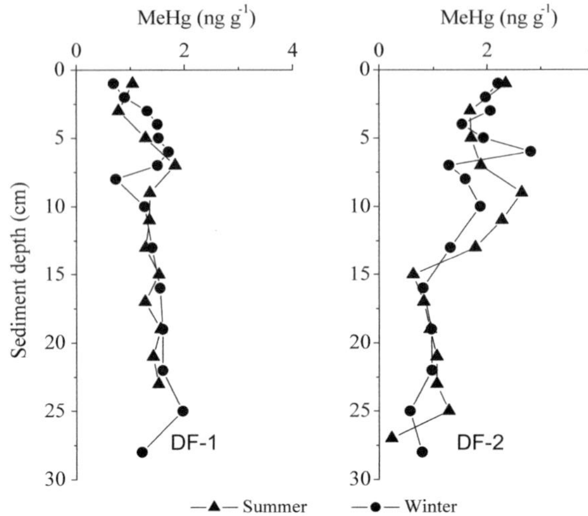

Fig. 6.57 Distributions of methlmercury (MeHg) in sediment profiles of the two sampling stations (DF-1, DF-2) in DF

(1) A large range of MeHg concentrations (0.75–6.64 ng g^{-1}), with several peak values within a depth of 1–12 cm, was observed in the vertical distribution of MeHg in sediments collected at PD-4 in summer. Furthermore, the MeHg value was fairly stable below a depth of 12 cm in the sediment profile of PD-4 in summer, suggesting that primary methylation occurred in sediments at depths of 1–12 cm compared to depths greater than 12 cm. (2) In contrast to sediments collected at PD-4 in summer, sediment samples collected from other sampling stations at PD showed a small range of MeHg concentrations (0.34–3.68 ng g^{-1}) in the vertical profiles. Generally, MeHg present in sediments originated from both external sources, e.g., input from surface runoff, and internal sources, e.g., Hg methylation in sediments. Based on the small concentration range, the vertical distribution of MeHg in sediments from PD-1, PD-2, and PD-3 suggested a stable input from surface runoff in the Wujiang watershed and limited vertical variation in the net Hg methylation in sediments. (3) Peak MeHg concentrations were observed in the sediment cores from PD-4 in summer and winter (6.64 ng g^{-1} in summer and 2.87 ng g^{-1} in winter), suggesting that active net Hg methylation occurred in the sediments at PD-4 compared with the other sampling sites. (4) Similar seasonal trends in the MeHg concentrations were observed in sediments from PD, with the exception of elevated MeHg concentrations in the surface sediment layer in summer compared with winter. (5) The methylation level of inorganic Hg is evaluated by calculating the ratio of MeHg to THg in sediments/soil. The ratios of MeHg to THg in sediments from PD ranged from 0.17 to 2.31% (mean: 0.87%) in summer and from 0.23 to 1.2% (mean: 0.71%) in winter, suggesting that higher net Hg methylation occurred in the sediment cores in summer than in winter.

As illustrated in Fig. 6.57, the distribution patterns of MeHg in sediment cores collected from DF are characterized below. (1) Peak MeHg concentrations in sediments were observed at DF-2 in both summer and winter (2.81 ng g^{-1} in summer and 2.64 ng g^{-1} in winter). (2) No clear seasonal variations in the MeHg concentrations in the sediment profiles at DF were observed for each of the sampling stations, with the exception of a slight peak in the MeHg concentrations in the surface sediment layer in summer. (3) Compared to the limited range of MeHg concentrations (0.69–1.97 ng g^{-1}) in sediment cores collected at DF-1, a wide range of MeHg concentrations (0.23–2.81 ng g^{-1}) was observed in sediment cores from DF-2, with two clear peaks within sediment depths of 0–10 cm; then, the MeHg concentrations decreased at sediment depth of greater than 10 cm and were fairly stable at depths greater than 15 cm. The ratios of MeHg to THg in the sediment profiles collected from DF ranged from 0.13 to 1.12% (mean: 0.73%) in summer, and 0.29–1.41% (mean: 0.73%) in winter, respectively.

The seasonal and spatial distribution patterns of MeHg in sediment profiles from both PD and DF showed higher values in the surface layer in summer than in winter. The possible reason for this phenomenon may be the favorable net Hg methylation conditions, including anaerobic environment, abundant OM, and higher temperature in summer. In contrast to winter, the stratification was more significant in both reservoirs in summer, leading to a relatively anaerobic environment in the surface sediments. The deposition of OM from the overlying water column into surface sediment later was more abundant in summer. In addition, the bottom water temperature during the summer was higher than in winter. The factors listed above could stimulate the activity of sulfate-reducing bacteria, resulting active net Hg methylation in summer. Compared to the PD-1 sampling station, clearer seasonal variations in MeHg levels were observed in the surface sediments at PD-3.

The MeHg concentrations in surface sediments (0–10 cm) at PD-4 and DF-2 were significantly higher than the concentrations in deeper sediment layers ($p < 0.01$), with the exception of PD-4 in winter. In contrast, no significant differences in the MeHg values in the surface sediment layer and deeper layer were observed at PD-1, PD-2, PD-3, and DF-1 for each of the sampling periods. Generally, most of the particulate matter from external sources that was suspended in the water body was rapidly deposited in upstream areas of the reservoir. In contrast, only a small portion was transported downstream. Therefore, the primary contributor to the MeHg levels in the sediments near the dam may be in situ Hg methylation. In addition, the deep water and slow water movement at the dam site resulted in significant water stratification and an anaerobic environment, particularly in the bottom water, which stimulated the net Hg methylation in the surface sediments.

The peak MeHg concentrations in sediments cores were observed within a depth of 10 cm for each of the reservoirs. Furthermore, the peak MeHg values in sediments from PD and DF were observed at the dam site. The peak MeHg levels in the sediment profiles suggested that the net Hg methylation predominantly occurred within a depth of 10 cm. As shown in a previous study, sulfate-reducing bacteria prefer to live in sediments within a depth of 7 cm, according to the distribution of 6 species of sulfate-reducing bacteria in sediments collected from Hongfeng reservoir (Liang et al. 2003). Iron-reducing bacteria can also participate in Hg methylation

(Fleming et al. 2006). In addition to the activity of iron/sulfate-reducing bacteria, many environmental physical, chemical and biological factors (e.g., THg concentration, RHg concentration, eH, pH, *T*, OM, sulfide concentration, and the circulation of Fe/Mn) impact Hg methylation in aquatic ecosystems (Ullrich et al. 2001). Significant positive correlations were observed between OM and MeHg concentrations in sediment at PD-4 in summer ($r = 0.61$, $p < 0.01$), in contrast to the insignificant correlation observed at DF. OM stimulates bacterial activity, which accelerates methylation. However, high OM may bind Hg^{2+} and decrease the bioavailable Hg concentrations, particularly in neutral pH environments (Miskimmin et al. 1992; Watras et al. 1995; Driscoll et al. 1995), preventing Hg methylation. In addition, non-bacteria-mediated methylation by humic acid has been reported (Weber 1993).

With the exception of the sediment profile at PD-4, the vertical distributions of MeHg in sediments collected at other sampling stations showed similar trends and small variations. The higher MeHg to THg ratios observed in sediments from PD in summer than in winter was attributed to the increased net Hg methylation and correspondingly higher MeHg concentrations at the dam site (PD-4). Furthermore, no significant differences in the MeHg to THg ratio in sediment from DF were observed between summer and winter during the sampling campaigns ($p > 0.05$). The MeHg to THg ratios in sediments from the two reservoirs were much lower than the ratio observed in sediments from reservoirs built above peatland/podzol, e.g., ratios of 10% in surface peat and 30% in podzol covered by water were observed over 10 years (Lucotte et al. 1999), which were related to the high OM (30–50%) in submerged soil. MeHg concentrations in sediment and fish tissues from newly constructed reservoirs were much higher than the concentrations in these sources from adjacent natural lakes (Abernathy and Cumbie 1977; Cox et al. 1979), and the flooded soil/vegetation were suggested to be the important sources of MeHg in fish tissues (Hecky et al. 1986; Jackson 1988). In this study, the MeHg concentrations in sediments from both PD and DF were slightly higher than the concentrations in sediments from Hongfeng lake (He 2007), and the MeHg to THg ratios in sediments from PD and DF were comparable with reported data from WJD (Jiang 2005).

2. Distribution of mercury species in sediment pore-water

(1) Filtered total mercury (DHg)

Hg adsorption and desorption between the solid and aqueous phases of sediment are complex physical-chemical processes. OM, clay minerals, iron oxides, and manganese oxides have very strong adsorption functions, and adsorb large amounts of Hg in sediment. Several simulation experiments suggested the following preference for Hg adsorption by other minerals, which is shown in a decreasing order: mercapto > illite > montmorillonite > amino > kaolinite > carboxyl > Sand (Reimers and Krenkel 1974). Hg adsorption/desorption processes in sediment are impacted by numerous factors, such as Hg concentrations, temperature, pH, redox conditions, and other varieties of complexations. The distribution of mercury between solid and aqueous phases in sediment is principally controlled by the cycling of S^{2-}, OM, and oxides (Fujiki and Tajima 1992). HgS precipitation is mainly produced in the environmental conditions of low pH and low S^{2-} concentrations. However, HgS precipitation will be transformed to soluble Hg sulfide compounds (e.g., HgS_2^{2-}) in a reducing environment in the presence of excess S^{2-} concentrations and high pH. These soluble Hg sulfide compounds subsequently enter the liquid phase from the solid phase. OM enhances the solubility of HgS and further transports Hg from the solid phase to the aqueous phase (Ravichandran et al. 1998).

The vertical distributions of DHg in the sediment pore water and overlying water columns of DF and PD are shown in Figs. 6.58 and 6.59. The mean DHg concentrations in sediment pore water at PD-1, PD-2, PD-3, and PD-4 in summer were 24.91 ± 18.08 ng L^{-1} (range: 8.79–71.56 ng L^{-1}), 20.61 ± 8.30 ng L^{-1} (range: 9.44–38.90 ng L^{-1}), 19.81 ± 8.00 ng L^{-1} (range: 7.89–33.52 ng L^{-1}) and 14.15 ± 5.86 ng L^{-1} (range: 6.04–31.26 ng L^{-1}), respectively. The corresponding concentrations at DF-1 and DF-2 were 11.60 ± 4.30 ng L^{-1} (range: 6.54–20.39 ng L^{-1}) and 7.32 ± 5.42 ng L^{-1} (range: 3.30–24.33 ng L^{-1}), respectively. During our sampling periods, the seasonal (summer) mean DHg concentration in sediment pore water at PD was 19.52 ± 11.31 ng L^{-1}, which was approximately 2 times higher than the concentration at DF (9.30 ± 5.19 ng L^{-1}).

In winter, the mean DHg concentrations in sediment pore water at the PD-1, PD-2, PD-3, and PD-4 sampling sites were 3.87 ± 1.06 ng L^{-1} (range: 2.12–5.55 ng L^{-1}), 3.32 ± 1.60 ng L^{-1} (range: 0.98–6.74 ng L^{-1}), 3.53 ± 1.56 ng L^{-1} (range: 1.17–7.49 ng L^{-1}), and 4.19 ± 4.58 ng L^{-1} (range: 2.00–20.38 ng L^{-1}), respectively. The corresponding concentrations at the DF-1 and DF-2 sampling sites were 10.95 ± 6.19 ng L^{-1} (range: 3.39–24.55 ng L^{-1}) and 8.06 ± 2.60 ng L^{-1} (range: 5.38–14.71 ng L^{-1}), respectively. Moreover, the

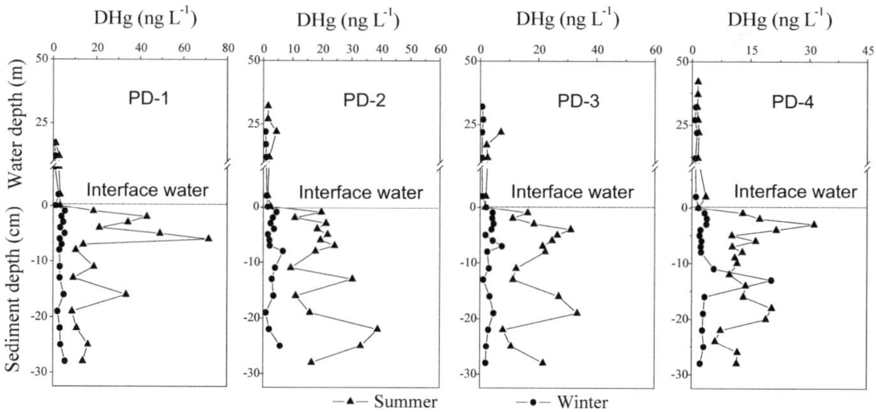

Fig. 6.58 Distributions of dissolved mercury (DHg) in water column and sediment pore water of four sampling stations (PD-1, PD-2, PD-3, PD-4) in PD

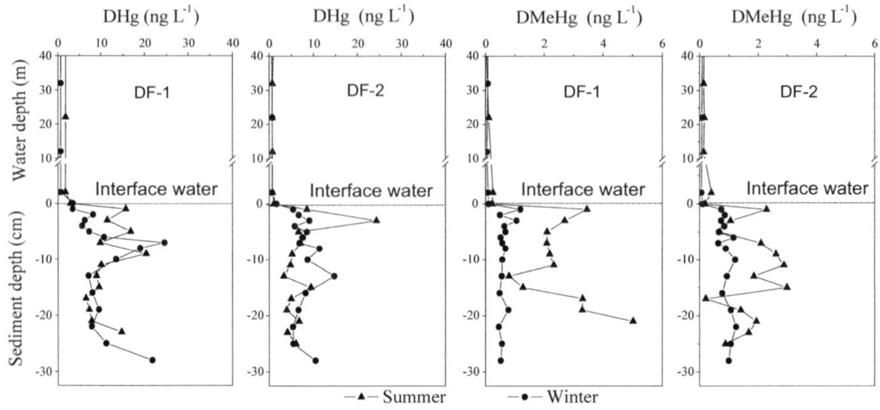

Fig. 6.59 Distributions of dissolved mercury (DHg) and dissolved methylmercury (DMeHg) in water column and sediment pore water of the two sampling stations (DF-1, DF-2) in DF

seasonal (winter) mean DHg concentration in sediment pore water at DF (9.50 ± 4.89 ng L^{-1}) was approximately 2.6 times higher than the concentration at PD (3.73 ± 2.57 ng L^{-1}).

The distribution patterns of pore water DHg at PD and DF exhibited greater variations than the THg concentrations in the solid phase of sediment, consistent with results from WJD and HF showing that DHg concentrations in the vertical profiles of pore water varied randomly without discernible trends. However, the DHg concentrations in pore water were generally higher than the concentrations in interface water, implying that the sediment was an important source of DHg for the water column. According to the statistical analysis, a significant correlation between the THg concentrations in the solid and aqueous phases (pore water) of sediment was not observed at either DF or PD for each of the seasons (summer and winter). Thus, the DHg concentrations in sediment pore water at PD and DF were controlled by factors other than the THg concentration in the solid phase. Furthermore, the DHg concentrations in sediment pore water at DF were the same in summer and winter: 9.45 and 9.50 ng L^{-1}, respectively. However, the DHg concentrations in sediment pore water at PD in summer were significantly higher than the concentrations in winter ($p < 0.01$), indicating that the Hg in sediment tended to exist in the liquid phase during summer.

The partitioning of Hg between the solid phase and aqueous phase is physically, chemically, or biologically controlled, and hence affected by a number of environmental parameters, such as pH, temperature, redox conditions, and bioturbation (Ullrich et al. 2001). Thermal stratification was observed in the water column of PD during the summer campaign; in contrast, thermal stratification did not occur during winter. Correspondingly, DO, pH and other chemical parameters exhibited significant differences between summer and winter ($p < 0.01$) in PD. Therefore, the seasonal changes in the physicochemical parameters of the water column were likely to be important factors controlling the seasonal distribution of DHg in sediment pore water of PD. However, the specific mechanism requires further study. The physicochemical parameters varied in summer and winter at DF as well. However, the variations were less pronounced compared with PD. Therefore, the DHg concentrations at DF were the same in summer and winter.

(2) Filtered methylmercury

The vertical distributions of DMeHg in sediment pore water and the overlying water column collected from PD and DF are illustrated in Figs. 6.59 and 6.60. The average (range) DMeHg concentrations in sediment pore water collected at PD-1, PD-2, PD-3, and PD-4 during the summer campaign were 0.51 ± 0.27 ng L^{-1} (range: 0.14–1.15 ng L^{-1}), 0.37 ± 0.20 ng L^{-1} (range: 0.15–1.00 ng L^{-1}), 0.65 ± 0.63 ng L^{-1} (range: 0.21–2.55 ng L^{-1}) and 0.26 ± 0.15 ng L^{-1} (range: 0.03–0.57 ng L^{-1}), respectively. The corresponding data collected from DF-1 and DF-2 in summer were 2.60 ± 1.15 ng L^{-1} (range: 0.80–5.02 ng L^{-1}) and 1.74 ± 0.86 ng L^{-1} (range: 0.21–2.99 ng L^{-1}), respectively. During our sampling periods, the seasonal (summer) mean concentration of DMeHg in sediment pore water at DF was 2.21 ng L^{-1}, which was approximately 5 times higher than the concentration at PD (0.47 ng L^{-1}).

The peak DMeHg concentration (2.55 ng L^{-1}) was observed in sediment pore water located in the surface sediment layer at PD-3 in summer, whereas for the other sampling sites (PD-1, PD-2, and PD-4), the maximum values occurred within depths ranging from 1 to 10 cm. The DMeHg concentrations in the surface sediment pore water (1–2 cm) from DF-1 and DF-2 were 2.29 and 3.46 ng L^{-1}, respectively, which were much higher than the concentrations in the corresponding overlying water. However, the peak DMeHg concentrations in sediment pore water at DF-1 and DF-2 were not located in the surface layer (1–2 cm). Greater

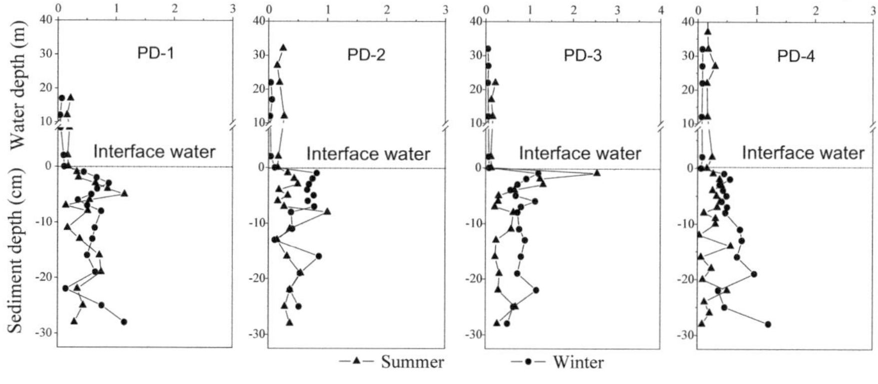

Fig. 6.60 Distributions of dissolved methylmercury (DMeHg) in water column and sediment pore water of the four sampling stations (PD-1, PD-2, PD-3, PD-4) in PD

fluctuations were observed in the vertical distribution of DMeHg in DF compared with PD.

Similarly, the mean (range) DMeHg concentrations in sediment pore water collected at PD-1, PD-2, PD-3, and PD-4 during the summer campaign were 0.62 ± 0.23 ng L^{-1} (range: 0.14–1.15 ng L^{-1}), 0.60 ± 0.21 ng L^{-1} (range: 0.11–0.86 ng L^{-1}), 0.82 ± 0.21 ng L^{-1} (range: 0.50–1.20 ng L^{-1}) and 0.60 ± 0.23 ng L^{-1} (range: 0.36–1.21 ng L^{-1}), respectively. The corresponding data collected at DF-1 and DF-2 in summer were 0.92 ± 0.19 ng L^{-1} (range: 0.64–1.24 ng L^{-1}) and 0.65 ± 0.21 ng L^{-1} (range: 0.45–1.18 ng L^{-1}), respectively. The seasonal (winter) mean DMeHg concentrations in sediment pore water at PD (0.63 ng L^{-1}) were slightly less than the concentrations at DF (0.84 ng L^{-1}) ($p > 0.05$).

During winter, the peak DMeHg concentration (1.20 ng L^{-1}) was observed in the sediment pore water located in the surface sediment layer (1–2 cm) at PD-3, whereas for the other sampling sites (PD-1, PD-2, and PD-4), the maximum values randomly occurred within depths ranging from 2 to 10 cm. Similarly, in winter, the maximum DMeHg concentration (1.19 ng L^{-1}) was observed in sediment pore water located in the surface layer at DF-1. For sampling station DF-2, the peak DMeHg concentration was observed at the bottom of the sediment profile. The fluctuations in the vertical distribution of DMeHg in both DF and PD were less pronounced in winter than in summer.

According to the statistical analysis, the DMeHg concentrations in sediment pore water from DF were significantly higher in summer than the concentrations in winter (K–S test, $p < 0.01$). However, a significant difference in the DMeHg concentrations at PD was not observed between winter and summer. Furthermore, significant relationships between the MeHg concentrations in the liquid and solid phases of the sediment cores were not observed for PD or DF during the whole sampling campaign. The DHg to DMeHg ratios in sediment pore water from PD ranged from 0.28 to 15.6% in summer, with a mean value of 2.94%, and from 3.68 to 77.2% in winter, with a mean value of 21.9%. The corresponding data from DF were 4.28–63.9% (mean value: 28.7%) in summer and 2.34–35.0% (mean value: 10.5%) in winter. The distribution patterns of MeHg in sediment pore water were more variable than the MeHg concentrations in the solid phase of sediment.

6.2.4 Diffusion Flux of Inorganic Mercury and Methylmercury to Water

The interface between bottom sediment and overlying water is not only a physical interface but also an important chemical interface in aquatic ecosystems (e.g., lakes, reservoirs, and rivers). Due to the different physicochemical characteristics between sediment and overlying water, active exchanges of energy and substances over the sediment–water interface is reasonable. Generally, four different migration and diffusion processes occur over the sediment–water interface, including diffusion, dispersion, bioturbation and hydrodynamic disturbance (Krom and Berner 1980). A previous study further confirmed that molecular diffusion is the most important method for the transport of dissolved components in sediment (Krom and Berner 1980). Driven by the concentration gradient, the dissolved components in sediment are diffused from high concentrations to low concentrations, which in turn impact the concentration and distribution of inorganic Hg and MeHg in the bottom water column and surface sediment pore water. The transport and diffusion of dissolved components from the surface sediment is one of the most important factors controlling the chemical characteristics of the overlying water column in aquatic systems. Furthermore, pollutants stored in sediment are eluted into the overlying water as secondary pollution. Therefore, the various physical and chemical behaviors of the sediment–water interface have received considerable attention by numerous researchers.

The diffusion fluxes of IHg and MeHg from sediment pore water to the water column were estimated in this study. We assumed that physical advection and bioirrigation are not important processes in the sediments and are insignificant compared to diffusion processes. The diffusive fluxes across the sediment–water interface are usually calculated based on the measured concentration gradient and Fick's first law, as described in the following equation (Feng et al. 2009a; Holmes and Lean 2006; Goulet et al. 2007):

$$F = -\left(\frac{\varphi D_w}{\theta^2}\right)\frac{\partial C}{\partial x} \tag{6.1}$$

where F is the diffusive flux of IHg or MeHg across the sediment–water interface (ng m^{-2} day^{-1}), φ is the sediment porosity, θ is the tortuosity (dimensionless), and D_w is the diffusion coefficient of IHg or MeHg in water in the absence of the sediment matrix (cm^{-2} s^{-1}). Tortuosity was estimated from porosity by Boudreau (1996) using the following equation:

$$\theta^2 = 1 - \ln\left(\varphi^2\right) \tag{6.2}$$

The diffusion coefficient depends on its specific ligand complex within the pore water. We assumed that Hg^{2+} exists as an anionic tetrachloro complex ($HgCl_4^-$) and MeHg exists as a neutrally charged chloride species ($MeHgCl^0$) (Gill et al. 1999). In the current study, the diffusion coefficients of IHg and MeHg were estimated to be 9.5×10^{-6} and 1.3×10^{-5} cm^{-2} s^{-1} at 25 °C, respectively (Gill et al. 1999; Covelli et al. 1999). The diffusion coefficients at 25 °C were corrected using the temperature of bottom water for each sampling period (Lerman 1979):

$$D_{T1} = D_{T2}(1 + 0.048\Delta t) \tag{6.3}$$

where Δt is the temperature difference in degrees centigrade?

The porosity was calculated using the following equation:

$$\varphi = 1 - \left[\frac{G}{(Vd)}\right] \tag{6.4}$$

where G is the dry weight of the sediment (g), V is the volume of fresh sediment (cm^3), and d is the density of the dry sediment (g cm^{-3}).

The concentration gradient was calculated based on the IHg or MeHg concentrations in the interface water and the pore water in the first sediment interval (1 cm). Although this calculation has a considerable uncertainty component, the estimated flux is only an approximate value and provides a relative basis for comparing Hg diffusion among the sampling sites.

1. Diffusion fluxes of inorganic mercury over the sediment–water interface

The spatial and seasonal distributions of dissolved inorganic Hg (IHg) in the water at the sediment–water interface and surface sediment pore water in PD and DF are listed in Table 6.12. The mean IHg concentrations in interface water in PD were 2.15 ± 0.71 and 1.54 ± 0.36 ng L^{-1} in summer and winter, respectively. The corresponding values in DF were 1.76 ± 1.24 n L^{-1} and 1.03 ± 0.69 L^{-1} in summer and winter, respectively. Meanwhile, the average IHg concentrations in surface sediment pore water in PD were 16.08 ± 3.29 and 3.68 ± 0.95 ng L^{-1} in summer and winter, respectively. The corresponding values in DF were 9.21 ± 4.20 and 3.44 ± 1.75 ng L^{-1} in summer and winter, respectively.

As shown in Table 6.12, the IHg concentrations in surface sediment pore water in PD and DF were generally higher than the concentrations in the corresponding interface water for each of the sampling stations and each of the seasons, with the exceptions of the concentrations at DF-1 in winter. The differences in the IHg concentrations between surface sediment pore water and interface water indicated that sediment was the potential source of IHg for the overlying water column. The IHg concentrations in surface sediment pore water at each of the sampling stations in PD and DF were significantly higher in summer than the concentrations in winter ($p < 0.01$). Similarly, the differences in the IHg concentrations between surface sediment pore water and interface water at each of the sampling stations in PD and DF were significantly higher in summer than the concentrations in winter

Table 6.12 Seasonal and spatial distribution of dissolved inorganic mercury (IHg) in surface sediment pore water and in interface water of PD and DF (ng L^{-1})

Sampling sites	Interface water				Surface soil pore water			
	PD		DF		PD		DF	
	Summer	Winter	Summer	Winter	Summer	Winter	Summer	Winter
PD-1/DF-1	3.02	1.15	2.64	3.34	18.27	4.98	12.18	2.20
PD-2	2.43	1.39	–	–	19.47	3.76	–	–
PD-3	1.65	2.02	–	–	13.75	3.13	–	–
PD-4/DF-2	1.50	1.61	0.88	1.52	12.81	2.85	6.24	4.67

($p < 0.01$), implying that the diffusion flux of IHg from surface sediment pore water to the overlying water column in summer was more active than in winter.

The diffusion fluxes of IHg from the sediment to the water column throughout the two sampling campaigns at PD and DF are shown in Fig. 6.61, and the corresponding data are summarized in Table 6.13.

Based on the calculated data, the diffusion fluxes of IHg over the sediment–water interface were positive for each of the sampling seasons and each of the sampling stations at PD and DF, with the exception of the data from DF-1 in winter, indicating steady diffusion of dissolved IHg from the surface sediment to the overlying water column. However, the IHg concentration in the interface water at DF-1 in winter was higher than the concentration in the corresponding surface sediment pore water, which in turn resulted in a negative diffusion flux of IHg over the sediment–water interface. This negative data indicated diffusion of IHg from interface water to the sediment, probably due to the extra source of IHg in the water column of DF-1 in winter (Rothenberg et al. 2008).

The spatial distribution of IHg fluxes over the sediment–water interface was less pronounced for each of the reservoirs (PD and DF). However, a clear seasonal trend in IHg fluxes over the sediment–water interface was observed at PD and DF during our sampling campaigns. As shown in Fig. 6.61, the diffusion fluxes of IHg over the sediment–water interface were significantly higher in summer than the fluxes in winter at each of the reservoirs ($p < 0.01$) (Table 6.13).

The seasonal trend in IHg fluxes over the sediment–water interface may be explained by the increased solubility of IHg under anoxic conditions in the summer (Benoit et al. 1998). Consequently, the IHg in sediment tended to exist in the liquid phase during summer, which in turn promoted the diffusion of IHg from the

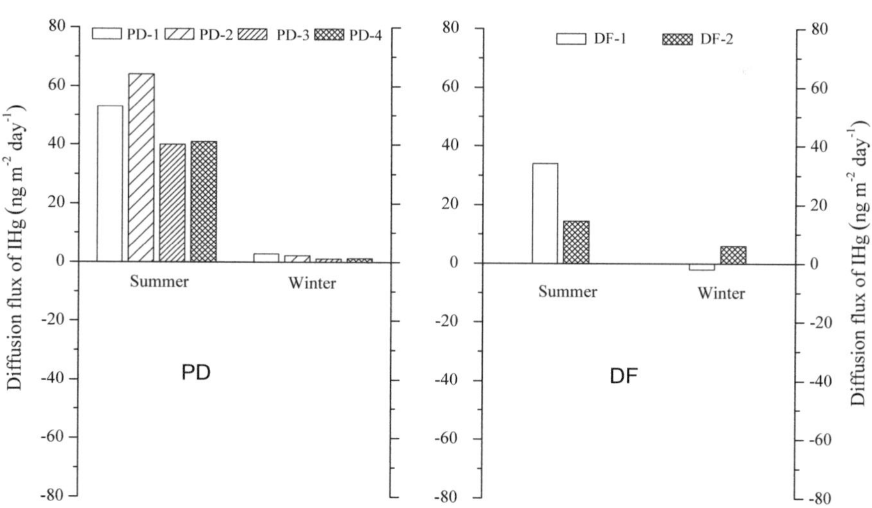

Fig. 6.61 Estimated diffusion fluxes of inorganic mercury (IHg) over the surface sediment and water column in PD and DF

Table 6.13 Estimated diffusion fluxes of inorganic mercury (IHg) and methylmercury (MeHg) across sediment to water column in PD and DF (ng m^{-2} day^{-1})

Sampling sites	IHg				MeHg			
	PD		DF		PD		DF	
	Summer	Winter	Summer	Winter	Summer	Winter	Summer	Winter
PD-1/DF-1	53.2	2.78	34.17	−2.07	0.72	0.34	11.9	2.79
PD-2	64.0	2.15	–	–	0.87	0.88	–	–
PD-3	40.2	1.06	–	–	11.0	1.46	–	–
PD-4/DF-2	41.3	1.27	14.5	5.95	0.55	0.56	10.3	1.57

sediment to the overlying water. However, the detailed mechanism underlying the seasonal trends in IHg fluxes over the sediment–water interface is very complex (physically, chemically, or biologically controlled) and requires further study.

2. Diffusion fluxes of methyl mercury over the sediment–water interface

The concentrations and distributions of MeHg in the water at the sediment–water interface and surface sediment pore water in PD and DF are listed in Table 6.14. The MeHg concentrations in the surface sediment pore water in PD and DF were generally higher than the concentrations in the corresponding interface water for each of the sampling stations and each of the seasons. The elevated MeHg concentrations in the surface sediment pore water indicated that MeHg in sediment was eluted to the overlying water and sediment was a very important source of MeHg for the water column.

The diffusion fluxes of MeHg from sediment to the overlying water at PD and DF are illustrated in Fig. 6.62, and the summary data are shown in Table 6.13. MeHg fluxes from the sediment to the interface water were all positive for each of the sampling stations (PD and DF) throughout the two sampling campaigns, indicating that the sediment was the net source of MeHg in the water column. Our calculated MeHg diffusion fluxes supported the finding that PD and DF are net sources of MeHg production (see the details in Chap. 8).

The estimated MeHg diffusion fluxes ranged from 0.34 to 11 ng m^{-2} day^{-1} at PD, with the maximum value observed at PD-4 in summer, and the minimum value observed at PD-1 in winter. We failed to observe any seasonal and spatial trends in

Table 6.14 Seasonal and spatial distribution of dissolved methylmercury in surface sediment pore water and in interface water of PD and DF (ng L^{-1})

Sampling sites	Interface water				Surface soil pore water			
	PD		DF		PD		DF	
	Summer	Winter	Summer	Winter	Summer	Winter	Summer	Winter
PD-1/DF-1	0.18	0.11	0.23	0.086	0.33	0.45	1.19	3.46
PD-2	0.16	0.11	–	–	0.33	0.82	–	–
PD-3	0.13	0.08	–	–	2.55	1.2	–	–
PD-4/DF-2	0.16	0.06	0.19	0.11	0.27	0.46	0.74	2.29

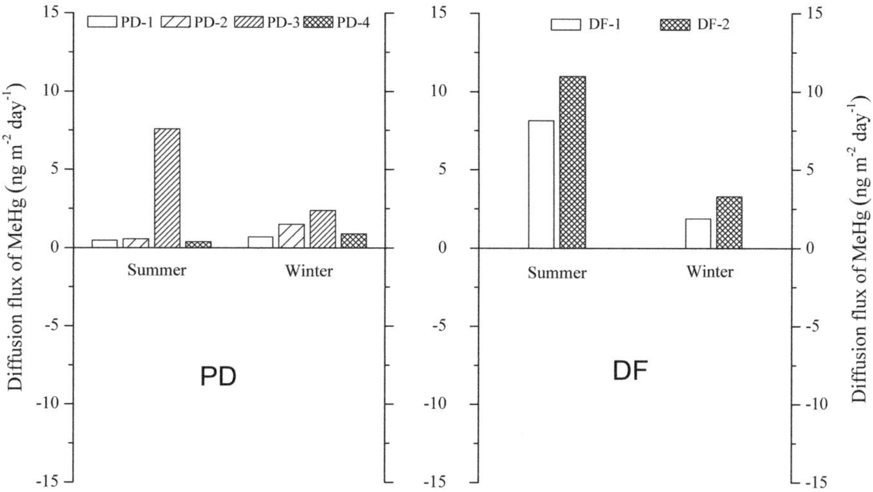

Fig. 6.62 Estimated diffusion fluxes of methylmercury (MeHg) over the surface sediment and water column in PD and DF

the MeHg diffusion fluxes at PD (see Fig. 6.62). The MeHg diffusion fluxes at DF ranged from 1.6 to 12 ng m^{-2} day^{-1}, with highest value observed at DF-1 in summer and the lowest value observed at DF-2 in winter (see Fig. 6.62). According to the statistical analyses, the MeHg fluxes were significantly higher in summer than the fluxes in winter throughout the DF sampling sites ($p < 0.01$), consistent with our previous study (Feng et al. 2009a) and the results of the Hg mass balance model (see the details in Chap. 7). Furthermore, the MeHg diffusion fluxes in summer and winter were higher at DF-1 than the values at DF-2.

The annual overall diffusion fluxes of IHg and MeHg from the sediment to the water column were quantified from our data using the following equation:

$$F_t = \sum \left[F_{(S,P)} D_{(S)} \left(\frac{A}{4} \right) \right] \times 10^{-9} \tag{6.5}$$

where F_t is the annual overall IHg or MeHg diffusion fluxes (g $year^{-1}$), F is the diffusive flux of IHg or MeHg (ng m^{-2} day^{-1}), S represents the sampling sites (PD: PD-1, PD-2, PD-3, and PD-4; DF: DF-1 and DF-2), P is the sampling period (summer: wet season; winter: dry season), A is the surface area of the reservoir (PD: 19.3×10^6 m^2; DF: 19.1×10^6 m^2), and $D_{(S)}$ is the number of days of each season (wet season: 180 days; dry season: 180 days). We assumed that the surface area of the reservoir is the same as the area of the surface sediment. For PD, each sampling site represents one-quarter of the surface area reservoir; for DF, each sampling site represents half of the surface area reservoir. Based on our calculated data, the overall MeHg diffusion fluxes in PD and DF were 14.2 and 45.61 g $year^{-1}$, respectively. Based on the mass balance calculation, the net MeHg fluxes reported in Chap. 7 (PD: 69.4 g $year^{-1}$; DF: 367.5 g $year^{-1}$) are generally consistent with our hypothesis that PD and DF were net sources of MeHg.

6.3 Biogeochemical Cycling of Mercury in Wujiangdu Reservoir

6.3.1 Sampling Sites Description, Sample Collection, Analytical Methods, and QA/QC

Four sampling stations spatially distributed from the upstream to the downstream of WJD were chosen as shown in Fig. 6.63. WJD-1 is located at the upper end of the reservoirs. WJD-2 is located in the middle part of the reservoirs. WJD-3 is situated in the downstream part of reservoirs. WJD-4 is located adjacent to the dam (within approximately 500 m). Water samples (water column profiles with 5–8 different depths), sediment cores (solid phase), sediment porewater (liquid phase), water–sediment interface water samples at each sampling station were collected each season in 2007.

Filtered water column samples were analyzed for DHg and DMeHg. THg, RHg, TMeHg, and TSS were analyzed in each of the unfiltered samples. Water–sediment interface water samples and sediment pore water samples were divided for DHg and DMeHg analysis. The freeze-dried sediment samples were grounded and homogenized for solid phase THg, MeHg, and organic matter concentrations analysis.

The sediment cores were immediately sliced using a plastic cutter in an oxygen-free glove box under argon. The first 10 cm was sectioned at 1 cm intervals and next 20 cm at 2 cm sections. The sediment samples were placed in acid-cleaned 50 mL plastic centrifuge tubes, capped and sealed with Parafilm. All samples were transported in an ice-cooled container to the lab within 24 h and stored at 3–4 °C for further laboratory processes.

Sediment samples were then centrifuged for 30 min at 3000 r min^{-1} at 5 °C to extract the pore water immediately after being transported to the laboratory. The pore water was then filtered through 0.45 μm disposable polycarbonate filter unit (Millipore) under argon in a glove box and placed in borosilicate glass bottles. At each sampling site, two sediment cores were collected, one for DHg analysis in pore water, and one for DMeHg analysis. Subsequently, the freeze-dried sediment samples were grounded and homogenized to a size of 150 meshes per inch with a mortar for solid-phase THg, MeHg, and organic matter concentrations analysis. Precautions were taken in order to avoid any cross-contamination during the sample processing. The grinder was thoroughly cleaned after each sample processing. The powdered samples were subsequently packed into plastic dishes, sealed in polyethylene bags and stored in a refrigerator within desiccators for further laboratory analysis.

All the water samples were acidified on site to 0.5% (v/v) with ultra-pure concentrated hydrochloric acid (HCl); the sample bottles were then capped, sealed with Parafilm, stored in a refrigerator at 3–4 °C in the dark. The analysis of concentrations of Hg species in water samples was conducted within 28 d after sampling. TSS was determined gravimetrically by filtering an aliquot of water (typically 1500 mL) through a pre-weighed 0.45 μm pore-size, 47 mm (diameter) polycarbonate membrane filter. Water quality parameters such as pH and temperature (*T*) were measured in situ using a portable analyzer (PD-501, Shanghai San-Xin Instrumentation Inc., China). Dissolved oxygen (DO) in water was monitored using an in situ DO probe (HI 7042S, HANNA Instruments® Inc. Italy). The concentrations of organic matter in the surface sediment samples were analyzed using KCr_2O_7 (Potassium dichromate) oxidation coupled with volumetric analysis.

Quality control for THg and MeHg determination was conducted by field blanks, matrix spikes, and duplicate samples. The method detection limit (3 × σ) was 0.02 ng L^{-1} for THg and 0.01 ng L^{-1} for MeHg in water samples. Limits of determination were 0.01 ng g^{-1} for THg and 0.002 ng g^{-1} for MeHg in sediment samples. The method blank was in each case less than the detection limit. Field blanks were 0.14 ± 0.04 ng L^{-1} for THg and 0.012 ng L^{-1} for MeHg. The relative standard deviations for duplicate sample analyses were <8.5% for THg and MeHg in water samples and were 7.8% for THg and MeHg in sediment samples. Recoveries for matrix spikes ranged from 87 to 113% and from 91 to 108% for THg and MeHg analysis in water samples. The average THg concentration of the geological standard of GBW07305 was 95 ± 7.0 ng g^{-1} (n = 15), which is comparable with the certified value of 100 ± 20 ng g^{-1}. The average MeHg concentration of 5.3 ± 0.50 ng g^{-1} (n = 15) was obtained from IAEA-405, with a certified value of 5.5 ± 0.53 ng g^{-1}.

Statistical evaluation was performed using SPSS 11.5 software. To reveal any relationship between the general sediment quality characteristics and Hg species, relationships between covariant sets of data were analyzed by regression analysis.

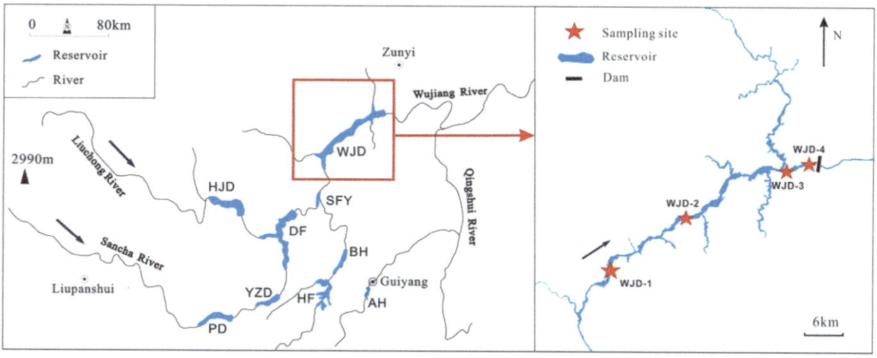

Fig. 6.63 Map of the study area and location of sampling stations at WJD (redrawed from Meng et al. 2016, with permission from John Wiley and Sons Inc.)

The correlation coefficient (*r*) and test of significance (*p*) were computed for correlation analysis. Correlation was significant at 0.05 (1-tailed). Linear regression fits were also processed to model the relationship between organic matter content and MeHg concentration in sediments. In addition, Kolmogorov–Smirnov (K–S) and Kruskal–Wallis (K–W) tests were performed to compare significant differences between 2 or more independent datasets. Differences were significant at $p < 0.05$.

6.3.2 Mercury Species in Water Column

1. General Water Quality Characteristics

Vertical profiles for *T*, pH, and DO from the WJD are displayed in Figs. 6.64, 6.65 and 6.66, and summary data is shown in Table 6.15.

Water temperature profiles exhibited expected seasonal and spatial patterns, with temperatures in the WJD ranging from 12 to 25 °C. The reservoirs were well mixed in fall and winter, and stratified in spring and summer, with stronger trends at the downstream stations of the reservoirs. TSS showed a seasonal pattern that was opposite of *T*, with the highest average TSS values in dry seasons (spring and winter) and the lowest values in wet seasons (summer and fall).

Due to the karstic geology of the Wujiang River basin, the water was slightly alkaline in most samples (pH = 7.7 ± 0.47) but slightly acidic in the bottom stratum at WJD-3 in summer (6.8) and at WJD-4 in winter (6.7) and summer (6.9), which may result from the formation of organic acids in the sediment (He et al. 2008a). The pH reached distinct peaks of 9.5 and 9.7 in the surface water at WJD-3

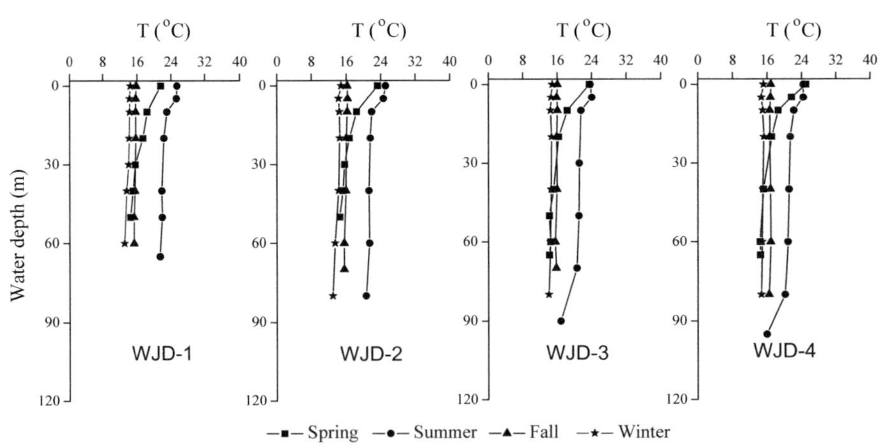

Fig. 6.64 Water column profiles of the temperature (*T*) at four sampling stations (WJD-1, WJD-2, WJD-3, WJD-4) in WJD (redrawed from Meng et al. 2010, with permission from The Alliance of Crop, Soil, and Environmental Science Societies; redrawed from Meng et al. 2011, with permission from Chinese Journal of Ecology)

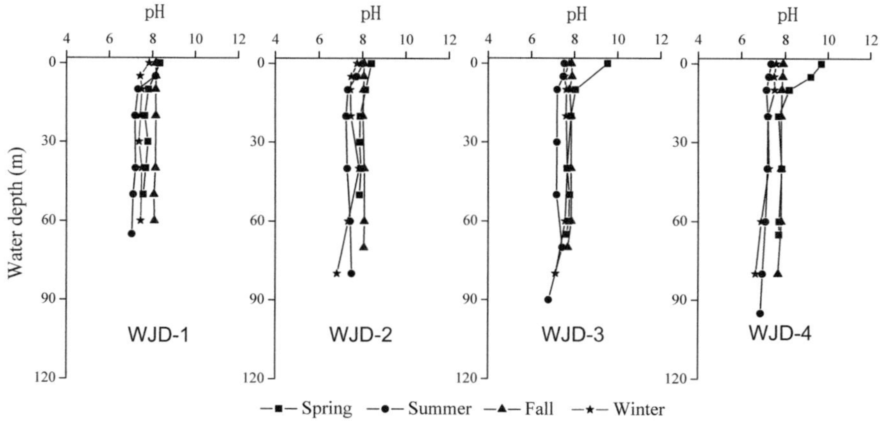

Fig. 6.65 Water column profiles of the pH at four sampling stations (WJD-1, WJD-2, WJD-3, WJD-4) in WJD (redrawed from Meng et al. 2010, with permission from The Alliance of Crop, Soil, and Environmental Science Societies; redrawed from Meng et al. 2011, with permission from Chinese Journal of Ecology)

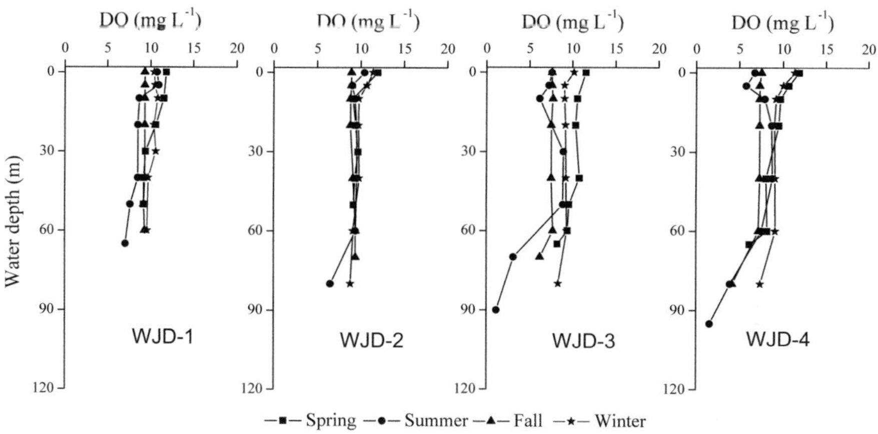

Fig. 6.66 Water column profiles of the dissolved oxygen (DO) at four sampling stations (WJD-1, WJD-2, WJD-3, WJD-4) in WJD (redrawed from Meng et al. 2010, with permission from The Alliance of Crop, Soil, and Environmental Science Societies; redrawed from Meng et al. 2011, with permission from Chinese Journal of Ecology)

and WJD-4 during spring, coinciding with maxima in DO (12 mg dm^{-3} at both sites) during algal blooms (He et al. 2008a). No significant variations of pH with depth were observed at WJD-1, WJD-2 during this study.

Vertical profiles of DO were pronounced and well-correlated with thermal stratification. Spatial and seasonal differences in DO levels were observed in the WJD. In the WJD, explicit deficiencies of DO were persistent in the bottom waters

Table 6.15 Distribution of routine sampling water parameters collected from the WJD [range (mean ± SD)]

Sampling site	Season	*T* (°C)	DO (mg dm^{-3})	pH	TSS (mg dm^{-3})
WJD-1	Spring	14–22 (17 ± 2.5)	9.2–12 (10 ± 1.1)	7.6–8.4 (7.8 ± 0.26)	1.2–8.8 (3.0 ± 2.6)
	Summer	21–25 (23 ± 1.7)	7.1–11 (8.9 ± 1.5)	7.0–8.2 (7.5 ± 0.51)	1.1–3.3 (1.9 ± 0.92)
	Fall	15–16 (16 ± 0.21)	9.2–9.3 (9.3 ± 0.038)	8.1–8.2 (8.1 ± 0.039)	1.6–3.3 (2.3 ± 0.64)
	Winter	13–14 (14 ± 0.49)	9.6–11 (10 ± 0.46)	7.4–7.9 (7.5 ± 0.16)	1.3–2.2 (1.8 ± 0.47)
WJD-2	Spring	15–23 (17 ± 3.3)	9.1–12 (9.8 ± 1.1)	7.9–8.4 (8.0 ± 0.22)	0.59–2.2 (1.3 ± 0.56)
	Summer	21–25 (22 ± 1.8)	6.5–10 (9.1 ± 1.2)	7.2–8.0 (7.5 ± 0.27)	0.84–2.1 (1.3 ± 0.47)
	Fall	16–16 (16 ± 0.33)	8.8–9.4 (9.0 ± 0.26)	8.0–8.0 (8.1 ± 0.019)	1.1–2.8 (1.6 ± 0.60)
	Winter	12–15 (14 ± 0.67)	8.8–11 (9.9 ± 0.88)	6.8–7.8 (7.5 ± 0.33)	0.79–2.4 (1.7 ± 0.55)
WJD-3	Spring	14–24 (17 ± 3.4)	8.3–12 (10 ± 1.1)	7.6–9.5 (8.0 ± 0.68)	0.83–5.4 (2.1 ± 1.7)
	Summer	17–24 (21 ± 2.4)	1.2–8.9 (6.2 ± 3.0)	6.8–7.5 (7.3 ± 0.25)	0.52–2.5 (1.2 ± 0.80)
	Fall	16–16 (16 ± 0.18)	6.2–7.7 (7.4 ± 0.54)	7.7–7.9 (7.8 ± 0.066)	0.32–1.6 (1.1 ± 0.41)
	Winter	14–15 (15 ± 0.24)	8.4–10 (9.2 ± 0.53)	7.1–7.8 (7.6 ± 0.20)	1.3–3.0 (2.2 ± 0.68)
WJD-4	Spring	14–25 (18 ± 4.0)	6.2–12 (9.1 ± 1.9)	7.7–9.7 (8.3 ± 0.81)	0.51–3.0 (1.6 ± 0.81)
	Summer	16–24 (21 ± 2.7)	1.6–8.8 (6.4 ± 2.5)	6.9–7.4 (7.1 ± 0.15)	0.26–1.6 (0.91 ± 0.63)
	Fall	16–17 (17 ± 0.16)	4.2–7.5 (6.9 ± 1.2)	7.7–7.9 (7.8 ± 0.071)	0.58–1.3 (0.86 ± 0.32)
	Winter	15–15 (15 ± 0.23)	7.4–11 (9.3 ± 1.2)	6.7–7.6 (7.2 ± 0.35)	1.1–2.7 (2.0 ± 0.63)

of WJD-3 and WJD-4 throughout the seasons without thermal stratification but were absent at WJD-1 and WJD-2. In summer, low DO concentrations (WJD-1: 7.1 mg dm^{-3}. WJD-2: 6.5 mg dm^{-3}. WJD-3: 1.2 mg dm^{-3}. WJD-4: 1.6 mg dm^{-3}) in the hypolimnion of the WJD were believed to be the result of intensive bacterial decomposition of settled degradable organic matter (He et al. 2008a). The spring maximum of DO at WJD-3 and WJD-4 at the surface was explained by algal blooms, as previously mentioned (He et al. 2008a). However, elevated DO in surface water was not observed at WJD-1 and WJD-2 during spring, possibly indicating the absence of an algal bloom there.

The WJD were completely stratified in summer especially at the downstream sites, but abundant algae were only present at WJD-3 and WJD-4 (Zhu et al. 2006; Dang 2008). The more pronounced chemical stratification at WJD-3 and WJD-4 compared to WJD-1 and WJD-2, is due to higher primary productivity levels at WJD-3 and WJD-4. Hence, primary productivity is the main factor controlling water column characteristics in stratified reservoirs.

2. Distribution of mercury species in water column

Spatial and seasonal distributions of THg, DHg, RHg, and PHg concentrations in the WJD are shown in Figs. 6.67, 6.68, 6.69 and 6.70, and summary data is shown in Table 6.16.

(1) Total Mercury

Annual mean concentrations of THg were 1.3 ± 0.56 ng L^{-1} in the WJD (range: 0.60–3.5 ng L^{-1}). THg concentrations in these reservoirs were significantly below the Chinese surface water standard of 50 ng L^{-1} (Environmental Quality Standards

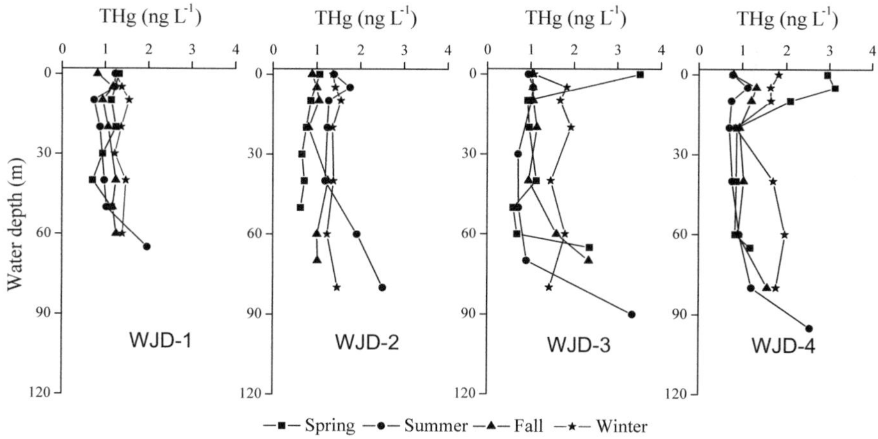

Fig. 6.67 Distributions of total mercury (THg) in water column profiles of the four sampling stations (WJD-1, WJD-2, WJD-3, WJD-4) in WJD (redrawed from Meng et al. 2010, with permission from The Alliance of Crop, Soil, and Environmental Science Societies; redrawed from Meng et al. 2011, with permission from Chinese Journal of Ecology)

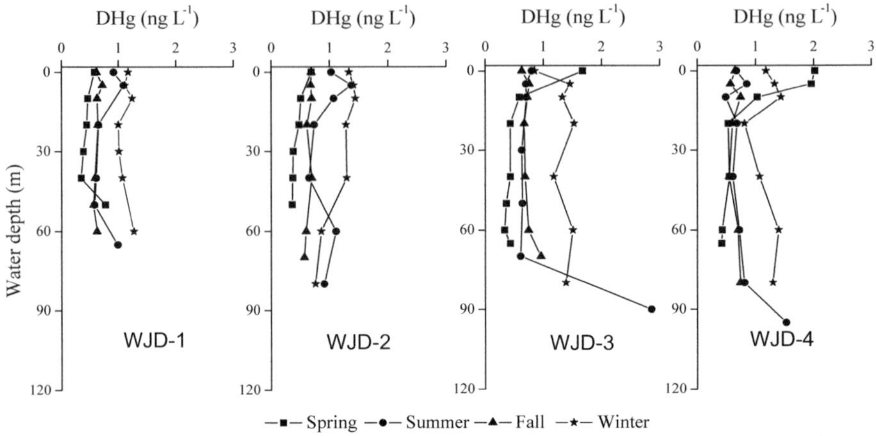

Fig. 6.68 Distributions of dissolved total mercury (DHg) in water column profiles of the four sampling stations (WJD-1, WJD-2, WJD-3, WJD-4) in WJD (redrawed from Meng et al. 2010, with permission from The Alliance of Crop, Soil, and Environmental Science Societies; redrawed from Meng et al. 2011, with permission from Chinese Journal of Ecology)

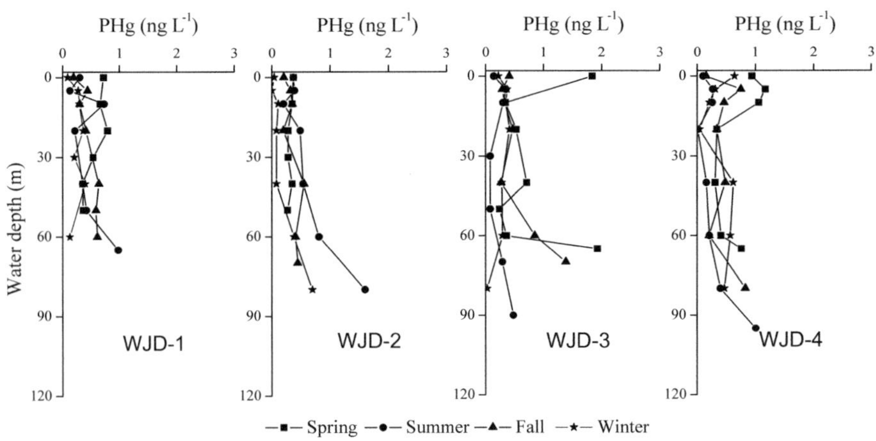

Fig. 6.69 Distributions of particulate mercury (PHg) in water column profiles of the four sampling stations (WJD-1, WJD-2, WJD-3, WJD-4) in WJD (redrawed from Meng et al. 2010, with permission from The Alliance of Crop, Soil, and Environmental Science Societies; redrawed from Meng et al. 2011, with permission from Chinese Journal of Ecology)

for Surface Water; GB3838-2002), and also below the 12 ng L^{-1} standard for THg recommended by the US-EPA to protect against adverse chronic effects on aquatic life (US-EPA 1992).

Peak levels of THg in the WJD were observed during spring in surface water at stations WJD-3 (3.6 ng L^{-1}) and WJD-4 (3.2 ng L^{-1}), coinciding temporally and spatially with peaks of TSS (WJD-3: 5.4 mg L^{-1}; WJD-4: 3.0 mg L^{-1}) and PHg

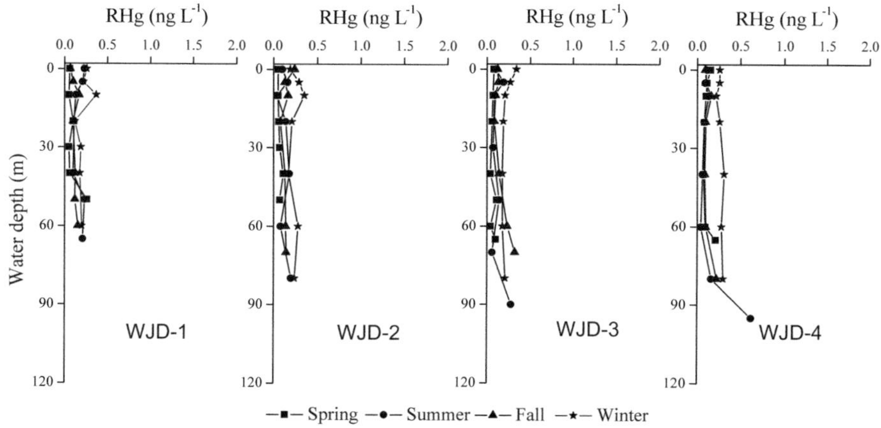

Fig. 6.70 Distributions of reactive mercury (RHg) in water column profiles of the four sampling stations (WJD-1, WJD-2, WJD-3, WJD-4) in WJD (redrawed from Meng et al. 2010, with permission from The Alliance of Crop, Soil, and Environmental Science Societies; redrawed from Meng et al. 2011, with permission from Chinese Journal of Ecology)

(WJD-3: 1.8 ng L^{-1}; WJD-4: 1.2 ng L^{-1}). These elevated levels of THg in the WJD may be attributed to seasonal increases in anthropogenic activities and the appearance of algal blooms (Zhu et al. 2006; Dang 2008). Algae have the capacity to bind Hg (Hurley et al. 1991), and the higher THg in the WJD compared to the newly constructed reservoir such as YZD, HJD, and SFY, could be due to the higher level of primary productivity in the WJD. Elevated levels of THg observed during summer in the bottom water of stations WJD-1 (2.0 ng L^{-1}), WJD-2 (2.6 ng L^{-1}), WJD-3 (3.4 ng L^{-1}) and WJD-4 (2.6 ng L^{-1}) coincided with elevated levels of PHg at these sites, perhaps suggesting that sediment resuspension was significant at that time.

No discernable difference in the annual mean concentration of THg was observed among WJD-1, WJD-2, WJD-3, WJD-4 (K–W test, $p = 0.99$). The concentrations of THg during the cold and dry seasons (spring and winter) in the WJD were higher than those in hot and wet seasons (summer and fall) (K–S test, $p = 0.001$).

(2) Dissolved Total Mercury

Spatial and temporal distributions of DHg in the water column in WJD is shown in Fig. 6.68.

Concentrations of DHg ranged from 0.33 to 2.8 ng L^{-1} (mean = 0.86 ± 0.42 ng L^{-1}) in the WJD. DHg represented 18–99% (mean = 67 ± 17%) of THg in the WJD. Regression analysis revealed a significant positive correlation between DHg and THg in the WJD ($r = 0.80$, $p < 0.0001$, $n = 110$), suggesting that DHg is the main Hg fraction present in the water column of the WJD.

Elevated concentrations of DHg were observed in bottom water in the WJD at WJD-3 (3.0 ng L^{-1}) and WJD-4 (1.6 ng L^{-1}) during summer, probably due to the

Table 6.16 Seasonal and spatial distribution of Hg species collected from the WJD [range (mean ± SD)]

Sampling site	Season	THg (ng L^{-1})	DHg (ng L^{-1})	RHg (ng L^{-1})	TMeHg (ng L^{-1})	DMeHg (ng L^{-1})
WJD-1	Spring	0.71–1.3 (1.1 ± 0.22)	0.35–0.78 (0.50 ± 0.16)	0.045–0.26 (0.094 ± 0.083)	0.13–0.21 (0.17 ± 0.029)	0.11–0.16 (0.13 ± 0.019)
	Summer	0.74–2.0 (1.2 ± 0.40)	0.59 ± 1.1 (0.81 ± 0.21)	0.10–0.23 (0.17 ± 0.058)	0.080–0.27 (0.16 ± 0.048)	0.051–0.13 (0.096 ± 0.030)
	Fall	0.82–1.3 (1.1 ± 0.17)	0.57–0.73 (0.64 ± 0.050)	0.067–0.17 (0.12 ± 0.035)	0.033–0.096 (0.050 ± 0.022)	0.030–0.046 (0.037 ± 0.010)
	Winter	1.2–1.6 (1.4 ± 0.12)	1.0–1.3 (1.1 ± 0.11)	0.12–0.37 (0.22 ± 0.078)	0.058–0.090 (0.076 ± 0.013)	0.046–0.087 (0.063 ± 0.014)
WJD-2	Spring	0.63–1.1 (0.78 ± 0.16)	0.36–0.69 (0.46 ± 0.13)	0.047–0.11 (0.069 ± 0.024)	0.13–0.22 (0.17 ± 0.031)	0.11–0.16 (0.13 ± 0.019)
	Summer	1.2–2.5 (1.6 ± 0.48)	0.65–1.4 (0.98 ± 0.24)	0.043–0.20 (0.13 ± 0.056)	0.11–0.35 (0.21 ± 0.091)	0.057–0.10 (0.085 ± 0.019)
	Fall	0.82–1.3 (1.0 ± 0.14)	0.57–0.70 (0.65 ± 0.053)	0.080–0.24 (0.15 ± 0.048)	0.053–0.081 (0.060 ± 0.010)	0.049–0.059 (0.052 ± 0.0039)
	Winter	1.2–1.6 (1.4 ± 0.10)	0.76–1.4 (1.2 ± 0.27)	0.17–0.35 (0.25 ± 0.065)	0.053–0.15 (0.082 ± 0.033)	0.041–0.066 (0.052 ± 0.010)
WJD-3	Spring	0.60–3.5 (1.5 ± 1.1)	0.33–1.7 (0.61 ± 0.48)	0.036–0.11 (0.070 ± 0.028)	0.15–0.28 (0.21 ± 0.054)	0.071–0.21 (0.15 ± 0.044)
	Summer	0.71–3.3 (1.2 ± 0.94)	0.61–2.9 (0.99 ± 0.82)	0.061–0.28 (0.13 ± 0.080)	0.14–2.9 (0.62 ± 1.0)	0.040–0.47 (0.11 ± 0.16)
	Fall	0.95–2.3 (1.3 ± 0.50)	0.63–0.96 (0.74 ± 0.11)	0.086–0.32 (0.16 ± 0.085)	0.033–0.096 (0.050 ± 0.022)	0.030–0.046 (0.037 ± 0.010)
	Winter	1.1–1.9 (1.6 ± 0.30)	0.85–1.5 (1.3 ± 0.24)	0.18–0.34 (0.22 ± 0.058)	0.11–0.15 (0.13 ± 0.016)	0.094–0.14 (0.11 ± 0.015)
WJD-4	Spring	0.85–3.1 (1.7 ± 1.0)	0.42–2.0 (0.99 ± 0.71)	0.072–0.22 (0.12 ± 0.050)	0.12–0.39 (0.22 ± 0.092)	0.089–0.21 (0.14 ± 0.044)
	Summer	0.70–2.5 (1.1 ± 0.61)	0.49–1.5 (0.80 ± 0.32)	0.040–0.62 (0.16 ± 0.19)	0.070–2.4 (0.43 ± 0.81)	0.026–1.4 (0.24 ± 0.49)
	Fall	0.80–1.6 (1.1 ± 0.27)	0.55–0.74 (0.65 ± 0.080)	0.086–0.22 (0.12 ± 0.051)	0.057–0.13 (0.086 ± 0.023)	0.039–0.095 (0.060 ± 0.022)
	Winter	0.85–2.0 (1.6 ± 0.36)	0.81–1.4 (1.2 ± 0.22)	0.22–0.31 (0.27 ± 0.030)	0.15–0.61 (0.27 ± 0.16)	0.095–0.44 (0.20 ± 0.12)

diffusion of Hg from sediment (Feng et al. 2009a). In addition, spring surface water samples at WJD-3 and WJD-4 were enriched in DHg This could be explained by atmospheric inputs combined with stratification of the water column. Except for these cases, none of the other sites exhibited spatial or seasonal trends in DHg (Fig. 6.68).

(3) Particulate Mercury

PHg concentrations ranged from 0.010 to 1.2 ng L^{-1} (mean = 0.44 ± 0.32 ng L^{-1}) in the WJD. Concentrations of both THg and PHg were elevated in surface water at WJD-3 and WJD-4 in spring (Fig. 6.69). The elevation of THg concentrations resulted from increased levels of PHg. TSS concentrations were also enhanced (WJD-3: 5.4 mg L^{-1}, WJD-4: 3.0 mg L^{-1}), primarily due to algal bloom formation in the surface water.

The elevated PHg levels may be due to an increased fraction of organic particles (Zhu et al. 2006; Dang 2008), which can adsorb more Hg than inorganic particles (Hurley et al. 1991). PHg concentrations in the bottom water at stations WJD-1 (summer), WJD-2 (summer), WJD-3 (spring and fall), and WJD-4 (summer) were all significantly higher than the corresponding overlying water, indicating the input of particulate Hg from sediment resuspension. In contrast, no discernible spatial or seasonal trends in PHg were observed in the newly constructed reservoirs (e.g., YZD, SFY, and HJD) because algal blooms and bottom sediments were absent there.

(4) Reactive Mercury

RHg concentrations ranged from 0.036 to 0.62 mg L^{-1} (mean = 0.15 ± 0.092 mg L^{-1}) in the WJD. We failed to observe any spatial or seasonal variation in RHg concentrations for the WJD (Fig. 6.70). The RHg concentration was at a maximum in the bottom water of the WJD (0.62 mg L^{-1}) during the summer. These elevated levels of RHg are possibly explained by diffusion of RHg from sediment, heightened by the lack of mixing. This explanation is supported in that no such maximum was observed in the newly constructed reservoirs (e.g., YZD), where a sediment layer is absent.

(5) Methylmercury

The distribution patterns and concentrations of total methylmercury (TMeHg), dissolved methylmercury (DMeHg), and particulate methylmercury (PMeHg) in the WJD are illustrated in Figs. 6.71, 6.72 and 6.73, and summary data is shown in Table 6.16.

Annual mean concentrations of TMeHg, DMeHg, and PMeHg are 0.26 ± 0.43 mg L^{-1} (range: 0.033–2.9 ng L^{-1}), 0.12 ± 0.09 mg L^{-1} (range: 0.03–1.4 ng L^{-1}), and 0.09 ± 0.16 mg L^{-1} (range: 0.01–0.98 ng L^{-1}), respectively. The corresponding mean ratios of DMeHg/TMeHg is 69 ± 22% in the WJD. Statistical analyses yielded significant positive correlations between DMeHg and TMeHg in the WJD ($r = 0.80$, $p < 0.0001$, $n = 111$). Furthermore, annual

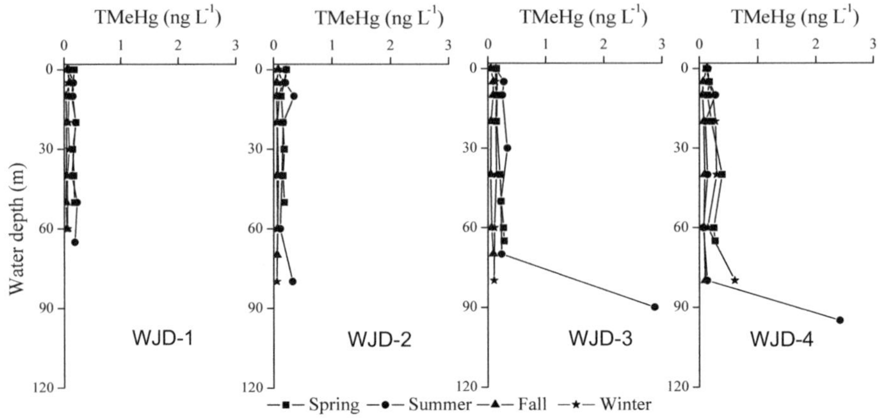

Fig. 6.71 Distributions of total methylmercury (TMeHg) in water column profiles of the four sampling stations (WJD-1, WJD-2, WJD-3, WJD-4) in WJD (redrawed from Meng et al. 2010, with permission from The Alliance of Crop, Soil, and Environmental Science Societies; redrawed from Meng et al. 2011, with permission from Chinese Journal of Ecology)

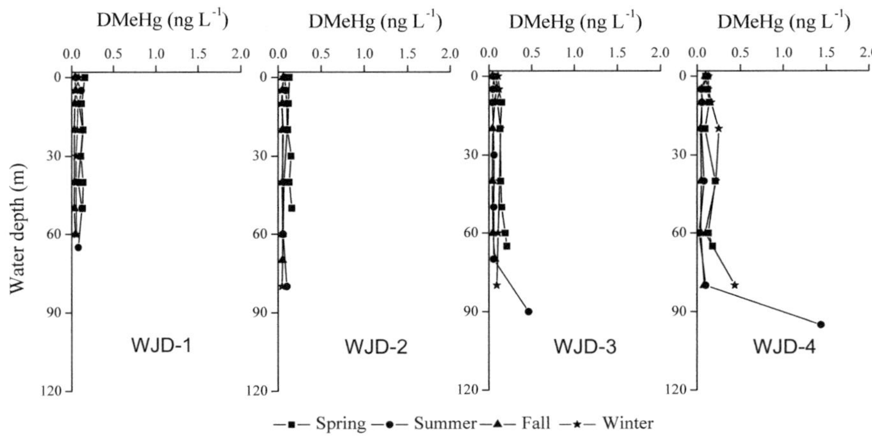

Fig. 6.72 Distributions of dissolved methylmercury (DMeHg) in water column profiles of the four sampling stations (WJD-1, WJD-2, WJD-3, WJD-4) in WJD (redrawed from Meng et al. 2010, with permission from The Alliance of Crop, Soil, and Environmental Science Societies; redrawed from Meng et al. 2011, with permission from Chinese Journal of Ecology)

average TMeHg concentrations showed statistically significant differences among the four sampling stations (WJD-1, WJD-2, WJD-3, WJD-4) in the WJD (K–W test, $p = 0.05$). In addition, annual average TMeHg concentrations at WJD-3 and WJD-4 were significantly higher than those at WJD-1 and WJD-2 (K–S test, $p = 0.01$). This observation suggests that considerably different levels of net Hg methylation exist between the upper and lower parts of the WJD. TMeHg

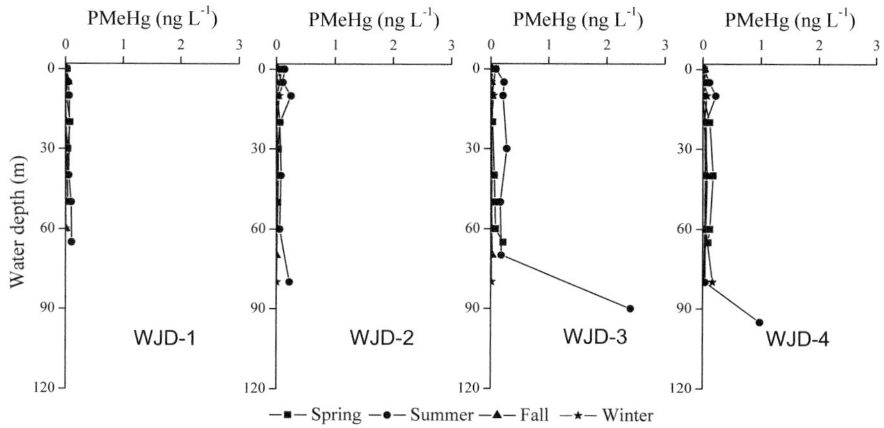

Fig. 6.73 Distributions of particulate methylmercury (PMeHg) in water column profiles of the four sampling stations (WJD-1, WJD-2, WJD-3, WJD-4) in WJD (redrawed from Meng et al. 2010, with permission from The Alliance of Crop, Soil, and Environmental Science Societies; redrawed from Meng et al. 2011, with permission from Chinese Journal of Ecology)

concentrations in WJD were significantly higher in summer compared to other seasons (K–S test, $p = 0.001$).

There are two possible sources of MeHg in the reservoirs: (1) in situ production being controlled by redox chemistry and/or settling particulate matter containing MeHg (Meili 1997; Eckley et al. 2005; He et al. 2008a), and (2) diffusion or resuspension, or both, of MeHg from underlying sediments (Mason and Sullivan 1999; Lawson et al. 2001). Gilmour and Henry (1991) showed that low pH and negative redox potential, not only increase methylation rates but also decrease demethylation rates, resulting in net production of MeHg.

In summer, TMeHg, DMeHg, and PMeHg were all low in the mixed layer but increased sharply to maxima in the low-oxygen region at downstream sites in the WJD. Similar observations were made in a stratified estuary in New England (Mason et al. 1993) and in a reservoir of southwestern China (He et al. 2008a). Our results indicate that active Hg methylation occurred at WJD-3 and WJD-4 during summer, in support of the conclusion by Ullrich et al. (2001) that high temperatures favor the Hg methylation process. In contrast, summer maxima of MeHg in bottom waters were not observed in the upper end of the WJD nor in any part of the newly constructed reservoirs (e.g., YZD, HJD, and SFY), plausibly ruling out sustained Hg methylation at those sites.

Feng et al. (2009b) reported that net annual Hg methylation was significantly higher in the WJD compared to DF because of the longer water residence time in the WJD. The water residence time in the WJD and the YZD is similar. However, the production of MeHg was much lower in the YZD. Hence, in this study, the key indicator for Hg methylation must not be the water residence time but rather the primary productivity. Another interesting outcome of this study is that Hg methylation was much more active in the lower end of the WJD compared to the upstream sites because of the primary productivity gradient, which supports the above conclusion.

The highest TMeHg concentrations during the summer stratification occurred at 90 m and 95 m at WJD-3 and WJD-4, respectively, where DO was lowest (WJD-3: 1.6 mg dm^{-3}, WJD-4: 1.2 mg dm^{-3}). Optimum conditions for Hg methylation were apparently reached at this depth during summer. The speciation and biochemical availability of Hg as well as environmental factors such as DO, pH, temperature, redox potential here favored net Hg methylation (c.g. Ullrich et al. 2001).

Regression analyses yielded inverse correlations when plotting TMeHg and DMeHg versus DO in the WJD, with Pearson correlation coefficients of −0.55 ($n = 111$, $p < 0.001$) and −0.40 ($n = 111$, $p < 0.001$), respectively. Similar relationships were also reported by He et al. (2008a). RHg and pH were significantly correlated with TMeHg (RHg: $r = 0.40$, $n = 111$, $p < 0.001$; pH: $r = -0.31$, $n = 111$, $p < 0.001$) as well as with DMeHg (RHg: $r = 0.50$, $n = 111$, $p < 0.001$; pH: $r = -0.24$, $n = 111$, $p = 0.01$). This suggests that low DO and pH and high RHg are prerequisites for elevated MeHg concentrations in the WJD. Hence, elevated MeHg in the bottom water may be plausibly attributed to in situ methylation and/or transfer from sediment to overlying waters; this is supported by the previous observation that sediment is the net source of MeHg to the water column (Feng et al. 2009a).

Spatial and seasonal distributions of PMeHg in the WJD showed little variation, with the exception of marked maxima in the bottom water at WJD-3 (2.4 ng L^{-1}) and WJD-4 (0.98 ng L^{-1}) in summer. These maxima may result from the release of MeHg from sediment resuspended into the water column. These results agree with previous observations that sediment resuspension may act as an additional MeHg source to water bodies (e.g., Mason and Sullivan 1999; Lawson et al. 2001).

Previous studies suggested that the ratio of MeHg to total Hg is recognized as a measure of Hg methylation efficiency (Sunderland et al. 2006; St. Louis et al. 1994; Rudd 1995; Gilmour et al. 1998). In our study, the TMeHg/THg ratio in water column in WJD (15 ± 14%) was slightly higher than those in the newly constructed reservoirs in Wujiang River (e.g., YZD, TMeHg/THg ratio = 13 ± 10%), but was approximately three to five times lower than those in the newly constructed reservoir in North American and North Europe (50–80%) (Kannan et al. 1998; St. Louis et al. 1994; Hall et al. 2005). These observations implied that the net Hg methylation in the newly constructed reservoir in North American and North Europe was much more active than that in WJD in Wujiang River.

Annual average concentrations of TMeHg, DMeHg, and PMeHg in the WJD were much higher than those in the newly constructed reservoir of YZD. This indicates that these two reservoirs, characterized by different levels of primary productivity, exhibit widely different strengths of Hg methylation and demethylation. A similar difference also appears to exist within the WJD, where only downstream stations (WJD-3 and WJD-4) demonstrate high TMeHg concentrations.

Previous studies have implied the presence of active MeHg production in newly constructed reservoirs and concluded that an enhanced methylation of Hg may last for more than 30 years after impoundment (St. Louis et al. 2004; Hall et al. 2005; Lucotte et al. 1999). Studies in North America showed that Hg methylation rates decrease with the age of the reservoir as a result of the decomposition of organic carbon in flooded soils (St. Louis et al. 2004; Hall et al. 2005; Lucotte et al. 1999). However, the observations obtained from the reservoirs in Wujiang River tell a different story. The WJD (an old reservoir) is characterized by a much more active net Hg methylation compared to the newly constructed reservoirs, such as YZD, SFY, and HJD. Furutani and Rudd (1980) found that organic material stimulates Hg methylation, hence diluting organic carbon concentrations in surface sediment by supplementing with inorganic material lowers MeHg production and bioaccumulation rates. Other studies have examined the differences in MeHg cycling in experimentally flooded wetland and upland catchments with varying carbon contents (Kelly et al. 1997; Bodaly et al. 2004).

Given the karstic environment of the Wujiang River basin, the organic carbon content (range: 2–5%) in submersed soils was very low (Jiang 2005). A recent study indicated that most of the organic matter in the newly constructed reservoirs (e.g., YZD) is derived from the watershed, with little autochthonous material evident (Jiang 2005). Primary productivity in the newly constructed reservoirs is currently much lower than that in the WJD, apparently due to the absence of cage culture fishing in the newly constructed reservoirs. In addition, lower organic carbon concentrations in the upland soils of the newly constructed reservoirs may inhibit methylating microorganisms from colonizing the newly constructed reservoirs or at least decrease their rate of metabolism. Hence, the low methylation rates in the newly constructed reservoirs are a result of the low organic carbon content in submersed soils and/or low primary productivity. However, it seems clear that Hg methylation will increase with increases in primary productivity.

(6) Comparisons with other Reservoirs

A comparison of THg and MeHg concentrations in water samples between the WJD and other areas are listed in Table 6.17. THg concentrations in water samples from the WJD were comparable to most literature data, which indicates that the WJD are apparently less impacted by local pollution sources.

However, the concentrations MeHg in water samples from WJD is slightly lower than that from the newly constructed reservoirs (<5 years old) in North America, but significantly higher than that from relatively old-aged reservoirs (>5 years old) in North America as well as from the newly constructed reservoirs in the same region (e.g., SFY, HJD, and YZD). These comparisons suggested that net Hg methylation in newly constructed reservoirs in North America was more active than that in the old-aged reservoirs. On the opposite, active net Hg methylation was only

Table 6.17 Comparison of levels of speciated mercury in the WJD with literature data

Sampling sites	Reservoir age (year)	THg (ng L^{-1})	TMeHg (ng L^{-1})	Reference
Flood-control imploundments, USA	2–4	0.74–6.97	0.06–6.6	Brigham et al. (2002)
Experimental reservoir, Canada	0.04	0.98–6.95	0.05–3.2	Kelly et al. (1997)
Experimental reservoir, Canada	3	1.1–6.0	0.1–2.1	Hall et al. (2005)
Quebec reservoir, Canada	3	<5	0.01–2	Lucotte et al. (1999)
Maryland reservoir, USA	12–133	0.4–6.8	0.048–0.38	Mason and Sveinsdottir (2003)
Narraguinnep reservoir, USA	16	0.47–1.06	0.010–0.043	Gray et al. (2005)
Caniapiscau Reservoir, Canada	17	1.19–1.69	0.06–0.09	Schetagne et al. (2000)
SFY, Wujiang River, China	3	0.4–4.9	0.030–0.22	This study
HJD, Wujiang River, China	4	0.3–6.6	0.05–0.17	This study
DF, Wujiang River, China	14	0.68–3.92	<0.05–0.50	This study
PD, Wujiang River, China	13	1.0–11.74	<0.05–0.51	This study
WJD, Wujiang River, China	28	0.60–3.5	0.033–2.9	This study
YZD, Wujiang River, China	6	0.40–1.9	0.0028–0.44	This study

observed in the old-aged reservoirs in Wujiang River, but not in the newly constructed reservoirs in the same region (e.g., SFY, HJD, and YZD), which implied that the net Hg methylation was gradually increased with the continuous evolution of the reservoirs. Therefore, in spite of the relatively low levels of MeHg in water in the newly constructed reservoirs from Wujiang River, we cannot exclude the possibility that the MeHg levels in water will increase with the evolution of the reservoir.

It is also interesting that concentrations of TMeHg in the water samples from downstream of the WJD appear higher than those reported for the other reservoirs and lakes worldwide, including Hongfeng reservoir, Caohai Lake, and Aha reservoir, which are located in the same region. Nevertheless, there is no discernable difference in MeHg levels between in the upstream and the middle of the WJD and uncontaminated global baseline waters. This further confirms that Hg methylation was amplified in the downstream of the WJD but not in the upstream and the middle of the WJD.

6.3.3 Mercury Species in Sediment Cores

1. General physical properties of sediment samples

Visual inspection of the sediment cores showed no macro-fauna or signs of bioturbation. Samples mainly consisted of fine particles, while sands and stones were virtually absent. As shown in Fig. 6.74, the water content in sediment cores from WJD showed the non-seasonal difference, but presented obviously spatial and vertical distribution characteristics for each of the sampling sites. The water content in sediment cores was the highest at the first 1 cm (WJD-1 80–85%; WJD-2 79–87%; WJD-3 89–93%; WJD-4 91–93%) and then decreased gradually to 60–70% at four sampling sites. Furthermore, the water content in surface sediment increased gradually from upstream (WJD-1 and WJD-2) to downstream (dam) of the WJD.

The vertical profiles of organic matter content in sediment cores from WJD are displayed in Fig. 6.75. The annual average concentrations of organic matter in sediment cores were 2.9 ± 0.57%, 3.1 ± 0.77%, 4.8 ± 2.3%, and 5.1 ± 2.0% at WJD-1, WJD-2, WJD-3, and WJD-4, respectively. Organic matter content in the sediments cores varied widely from 2.3 to 11% at WJD-3, and from 2.7 to 11% at WJD-4, respectively, but remained nearly monotonic at upstream sites (ranging from 2.2 to 5.2% at WJD-1 and from 2.2 to 5.5% at WJD-2, respectively). The maximum values of organic matter content in sediment at four sampling sites in WJD were observed at the surface sediment. Statistical analysis revealed that organic matter contents at depth of 1–5 cm in sediment were significantly higher than those in the corresponding sediment at depth of 6–30 cm across the four sampling sites (K–S test, $p < 0.01$). Furthermore, the organic matter contents in the

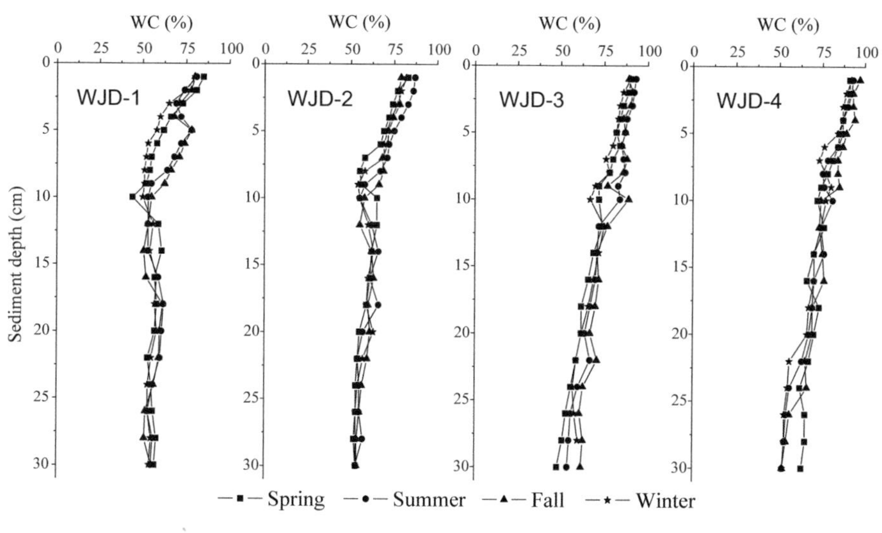

Fig. 6.74 Distributions of water content (WC) in sediment profiles of the four sampling stations (WJD-1, WJD-2, WJD-3, WJD-4) in WJD

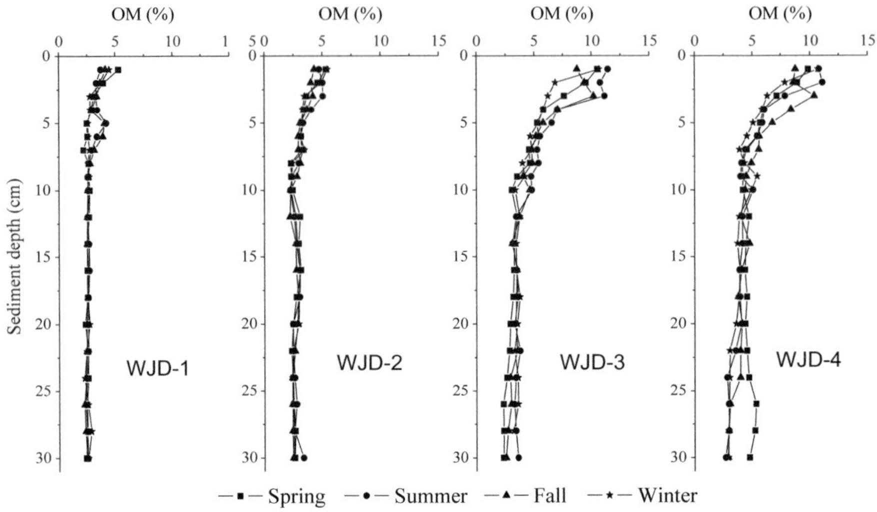

Fig. 6.75 Distributions of organic matter (OM) in sediment profiles of the four sampling stations (WJD-1, WJD-2, WJD-3, WJD-4) in WJD (redrawed from Meng et al. 2016, with permission from John Wiley and Sons Inc.)

first 1 cm of sediment at downstream sites (WJD-3 and WJD-4) (ranging from 8.8 to11%) were significantly higher than those at upstream sites (ranging from 3.7 to 5.5%) throughout the four sampling campaigns (K–S test, $p < 0.01$, Table 6.18). Temporal differences in organic matter content in sediment are not detected for all sampling sites.

Concentrations of TN, TP, chlorophyll, and phytoplankton cell density/biomass in WJD increase from the upstream to the downstream during our sampling periods (Dang 2008). It was showed that the water at the downstream of WJD is at a state of hypereutrophic, while the upper and middle reach of WJD is oligotrophic–mesotrophic during our sampling periods (Dang 2008; Kimmel and Groeger 1984). Our field investigation showed that the cage aquaculture activities are absent at WJD-1, appeared to be sporadic at WJD-2, and rather pervasive across the sampling sites of WJD-3 and WJD-4. Owing to the high intensities of cage aquaculture activities at the downstream of WJD, the fish feeds and fish feces were potentially the main sources of organic matter inputs to sediments there, which also resulted in higher primary productivity compared to the upstream of WJD. The organic matter content ranged from 0.4 to 6.9% in submersed soils in the catchment of WJD (see detail in Chap. 6, Sect. 6.1), which was comparable with the mean (range) values of organic matter in sediment at sampling sites of WJD-1. Therefore, it is reasonable to believe that the organic matters in sediment at the upstream of WJD were mainly derived from the watershed, such as soil erosion and surface runoff inputs with little contribution from the autochthonous sources.

2. Distribution of mercury species in sediment cores

(1) Total mercury

The distribution patterns of THg concentrations in sediment profiles are illustrated in Fig. 6.76. Overall, no discernable seasonal trends in the distribution of THg were observed in the sediment profiles of WJD during our sampling periods. However, an obvious spatial variation was observed between WJD-1 and other sampling sites. The concentrations of THg in sediment cores at WJD-2, WJD-3, and WJD-4 exhibited a very narrow range (106–494 ng g^{-1}), while the concentrations of THg in sediment at WJD-1 ranged widely from 128 to 1376 ng g^{-1}. Furthermore, the vertical distributions of THg in sediment cores of WJD showed a small variation, with the exception of sharp peaks in the depth of 10–15 cm at WJD-1 close to the inflow of WJD. The solid-phase THg record can be used to reconstruct the evolution of the anthropogenic Hg deposition. WJD-1 is located at the downstream from an Hg mining area. Therefore, the sharp peaks at depth of 10–15 cm in sediment cores throughout the four sampling campaigns indicated the direct input from the point source of the nearby Hg mining activities in the past.

The annual average THg concentrations in sediment cores were 389 ± 259 ng g^{-1}, 238 ± 69 ng g^{-1}, 300 ± 85 ng g^{-1}, and 268 ± 74 ng g^{-1} at WJD-1, WJD-2, WJD-3, and WJD-4, respectively. These values were comparable with the background level of THg (260 ng g^{-1}) in the soil in Guizhou province (Wang et al. 1992). Generally, uncontaminated sediments in reservoir or lake were suggested to have mean THg concentrations ranging from 50 to 300 ng g^{-1} in the

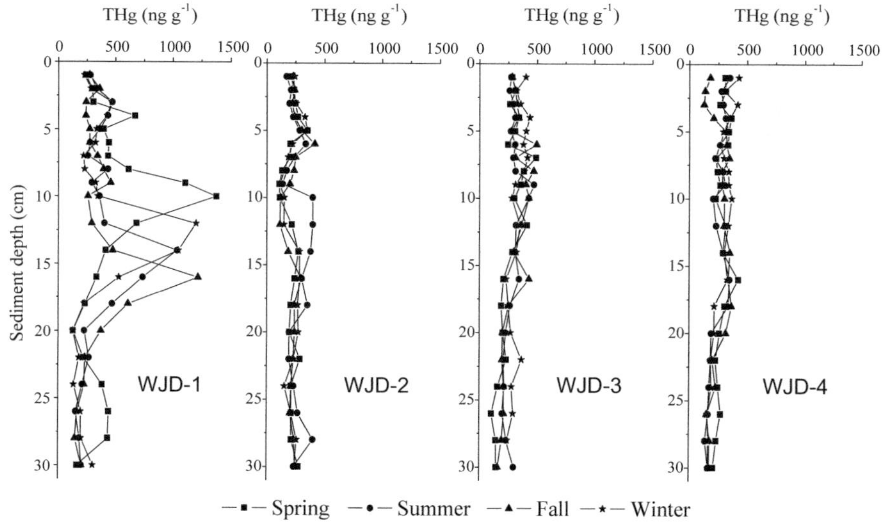

Fig. 6.76 Distributions of total mercury (THg) in sediment profiles of the four sampling stations (WJD-1, WJD-2, WJD-3, WJD-4) in WJD (redrawed from Meng et al. 2016, with permission from John Wiley and Sons Inc.)

study region (Yan et al. 2008). The average THg concentration in sediment collected from WJD-2, WJD-3 and WJD-4 suggested that sediment cores at the middle and downstream of WJD have not been impacted severely with Hg contamination. However, the mean THg concentration in sediment at WJD-1 was higher than the background level of THg in soil in the study region (Wang et al. 1992), and the elevation of the average THg concentration in sediment at WJD-1 resulted from the sharp peak at the depth of 10–15 cm in sediment profile. Overall, the observed THg concentrations in sediment in WJD were much higher than those observed in North American and North Europe. For example, French et al. (1999) reported the mean THg concentration in sediment collected from 34 reservoirs in Newfoundland, Canada was 39 ng g^{-1}.

(2) Methyl mercury

In comparison with distribution patterns of THg in sediment profiles, different vertical and spatial trends of MeHg in sediment cores were observed (Fig. 6.77).

The spatial distribution of MeHg concentrations in sediment reveals that the MeHg concentrations were relatively higher at the middle and downstream of WJD (Sampling sites of WJD-2, WJD-3, and WJD-4) compared to that at the upstream (WJD-1) of WJD. Statistically significant differences of MeHg levels in sediment were found among the WJD-1, WJD-2, WJD-3, and WJD-4 throughout the four sampling campaigns (K-W test, $p = 0.001$). Furthermore, MeHg concentrations in sediment collected from WJD-3 and WJD-4 were significantly higher than those from WJD-1 and WJD-2 (K–S test, $p = 0.003$). These observations suggested

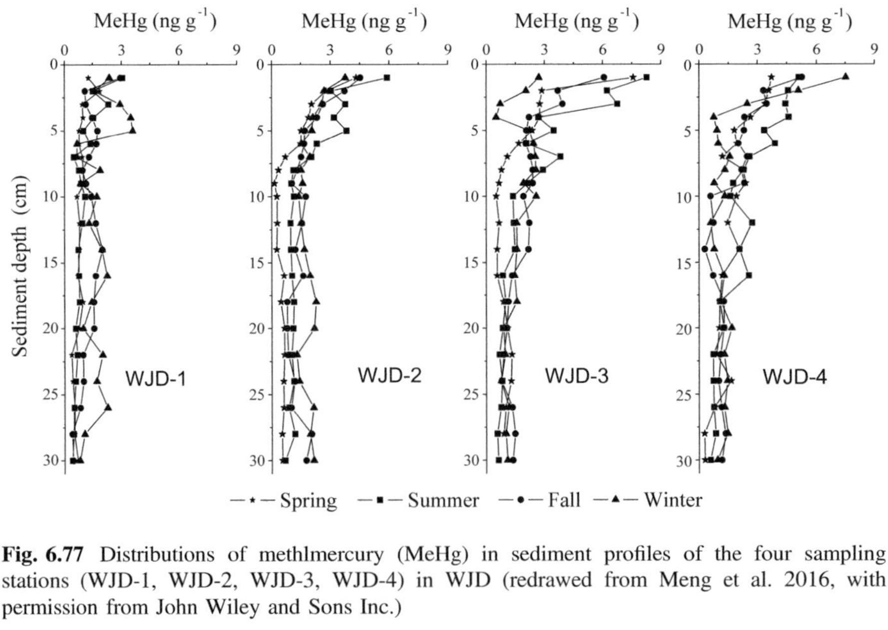

Fig. 6.77 Distributions of methlmercury (MeHg) in sediment profiles of the four sampling stations (WJD-1, WJD-2, WJD-3, WJD-4) in WJD (redrawed from Meng et al. 2016, with permission from John Wiley and Sons Inc.)

considerably different net Hg methylation among WJD-1, WJD-2, WJD-3, and WJD-4.

Percent THg as MeHg (MeHg %) is recognized as an indicator of net Hg methylation in substrates (Sunderland et al. 2006). Generally, MeHg concentrations accounted for approximately 1.0 to 1.5% of THg in sediments and these ratios tend to be lower (typically <0.5%) in estuarine environments (Ullrich et al. 2001). However, the MeHg% in sediment in WJD can reach up to 2.8, 2.9, and 4.2% at WJD-2, WJD-3, and WJD-4, respectively (Table 6.18), which were much higher than those in the greater depth of sediment cores as well as the other reservoirs in the same area (Yan et al. 2008). Moreover, the highest values of MeHg% were all observed at the surface sediment at WJD-2, WJD-3, and WJD-4 throughout the four sampling campaigns, indicating the active net Hg methylation occurred in this sediment layer. However, MeHg% fluctuated throughout the sediment cores at WJD-1 in each season with the highest value (1.5%) observed in the depth of 4 cm in fall.

The major differences among the four sampling stations are that both water depth and the organic matter content in the sediment increases from the upstream to the downstream of WJD. The distribution patterns of MeHg and MeHg% in sediment cores were mirrored with the organic matter content in sediment throughout the sampling sites, suggesting that relatively high production of MeHg is related to high organic matter content in sediment. Moreover, the regression analyses yielded significantly positive correlations when plotting MeHg versus organic matter contents in sediment for each of the sampling sites (Fig. 6.78), further implying that

Table 6.18 Seasonal and spatial distribution of organic matter content (OM), total mercury (THg), methylmercury (MeHg), and MeHg (%) in surface sediment of WJD

Sampling sites	Seasons	OM (%)	THg (ng g^{-1})	MeHg (ng g^{-1})	MeHg/THg (%)
WJD-1	Spring	5.3	240	3.0	1.3
	Summer	3.7	280	3.0	1.1
	Fall	4.2	273	2.4	0.87
	Winter	4.5	228	1.3	0.56
WJD-2	Spring	5.4	209	5.9	2.8
	Summer	4.8	170	4.5	2.7
	Fall	4.3	229	3.8	1.6
	Winter	5.5	241	4.3	1.8
WJD-3	Spring	11	282	8.3	2.9
	Summer	11	273	6.1	2.2
	Fall	8.8	282	2.7	1.0
	Winter	11	401	7.6	1.9
WJD-4	Spring	9.9	307	5.1	1.7
	Summer	11	350	5.3	1.5
	Fall	8.8	179	7.5	4.2
	Winter	11	428	3.7	0.87

organic matter in sediment plays an important role in the methylation of IHg in WJD. Moreover, Pearson Correlation Coefficients in WJD-1 ($r = 0.52$) was much lower when comparing with the data in WJD-2 ($r = 0.77$), WJD-3 ($r = 75$), and WJD-4 ($r = 70$). It seems to be other factors, apart from organic matter content, control the MeHg production in the sediment at WJD-1. The relationship between MeHg concentration and organic matter in sediment was also observed by other studies (Feng et al. 2009a; Graham et al. 2012). However, there is no such a significant correlation between the MeHg concentrations and THg concentrations in sediment, which is in agreement with a previous study (Kelly et al. 1995). This indicated that THg concentration is not a useful indicator for predicting MeHg concentrations in WJD.

The absence of obvious peak for organic matter in surface sediment at WJD-1 in all seasons, consistent with the low levels of MeHg, suggested that the production of MeHg in this sediment layer is limited. As described, in general physical properties of sediment samples, watershed soil erosion, and surface runoff is the primary source of organic matter to sediment at WJD-1. Bishop and Lee (1997) reported that the strong association of Hg with humic matter implies the watershed transport of Hg. Therefore, we suggested that transport of terrestrial organic matter with surface runoff could be an important source of MeHg to sediment at WJD-1. The remarkably higher values of MeHg and MeHg% at depth of 4 cm at WJD-1, also implied the existence of additional factors controlling MeHg production there.

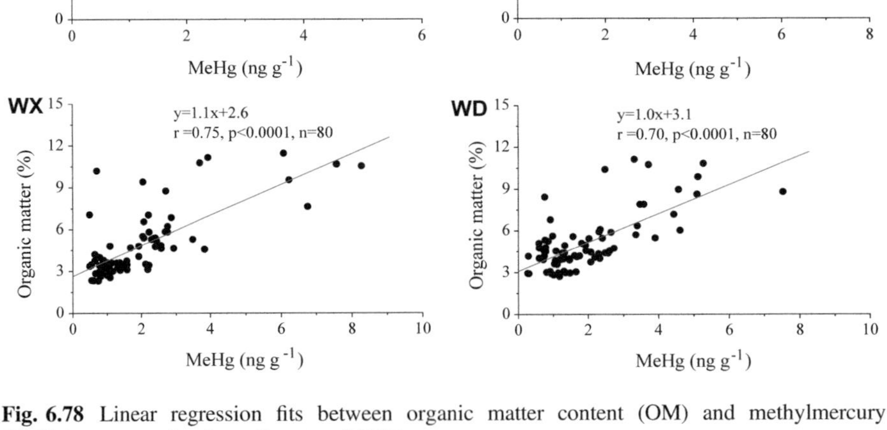

Fig. 6.78 Linear regression fits between organic matter content (OM) and methylmercury (MeHg) concentration in sediments of WJD

Studies in North American have implied the presence of active MeHg production in newly constructed reservoirs and concluded that an enhanced Hg methylation may last for >30 years after impoundment (St. Louis et al. 2004; Hall et al. 2005; Lucotte et al. 1999). Hg methylation rates decreased with the age of the reservoir, as a result of the decomposition of organic matters in flooded soil (St. Louis et al. 2004; Hall et al. 2005; Lucotte et al. 1999). Our current study showed that the reservoirs located in Southwestern China may have different Hg biogeochemical dynamics from reservoirs in Europe and North America. The organic matter contents in submersed soil were much lower than the organic matter concentrations (varied from 30 to 50%) in submerged soil from the boreal forest or wetland in North America and Europe (St. Louis et al. 2004; Hall et al. 2005; Lucotte et al. 1999). The major source of organic matter in newly constructed reservoir in Wujiang River is mainly derived from the watershed input, with little autochthonous contribution due to the low primary productivity. Recent studies concluded that the low organic matter contents in the submerged upland soil of the Wujiang River Basin may inhibit methylating microorganisms or at least decrease their rate of metabolism. Therefore, due to the low organic matter contents in submersed soil, the newly constructed reservoirs such as SFY, HJD, and YZD reservoirs in Wujiang River did not show net production of MeHg in reservoir systems.

Phytoplankton-derived organic matters jointly with the fish feeds and fish feces were potentially the significant sources of organic matter input to surface sediments of the downstream of WJD. Therefore, it may imply that the organic matters in surface sediment originated from cage aquaculture activities in WJD were easily decomposed by microorganisms which mediated Hg methylation processes in the sediments. It is well known that organic matters in sediment play an important role in the methylation of IHg (Andersson et al. 1990). Numerous studies indicated that elevated MeHg concentrations in sediments were observed with elevated organic matter contents in sediments (Fjeld and Rognerud 1993), which are attributed to the stimulation effect of organic nutrients on microbial methylation activity. Cossa and Gobeil (2000) explained that increased oxygen consumption during organic waste degradation causes progressively more anoxic conditions at the sediment/water interface, which may lead to active methylation process.

The vertical profiles for pH and dissolved oxygen in the water column of WJD were documented in Sect. 6.3.2. In brief, the water was slightly acidic in the bottom stratum, as a result of the formation of organic acids in the sediment. Moreover, explicit deficiencies of dissolved oxygen were persistent at the bottom water of WJD-3 and WJD-4 sites throughout the sampling seasons, but were less pronounced at WJD-3 site and were absent at WJD-1 site. It is generally accepted that the aquatic environment with low dissolved oxygen and pH favored net Hg methylation (Ullrich et al. 2001). The intensive cage aquaculture activities contributed to the high primary productivity in downstream of WJD. It can be seen that the contribution of organic matters to sediment from cage aquaculture activities is the key factor to explain the different MeHg production among WJD-1, WJD-2, WJD-3, and WJD-4 sites.

Moreover, it is accepted that supply of oxygen to the surface sediment tends to decrease with the increase of the water depth. The water depths became deeper and deeper from the upstream to the downstream of WJD. Therefore, the limit of oxygen supply to the epilimnion of water column also may play an important role in influencing the variation of MeHg production in sediment in WJD (Lambertsson and Nilsson 2006). The lower levels of MeHg at WJD-1 site compared to the other sampling sites suggested that the influence of organic matter originated from runoff and soil erosion from the catchment on Hg methylation may be minor.

MeHg concentrations in sediment cores at WJD-2, WJD-3, and WJD-4 showed definite maxima just blow the water–sediment interface and decreased with depth throughout the four sampling campaigns, which is different from that at WJD-1 site. This is in a good agreement with the previous observation that MeHg concentration often reached the maximum value in the anaerobic surface sediments and then sharply declines with increasing sediment depth (Bloom et al. 1999). MeHg concentrations fluctuated throughout the sediment cores at WJD-1 site in four seasons. This fluctuation could be explained by the different intensity of river erosion and surface runoff which act as potential sources of MeHg to sediment at WJD-1. Previous studies also observed the seasonal variation of maximum MeHg concentrations in sediment profile. For instance, Feng et al. (2009a) reported that MeHg concentrations in sediment cores were the highest in July campaign and the lowest in the December campaign. However, Bloom et al. (1999) found a sharp peak of MeHg concentration in sediment in early spring, following a decrease during the remaining seasons. Seasonal variations in MeHg production and demethylation were controlled by numerous factors, such as temperature, seasonal change in productivity/nutrient supply, pH, and redox conditions (Ullrich et al. 2001). As discussed in section general physical properties of sediment samples, the amount of seasonal input of fresh organic matter to the sediment surface is relatively minor in comparison to the total amount of organic matter already existed in the sediment; consequently, the temporal differences in organic matter contents in sediment are therefore not seen in the current study. Therefore, organic matter content in sediment alone could not explain the seasonal variations of MeHg in surface sediment. We speculated that the seasonal variation of redox conditions in the surface sediment of WJD may play an important role in controlling the temporal tread of MeHg (Feng et al. 2009a). Certainly, other reasons may also affect methylation process; thus, further work is urgent.

3. Distribution of mercury species in sediment pore-water

(1) Filtered total mercury

(I) Interface water

The spatial and seasonal patterns of DHg in the interface water of the WJD are shown in Figs. 6.79. The annual mean concentrations of DHg in the interface water at WJD-1, WJD-2, WJD-3, and WJD-4 were 1.1 ± 0.35, 1.3 ± 0.40, 2.2 ± 1.7 and 3.3 ± 1.9 ng L^{-1}, respectively.

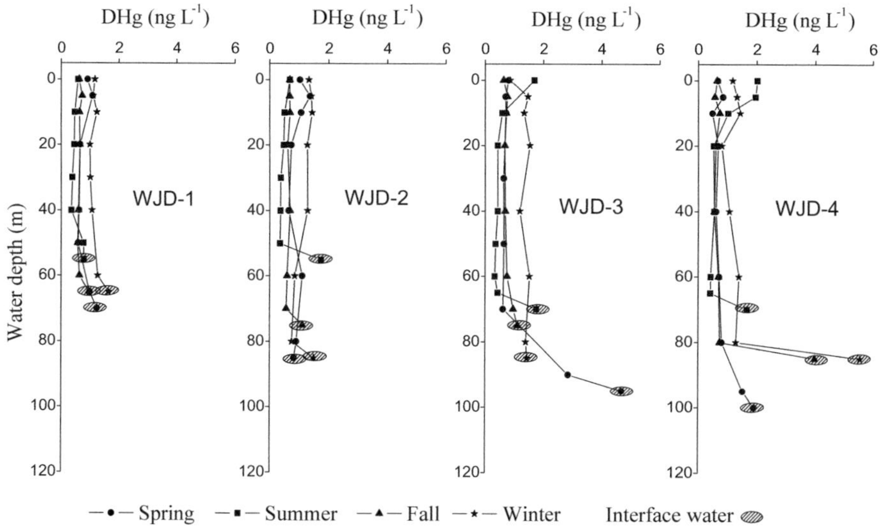

Fig. 6.79 Distributions of the dissolved mercury (DHg) in water column and interface water at four sampling stations (WJD-1, WJD-2, WJD-3, WJD-4) in WJD (redrawed from Meng et al. 2010, with permission from The Alliance of Crop, Soil, and Environmental Science Societies; redrawed from Meng et al. 2011, with permission from Chinese Journal of Ecology)

Interface water DHg concentration downstream of the WJD (WJD-3 and WJD-4) was elevated compared to the upper stream of the WJD (WJD-1 and WJD-2). As shown in Fig. 6.79, concentrations of DHg in interface water at WJD were significantly higher than those in the corresponding overlying water at each of the sampling stations and each of the sampling campaigns. The interface water DHg is plausibly perpetrated by diffusion of DHg from sediment–water due to a continuous concentration gradient (Jiang 2005; He et al. 2008b).

(II) Sediment pore water

The seasonal and spatial distributions of DHg in pore water in WJD were shown in Fig. 6.80. The annual mean concentrations of DHg in the sediment pore water at WJD-1, WJD-2, WJD-3, and WJD-4 were 7.1 ± 3.9 ng L^{-1} (2.2–23 ng L^{-1}), 6.3 ± 4.4 ng L^{-1} (2.0–28 ng L^{-1}), 5.8 ± 4.3 ng L^{-1} (1.2–23 ng L^{-1}) and 5.6 ± 4.7 ng L^{-1} (1.2–29 ng L^{-1}), respectively.

The distribution patterns of pore water DHg were completely different from those of pore water DMeHg in WJD, and were more variable than the THg concentrations in the solid phase of sediment. This is in a good agreement with our previous study (Feng et al. 2009a) that concentrations of DHg in vertical profiles of pore water varied randomly without discernible trends throughout the four sampling stations in each season. However, DHg concentrations in pore water were generally higher than those in interface water, implying that the sediment was an important source of DHg to water column. Moreover, DHg concentrations in pore water in

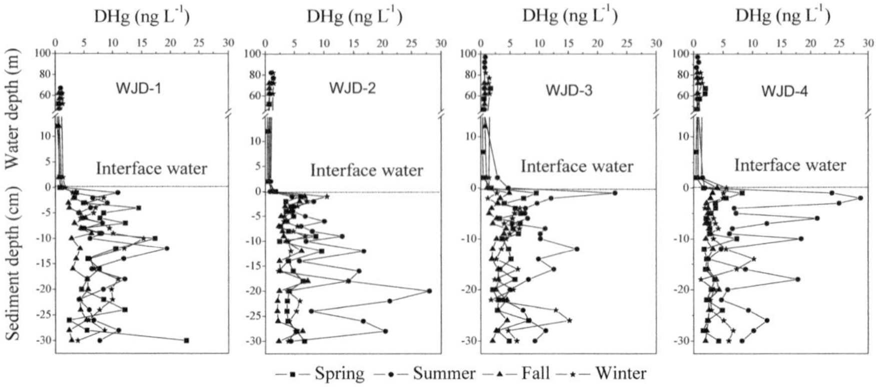

Fig. 6.80 Distributions of dissolved mercury (DHg) in water column and sediment pore water of the four sampling stations (WJD-1, WJD-2, WJD-3, WJD-4) in WJD (redrawed from Meng et al. 2016, with permission from John Wiley and Sons Inc.)

summer were significantly higher than those in the other seasons for all sampling sites (K–S test, $p < 0.01$), indicating that the Hg in sediment tends to exist in the liquid phase during summer. The seasonal trend of DHg in sediment pore water may be explained by the increased solubility of Hg under anoxic condition during summer (Benoit et al. 1998). However, the partition of Hg between the solid phase and aqueous phase is physically, chemically, or biologically controlled, and hence affected by a number of environmental parameters such as pH, temperature, redox conditions, and bioturbation (Ullrich et al. 2001).

(2) Filtered methylmercury

(I) Interface water

The spatial and seasonal patterns of DMeHg in the interface water of the WJD are shown in Figs. 6.81. The annual mean concentrations of DMeHg in the interface water at WJD-1, WJD-2, WJD-3, and WJD-4 were 0.17 ± 0.074, 0.14 ± 0.085, 0.98 ± 1.5 and 2.5 ± 1.7 ng L^{-1}, respectively. The corresponding ratio of DHg as DMeHg (DMeHg/DHg) at WJD-1, WJD-2, WJD-3, and WJD-4 were 16 ± 8.2%, 10 ± 3.6%, 28 ± 28%, and 72 ± 23%, respectively. Generally, the interface water DMeHg concentrations and DMeHg/DHg were gradually increased from upstream (WJD-1 and WJD-2) to downstream of the WJD (WJD-3 and WJD-4).

There was no discernible difference in DMeHg levels between the interface water and the overlying water at WJD-1 and WJD-2 sites. The levels of DMeHg in the interface water represent maxima concentrations in seasonal vertical profiles at WJD-3 and WJD-4 sites, with the summer data of WJD-4 as an exception. Again, the baseline MeHg levels in interface water are probably due to the diffusion of MeHg from the surface layer of sediment (Furutani and Rudd 1980; Mason et al. 1993). However, the enhancement in DMeHg at WJD-3 during the entire period of sampling and at WJD-4 during winter, fall and spring demand for an additional

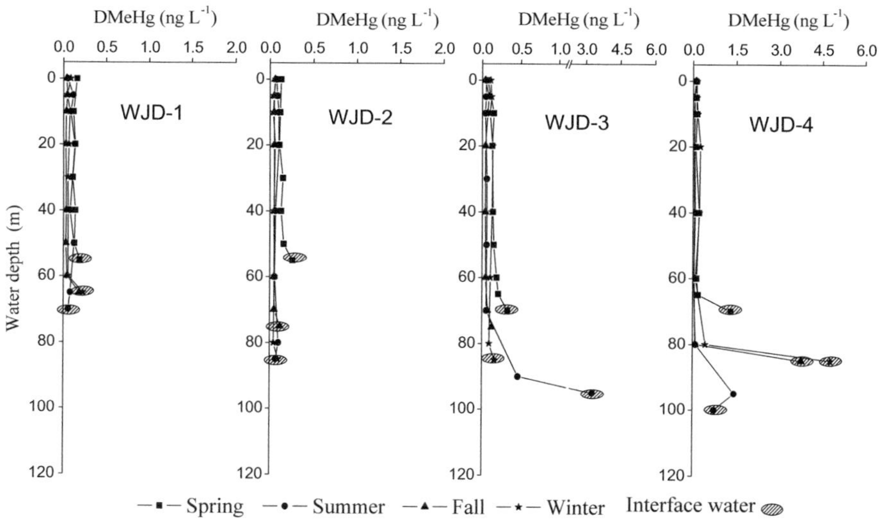

Fig. 6.81 Distributions of the dissolved methylmercury (DMeHg) in water column and interface water at four sampling stations (WJD-1, WJD-2, WJD-3, WJD-4) in WJD (redrawed from Meng et al. 2010, with permission from The Alliance of Crop, Soil, and Environmental Science Societies; redrawed from Meng et al. 2011, with permission from Chinese Journal of Ecology)

source, such as the in situ net Hg methylation and the diffusion of MeHg from the surface layer of sediment.

The maximum DMeHg were not positioned at interface water but at the bottom of water column (95 m) at WJD-4 station during summer. The implication is that MeHg in this region is not only due to diffusion of MeHg from surface sediment and/or the accumulation of settling particulate matter, but also from in situ methylation in anoxic water (Meili 1997; Eckley et al. 2005; He et al. 2008a). Diffusion of MeHg from the anoxic region (95 m) into the deep water is implied in this case.

MeHg data indicated that Hg methylation was present during all sampling seasons in WJD-4, which is in agreement with earlier observation (Guo 2008; Guo et al. 2008a, b; Feng et al. 2009b). The ratios of DMeHg/DHg in interface water (spring: 79%; summer: 39%; fall: 94%; winter: 86%) in WJD-4 were elevated compared to overlying water (18 $\pm$ 18%) with the exception of a summer maximum of 94% presenting at 95 m. The elevated proportions of DMeHg/DHg in interface water are probably a result of active Hg methylation and/or the dissolution of oxides and anaerobic decomposition of particulate organic matter (Eckley et al. 2005; He et al. 2008a).

(II) Sediment pore water

The spatial and seasonal patterns of DMeHg in the sediment pore water of the WJD are shown in Figs. 6.82.

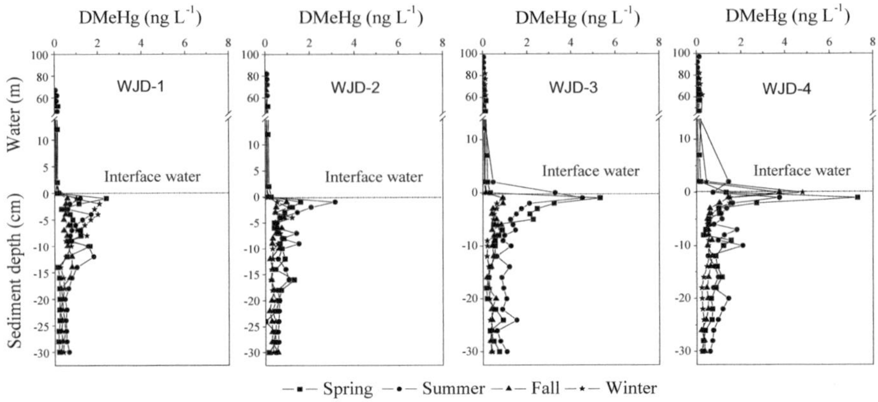

Fig. 6.82 Distributions of dissolved methylmercury (DMeHg) in water column and sediment pore water of the four sampling stations (WJD-1, WJD-2, WJD-3, WJD-4) in WJD (redrawed from Meng et al. 2016, with permission from John Wiley and Sons Inc.)

Similar to the distribution patterns of MeHg in the solid phase, DMeHg concentrations in pore water varied widely from 0.15 to 3.1 ng L^{-1}, from 0.15 to 5.3 ng L^{-1}, from 0.21 to 7.3 ng L^{-1}, and from 0.15 to 2.3 ng L^{-1} at WJD-2, WJD-3, WJD-4, and WJD-1 sites, respectively. The levels of DMeHg in pore water generally represented the highest concentration of surface sediment at WJD-2, WJD-3, and WJD-4 sites throughout the four sampling campaigns, typically coinciding with the peak concentrations of MeHg in the solid phase, and then declined gradually with the depth. This again suggested that the active net Hg methylation occurred in surface sediment. Moreover, considerable different levels of net Hg methylation process among WJD-1, WJD-2, WJD-3, and WJD-4 were further confirmed by the significant difference of pore water DMeHg at the surface layer. Seasonal distribution of DMeHg in pore water showed that the DMeHg concentrations in spring and summer were significantly higher than those in fall and winter (K–S test, $p < 0.001$) at WJD-2, WJD-3, and WJD-4. In comparison, DMeHg fluctuated throughout the sediment cores at WJD-1 site, without any clear distribution trends. Previous studies showed that oxic and alkaline conditions generally favor sediment uptake of MeHg, whereas anoxic and acidic conditions favor MeHg release (Ullrich et al. 2001). The solubility of MeHg in sediment under anoxic environment can be increased as a result of the formation of soluble sulfide complexes (Benoit et al. 1998). Therefore, it is reasonable to assume that the water characteristics in the stratified reservoir could be one of the important factors controlling the distribution of DMeHg in sediment between solid and liquid phase. Hence, we believed that seasonal distributions of DMeHg in pore waters at WJD-2, WJD-3, and WJD-4 may be linked with redox condition changes due to the seasonal stratification of WJD (Gill et al. 1999).

It is clear that DMeHg concentrations in interface water at all sampling sites were much lower than those in the pore water in surface sediment throughout the four sampling campaigns, with the data at WJD-4 in fall and winter as an exception. The slope of the DMeHg in vertical profile between surface sediment pore water and interface water indicated the positive diffusion of MeHg from sediment to water column. MeHg concentration in sediment pore water is usually much higher than that in the overlying water column (Ullrich et al. 2001). However, the values of DMeHg in interface water during fall and winter at WJD-4 were approximately 2.5–3 times higher than those in pore water at surface sediment. Moreover, it is interesting to note that the concentration of DMeHg in bottom water (95 m) at WJD-4 during summer was clearly higher than that in the interface water (sediment–water interface) as well as the overlying water, which agreed with the previous observation that MeHg levels in hypolimnetic waters of seasonally stratified reservoirs generally increase during summer stratification (Yan et al. 2013). These results implied that Hg methylation processes were much active above the sediment–water interface at WJD-4 sites during fall and winter.

There are two possible sources of MeHg contributed to the peak levels of DMeHg in the water column: in situ production controlled by redox chemistry and/or setting particulate matter containing MeHg (Meili 1997), and diffusion or resuspention, or both, of MeHg from underlying sediments (Lawson et al. 2001). The increased decomposition of organic matters in summer results in more anoxic conditions at surface sediments and hypolimnetic waters or/and interface water, which favored the net Hg methylation (Ullrich et al. 2001). Decreased dissolved oxygen concentrations and low pH in hypolimnetic water were detected during summer at WJD-4. Therefore, we speculated that the elevated MeHg concentrations in interface water during fall and winter as well as in bottom water during summer at WJD-4 was not only due to the redox-controlled release of MeHg from bottom sediments or/and the accumulation of settling particulate matters, but also related to in situ Hg methylation process in anoxic water (Bravo et al. 2014). Obvious peaks of MeHg concentrations in these layers suggested that the net MeHg production occurred both in the surface sediment and anoxic water layers (the bottom water and interface water). Maximum methylation rates usually occurred at the boundary between oxidized and anoxic conditions, which may vary seasonally and frequently coincide with the sediment–water interface at WJD-4 in WJD.

6.3.4 *Diffusion Flux of Inorganic Mercury and Methylmercury to Water*

The spatial and seasonal distributions of dissolved inorganic Hg (IHg) and dissolved methylmercury (DMeHg) in the water at the sediment–water interface and surface sediment pore water in WJD are listed in Tables 6.19 and 6.20.

The diffusion fluxes of both IHg and MeHg from sediment to water column throughout the four sapling campaigns were calculated and presented in Figs. 6.83 and 6.84, and the summary data are shown in Table 6.21. Estimated diffusion fluxes

Table 6.19 Seasonal and spatial distribution of dissolved inorganic mercury (IHg) in surface sediment pore water and in interface water of WJD (ng L^{-1})

Sampling sites	Interface water				Surface sediment porewater			
	Spring	Summer	Fall	Winter	Spring	Summer	Fall	Winter
WJD-1	0.60	1.14	0.77	1.37	1.22	9.71	2.63	1.96
WJD-2	1.48	0.76	1.00	1.40	4.46	1.43	5.78	9.96
WJD-3	1.42	1.40	0.98	1.24	4.16	18.45	3.97	1.86
WJD-4	0.35	1.17	0.23	1.30	0.94	19.77	1.78	3.56

Table 6.20 Seasonal and spatial distribution of dissolved methylmercury (DMeHg) in surface sediment pore water and in interface water of WJD (ng L^{-1})

Sampling sites	Interface water				Surface sediment porewater			
	Spring	Summer	Fall	Winter	Spring	Summer	Fall	Winter
WJD-1	0.19	0.06	0.19	0.23	2.4	1.18	0.58	1.00
WJD-2	0.26	0.07	0.12	0.098	1.58	3.13	0.95	0.52
WJD-3	0.33	3.30	0.13	0.16	5.32	4.52	0.89	0.88
WJD-4	1.32	0.74	3.7	4.78	7.31	3.96	1.42	1.50

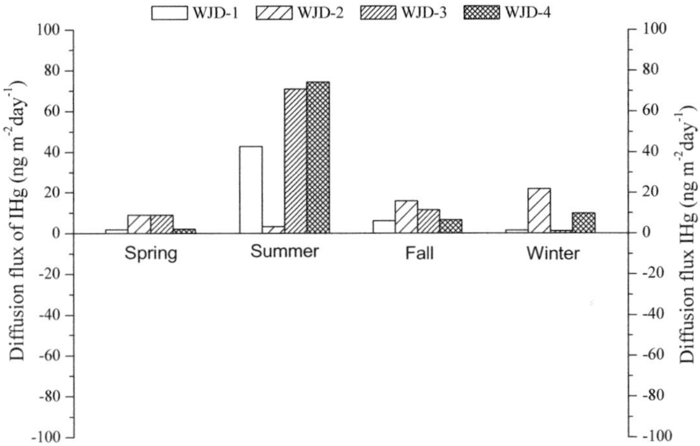

Fig. 6.83 Estimated diffusion fluxes of inorganic mercury (IHg) over the surface sediment and water column in WJD

of IHg and MeHg ranged from 1.5 to 73 ng m^{-2} day^{-1} and from −19 to 29 ng m^{-2} day^{-1}, respectively. The IHg flux from sediment to water column in WJD was comparable with the reported values from DF (ranging from 41 to 63 ng m^{-2} day^{-1}) and WJD (ranging from 44 to 65 ng m^{-2} day^{-1}) reservoirs estimated in 2004 (Feng et al. 2009a) as well as at Mugu Lagoon (ranging from −0.49 to 75 ng m^{-2} day^{-1}; Rothenberg et al. 2008), while these values were lower than IHg fluxes in Lavaca bay (ranging from 0.1 to 140 ng m^{-2} day^{-1}; Gill et al. 1999).

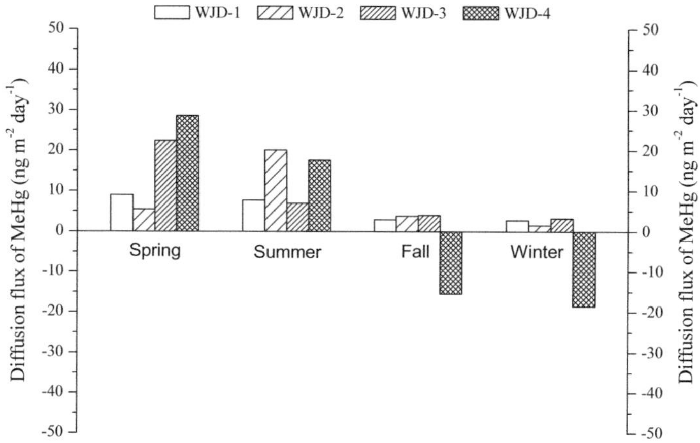

Fig. 6.84 Estimated diffusion fluxes of methylmercury (MeHg) over the surface sediment and water column in WJD

Table 6.21 Estimated diffusion fluxes of inorganic mercury (IHg) and methylmercury (MeHg) across sediment to water column in WJD (ng m^{-2} day^{-1})

Sampling sites	IHg				MeHg			
	Spring	Summer	Fall	Winter	Spring	Summer	Fall	Winter
WJD-1	1.84	42.8	6.00	1.51	8.97	7.68	1.73	2.7
WJD-2	8.98	3.21	15.8	21.8	5.45	20.1	3.76	1.47
WJD-3	8.99	71.0	11.4	2.01	22.4	6.95	3.96	3.2
WJD-4	2.06	74.3	6.46	9.75	28.6	17.6	−15.5	−18.6

IHg fluxes from sediment to interface water at WJD-1, WJD-2, WJD-3, and WJD-4 were all positive throughout the four sampling campaigns, indicating that the sediment was the net source of IHg to water column. The highest value of IHg flux was observed in summer, with the exception of WJD-3 site. This may be resulted from the increased solubility of IHg during summer. However, the spatial distribution trends of IHg flux at WJD were not consistent.

MeHg fluxes reported at WJD were higher than the values observed at Mugu Lagoon (ranging from 0.14 to 5.3 ng m^{-2} day^{-1}; Rothenberg et al. 2008) and in four wetlands along the St. Lawrence River, Canada (ranging from −1.6 to 10 ng m^{-2} day^{-1}; Holmes and Lean 2006), but much lower than those reported for other system, for example, at the Gulf of Trieste (mean values of 380 ng m^{-2} day^{-1}; Covelli et al. 1999) and at the Lavaca Bay, Texas (ranging from 0.1 to 1700 ng m^{-2} day^{-1}; Gill et al. 1999). Negative values were observed at WJD-4 during fall and winter campaigns, indicating the concentration of MeHg in interface water were higher than those in surface sediment pore water (Table 6.20). This phenomenon can be explained by the occurrence of active Hg methylation above the sediment–water interface. Due to the fact that the concentrations of MeHg in interface water at WJD-4 in fall and winter were higher than those in overlying water, MeHg can be diffused from interface water to overlying water as a result of concentration gradients. If MeHg fluxes were estimated based on the concentration gradients between MeHg concentrations in the interface water and the overlying water (bottom of water column), the values of MeHg flux across the interface water to water column at WJD-4 can reach up to 24 and 25 ng m^{-2} day^{-1} during fall and winter, respectively. Hence, we can see that the surface sediments at WJD-1, WJD-2, and WJD-3 were a net source of MeHg to the overlying water throughout the four sampling campaigns.

For the sampling site of WJD-4, we can see that MeHg can not only be produced in sediment but also in the hypolimnetic water (especially the interface water). The maximum MeHg fluxes at each of the sampling sites were all observed in spring or summer. Statistical analyses showed that the MeHg fluxes in spring and summer were significantly higher than those in fall and winter throughout the four sampling sites (K–S test, $p < 0.001$), which is in an agreement with our earlier studies (Feng et al. 2009a; Guo 2008). Moreover, the average diffusion fluxes of MeHg in spring and summer were higher in WJD-3 and WJD-4 than those in WJD-1 and WJD-2, indicating that more active net Hg methylation occurred at downstream of WJD than at the upstream sections. MeHg diffusion from the sediment is an important source of MeHg to the water column which supported the finding that WJD is a net MeHg production source (Guo 2008). Moreover, the diffusion flux data indicated that Hg methylation was present during all the sampling seasons at WJD-4, which is also supported by our previous observations (Guo 2008; Feng et al. 2009a, b).

The annual overall diffusion fluxes of IHg and MeHg from sediment to the water column in WJD were qualified. The surface area of WJD is 47.8×10^6 m^2. We assumed that each sampling site represents a quarter of the surface area of the reservoir, and the surface area of the reservoir is the same as the area of the surface sediment. The annual overall IHg and MeHg diffusion fluxes in WJD are 314 and 109 g $year^{-1}$, respectively, which were somewhat different from the mass balance calculation (THg 156 g $year^{-1}$; MeHg 872 g $year^{-1}$) concluded in Chap. 7, but the differences were within one order of magnitude.

The net MeHg yields based on the mass balance calculation is about eight times higher than the annual overall MeHg diffusion fluxes data. On one hand, the WJD is located in the karstic environment of the Wujiang River Basin, which is a typical bioirrigation gorge. Hence, the total area of surface sediment may be much larger than the surface area of the reservoir. On the other hand, the active Hg methylation occurred predominantly in surface sediments and extended to the bottom of the water column. It is, therefore, should be emphasized that water column methylation is potentially very important. Moreover, the overall diffusive flux of MeHg could be enhanced if bioirrigation and bioturbation of pore water are considered. Elevated THg concentrations at bottom water in WJD were observed throughout the four sampling sites, which may be attributed to the positive value of IHg flux. However, probably due to the large water volume (2.3×10^8 m^3) and short water residence time (81 days, see detail in Chap. 7) of WJD, the influence of IHg flux on the distributions of THg in water is not significant. In all, our study confirmed the active Hg methylation occurred in WJD, especially in downstream of reservoir, which may pose a potential threat to the reservoir system and downstream aquatic ecosystems.

References

Abernathy AR, Cumbie PM (1977) Mercury accumulation by largemouth bass (Micropterus salmoides) in recently impounded reservoirs. Bull Environ Contam Toxicol 17:595–602

Amyot M, Lean D, Mierle G (1997) Photochemical formation of volatile mercury in high Arctic lakes. Environ Toxicol Chem 16:2054–2063

Andersson I, Parkman H, Jernelov A (1990) The role of sediments as sink or source for environmental contaminants: a case study of mercury and chlorinated organic compounds. Limnologica 20:347–359

Bai WY (2006) The primary study on the distributions and transformations of the different species mercury in Aha reservoir. Institute of Geochemistry Chinese Academy of Sciences, Beijing, China (in Chinese, with English abstract)

Barkay T, Gillman M, Turner RR (1997) Effects of dissolved organic carbon and salinity on bioavailability of mercury. Appl Environ Microbiol 63:4267–4271

Benoit JM, Gilmour CC, Mason RP, Riedel GS, Riedel GF (1998) Behavior of mercury in the Patuxent River estuary. Biogeochemistry 40:249–256

Bishop KH, Lee YH (1997) Catchments as a source of mercury/methylmercury in boreal surface waters. In: Sigel A, Sigel H (eds) Metal ions in biological systems, vol 34—mercury and its effect on environment and biology. Marcel Dekker, New York, NY, USA, pp 113–130

Bloom NS (1992) On the chemical form of mercury in edible fish and marine invertebrate tissue. Can J Fish Aquat Sci 49:1010–1017

Bloom NS, Effler SW (1990) Seasonal variability in mercury speciation of Onondaga Lake. Water Air Soil Pollut 53:251–265

Bloom NS, Gill GA, Cappellino S, Dobbs C, Mcshea L, Driscoll C (1999) Speciation and cycling of mercury in Lavaca Bay, Texas, sediments. Environ Sci Technol 33:7–13

Bodaly RA, Hecky RE, Fudge RJP (1984) Increases in fish mercury levels in lakes flooded by the Churchill River diversion, northern Manitoba. Can J Fish Aquat Sci 41:682–691

Bodaly RA, Rudd JWM, Fudge RJP, Kelly CA (1993) Mercury concentrations in fish related to size of remote Canadian Shield lakes. Can J Fish Aquat Sci 50:980–987

Bodaly RA, Beaty KG, Hendzel LH, Majewski AR, Paterson MJ, Rolfhus KR, Penn AF, St. Louis VL, Hall BD, Matthews CJD, Cherewyk KA, Mailman M, Hurley JP, Schiff SL, Venkiteswaran JJ (2004) Experimenting with hydroelectric reservoirs. Environ Sci Technol 38:347–352

Bonzongo JC, Heim KJ, Warwick JJ, Lyons WB (1996) Mercury levels in surface waters of the Carson river-Lahontan Reservoir system, Nevada: influence of historic mining activities. Environ Pollut 92(2):193–201

Boudreau BP (1996) The diffusive tortuosity of fine-grained unlithified sediments. Geochim Cosmochim Acta 60:3139–3142

Bravo AG, Cosio C, Amourou D, Zopfi J, Cheualley PA, Spangenberg JE, Ungureanu VG, Dominik J (2014) Extremely elevated methyl mercury levels in water, sediment and organisms in a Romanian reservoir affected by release of mercury from a chlor-alkali plant. Water Res 49:391–405

Brigham ME, Krabbenhoft DP, Olson ML, Dewild JF (2002) Methylmercury in flood-control inpoundment and natural waters of Northwestern Minnesota, 1997–1999. Water Air Soil Pollut 138:61–78

China National Environmental Monitoring Center, National Environmental Protection Agency (1990) The Atlas of Soil Environmental Background Value in the People's Republic of China. China Environmental Science, Beijing, China
China Environmental Monitoring Station editor (1992) Presided over by State Environmental Protection Administration. Background value of soil element in China. China Environmental Science Press, Beijing, pp 1–195 (in Chinese)
Cossa D, Gobeil C (2000) Mercury speciation in the Lower St. Lawrence estuary. Can J Fish Aquat Sci 57:38–147
Covelli S, Faganeli J, Horvat M, Brambati A (1999) Pore water distribution and benthic flux measurements of mercury and methylmercury in the Gulf of Trieste (Northern Adriatic Sea). Estuar Coast Shelf Sci 48:415–428
Cox JA, Carnahan J, Dinuzio J, Mccoy J, Meister J (1979) Source of mercury in fish in new impoundments. Bull Environ Contam Toxicol 23:779–783
Dang AZ (2008) The impact of cascade exploitation on the phytoplankton in the upper reaches of the Wujiang River. Master dissertation. Guizhou Normal University, Guiyang, China (in Chinese, with English abstract)
Driscoll CT, Blette V, Yan C, Schofield CL, Munson R, Holsapple J (1995) The role of dissolved organic carbon in the chemistry and bioavailability of mercury in remote Adirondack lakes. Water Air Soil Pollut 80:499–508
Eckley CS, Hintelmann H (2006) Determination of mercury methylation potentials in the water column of lakes across Canada. Sci Total Environ 368:111–125
Eckley CS, Watras CJ, Hintelmann H (2005) Mercury methylation in the hypolimnetic waters of lakes with and without connection to wetlands in northern Wisconsin. Can J Fish Aquat Sci 62:400–411
Environmental Impact Assessment Report (1987) Guiyang Engineering Corporation Limited, pp 1–98 (in Chinese)
Feng XB, Li GH, Qiu GL (2006) A preliminary study on mercury contamination to the environment from artisanal zinc smelting using indigenous methods in Hezhang country, Guizhou, China: part 2. Mercury contaminations to soil and crop. Sci Total Environ 368:47–55
Feng XB, Jiang HM, Qiu GL, Yan HY, Li GH, Li ZG (2009a) Geochemical processes of mercury in Wujiangdu and Dongfeng reservoirs, Guizhou, China. Environ Pollut 157:2970–2984
Feng XB, Jiang HM, Qiu GL, Yan HY, Li GH, Li ZG (2009b) Mercury mass balance study in Wujiangdu and Dongfeng reservoirs, Guizhou, China. Environ Pollut 157:2594–2607
Fjeld E, Rognerud S (1993) Use of path-analysis to investigate mercury accumulation in brown trout (Salmo trutta) in Norway and the influence of environmental factors. Can J Fish Aquat Sci 50:1158–1167
Fleming EJ, Mack EE, Green PG, Nelson DC (2006) Mercury methylation from unexpected sources: molybdate-inhibited freshwater sediments and an iron-reducing bacterium. Appl Environ Microbiol 72:457–464
French KJ, Scruton DA, Anderson MR (1999) Influence of physical and chemical characteristics on mercury in aquatic sediments. Water Air Soil Pollut 110:347–362
Fujiki M, Tajima S (1992) The pollution of Minamata Bay by mercury. Water Technol 25: 133–140
Furutani A, Rudd JW (1980) Measurement of mercury methylation in lake water and sediment samples. Appl Environ Microbiol 40:770–776
Gill GA, Bruland KW (1990) Mercury speciation in surface freshwater systems in California and other areas. Environ Sci Technol 24:1392–1400
Gill GA, Bloom NS, Cappellino S, Driscoll CT (1999) Sediment-water fluxes of mercury inLavaca Bay, Texas. Environ Sci Technol 33:663–669
Gilmour CC, Henry EA (1991) Mercury methylation in aquatic systems affected by acid deposition. Environ Pollut 71:131–169
Gilmour CC, Riedel GS, Ederington MC, Bell JT, Benoit JM, Gill GA, Stordal MC (1998) Methylmercury concentrations and production rates across a trophic gradient in the northern Everglades. Biogeochemistry 40:327–345

Goulet RR, Holmes J, Page B, Poissant L, Siciliano SD, Lean DRS, Wang F, Amyot M, Tessier A (2007) Mercury transformations and fluxes in sediments of a riverine wetland. Geochim Cosmochim Acta 71:3393–3406

Grace MV, Mason RP, Fitzgerald WF (1991) Cycling of volatile mercury in temperate lake. Water Air Soil Pollut 56:791–807

Graham AM, Aiken GR, Gilmour CC (2012) Dissolved organic matter enhances microbial mercury methylation under sulfidic conditions. Environ Sci Technol 46:2715–2723

Gray JE, Fey DL, Holmes CW, Lasorsa BK (2005) Historical deposition and fluxes of mercury in Narraguinnep Reservoir, southwestern Colorado, USA. Appl Geochem 20:207–220

Guizhou Statistical Yearbook (2008) National Bureau of statistics of the People's Republic of China. China Statistics Press, Beijing, pp 1–564 (in Chinese)

Guo YN (2008) Input and output fluxes of mercury in different evolutive reservoirs in Wujiang River Basin. Ph.D. dissertation. Graduate School of the Chinese Academy of Sciences, Beijing (in Chinese, with English abstract)

Guo YN, Feng XB, He TR, Zhang JF, Liang P, Meng B, Yao H (2008a) Temporal and spatial distribution of different mercury species in precipitation of Wujiang River Basin. Acta Sci Circumst 28:1441–1446 (in Chinese, with English abstract)

Guo YN, Feng XB, Li ZG, He TR, Yan HY, Meng B, Zhang JF, Qiu GL (2008b) Distribution and wet deposition fluxes of total and methylmercury in Wujiang River Basin, Guizhou, China. Atmos Environ 42:7096–7103

Hall BD, St. Louis VL, Rolfhus KR, Bodaly RA, Beaty KG, Paterson MJ, Cherewyk KAP (2005) Impact of reservoir creation on the biogeochemical cycling of methyl and total mercury in boreal upland forests. Ecosystems 8:248–266

He TR (2007) Biogeochemical cycling of mercury in Hongfeng Reservior, Guizhou, China. Ph.D. dissertation in the Graduate School of the Chinese Academy of Sciences (GSCAS), Beijing (in Chinese, with English abstract)

He TR, Feng XB, Guo YN, Li ZG, Qiu GL, Liang L (2008a) Geochemical cycling of mercury in the sediment of Hongfeng Reservoir. Environ Sci 29:1768–1774 (in Chinese, with English abstract)

He TR, Feng XB, Guo YN, Qiu GL, Li ZG, Liang L, Julia L (2008b) The impact of eutrophication on the biogeochemical cycling of mercury species in a reservoir: a case study from Hongfeng Reservoir, Guizhou, China. Environ Pollut 154:56–67

Hecky RE, Bodaly RA, Ramsey DJ (1986) Enhancement of mercury bioaccumulation in fish by flooded terrestrial material ecosystem. Appendix no. 6. Canada-Manitoba agreement on the study and monitoring of mercury in the Churchill River diversion, Winnipeg, Manitoba

Hecky RE, Ramsey DJ, Bodaly RA, Strange NE (1991) Increased methylmercury contamination in fish in newly formed freshwater reservoirs. In: Suzuki T, Imura N, Clarkson TW (eds) Advances in mercury toxicology. Plenum Press, New York, pp 33–52

Hines NA, Brezonik PL, Engstrom DR (2004) Sediment and porewater profiles and fluxes of mercury and methylmercury in a small seepage lake in northern Minnesota. Environ Sci Technol 38:6610–6617

Holmes J, Lean D (2006) Factors that influence methylmercury flux rates from wetland sediments. Sci Total Environ 368:306–319

Hudson RJM, Gherini SA, Watras CJ, Porcella DP (1994) Modeling the biogeochemical cycle of mercury in lakes: the Mercury Cycling Model (MCM) and its application to the MTL study lakes. In: Watras CJ, Huchabee JW (eds) Mercury pollution-integration and synthesis. Lewis Publishers, Boca Raton, FL

Hurley JP, Watras CJ, Bloom NS (1991) Mecury cycling in a northern Wisconin seepage lake—the role of particulate matter in vertical transport. Water Air Soil Pollut 56:543–551

Hurley JP, Watras CJ, Bloom NS (1994) Distribution and flux of particulate mercury in four stratified seepage lakes. In: Mercury pollution-integration and synthesis, pp 69–82

Jackson TA (1988) The mercury problem in recently formed reservoirs of northern Manitoba (Canada): effects of impoundment and other factors on the production of methylmercury by microorganisms in sediments. Can J Fish Aquat Sci 45:97–121

Jiang HM (2005) Effects of hydroelectric reservoir on the biogeochemical cycle of mercury in the Wujiang River. Ph.D. dissertation. Graduate School of the Chinese Academy of Sciences, Beijing (in Chinese, with English abstract)

Jiang YK (2007) Sulfur isotope geochemistry and carbonate weathering in karst catchment. Ph.D. dissertation. Graduate School of the Chinese Academy of Sciences, Beijing (in Chinese, with English abstract)

Jiang YK, Liu CQ, Tao FX (2006) Sulfur isotopic compositions of Wujiang River water in Guizhou Province during low-flow period. Geochimica 35:623–628 (in Chinese, with English abstract)

Kannan K, Smith RG Jr, Lee RF, Windom HL, Heitmuller PT, Macauley JM, Summers JK (1998) Distribution of total mercury and methylmercury in water, sediment, and fish from South Florida Estuaries. Arch Environ Contam Toxicol 34:109–118

Kelly CA, Rudd JWM, Louis VL, Heyes A (1995) Is total mercury concentration a good predictor of methyl mercury concentration in aquatic systems? Water Air Soil Pollut 80:715–724

Kelly CA, Rudd JWM, Bodaly RA, Roulet NP, St. Louis VL, Heyes A, Moore TR, Schiff S, Aravena R, Scott KJ, Dyck B, Harris R, Warner B, Edwards G (1997) Increases in fluxes of greenhouse gases and methylmercury following flooding of an experimental reservoir. Environ Sci Technol 31:1334–1344

Kimmel BL, Groeger AW (1984) Factors controlling primary production in lakes and reservoirs: a perspective. Lake Reservoir Manage 1:277–281

Kotnik J, Horvat M, Tessier E, Ogrinc N, Monperrus M, Amouroux D, Fajon V, Gibičar N, Žižek S, Sprovieri F, Pirrone N (2007) Mercury speciation in surface and deep waters of the Mediterranean Sea. Mar Chem 107:13–30

Krom MD, Berner RA (1980) Adsorption of phosphate in anoxic marine sediment. Geochim Cosmochim Acta 45:207–216

Lambertsson L, Nilsson M (2006) Organic material: The primary control on mercury methylation and ambient methyl mercury concentrations in estuarine sediments. Environ Sci Technol 40:1822–1829

Lanzillotta E, Ceccarini C, Ferrara R (2002) Photo-induced formation of dissolved gaseous mercury in coastal and offshore seawater of the Mediterranean basin. Sci Total Environ 300:179–187

Lawson NM, Mason RP, Laporte JM (2001) The fate and transport of mercury, methylmercury, and other trace metals in Chesapeake Bay Tributaries. Water Res 35:501–515

Lee YH, Hultberg H (1990) Methylmercury in some Swedish surface waters. Environ Toxicol Chem 9:833–841

Lerman A (1979) Geochemical processes-water and sediment environments. Wiley, New York, USA

Liang XB, Zhu JM, Liu CQ, Wei ZQ, Wang FS, Wan GJ, Huang RG (2003) Enzymatic and microbial degreadation of organic matter in Lake Hongfeng of Guizhou Province. Quat Sci 23:565–572 (in Chinese, with English abstract)

Lindqvist O, Johansson K, Aastrup M, Andersson A, Bringmark L, Hovsenius G, Hikanson L, Iverfeldt A, Mcili M, Timm B (1991) Mercury in the Swedish environment-recent research on causes, consequences and corrective methods. Water Air Soil Pollut 55:143–177

Liu GL, Cai Y, Philippi T, Kalla P, Scheidt D, Richards J, Scinto L, Appleby C (2008) Distribution of total and methylmercury in different ecosystem compartments in the everglades: implications mercury bioaccumulation. Environ Pollut 153:257–265

Liu et al (2009) Cycling of nutrients in soil-plant systems of karstic environment, southwest China. Science Press Ltd., Beijing, pp 1–90 (in Chinese, with English abstract)

Liu CQ (2009) Biogeochemical processes and cycling of nutrients in the earth's surface: cycling of nutrients in soil–plant systems of karstic environments, Southwest China. Science Press, Beijing, China

Lu HS, Liang LB, Xu YQ (1999) A study of seasonal change of phytoplankton in Fujian Gutian Reservoir and evaluation of its water quality. J Fujian Teach Univ 15:95–99 (in Chinese, with English abstract)

Lucotte M, Schetagne R, Therien N, Langlois C, Tremblay A (1999) Mercury in the biogeochemical cycle—natural environments and hydroelectric reservoirs of Northern Quebec (Canada). Springer, pp 1–334

Lyons WB, Fitzgibbon TO, Welch KA, Carey AE (2006) Mercury geochemistry of the Scioto River, Ohio: impact of agriculture and urbanization. Appl Geochem 21:1880–1888

Mason RP, Fitzgerald WF (1990) Alkylmercury species in the equatorial Pacific. Nature 347: 457–459

Mason RP, Sullivan KA (1999) The distribution and speciation of mercury in the South and equatorial Atlantic. Deep Sea Res Part II Top Stud Oceanogr 46:937–956

Mason RP, Sveinsdottir AY (2003) Mercury and methylmercury concentrations in water and largemouth bass in Maryland Reservoirs. Maryland Department of Natural Resources, Annapolis, MD

Mason RP, Fitzgerald WF, Hurley JP, Hanson JAK, Donaghay PL, Sieburth JM (1993) Mercury biogeochemical cycling in a stratified estuary. Limnol Oceanogr 38:1227–1241

McMurtry MJ, Wales DL, Scheider WA, Beggs GL, Dimond PE (1989) Relationship of mercury concentrations in lake trout and smallmouth bass to the physical and chemical characteristics of Ontario lakes. Can J Fish Aquat Sci 46:426–434

Meili M (1997) Mercury in lakes and Rivers. In: Sigel A, Sigel H (eds) Metal ions in biological systems: mercury and its effect on environment and biology, vol 34. Marcel Dekker Inc., New York, pp 21–51

Meng B, Feng XB, Chen CC, Qiu GL, Sommar J, Guo YN, Liang P, Wan Q (2010) Influence of eutrophication on the distribution of total mercury and methylmercury in hydroelectric reservoirs. J Environ Qual 39:1624–1635

Meng B, Feng XB, Chen CX, Qiu GL, Guo YN, Liu K, Yao H, Zhang JF (2011) Distribution of total mercury and methylmercury in two hydroelectric reservoirs of Guizhou Province, China. Chin J Ecol 30:951–960 (in Chinese, with English abstract)

Meng B, Feng XB, Qiu GL, Li ZG, Yao H, Shang LH, Yan HY (2016) The impacts of organic matter on the distribution and methylation of mercury in a hydroelectric reservoir in Wujiang River, Southwest China. Environ Toxicol Chem 35:191–199

Meuleman C, Leermakers M, Baeyens W (1995) Mercury speciation in Lake Baikal. Water Air Soil Pollut 80:539–551

Miskimmin BM, Rudd JWM, Kelly CA (1992) Influence of dissolved organic carbon, pH, and microbial respiration rates on mercury methylation and demethylation in lake water. Can J Fish Aquat Sci 49:17–22

Mitchell CPJ, Branfireun BA, Kolka RK (2008) Spatial characteristics of net methylmercury production hot spots in peatlands. Environ Sci Technol 42:1010–1016

Porvari P, Verta M (2003) Total and methyl mercury concentrations and fluxes from small boreal forest catchments in Finland. Environ Pollut 123:181–191

Ramlal PS, Anema C, Furutani A, Hecky RE, Rudd JWM (1987) Mercury methylation and demethylation studies at Southern Indian Lake, Manitoba: 1981–1983. Can Tech Rep Fish Aquat Sci, vol 1490

Ramlal PS, Kelly CA, Rudd JW, Furutani A (1993) Sites of methyl mercury production in remote Canadian Shield lakes. Can J Fish Aquat Sci 50:972–979

Ramsey DJ. (1990) Experimental studies of mercury dynamics in the Churchill River diversion, Manitoba, CSEB, 1990—managing the effects of hydroelectric developments. In: Delisie CE, Bouchard MA (eds) Proceeding of a symposium. Collection Environment et Géologie, vol 9. Université de Montréal, Montréal, Québec, 6–7 Apr 1989, p 650

Ravichandran M, Aiken GR, Reddy MM, Ryan JN (1998) Enhanced dissolution of innabar (mercury sulfide) by dissolved. organic matter isolated from the Florida Everglades. Environ Sci Technol 32:3305–3311

Reimers RS, Krenkel PA (1974) Kinetics of mercury adsorption and desorption in. sediments. J Water Pollut Control Fed 46:352–365

Rothenberg SE, Ambrose RF, Jay JA (2008) Mercury cycling in surface water, pore water and sediments of Mugu Lagoon, CA, USA. Environ Pollut 154:32–45

Rudd JWM (1995) Sources of methyl mercury to freshwater ecosystems: a review. Water Air Soil Pollut 80:697–713
Schetagne R, Doyo JF, Fournier JJ (2000) Export of mercury downstream from reservoirs. Sci Total Environ 260:135–145
St. Louis VL, Rudd JWM, Kelly CA, Beaty KG, Bloom NS, Flett RJ (1994) Importance of wetlands as sources of methyl mercury to boreal forest ecosystems. Can J Fish Aquat Sci 51: 1065–1076
St. Louis VL, Rudd JWM, Kelly CA, Bodaly RAD, Paterson MJ, Beaty KG, Hesslein RH, Heyes AW, Majewski AR (2004) The rise and fall of mercury methylation in an experimental reservoir. Environ Sci Technol 38:1348–1358
Steven DS, Nelson JOD, Lean RS (2002) Microbial reduction and oxidation of mercury in freshwater lakes. Environ Sci Technol 36:3064–3068
Sunderland EM, Gobas FAPC, Branfireun BA, Heyes A (2006) Environmental controls on the speciation and distribution of mercury in coastal sediments. Mar Chem 102:111–123
Tremblay A, Lucotte M, Rheault I (1996) Methylmercury in a benthic food web of two hydroelectric reservoirs and a natural lake of northern Quebec (Canada). Water Air Soil Pollut 91:255–269
Tseng CM, Lamborg C, Fitzgerald WF, Engstromd R (2004) Cycling of dissolved elemental mercury in Arctic Alaskan lakes. Geochim Cosmochim Acta 68:1173–1184
Ullrich SM, Tanton TW, Abdrashitova SA (2001) Mercury in the aquatic environment: a review of factors affecting methylation. Crit Rev Environ Sci Technol 31:241–293
U.S. EPA (1992) Water quality standards; establishment of numeric criteria for priority toxic pollutants; states' compliance; final rule. Fed. Regist. USEPA, 40 CFR Part 131, 57/246, 60847–60916
Verdon R, Brouard D, Demers C, Lalumiere R, Laperle M, Schetagne R (1991) Mercury evolution (1978–1988) in fishes of the La Grande hydroelectric complex, Quebec, Canada. Water Air Soil Pollut 56:405–417
Wang YC (2005) Thermal stratification and paroxysmal deterioration of water quality in Canyon-reservoir, Southwestern China. J Lake Sci 17:54–60 (in Chinese, with English abstract)
Wang PJ, Gu YH, Li DH (1992) The discussion on the state of mercury in a water body in Baihua Lake Reservoir. J Guizhou Normal Univ 10:29–32 (in Chinese, with English abstract)
Watras CJ, Morrison KA, Host JS (1995) Bloom N.S. concentration of mercury species in relationship to other site-specific factors in the surface waters of northern Wisconsin lakes. Limnol Oceanogr 40:556–565
Weber JH (1993) Review of possible paths for abiotic methylation of mercury (II) in the aquatic environment. Chemosphere 26:2063–2077
Winfrey MR, Rudd JWM (1990) Environmental factors affecting the formation of methylmercury in low pH lakes. Environ Toxicol Chem 9:853–869
Yan HY (2005) Establishment of analytical methods for different mercury species in environmental samples, and a preliminary study on the biogeochemical cycling of mercury in Baihua Lake, Guizhou Province. Ph.D. dissertation in the Graduate School of the Chinese Academy of Sciences (GSCAS), Beijing (in Chinese, with English abstract)
Yan HY, Feng XB, Shang LH, Qiu GL, Dai QJ, Wang SF, Hou YM (2008) The variations of mercury in sediment profiles from a historically mercury-contaminated reservoir, Guizhou province, China. Sci Total Environ 407:497–506
Yan HY, Li QH, Meng B, Wang CP, Feng XB, He TR, Dominik J (2013) Spatial distribution and methylation of mercury in a eutrophic reservoir heavily contaminated by mercury in Southwest China. Appl Geochem 33:182–190
Yao H, Feng XB, Guo YN, Meng B (2009) Spatial and temporal distribution of methylmercury in two new reservoirs on the upper course of Wujiang River. Resour Environ Yangtze Basin 18:343–349 (in Chinese, with English abstract)
Yao H, Feng XB, Qiu GL, Yan HY (2011) Distribution of different mercury forms of Hongjiadu Reservoir from Guizhou Province. Chin J Ecol 30:961–968 (in Chinese, with English abstract)

Zhang JF, Feng XB, Yan HY, Guo YN, Yao H, Meng B, Liu K (2009a) Seasonal distributions of mercury species and their relationship to some physicochemical factors in Puding Reservoir, Guizhou, China. Sci Total Environ 408:122–129

Zhang JF, Feng XB, Yan HY, Li P, Guo YN, Yao H, Meng B (2009b) Preliminary study on methylmercury distribution in Yelanghu Reservoir. Earth Environ 37:293–298 (in Chinese, with English abstract)

Zhu J, Liu CQ, Wang YC, Li SL, Li J (2006) Spatial-temporal variation of dissolved silicon in Wujiangdu reservoir. Adv Water Sci 17:330–333 (in Chinese, with English abstract)

Chapter 7
Biogeochemical Cycling of Mercury in the Hongfeng, Baihua, and Aha Reservoirs

Abstract To understand the biogeochemical process of mercury (Hg) in reservoir in Wujiang River Basin, Southwest China, three reservoirs including Hongfeng Reservoir (HF), Baihua Reservoir (BH), Aha Reservoir (AH), and Wujiang River were selected in this study. The primary objectives of this chapter were seasonal variations of Hg species in different sectors of reservoirs (e.g., water column, river water, sediment, and pore water), processes of Hg methylation and their possible controlling factors in the reservoirs in the Wujiang River Basin, in the branch of, (1) to investigate spatial and (2) to reveal the Southwest China.

Keywords Reservoir · Mercury distributions · Drinking water area · Inflows/outflow

7.1 Introduction

The Hongfeng Reservoir (HF) and Baihua Reservoir (BH) were built on the Maotiao River, a major area of the south bank of the Wujiang River. The Aha Reservoir (AH) was built on the Nanming River in the secondary upper reaches of the south bank. These water bodies serve as important water sources for the city of Guiyang and are referred to as "two lakes and one reservoir," supplying 60% of all drinking water to 1.2 million people (Fig. 7.1). Human activities (e.g., industry, agriculture, and municipal sewage discharge) have led to declining water quality levels and to gradual eutrophication. Currently, water quality levels only reach Grade III in the AH. In some sections of the HF and BH, algae grow frequently and the water quality levels have declined considerably, spurring widespread concern among residents and government officials.

The Hongfeng Reservoir, which has a surface area of 57.2 km^2 and volume of 6.01×10^8 m^3, was dammed in 1960 and is located 28 km into the suburbs of the city of Guiyang, Guizhou Province, southwestern China (Fig. 7.1). The reservoir was constructed for hydroelectric power generation, flood control, tourism, drinking water, and fishing. There are nearly two-dozen factories in the drainage area of the

X. Feng et al., *Biogeochemical Cycle of Mercury in Reservoir Systems in Wujiang River Basin, Southwest China*, https://doi.org/10.1007/978-981-10-6719-8_7

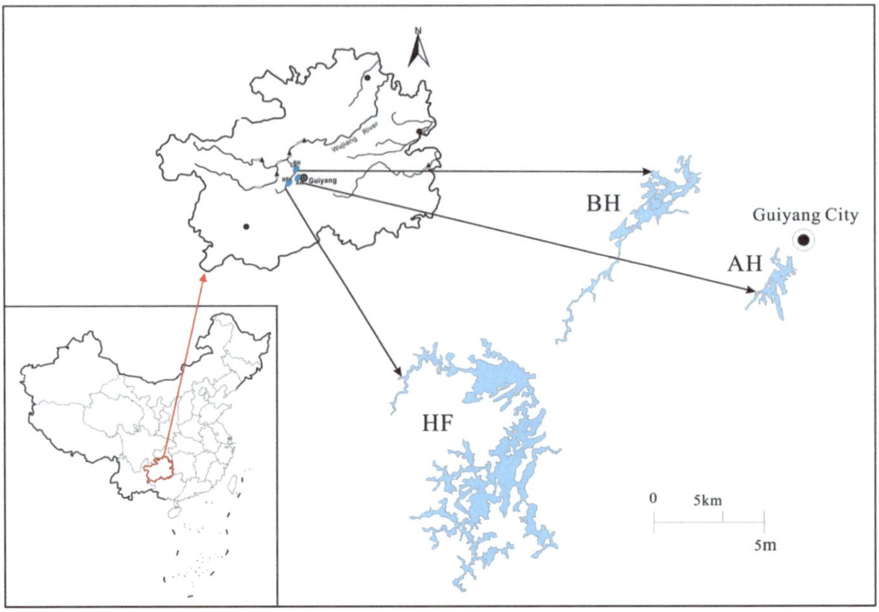

Fig. 7.1 Locations of Hongfeng Reservoir (HF), Baihua Reservoir (BH), Aha Reservoir (AH) in Guizhou Province Southwest China

HF. The reservoir thus also holds a large volume of agricultural, domestic, and industrial wastewater, transforming the reservoir into a hyper-eutrophic reservoir (Zhang 1999). Considerable drainage has resulted in a series of series environmental problems in the HF. Numerous studies have been conducted on the biogeochemical cycling of nutrients as well as on eutrophication processes in this reservoir (Xiao and Liu 2004; Liang et al. 2004). The Yangchang, Maxian, Houliu, and Maibao Rivers flow into this reservoir.

The Baihua Reservoir, which has an area of 14.5 km^2 and volume of 1.91×10^8 m^3, was dammed in 1966 and is situated 18 km northwest of Guiyang, the capital of Guizhou Province, southwestern China. The reservoir was constructed for hydroelectric power generation, flood control, tourism, drinking water, and fishing. The Maotiao, Dongmenqiao, Changchong, and Maixi Rivers as well as the Banpotang Stream flow into this reservoir. The Guizhou Organic Chemical Plant (GOCP), the only plant in China to use metallic Hg as a catalyst for producing acetic acid, is located in the upper reaches of the BH (Fig. 7.1). Wastewater from the GOCP serves as an important mercury source for the reservoir. In addition, seven small coal mines, one iron mine, and one Guizhou aluminum plant are located along the upper reaches of the reservoir. These pollution sites constitute Hg emission sources.

The Aha Reservoir is located roughly 8 km southwest of Guiyang, the capital of Guizhou Province. The reservoir was impounded in 1960, and its impounded area was enlarged in 1982. The reservoir covers approximately 4.5 km^2 and has a total water volume of 5.42×10^7 m^3. The reservoir was constructed for drinking water supplies and flood control. Five rivers (the Youyu, Caichong, Lanni, Sha, and Baiyan Rivers) flow into the reservoir, as is shown in Fig. 7.1. The AH is polluted with industrial and domestic wastewater in the catchment. Most coal mines in the area are closed in 2011.

7.1.1 Biogeochemical Cycling of Mercury in the Hongfeng Reservoir

7.1.1.1 Sampling Sites Description and Sample Collection

Four sampling sites in the reservoir, six sampling sites in the inflows, and one sampling site in the outflow (Maotiao River—MTH) are shown in Fig. 7.2. From 2003 to 2004, water, pore water, and sediment profile samples were collected from the DB and HW of HF, and surface water samples were collected from the inlets of six rivers: the Yangchang River (YHH), Houliu River (HLH), Maxian River (MXH), Taoyuan River (TYH), Maibao River (MBH), and sewage ditch of the Guizhou Organic Chemical Plant (GOCP) in the Maotiao River (MTH). To further examine mercury geochemical cycling in the HF aquatic system, water column, pore water, sediment profile, and BH surface water samples were collected in June, July, and September 2006. In addition, to enhance the representativeness of the sampling sites, another two sampling sites at Yaodong and Jiangjundong were added. These sampling sites are shown in Fig. 7.2.

Unfiltered and filtered water samples from HF were collected using trace metal clean protocols. Four sampling campaigns were conducted in November 2003, and February, May, and September 2004, representing autumn, winter, spring, and summer seasons, respectively, according to the local climate conditions. A part of samples was collected as a supplementary in 2006. Water samples in the reservoir were taken from different depths throughout the entire water columns (0, 4, 8, 12, 16, 20, 24, 28, and 30 m), while only surface water was sampled in the inflows and outflow of the reservoir.

Total Hg (THg), reactive Hg (RHg), dissolved Hg (DHg), dissolved gaseous mercury (DGM), total methylmercury (TMeHg), and dissolved methylmercury (DMeHg) were analyzed for each sample. The analytical methods used for Hg speciation in water have been described in detail in Chap. 3.

Dissolved organic carbon (DOC), total suspended particles (TSP), and water quality parameters such as pH, temperature (T), dissolved oxygen (DO), and total dissolved solid (TDS) were also measured. DOC was measured by the high-temperature combustion method. Water quality parameters such as T, DO, pH, and TDS were measured by a portable analyzer (Radiometer Analytical) on site.

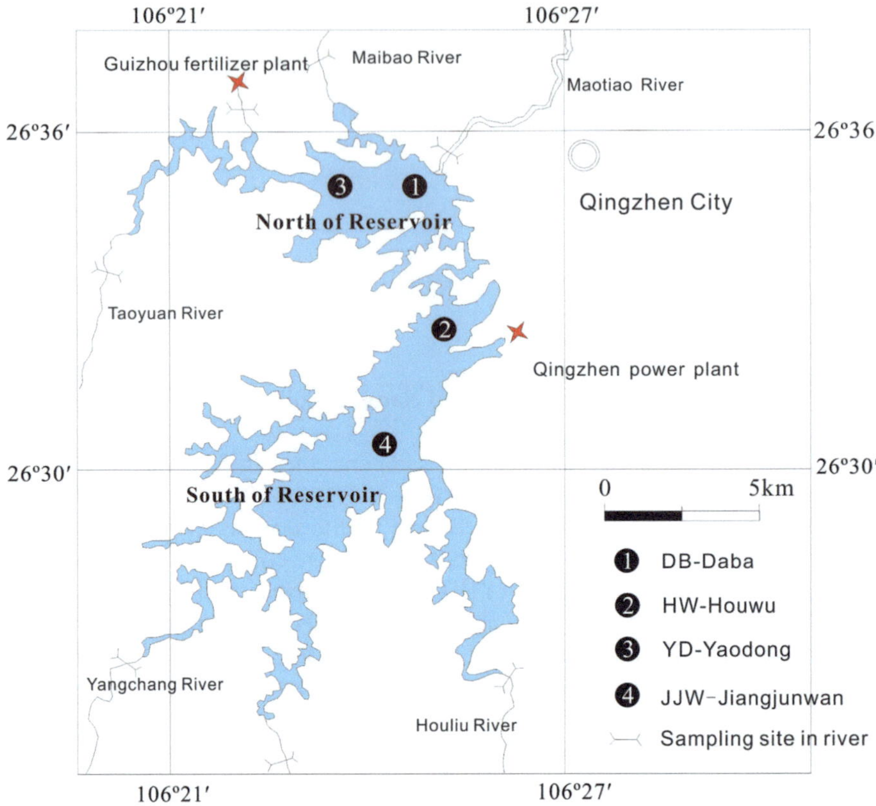

Fig. 7.2 Map of the sampling sites and Hongfeng Reservoir (revised from Zhang 1999)

For water samples, quality assurance and quality control of the analytical process were carried out using duplicates, method blanks, field blanks, and matrix spikes. Field blanks and duplicates were taken regularly (>10% of samples) throughout each sampling campaign. Detection limits were estimated as three times the standard deviation of the blank measurement and are, respectively, 0.004 ng L^{-1} for DGM, 0.02 ng L^{-1} for RHg, 0.10 ng L^{-1} for DHg, and 0.009 ng L^{-1} for DMeHg. The relative standard deviations (RSD) on precision tests for the duplicate samples varied from 1.1 to 12.5% for MeHg analysis, and were <8% for inorganic mercury species analysis. Recoveries for matrix spikes were in the range of 88.2–110% for MeHg analysis, and 86.1–110.3% for inorganic mercury analysis. The estimated detection limit is 0.10 mg L^{-1} for SO_4^{2-}, 0.002 mg L^{-1} for iron, and 0.001 mg L^{-1} for Mn. The relative average deviations on precision tests for the duplicate samples vary from 1.3 to 4.0% for SO_4^{2-}, from 3.5 to 9.2% for sulfide, from 1.4 to 7.1% for iron, and from 0.4 to 6.4% for Mn in filtered water. For sediment samples, quality assurance and quality control of the analytical process were carried out using duplicates, method, and certified reference materials (CC 580, marine sediment). The mean MeHg concentration of 74.7 ± 5.0 ng g^{-1} (n = 10) was obtained from CC 580 with a certified value of 75.5 ± 4.0 ng g^{-1}.

7.1.1.2 Mercury Species in the Water Columns

1. General water quality characteristics

Figure 7.3 shows the distributions of the main water quality parameters for the Houwu and Daba. We found no discernible differences in water temperature, pH, or DO distributions in the water column, demonstrating that the water in the reservoir is thoroughly mixed due to thermal stratification that occurred in November of 2003 and February 2004 (Fig. 7.3).

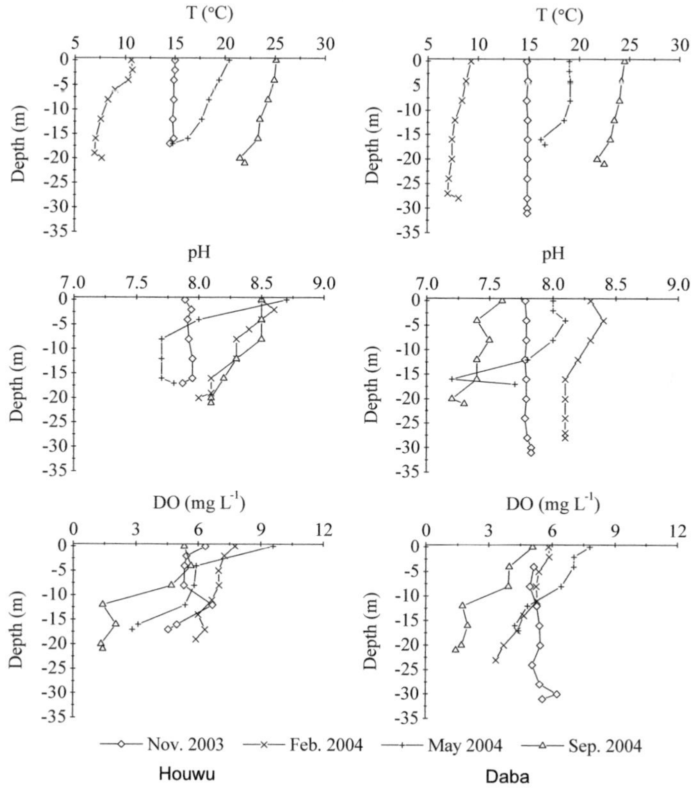

Fig. 7.3 Seasonal distributions of T, pH, and DO in the Hongfeng Reservoir (redrawed from He et al. 2008c, with permission from Elsevier; redrawed from He et al. 2008d, with permission from Chinese Journal of Geochemistry)

However, in May 2003, an anaerobic layer developed in the deepwater area. The pH and DO distributions showed significant differences in their vertical profiles. Both the pH and DO values reached maximum levels at the surface as a result of algae formation but began to decrease in deepwater areas due to stratification in the water column. In particular, the dissolved oxygen levels reached up to 9.6 mg L^{-1} in Houwu as a result of oxygen production from algae growth.

During the September study period, as decomposing algae depleted dissolved oxygen, the DO concentrations decreased in the entire water column, especially in the hypolimnion. A sharp decrease in DO was found from 8 m to 12 m in depth, showing that the reservoir was well stratified. Seasonal distributions of TSP and DOC at the two sampling sites are presented in Fig. 7.4. The TSP concentrations were generally low, ranging from 0.8 to 5.8 mg L^{-1}, with an average concentration of 2.1 mg L^{-1} in all of the samples, except in the sample from May 2004 in the Houwu. However, as a result of algal growth, highly elevated average concentrations of TSP (up to 15.67 mg L^{-1}) were observed in the Houwu in May 2004. The DOC concentrations ranged from 1.74 to 3.23 mg L^{-1}, which are not as high as those found in certain bog lakes in North America (Hines et al. 2004).

The distributions of the physical and chemical characteristics of the water columns at Houwu and Daba showed spatial variations, as there are many different internal and external contamination sources in the HF. During warm seasons, the average concentrations of TSP, TDP, chlorophyll, and DOC in the Houwu were

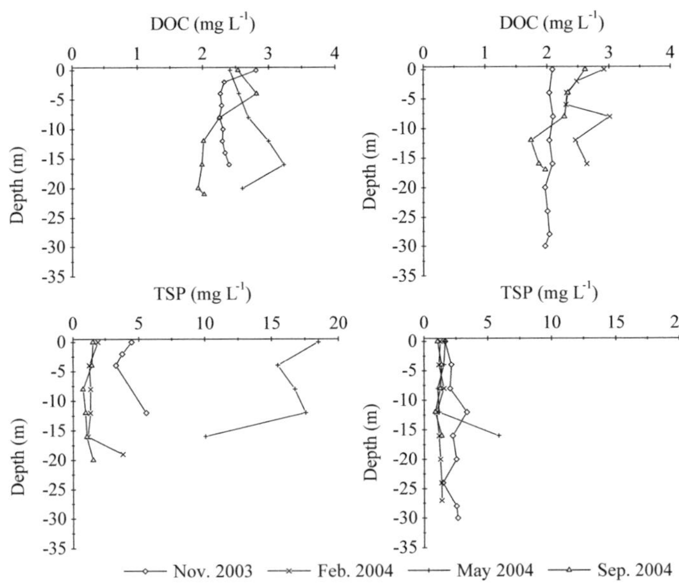

Fig. 7.4 Seasonal distributions of DOC and TSP in the Hongfeng Reservoir from 2003 to 2004 (redrawed from He et al. 2008c, with permission from Elsevier)

higher than those at Daba, while the DO and pH levels in the hypolimnion of the Houwu were lower than those at Daba. These variations show that eutrophication in the Houwu is more developed than it is at Daba.

Compared to the water pH levels found in 2004, the average decreasing pH value with water depth was found to be 7.9 across the three sampling periods, with no significant difference for 2006 (Fig. 7.5). The DO levels in water were similar in 2004 and 2006 and were saturated in surface water, but were anoxic in deeper water areas. Despite the significantly different distributions of DO levels found in 2004, the dissolved oxygen levels were very low, reaching a maximum of only 5.6 mg L^{-1} in surface water. These levels saturated in the summer of 2006, as the water quality levels improved after the prohibition of fish farming in the HF.

2. Distribution of mercury species in water columns
(1) THg, DHg, and PHg in water columns

Spatial and temporal distributions of THg and DHg at Houwu and Daba are presented in Fig. 7.6. The THg concentrations ranged from 2.49 to 13.9 ng L^{-1}, with an average concentration of 6.89 ng L^{-1}. The DHg concentrations ranged from 1.19 to 7.96 ng L^{-1}, with an average of 3.98 ng L^{-1}. The mercury concentrations in the HF water were distinctly higher than those in other natural waters reported for Europe and North America (e.g., Bloom et al. 2004; Sullivan and Mason 1998). THg and DHg showed no discernible vertical distribution trends throughout all of the water columns across the sampling areas. However, we found spatial and seasonal variations in the THg and DHg concentrations in the water column of the reservoir. These spatial variations suggest that the two basins of the reservoir have been affected by different mercury contamination sources. The highest average concentrations of THg and PHg were observed at Houwu in May 2004 (up to 11.43 and 5.95 ng L^{-1}, respectively), while lower average concentrations of THg and

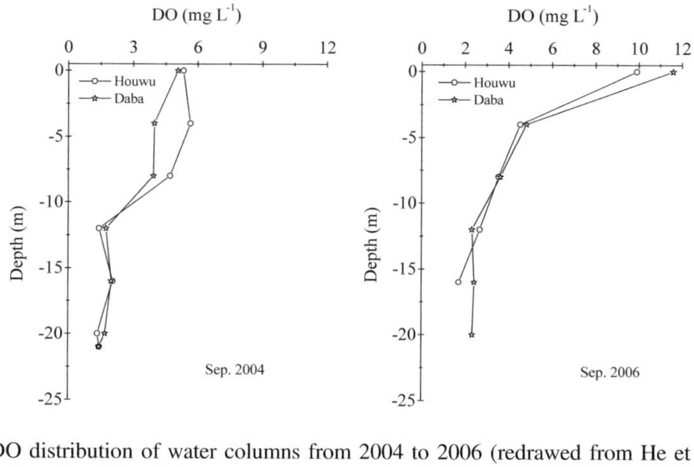

Fig. 7.5 DO distribution of water columns from 2004 to 2006 (redrawed from He et al. 2008c, with permission from Elsevier)

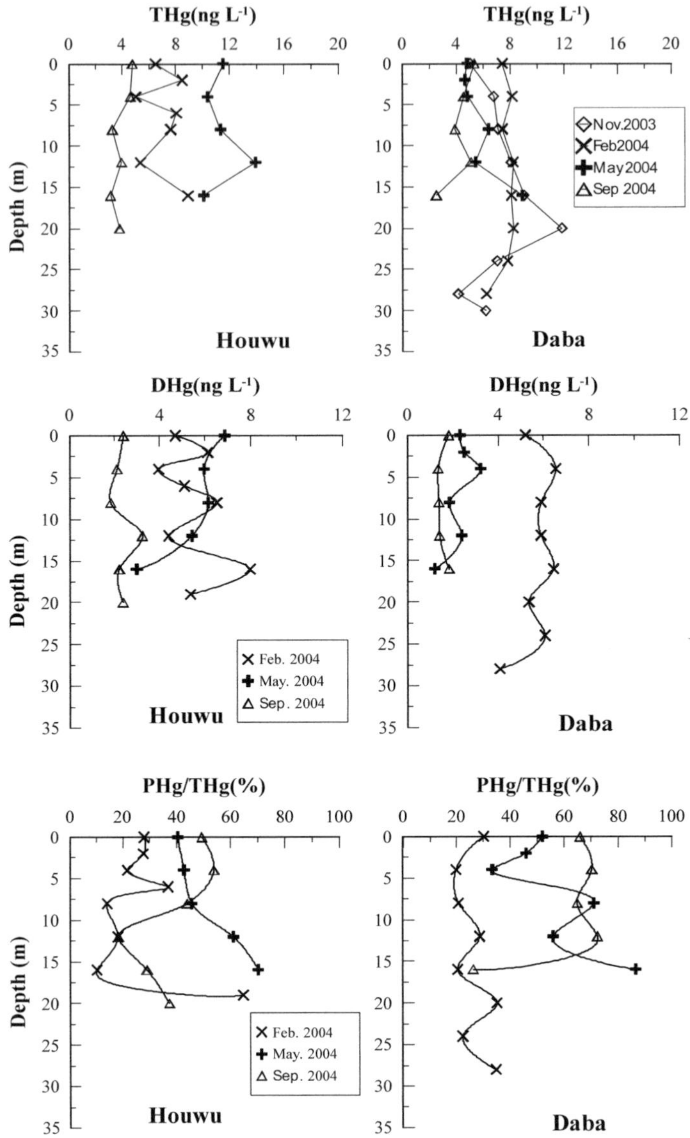

Fig. 7.6 THg, DHg, and PHg in water columns of the Hongfeng Reservoir from 2003 to 2004 (redrawed from He et al. 2008c, with permission from Elsevier; redrawed from He et al. 2010, with permission from Journal of Lake Science; redrawed from He et al. 2008d, with permission from Chinese Journal of Geochemistry)

PHg (6.06 and 3.84 ng L^{-1}, respectively) were found at Daba at the same time. Similarly, elevated average concentrations of TSP were found at Houwu in May; the TSP levels averaged at 15.67 mg L^{-1} compared to a value of 2.1 mg L^{-1} found for the Daba at the same time. This difference is attributed to the appearance of large algae populations at Houwu in May due to aquaculture activity and the addition of wastewater enriched with N and P. The distribution of mercury found also shows that macroalgae may be able to bind mercury and may represent a substantial pool of mercury in the aquatic system. The levels of total and dissolved mercury recorded in September were fairly low relative to those recorded in November and February. This seasonal distribution of mercury could be a result of wastewater contamination. In May and September, high runoff volumes were due to abundant precipitation-diluted mercury concentrations in the water, whereas there was very little precipitation in February and November, thus causing mercury concentrations in the water to increase.

The PHg/THg ratios were high during May and September at 57 and 49%, respectively, while the PHg/THg ratio was only 26% in February (Fig. 7.6). However, increased TSP levels were not observed in May and September, with the exception of a high TSP value in the Houwu in May due to algal growth. Therefore, the PHg proportions likely increased due to the increased fraction of organic particles, which can absorb more mercury than inorganic particles, in May and September. In September, however, the DHg levels in the hypolimnion increased once again, especially in the Houwu, with the highest proportion reaching 82%. This is likely attributable to the anaerobic decomposition of particulate organic matter in the hypolimnion, which resulted in an increase in dissolved organic matter in anoxic water, which can absorb more mercury than chloride and hydroxide complexes (Coquery et al. 1997). Moreover, the dissolution of iron and manganese oxide in anoxic water likely also contributed to the increase in the dissolved mercury levels (Regnell et al. 2001).

The mercury level range and average values for 2006 are shown in Table 7.1. The results show that the THg levels were significantly higher in June than in July, exceeding 4 ng L^{-1} at the Houwu site. The THg levels at Daba were also higher than those in July, reaching nearly 2 ng L^{-1}. This abnormal increase in mercury concentrations is attributed to runoff flows resulting from continuous rain prior to sampling.

Table 7.1 Range of the THg and DHg values in the Hongfeng Reservoir in 2006

Date	Site	Range		Average	
		THg	DHg	THg	DHg
2006.06	DB	3.7–5.4	2.9–4.3	4.7	3.7
	HW	4.5–6.5	2.5–4.1	6.2	3.4
2006.07	DB	2.0–3.4	1.0–1.5	2.7	1.4
	HW	2.0–3.0	0.5–1.9	2.5	1.4
	JJW	2.5–4.2	1.0–2.0	3.3	1.6
	YD	1.5–5.4	0.7–1.9	3.3	1.3

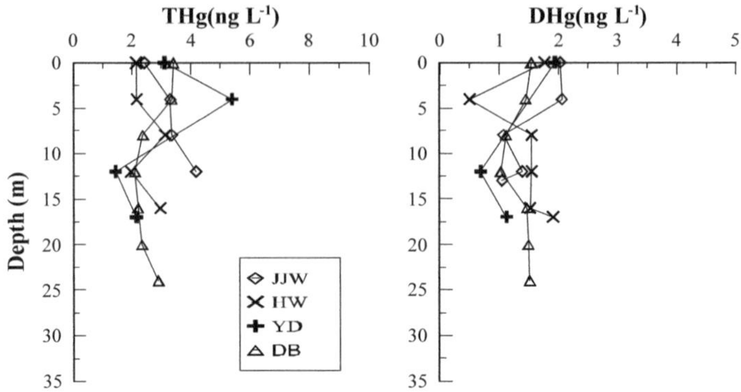

(JJW: Jiangjunwan; HW: Houwu; YD: Yaodong; DB: Daba)

Fig. 7.7 The THg and DHg levels in the water columns of the Hongfeng Reservoir in 2006

We found no significant differences in the THg levels in the water columns at the sampling sites in 2006 (Fig. 7.7). The THg levels in the water were lower in June 2006 than in May 2004, but were higher than those recorded in September 2004. In July 2006, the THg levels were lower than those recorded in the summer and spring of 2004.

Generally speaking, mercury concentrations in a nonpolluted water body are often higher in rainy spring and summer seasons than in winter due to runoff-related inputs. In the HF, the mercury concentrations in the water columns of the four sites were found to be lower in spring and summer than in autumn and winter due to rain dilution effects. However, different results were found at the Houwu site due to high levels of eutrophication in the water body resulting from anthropogenic activities. In spring, most mercury was absorbed by algae in the Houwu, forming a potential mercury pool and changing the transportation and migration of mercury into the reservoir. In conclusion, anthropogenic sources affected spatial and seasonal variations of mercury in the HF more than natural sources.

The average ratio of THg as PHg found in water columns was 29% in June 2006, which is lower than that recorded in the spring and summer of 2004. In July of 2006, the average %PHg value was 50%, which is similar to that of the corresponding period for 2004. Vertical distribution variations of DHg for all of the sampling sites show slight discrepancies for June and July 2006 (Fig. 7.7). These results suggest that it may be possible to limit algal growth by improving the environmental quality. The annual variations of the %PHg found may be attributed to improvements in the HF water quality levels. For instance, the organic particulate matter levels declined with a significant decrease in algae populations with the prohibition of fish farming in the HF in 2005. However, the above phenomenon did not exclude the high proportion of DHg found in June, which was attributed to DHg inputs from other sources.

Overall, the mercury concentrations in water were lower and the water quality levels were higher in 2006 than in 2004 due to the prohibition of fish farming in

2005. In addition, according to the mercury distributions in the input rivers, the THg and DHg levels in the sewage ditch of the Guizhou Organic Chemical Industrial Plant (GOCP) declined significantly in June and July 2006 relative to the 2004 levels. These results show that the mercury concentrations were influenced by human activities. However, this occurred only occasionally due to short water retention periods (0.325 years) and fast water flow rates. For example, the water THg levels were abnormal at Houwu in the spring of 2004 as well as June 2006 and reverted back to normal concentrations in the following sampling period. The above results and discussion show that the partitioning of mercury between water and particulates in the HF is mainly dependent on the effects of endogenous organic matter and redox conditions. The binding capacity of mercury with high organic fraction particles was found to be much higher than that with inorganic particles and particulate-bound mercury dissolved in water.

(2) RHg in water columns

Seasonal distribution patterns of RHg in the reservoir are presented in Fig. 7.8. The RHg levels ranged from 0.14 to 2.17 ng L^{-1}, with an average concentration of 0.64 ng L^{-1}. The RHg concentrations along the water surface were lower than those underwater in all of the vertical profiles, with the exception of those for the Daba in September. This is likely attributable to intense particulate scavenging and/or to a biological reduction of Hg^{2+} near the water surface and to a subsequent release of Hg^0 into the atmosphere (Gill 1986; Kim and Fitzgerald 1986; Dalziel 1995). The unusually high RHg value recorded from the Daba in September may be attributed to contamination from a nearby chemical fertilizer plant. This theory is supported by the presence of very high RHg concentrations (up to 81.49 ng L^{-1}) observed in the fertilizer plant drain, which was contaminated by the chemical fertilizer plant.

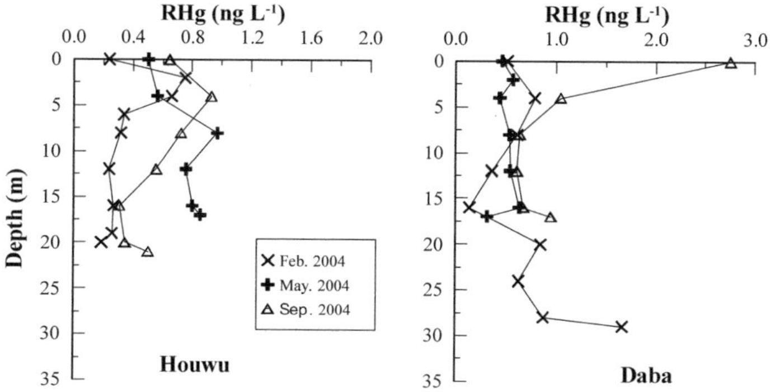

Fig. 7.8 Seasonal distribution of RHg in the Hongfeng Reservoir (redrawed from He et al. 2008c, with permission from Elsevier; redrawed from He et al. 2008a, with permission from Journal of Research of Environmental Sciences; redrawed from He et al. 2008d, with permission from Chinese Journal of Geochemistry)

The RHg levels decreased from the subsurface to deep water areas in most of the vertical profiles, potentially due to the methylation of Hg^{2+} or presence of S^{2-} in deep anoxic water. As RHg is the main species that can be reduced to Hg^0 or methylated through a bacterially mediated process (Ullrich et al. 2001), reduction, and methylation processes likely controlled the RHg concentrations. However, the above-listed distribution patterns of RHg in the water column were not observed in the Houwu in May or at Daba in February and were likely due to contamination from the chemical fertilizer plant at the Daba as well as excessive algae formation in the Houwu, as discussed above.

(3) DGM in water columns

For our comparisons between vertical profiles of DGM, samples of each vertical profile were collected midday as water DGM concentrations. Surface water, in particular, was found to be significantly affected by sunlight. The DGM concentrations ranged from 0.02 to 0.11 ng L^{-1} during the four sampling periods (Fig. 7.9). The DGM concentrations were found to be highest along the water surface and decreased with water depth. The average concentrations of DGM in September (0.08 ng L^{-1} at Daba; 0.07 ng L^{-1} in the Houwu) were higher than those recorded in February (0.04 ng L^{-1} at Daba; 0.05 ng L^{-1} in the Houwu). This DGM distribution pattern is consistent with the hypotheses that the photoreduction of Hg (II) complexes constitutes the main source of Hg^0 formation in water and that temperature plays an important role in the photoreduction process. We found a sharp decrease in DGM concentrations at 8–12 m in depth in September 2004. This suggests that seasonal stratification may also affect the vertical DGM distributions. The lowest average DGM concentration (0.04 ng L^{-1}) was observed in the Houwu in May rather than in February, whereas the DGM concentrations at Daba were as high in May (0.08 ng L^{-1}) as they were in September. This suggests

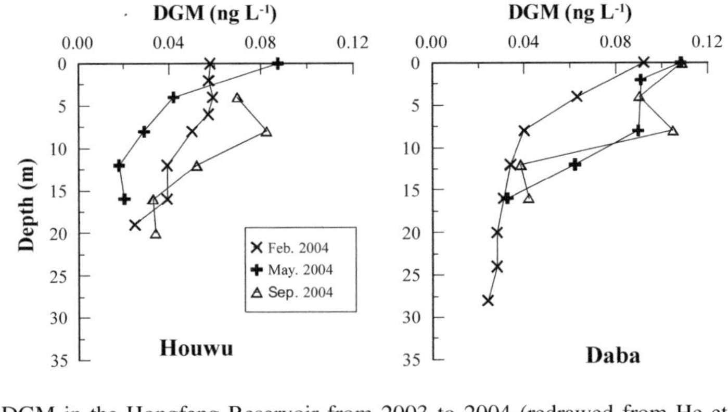

Fig. 7.9 DGM in the Hongfeng Reservoir from 2003 to 2004 (redrawed from He et al. 2008c, with permission from Elsevier; redrawed from He et al. 2008a, with permission from Journal of Research of Environmental Sciences; redrawed from He et al. 2010, with permission from Journal of Lake Science; redrawed from He et al. 2008d, with permission from Chinese Journal of Geochemistry)

that other processes may also control DGM production. The lowest average DGM concentration observed in the Houwu in May could be attributed to the presence of algae, which can block sunlight. Early studies suggest that bacteria and some eukaryotic microorganisms, such as algae, reduce Hg(II) levels. Ben-Bassat and Mayer (1987) found that Hg^0 formation decreases as a function of the inhibition of photosynthesis in cultures of green algae chlorella.

The lowest average water DGM value was found in the Houwu in spring. The average DGM value recorded from the Daba in summer was similar to that recorded in spring. This result implies that primarily solar radiation, temperature, and other effects control DGM production (e.g., plankton and biotas in water bodies). The highest Hg(0) production rate was observed during periods of algal growth, as plant photosynthesis can generate Hg(II) reductant (Vandal et al. 1991). Mason et al. (1995) found that phytoplankton, especially that is smaller than 3 μm, is the main cause of Hg(II) biotic reduction. The above results show that plants facilitate elemental Hg formation. However, the DGM concentrations remained low throughout the year despite the presence of green algae in the Houwu. The large algae population may have prevented light from penetrating water, thereby reducing the rate of Hg(II) photoreduction during alga, growth. Another study found that the formation of elemental mercury in water is significantly related to DOC levels (Jiang 2005). However, no significant correlation was found between DGM and DOC in the HF ($R = 0.0049$, $p = 0.98$, Fig. 7.10).

Temperature levels may also affect Hg(0) formation. Sullivan and Mason (1998) found that the lowest elemental mercury production rate occurred at the lowest temperatures through board incubation experiments, and Jiang (2005) found a significant correlation between temperature and DGM in the Wujiangdu Reservoir (WJD) and Dongfeng Reservoir (DF).

While the ways in which temperature affects microbial activity and controls Hg(0) formation remains contested (Krabbenhoft et al. 1998), some studies have shown that microbial activity has a major effect on atomic mercury formation. Higher sample

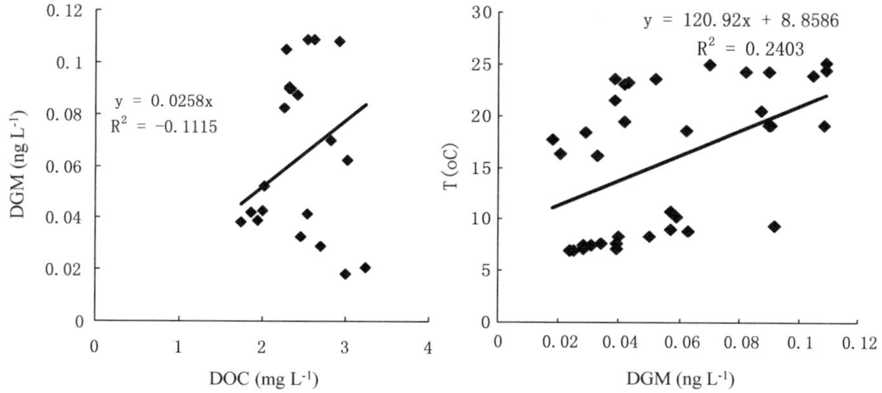

Fig. 7.10 Correlation between DGM and DOC and temperature

temperatures denote radiation strength grades rather than water temperatures. We found a slight correlation between the temperatures and DGM levels for the HF ($r = 0.49$, $p = 0.0021$, in Fig. 7.10), implying that temperature should not significantly affect DGM production.

(4) MeHg in water columns

Seasonal distributions of TMeHg in the water columns are shown in Fig. 7.11. An analysis of variance revealed no statistically significant differences ($F = 3.01$, $p = 0.059$) between the TMeHg distributions in the water columns at both sampling sites for the May, November, and February study periods. The TMeHg concentrations in these campaigns varied from 0.05 to 0.33 ng L^{-1} and increased slightly with water column depth during the February and November study periods.

In fall and winter, the TMeHg levels at the two sampling sites increased slightly from the surface to the bottom layer, implying that a methylmercury source from sediment mercury emissions or factors improved methylation (e.g., low DO and pH). However, an initially decreasing then increasing trend was found at two sites in spring, and the highest TMeHg value found in surface water areas can be attributed to particular matter that bounded mercury input during a rainstorm (Fig. 7.11).

The MeHg concentrations in summer (September) were statistically elevated relative to those found in the other three sampling campaigns ($F = 9.48$, $p < 0.001$). The highest value of 0.92 ng L^{-1} occurred in the Houwu and was 2.5 times higher than the highest value recorded during the other seasons. We found a distinct vertical distribution pattern of MeHg in the water column. The TMeHg levels increased from 0.15 ng L^{-1} at the surface to 0.92 ng L^{-1} in the hypolimnion of the Houwu, while the TMeHg levels increased from 0.08 at the surface to 0.81 ng L^{-1} at the base of the Daba. We found a sharp increase in TMeHg concentrations at depths of 8–12 m that corresponded to a sharp decrease in the

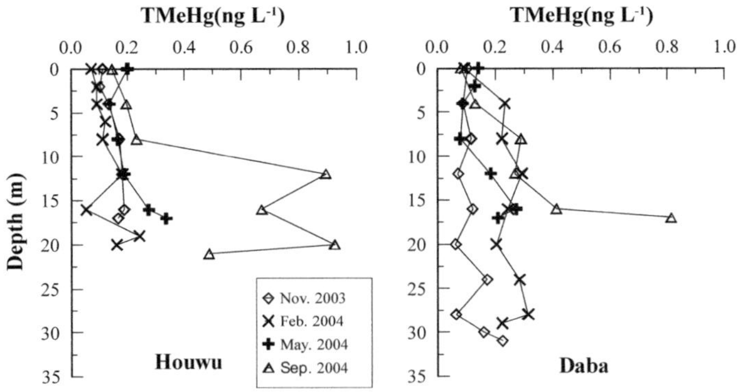

Fig. 7.11 Seasonal distribution of TMeHg in the Hongfeng Reservoir water from 2003 to 2004 (redrawed from He et al. 2008c, with permission from Elsevier redrawed from He et al. 2010, with permission from Journal of Lake Science; redrawed from He et al. 2008d, with permission from Chinese Journal of Geochemistry)

dissolved oxygen concentrations at the same depth. The spatial and temporal distributions of MeHg show that the MeHg levels increased significantly in the hypolimnion in September, especially in the Houwu.

The MeHg content in water is influenced by a wide variety of environmental factors, such as the total and reactive mercury content, temperatures, redox potential levels, pH levels, and inorganic and organic solutes in water (Ullrich et al. 2001). However, these factors cannot be measured independently of one another, as they often interact, forming a complex system of synergistic and antagonistic effects. It is generally believed that Hg methylation is predominantly a microbial-mediated process, and some studies have shown that methylation is carried out by sulfate-reducing bacteria in the water column (Gilmour and Henry 1991; Berman and Bartha 1986). The methylation rates appear to be enhanced under anaerobic conditions due to increased anaerobic sulfate-reducing bacterial activity. According to our investigation, TMeHg has a strong negative relationship with DO, with a Pearson correlation coefficient of $-0.74(n = 78, p < 0.0001$, Fig. 7.12). As the pH, DOC, and salinity levels in all of the samples varied within a narrow range, no significant correlations between methylmercury and these parameters were observed ($r = 0.33, p < 0.01$).

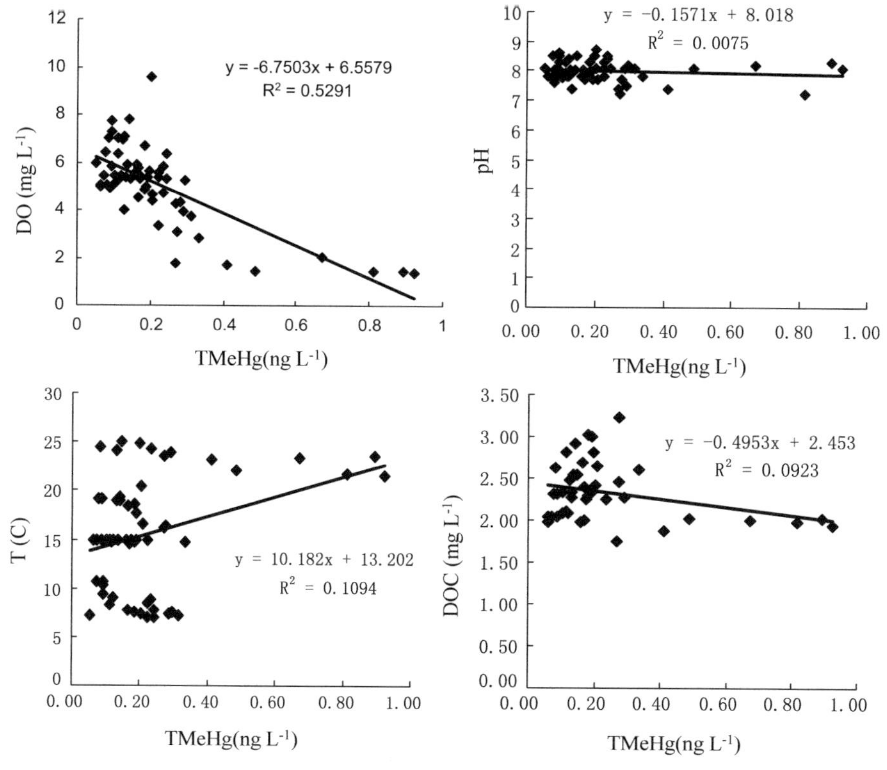

Fig. 7.12 Correlation analysis of TMeHg and the water parameters

Gilmour and Henry (1991) suggests that both low pH and negative redox potential levels, which are common to anoxic hypolimnia, not only increase methylation rates but also decrease demethylation rates, thus resulting in a net increase in MeHg. Eckley et al (2005) showed that the methylation rates in hypolimnetic water account for the observed accumulation of MeHg in hypolimnetic water in summer in two pristine Wisconsin lakes. Some studies have also shown that the accumulation of settling particulate matter from the epilimnion, such as hydrous ferric and manganese oxides, can bind MeHg and that their dissolution in the hypolimnion contributes to the high concentrations of MeHg found in deep water areas (Meili 1997). Other studies have indicated that increased MeHg levels mostly result from the release of MeHg from sediments, especially in highly contaminated sites (e.g., Regnell and Ewald 1997; Verta and Matilainen 1995). Moreover, many studies have shown that MeHg increasingly releases flux from sediments among decreased levels of pH and DO (e.g., Ullrich et al. 2001).

The highest values of DHg and DMeHg in the Houwu in September were not found at the sediment–water interface but at depths of 12 and 20 m. This suggests that MeHg in the water column was not produced from the release of MeHg in sediment but from in situ methylation in anoxic water or from the deposition of particular methylmercury in the surface water (Fig. 7.11). In the Houwu in summer, the decomposition of dead algae consuming considerable amounts of oxygen in water led to the formation of the lowest levels of dissolved oxygen in bottom water areas across all four seasons, facilitating methylation.

At Daba in September, however, the MeHg levels exhibited a strong increasing gradient toward the sediment, indicating that MeHg released from sediment had a strong effect on the MeHg depth profile of the Daba. Despite an increase in MeHg in deep water areas in the Daba in September, the Wilcoxon rank sum test results showed that the MeHg concentrations in the Houwu were much higher than those in the Daba at the same depths in September ($p < 0.05$), especially for the hypolimnion layer (Fig. 7.11). This finding suggests that MeHg is formed in the hypolimnion layer of the Houwu in September. The decomposition of a large volume of algae induced by high nutrient concentrations in the Houwu led to low DO and pH levels, which may have accelerated Hg methylation. On the other hand, seasonal variations of methylmercury in surface sediments in the pore water of the Houwu also show that the release of methylmercury does not constitute the main source of water in summer. A comparison between the MeHg concentrations at Daba and Houwu shows that the average TMeHg levels in the Houwu are higher than those in the Daba during all seasons except for winter. However, the TMeHg levels in the Daba are higher than those in the Houwu in winter, as there are many different sources of pollution in the HF. In the Houwu, warm power plant discharge water constitutes the main source of aquaculture water and endogenous pollution in the HF. These pollutants exacerbate water body eutrophication and large algae breeding, causing further water quality deterioration. According to the water parameters, the dissolved oxygen levels in the bottom water layers are lower and the DOC levels are slightly higher in the Houwu than in the Daba. These changes create favorable conditions for the methylation of mercury.

In the Daba, all pollution sources, including mercury, mainly derive from the Guizhou Chemical Fertilizer Plant (GCFP). However, water discharge and runoff occurring during the rainy season diluted these pollutants in the Daba. There is thus no obvious difference in the mercury concentrations between the Houwu and Dabas. However, in winter, fewer fish farming activities, weak runoff dilution, and hydrodynamic conditions enhance the effects of GCFP on the mercury concentrations in Houwu water. Xiao's research (2002) confirms that chemical plants mainly affect mercury distribution profiles in winter. As the main species of mercury methylation, the RHg levels are 2 times higher in the Daba than those in the Houwu in winter due to the serious pollution levels in the Daba. The RHg data also show that the pollution levels are higher in the Daba than in the Houwu in winter and that the RHg concentrations in the Daba are twice as high as those in the Houwu. Overall, we believe that RHg is converted into MeHg.

Seasonal variations in DMeHg in the reservoir show that (Fig. 7.13) the DMeHg levels range from 0.01 to 0.11 ng L^{-1}, with an average of 0.05 ng L^{-1} in the Daba, and from 0.03 to 0.13 ng L^{-1}, with an average of 0.08 ng L^{-1} in the Houwu in autumn. The DMeHg levels range from 0.08 to 0.13 ng L^{-1} in winter, with an average value of 0.11 ng L^{-1} in the Daba, and from 0.03 to 0.11 ng L^{-1}, with an average of 0.05 ng L^{-1} in the Houwu. The DMeHg levels range from 0.03 to 0.08 ng L^{-1}, with an average value of 0.05 ng L^{-1} in the Daba, and from 0.04 to 0.20 ng L^{-1}, with an average of 0.10 ng L^{-1} in the Houwu in spring. The DMeHg levels range from 0.03 to 0.53 ng L^{-1}, with an average of 0.10 ng L^{-1} in the Daba, and from 0.08 to 0.91 ng L^{-1}, with an average of 0.37 ng L^{-1} in the Houwu in summer. No significant difference in the water DMeHg to TMeHg proportions was observed during the November, February, and May study periods; the average value was roughly 43%. In September, however, the DMeHg proportions increased significantly to an average of 65% and to a maximum of 98% at a depth of 20 m in the Houwu. The DMeHg proportions were also elevated (73%) in the outflow area

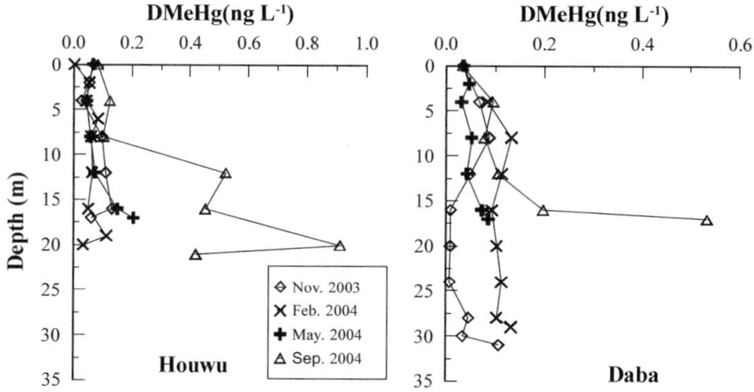

Fig. 7.13 DMeHg in water columns of the Houwu and Dabas in the Hongfeng Reservoir during 2003–2004 (redrawed from He et al. 2008c, with permission from Elsevier; redrawed from He et al. 2008d, with permission from Chinese Journal of Geochemistry)

of the reservoir, which is composed of hypolimnion water. Eckley et al. (2005) and Baeyens and Meuleman (1998) also observed elevated proportions of DMeHg in the hypolimnion of seasonally stratified lakes. Hydrous ferric and manganese oxides and organic particles have strong MeHg and Hg(II) binding capacities. On one hand, under anoxic conditions, the mercury methylation rates increased significantly, resulting in an increase in the MeHg concentrations. On the other hand, the dissolution of oxides and anaerobic decomposition of particulate organic matter may have spurred an increase in the DMeHg proportions in the hypolimnion (Regnell et al. 2001).

The TMeHg concentrations in Houwu water were statistically higher than those in the Daba for all of the study periods, with the exception of the February period (Wilcoxon rank sum test, $p < 0.05$) (Fig. 7.14). The average TMeHg concentrations in the Houwu in May, September, and November reached 0.22, 0.50, and 0.15 ng L^{-1}, respectively, while corresponding values for the Daba were recorded as 0.16, 0.34, and 0.12 ng L^{-1}, respectively. In February, however, the MeHg levels in the Daba (0.23 ng L^{-1}) were higher than those recorded from the Houwu (0.13 ng L^{-1}). The different contamination sources could be responsible for these seasonal and spatial variations. For the Houwu, aquaculture activities constituted the main contamination source. In the Daba, the main sources of contamination were the nearby chemical fertilizer plant and domestic wastewater inputs. Fish farms in the Houwu contributed large volumes of N and P to the water, resulting in the formation of larger algae populations than those in the Daba. The decomposition of algae caused the DO and pH levels to decrease, favoring the methylation of mercury. In winter, contamination in the Houwu decreased significantly with tempered fish farm activity, but the contamination levels in the Daba remained high as pollutants were derived from the chemical fertilizer plant and from domestic wastewater. The reactive Hg levels in the Daba (0.72 ng L^{-1}) were also higher than those in the Houwu (0.36 ng L^{-1}), while the DO levels in the Daba (4.9 mg L^{-1}) were lower than those recorded in the Houwu (6.7 mg L^{-1}) in February. All of these factors resulted in higher mercury methylation rates in the Daba in winter.

Vertical distributions of TMeHg and DMeHg in the water column of the HF for 2006 are shown in Fig. 7.14. The highest DMeHg and TMeHg values found in each sampling site are located at the water–sediment interface and gradually decrease from the surface to the base. This suggests that elevated methylmercury levels in deeper water areas in these sites are mainly attributed to the release of methylmercury from sediments or to the resuspension of sediments rather than from methylmercury generation in the water. The maximum and average values of TMeHg and DMeHg in the water increased in July, but decreased in September (shown in Table 7.2). In particular, the methylmercury levels in the pore water are clearly higher in July, suggesting that upward pore water mercury diffusion and sediment resuspension reached their maximum levels.

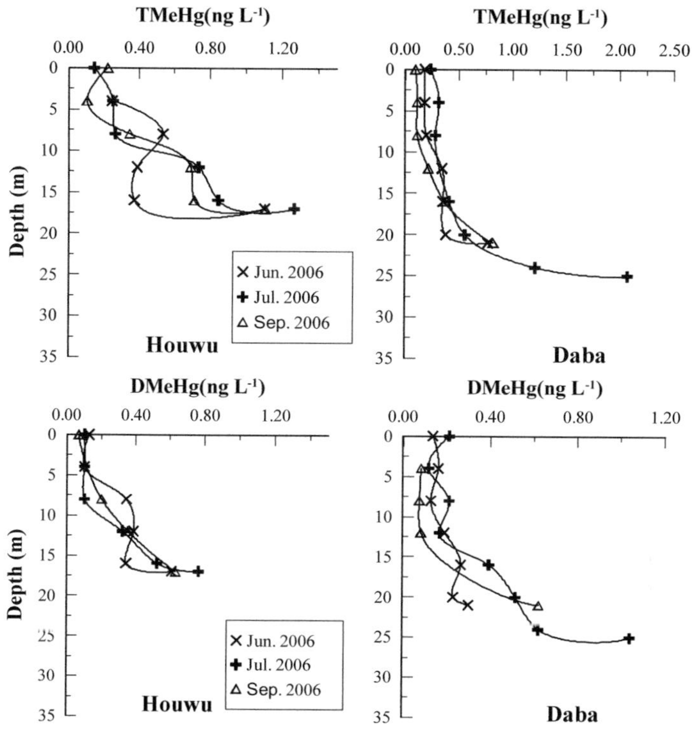

Fig. 7.14 TMeHg and DMeHg profiles for the Daba and Houwu of the Hongfeng Reservoir

Table 7.2 MeHg in the Hongfeng Reservoir in 2006

Date	Sampling Site	Range (ng L^{-1})		Average (ng L^{-1})	
		TMeHg	DMeHg	TMeHg	DMeHg
2006.06	DB	0.18–0.78	0.13–0.31	0.34	0.20
	HW	0.25–1.1	0.10–0.61	0.53	0.32
2006.07	DB	0.23–2.0	0.12–1.04	0.72	0.41
	HW	0.14–1.3	0.10–0.76	0.58	0.32
	JJW	0.17–2.2	0.10–1.5	0.87	0.48
	YD	0.15–0.70	0.08–0.27	0.39	0.15
2006.09	DB	0.09–0.82	0.07–0.62	0.28	0.21
	HW	0.22–1.1	0.06–0.63	0.53	0.32
	JJW	0.28–1.6	0.09–1.0	0.51	0.31

Table 7.3 Main characteristic parameters of each river (Zhang 1999)

Parameters	Taoyuan River	Yangchang River	Maxian River	Houliu River	Maibao River
Length (km)	51.8	72.0	35.4	23.3	6.4
Watershed area (km^2)	205	817	252	88	–
Natural fall of river (m)	220	175	20	50	–
Annual average flow ($m^3 s^{-1}$)	4.14	12.67	5.31	1.86	–

3. Mercury in inflows and outflows
(1) Standard physical and chemical parameters

The HF inflows include the Yangchang, Taohuayuan, Maxian, Liuguang, and Maibao Rivers (see Fig. 7.2). HF water comes from rainwater, and thus, the reservoir's volume is dependent on rainfall patterns and increases and decreases rapidly. As industrial and mining enterprises and villages are positioned near the bank of the reservoir, these rivers are subjected to differing degrees of pollution.

The dissolved oxygen and pH distributions for these rivers are shown in Table 7.4. It is evident that the DO and pH levels in the Maotiao River are significantly lower than those of these rivers in terms of inflows. This suggests that the water quality levels change significantly as it moves through the reservoir. In particular, the dissolved oxygen levels decrease to 1.8 mg L^{-1} in summer. Overall, dissolved oxygen in the rivers is abundant and the dissolved oxygen levels in summer are lower than those recorded in autumn and spring, which may be due to high summer temperatures and to low pressure levels along the river's surface.

Table 7.4 DO, DOC, and pH levels in the Hongfeng Reservoir inflows and outflows (redrawed from He et al. 2008c, with permission from Elsevier)

Inflow and outflow	Nov-03		May-04		Sep-09		
	DO (mg L^{-1})	pH	DO (mg L^{-1})	pH	DO (mg L^{-1})	pH	DOC (mg L^{-1})
Maxian River (inflow)	8.0	8.1	8.4	7.9	5.4	8.0	2.39
Yangchang River (inflow)	8.5	8.0	8.9	7.6	6.7	7.9	2.12
Taoyuan River (inflow)	8.3	8.4	–	–	7.2	8.2	0.66
Houliu River (inflow)	7.4	7.8	8.3	7.6	5.3	7.6	2.47
Maibao River (inflow)	7.7	7.7	8.5	7.7	8.0	7.7	0.54
Chemical plant (inflow)	7.6	7.6	5.5	8.0	5.3	7.5	–
Maotiao River (outflow)	4.7	6.9	3.2	7.4	1.8	7.4	1.87

(2) Mercury in inflows and outflows

Distributions and seasonal variations in THg concentrations for rivers flowing into the HF are shown in Fig. 7.15. The THg concentrations in these rivers range from 2.2 to 350 ng L^{-1}, with an average of 51 ng L^{-1}. The average annual distribution of the THg concentrations in the different sites is as follows: sewage ditch (GOCP) > Maobao River > Yangchang River > Houliu River > Maxian River > Taohuayuan River. Although flows through the sewage ditch (GOCP) and Maibao River are limited, local industrial and domestic wastewater flows constitute the main water source. Therefore, the mercury concentrations here are an order of magnitude higher than those of other rivers. For the HF, the mercury concentrations found in the Yangchang River are highest among those of several water supply rivers. The Yangchang River is one of the largest rivers among several other water supply rivers in the HF. As the main water source of the HF, the Yangchang River receives domestic sewage from Pingba County and industrial and domestic wastewater from the Anshun Chemical Fertilizer Plant, Pingba Distillery, Liyang Company, Huanyu Machinery Factory, and Pingba Fertilizer Plant. The Houliu River is mainly affected by rural farmland drainage pollution, but the THg concentrations remain high due to its limited flows and self-purification capacities.

The THg, DHg, and RHg concentrations measured from the inflows and outflows during the four study periods are presented in Table 7.3. Relatively high concentrations of mercury were observed in the Yangchang and Maibao Rivers as well as in the Fertilizer Plant drainage area, which likely result from industrial and domestic wastewater sources of mercury in the drainage area of the HF. PHg, with proportions relative to THg reaching 58, 64, and 84% in the Yangchang and Maibao Rivers as well as Fertilizer Plant Drain, respectively, contributes the most THg to these rivers. The THg and DHg concentrations in the outflows were much lower than those found in contaminated inflows. While the water flows were not measured in this study, we estimated rough annual inputs and outputs of mercury species based on the average concentrations of mercury species across the four study periods and based on long-term average annual flows for the rivers reported

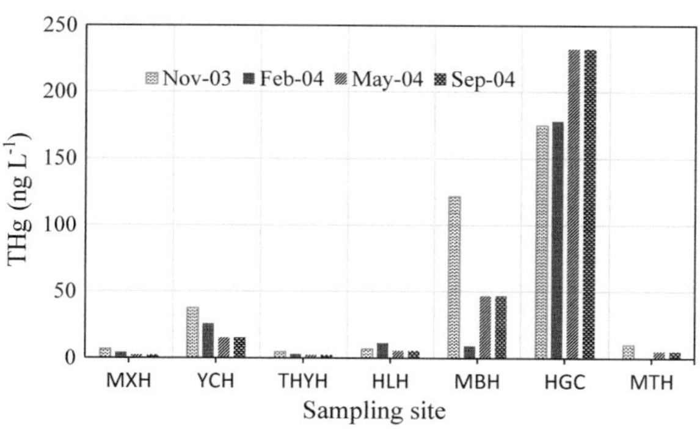

Fig. 7.15 Concentration distributions and seasonal variations of THg in the Hongfeng Reservoir inflows

by Zhang (1999) (Table 7.3). These estimates suggest that more than 50% of THg from the inflows was removed by the reservoir and that most of this THg was assumed to be buried in sediments.

In different seasons, we found that the mercury concentrations in the Yangchang River resulting from industrial activities are higher in autumn and winter than those recorded in spring and summer. This indicates that mercury in the water is diluted by rain. In the Majian and Houliu Rivers, which mainly receive farmland wastewater discharge, the highest concentrations of THg and DHg were found in spring. Prior to sampling in spring, a large volume of surface runoff from the surrounding mountains and farmland flowed into the river as a result of heavy rainfall in the HF watershed. This may have increased the mercury content in the river.

The seasonal distribution of DHg in the inflows and outflows is presented in Fig. 7.16. The average annual concentrations of DHg in the river were consistent with those of THg, and the DHg levels correlated with the THg levels in each season. The complex correlation coefficients and complex determinants were recorded as 0.85 and 0.73, respectively ($p < 0.001$). We found a significant correlation between the THg and PHg levels in the rivers (complex and coefficient coefficients of 0.99 and 0.99, respectively (Fig. 7.17a), suggesting that the distribution of PHg in the river accounts for 99% of the THg levels. At the same time, we found a significant correlation between the total suspended and total particulate mercury in the river ($r = 0.66$, $p < 0.001$, Fig. 7.17b).

The THg and DHg levels in the Maotiao River in the summer and autumn were much lower than those recorded in the contaminated inflows, but were slightly higher than those recorded in the Maxian and Taoyuan Rivers. In spring, the THg and DHg levels in the Maotiao River were lower than those recorded in the inflows. Thus, the mercury in the inflows declined, with only a small amount of mercury continuing to flow out. The concentrations of THg and total suspended solids (TSP) in the HF in June and July of 2006 were not significantly different from those recorded in slightly contaminated rivers in 2004. As one heavily polluted river, the THg concentrations in the GCOP's ditch significantly decreased in 2006, but an exceptionally high THg concentration of 150 ng L^{-1} was observed in the

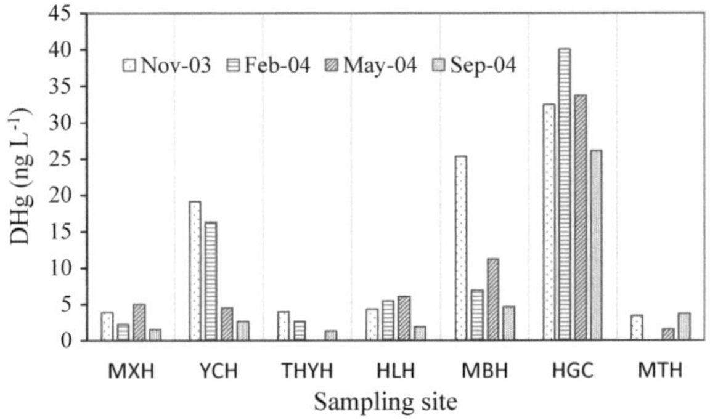

Fig. 7.16 Concentration distributions and seasonal variations of DHg in Hongfeng Reservoir inflows

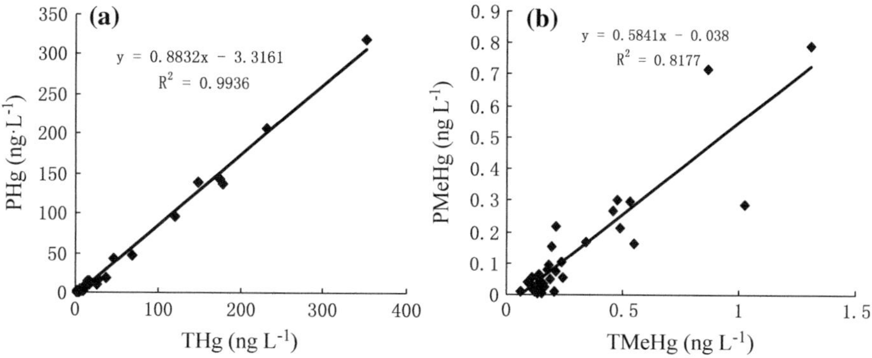

Fig. 7.17 Correlations between THg and PHg **a**, TMeHg, and PMeHg levels **b** in the rivers

Table 7.5 Distribution of mercury and TSP in Hongfeng Reservoir in June and July

Date	Parameters	MXH	YCH	TYH	HLH	MBH	GOCP	MTH
2006.06	TSP	74	24	17	26	12	81	–
	THg	3.9	150	2.4	2.8	15	–	4.3
	DHg	1.2	10	1.2	1.8	1.1	7.1	1.3
	TMeHg	0.48	0.86	0.13	0.22	0.49	–	0.55
	DMeHg	0.18	0.15	0.13	–	0.27	–	0.38
2006.07	TSP	2.9	4.5	4.1	12	11	21	1.3
	THg	2.4	14	1.2	1.5	16	68	3.1
	DHg	1.1	1.7	0.50	1.1	2.6	21	1.8

Yangchang River in June of 2006 as a result of human activities. The THg levels recorded in June of 2006 were slightly higher than those recorded in July, possibly due to heavy rainfall and surface runoff patterns occurring prior to the June sampling period (Table 7.5).

(3) MeHg levels in inflows and outflows

Figures 7.18 and 7.19 show the seasonal distribution of the MeHg levels in rivers discharging into the HF. We found no significant differences in the annual average TMeHg and DMeHg levels in all of the inflows of the HF except in the Taohuayuan River. We found that the MeHg concentrations were not high in rivers with high THg levels. Therefore, we infer that inorganic mercury is not a main influencing factor of mercury methylation. TMeHg, the DMeHg, and DOC levels in the Taohuayuan River are low throughout the year, potentially due to infertile soil around the river. Due to seasonal variations in TMeHg and DMeHg, the levels were higher in spring than in other seasons, though this was not the case for rivers polluted with industrial and domestic wastewater. This shows that MeHg in surface soil is transported into rivers via runoff during spring rainstorms. We also found significant correlations between TMeHg and PMeHg, with complex correlation

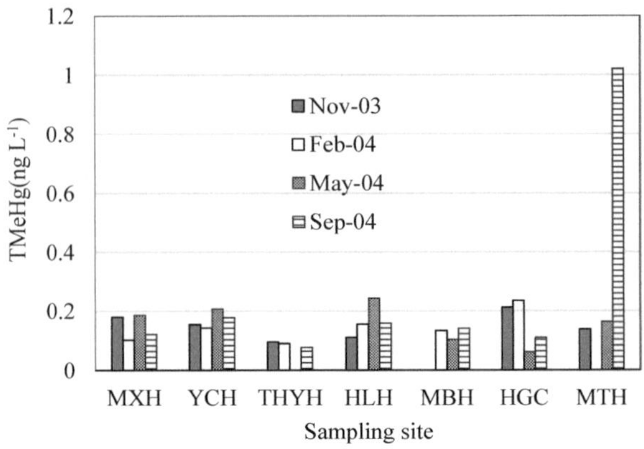

Fig. 7.18 Concentrations and seasonal variations of TMeHg in Hongfeng Reservoir inflows/outflows

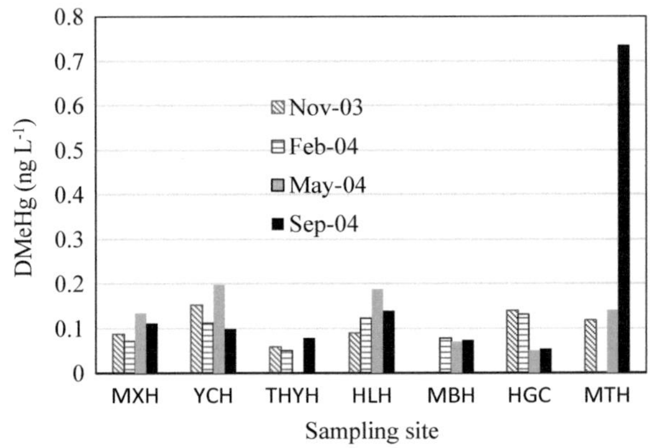

Fig. 7.19 Concentrations and seasonal variations of DMeHg in Hongfeng Reservoir inflows/outflows

coefficients and complex determinants of 0.90 and 0.82, respectively ($p < 0.001$). However, we found no correlation between PMeHg and TSP ($r = 0.006$), indicating that PMeHg is not controlled by TSP but is rather related to particulate matter properties, such as the particulate matter compositions. Hurley et al. (1994) found that inorganic mercury is more likely to bind to mineral particles and rock fragments, whereas methylmercury is more readily bound to organic particle matter.

We found no significant differences in the DMeHg levels in the HF inflows/outflows in spring or autumn. However, the TMeHg and DMeHg levels in the Maotiao River in summer increased significantly to 1.0 and 0.74 ng L^{-1}, respectively. These results show that the MeHg levels increase considerably after the river flows through the reservoir in summer, suggesting that the reservoir may serve as an important source of MeHg for the downstream river area in summer.

7.1.1.3 Mercury Species in the Sediment Cores

1. General physical properties of the sediment samples

Concentrations and distributions of OM% and water content in the HF sediment cores area are shown in Fig. 7.20a, b. The OM levels range from 2.2 to 8.8% (with an average value of 5.1%) in the Daba and from 2.7 to 7.6% (with an average value of 5.1%) at Houwu. This indicates that there are no significant differences between the OM levels in the two sites. At Houwu, the OM levels reach a maximum value at a depth of 5 cm, decrease within the upper 4 cm sediment layer, and then decline with depth. The OM levels at Daba first increase in the upper 4 cm sediment layer and then exhibit a pattern similar to that found at Houwu. The fact that the OM levels in the HF are higher than those of the DF (in the same river basin) indicates that the HF has accumulated large quantities of endogenous OM with the eutrophication and afflux of large amounts of anthropogenic OM from local sources.

However, the organic matter levels are much lower in HF sediment than those found in peat and ash soil podzolic soil sediment, ranging from 30 to 50% (Lucotte et al. 1999). Regarding the distribution of the water content in the sediment, no significant differences were found between the Houwu and Daba in the HF, with both sites showing a gradual decreasing trend from more than 90% at the surface to roughly 60% at the base.

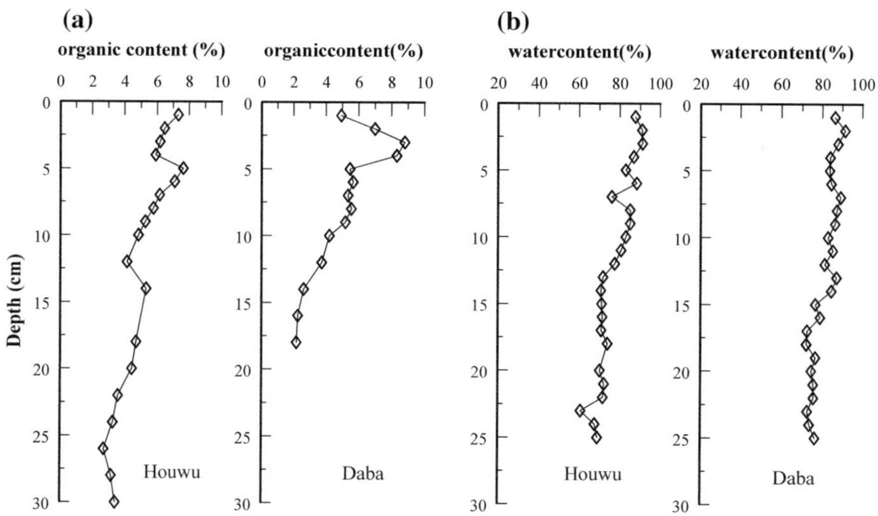

Fig. 7.20 Organic matter (OM%) and water content (WC%) in Daba and Houwu sediments in the Hongfeng Reservoir

2. Distribution of mercury species in the sediment cores
(1) Concentration distributions and seasonal variations of THg in the sediment cores

Vertical distributions and seasonal variations of THg in sediment cores of the Houwu and Daba are illustrated in Fig. 7.21. In winter, the THg levels range from 0.36 to 0.57 μg g^{-1} (with an average value of 0.43 μg g^{-1}) in the Daba and from 0.21 to 0.44 μg g^{-1} (with an average value of 0.35 μg g^{-1}) in the Houwu. In summer, THg values range from 0.33 to 0.60 μg g^{-1} (with an average value of 0.42 μg g^{-1}) in the Daba and from 0.28 to 0.52 μg g^{-1} (with an average value of 0.37 μg g^{-1}) in the Houwu. In the HF, the THg levels are significantly higher than those in other uncontaminated sediments. French et al. (1999) reported that the THg levels in sediments sampled from 34 reservoirs across Newfoundland in Canada averaged 0.039 μg g^{-1}. However, the average concentrations of THg in the WJD and DF, which are located in the same basin of the HF, were recorded at 0.25 and 0.17 μg g^{-1}, respectively. These findings indicate that sediments in the HF are contaminated by mercury due to the release of wastewater.

Relationships between mercury and organic matter content in HF sediments are shown in Fig. 7.22. Lindberg and Harriss (1974) found that organic matter mineralization may cause Hg concentration decline with sediment depth. During the degradation of organic matter, Hg associated with organic matter is released into pore water and is brought to the sediment surface where it is incorporated into newly deposited organic matter. The sediment depths in the HF are converted into sedimentary chronological sequences based on the mass depth and average sedimentation rate (0.17 ± 0.1 g cm^{-2} a^{-1}) (Wan et al. 2000). The history of

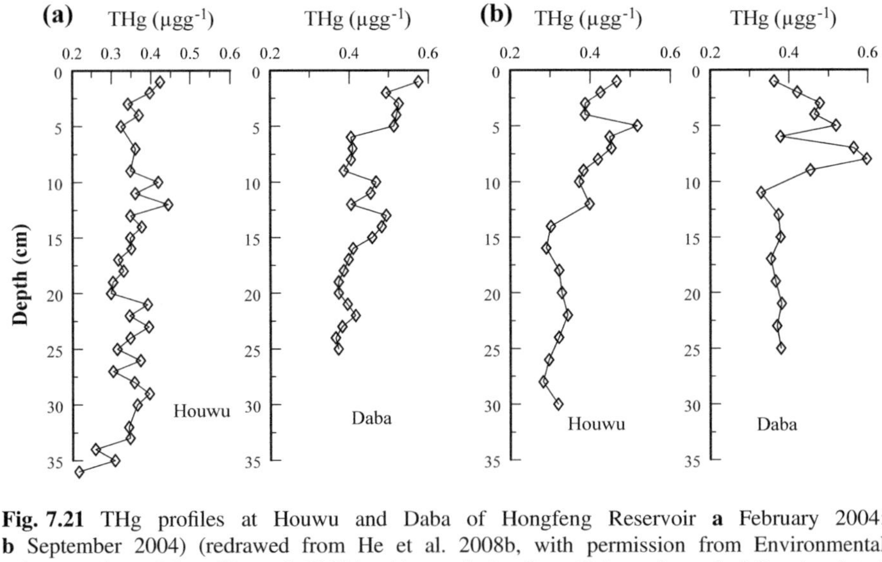

Fig. 7.21 THg profiles at Houwu and Daba of Hongfeng Reservoir **a** February 2004; **b** September 2004) (redrawed from He et al. 2008b, with permission from Environmental science; redrawed from He et al. 2008d, with permission from Chinese Journal of Geochemistry)

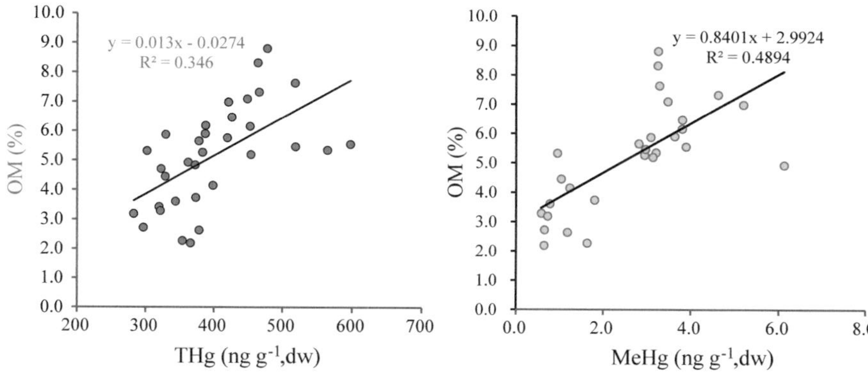

Fig. 7.22 Relationships between mercury and organic matter content in Hongfeng Reservoir sediments

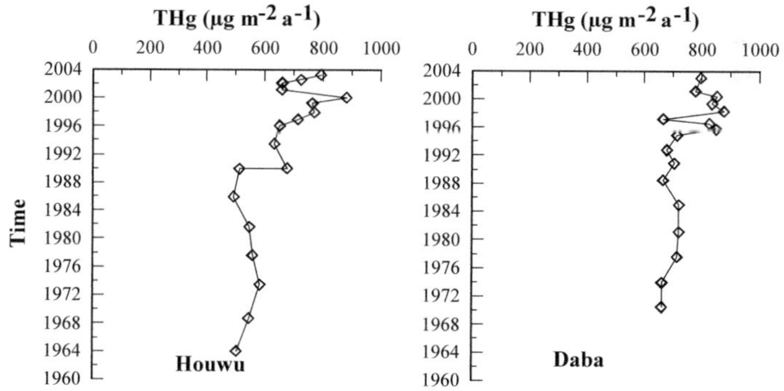

Fig. 7.23 Historical deposition rate of THg in DB and HW sediment in the Hongfeng Reservoir

mercury deposition in a time series is calculated based on the mercury concentrations and average deposition rate (Fig. 7.23). The depositional profiles of the Daba and Houwu show similar tendencies, which are higher at the surface and remain constant with depth. Regarding the deposition time, the years 1990 and 1994, respectively, mark divisions in the mercury deposition rates at Houwu and Daba. These changes in the THg depositional profiles with time may reflect two factors. First, early diagenetic alterations of sediments (i.e., the re-migration of mercury in sediments) may influence the history of mercury deposition, with organic matter degradation constantly enriching mercury in surficial sediments. On the other hand, this historical distribution of mercury may be related to increasingly severe mercury contamination levels. According to water monitoring reports, while water in the HF prior to 1990 was of good quality, rapid economic development has spurred a series of pollution incidents in the local area since the 1990s.

(2) Concentration distribution and seasonal variation of MeHg in sediment

MeHg at the two sites is shown in Table 7.6. As can be seen, average concentrations of MeHg in samples collected at both sites appears the highest in spring, especially at Daba, where MeHg in sediment remains significantly higher in spring than in any other seasons, whereas the MeHg concentration in sediment in summer in favor of mercury methylation (Kotnik et al. 2002) remains at a similar level to that in autumn and winter. These findings differ significantly from that reported by Jiang (2005) on WJD and DF.

From the profile distribution of sediment MeHg (Figs. 7.24 and 7.25), we can see that higher MeHg concentrations occur at 0 to 10-cm-depth sediment, whereas the peak value of MeHg in sediment profiles occur at different depths for each sediment core. The maximum MeHg concentrations at both of two sites in autumn and winter do not occur at the surface, but at 2 cm depth. In spring, maximum values at the two sites are both observed in surficial sediment and are much higher

Table 7.6 Concentration distribution and seasonal variation of sediment MeHg in Daba and Houwu of Hongfeng Reservoir

Sampling site	Date of sampling	Min (ng g^{-1})	Max (ng g^{-1})	Average (ng g^{-1})
Daba	2003.11	0.54	5.3	2.7
	2004.02	0.84	6.2	2.5
	2004.05	1.0	8.4	4.6
	2004.09	0.24	6.1	2.6
Houwu	2003.11	0.36	5.7	2.4
	2004.02	0.61	6.2	2.2
	2004.05	0.19	7.6	2.6
	2004.09	0.57	4.6	2.2

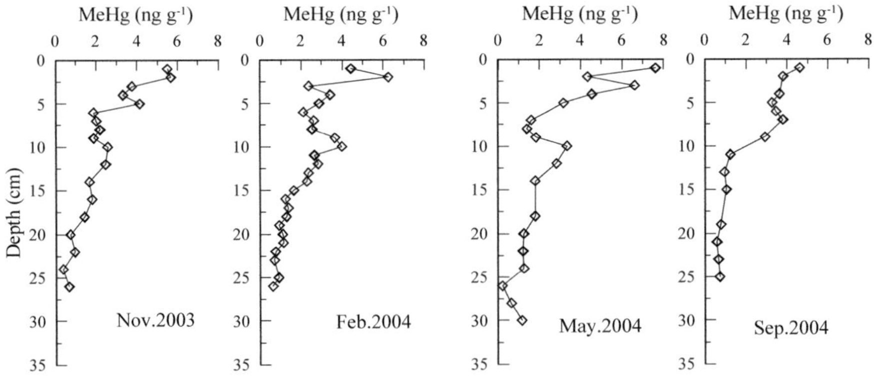

Fig. 7.24 Concentration distribution and seasonal variation of sediment MeHg at Houwu in Hongfeng Reservoir (redrawed from He et al. 2008b, with permission from Environmental science; redrawed from He et al. 2008d, with permission from Chinese Journal of Geochemistry)

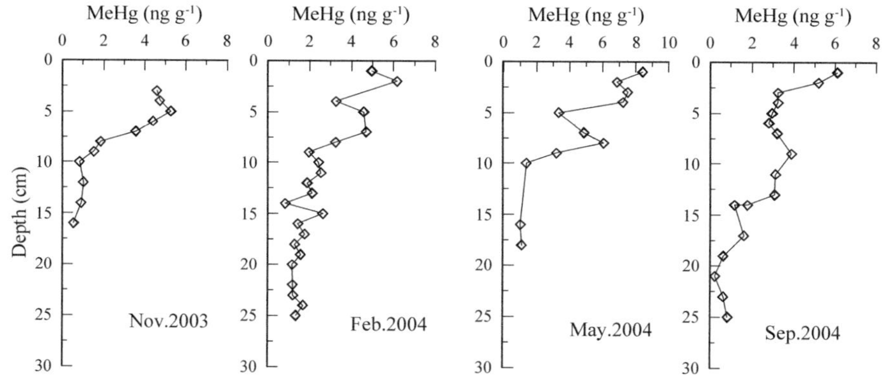

Fig. 7.25 Distribution and seasonal variation of MeHg in sediment at Daba in Hongfeng Reservoir (redrawed from He et al. 2008b, with permission from Environmental science; redrawed from He et al. 2008d, with permission from Chinese Journal of Geochemistry)

than in autumn and winter. Maximum values also observed in surficial sediments in summer are lower than in spring and remain as low as in autumn and winter.

Mercury methylation is mainly related to the microbial activity process, during which methylcobalamin is the main supplier of environmental methyl. As methylation mostly occurs under anaerobic conditions, hence, an aerobic environment is more conducive to demethylation. Numerous studies show that sulfate-reducing bacteria (SRB) are the major mercury-methylating bacteria (Gilmour and Henry 1991). Korthals and Winfrey (1987) observed that highest methylation rates occur in redox interface which is also the main activity zone of SRB. In seasonally stratified water body, the redox interface migrates with seasonal changes. In a study on redox boundary migration in Lugu Lake, Ma et al. (2000) found that redox boundary in winter coincides with sediments–water interface and migrate up into the overlying water in summer, with no obvious release of Mn, Fe, and other heavy metals.

In autumn and winter, when the entire overlying water in HF is in good oxygenation state, dissolved oxygen in pore water reaches up to 5 mg L^{-1}. The yellow oxide layer on the sediment surface and the clear pore water indicate that surfacial sediments are in aerobic state; meantime, the redox boundary below surfacial sediments coincides well with the concentration distribution of sediment MeHg. The highest MeHg concentration occurs at the depth of 2 cm below the sediment surface rather than at the surface, remaining consistent with the redox boundary (Figs. 7.24 and 7.25). While in spring, when the lake stratifies, dissolved oxygen of the pore water decreases to 3 mg L^{-1}, the yellow layer in the sediment surface disappears, and the pore water becomes turbid, showing that surfacial sediments start to enter into a hypoxic state, overlapping the redox boundary. Meanwhile, methylation rate in the surficial sediments reaches its maximum, and thus in spring, the highest MeHg concentration occurs at the water–sediment interface and remains significantly higher than in other seasons, especially at Daba where it reaches a maximum of 8.4 ng g^{-1} (Figs. 7.24 and 7.25). According to Wang (2003), in winter, obvious reduction of sulfate is only observed below the sediment surface. While in late spring, it takes place in the first 10 cm water column above the sediment surface. As the lake further stratifies in summer, when severe hypoxia occurs at the bottom, especially at Houwu, the redox boundary migrates to the bottom of the lake where the highest rate of methylation also appears. MeHg concentration and distribution in water aforesaid support that methylation takes place at the bottom of the water column at Houwu. Maximum concentration of MeHg in sediments is observed at the surface; however, it is significantly lower than in spring. Especially at Houwu, the maximum MeHg reaches only 4.6 ng g^{-1} in summer, suggesting that the maximum methylation rate in summer does not occur in sediments but in the water column as the redox boundary migrates.

From the distribution of sediment MeHg maximum values in HF, it can be concluded that methylation occurs within a zone of about 0–8 cm. Liang et al. (2003) analyzed the distribution of SRB of six species in sediments in HF using PCR method; the results show that SRB is mainly distributed at the depth of 7 cm in surfacial sediments. According to latest researches, apart from SRB, iron-reducing bacteria (FeRB) can also improve methylation of mercury. Fleming et al. (2006) observed that when SRB in sediments were all inhibited by molybdate, methylation still occurs with a considerable rate. Studies of Wang (2003) showed that the activity of FeRB in sediments in HF reaches its maximum below the depth of 10 cm, with the peak appearing at depths varying between 15 and 20 cm. Hence, the contribution of FeRB to the methylation of the sediment mercury is minor.

In addition, the methylation rate in environment is not only affected by bacterial activity, but also by multiple physical, chemical, and biological factors, including THg concentrations, mercury activity, redox conditions, pH, temperature, organic content, sulfide content, and ferromanganese circulation. There is certain correlation between the organic matter and the distribution of MeHg in sediment cores in HF ($r = 0.70$, $p < 0.001$, Fig. 7.22). Though what role organic matter plays in methylation is not known yet, organic nutrients in general can stimulate the microbial activity and thus increase the methylation rate. However, other studies also show that as organic matter can bound with divalent mercury. This leads to its decrease of the bioavailable mercury concentration, and thus the methylation is limited by the high concentration of DOC, especially in neutral pH environment (Miskimmin et al. 1992; Watras et al. 1995; Driscoll et al. 1995). Besides, abiotic methylation directly through humic acid has also been reported (Weber 1993).

The proportion of sediment MeHg as THg (% MeHg) ranges from 0.2 to 1.2% at Daba in winter, with an average value of 0.6% and between 0.1 and 1.6% in summer, with an average value of 0.6%, while it ranges from 0.20 to 1.6% at Houwu in winter, with a mean value of 0.5% and from 0.20 to 1.0% in summer, with an average value of 0.5%. Data show that %MeHg in sediments is tantamount to the methylation rate measured using mercury isotope approach (Benoit et al. 2003; Sunderland et al. 2004). The results above show that there is no significant difference in methylation rates in winter and summer. The %MeHg in HF is far lower than in sediments in reservoirs built on peatlands and in podzolic soil.

Lucotte et al (1999) reported that the proportion of MeHg reaches up to 10% at 0–10 cm depth in peatlands inundated over 10 years, while that at the humus layer in podzolic soil reaches as high as 30%. This may be related to the reservoir effects and its rich organic content (30–50%). Several studies show that MeHg in sediments and fish in new reservoirs are higher than in adjacent natural lakes (Abernathy and Cumbie 1977; Cox et al. 1979). According to Jackson, inundated soil and vegetation are important sources that increase the MeHg content in fish (Jackson 1988). These exceptionally high MeHg levels will slowly return to the level in natural lakes over time. (Porvari 1998; Schetagne et al. 1995) MeHg in sediments of HF is slightly lower than that reported by Jiang (2005) in the WJD (in the same river basin as HF). The 18-year younger WJD not only has as abundant endogenous organic matter as HF (4.4% at the surface), but also stratifies seasonally.

As HF is a seasonal hypoxia reservoir, by analyzing profile distribution features and seasonal variation of sediment MeHg in it, we can see that sediment MeHg distribution is mainly controlled by seasonal migration of redox zones. The concentration distribution of MeHg is also affected by organic matter to some degree. Seasonal changes in sediment MeHg concentrations and the maxima of profile distribution are mainly controlled by seasonal migration of the redox zone. Sediment MeHg concentration features obvious seasonal variations, and MeHg content remains the highest in spring and keeps at similar levels in summer, autumn, and winter; generally, its profile distribution decreases with depth. Its maxima appear at the subsurface in autumn and winter and at the surface in spring and summer. MeHg concentration peaks occur mainly within the active zone of the SRB. Organic matter content to a certain extent also affects the concentration and distribution of MeHg, while other factors such as the temperature and temperature changes have relatively minor impact on the MeHg concentration and distribution in HF.

3. Distribution of mercury species in pore water
(1) Distribution and seasonal variation of DHg in pore water

The organic matter, clay minerals, and Fe and Mn oxides in sediment have a strong adsorption capacity for mercury in sediment. Adsorption and desorption are very complex physical and chemical processes. According to some simulation experiments, mercury affinity shows an order as follows: mercapto > illite > montmorillonite > amino > kaolinite > carboxyl > Sand (Reimers and Krenkel 1974). Mercury adsorption and desorption in sediments are affected by a number of factors such as mercury concentration, temperature, pH, redox conditions, and various complex coordination reactions. Studies have shown that the physicochemical cycle of mercury in sediments between the solid and liquid phases is mainly controlled by S^{2-}, OM% and oxide contents (Fujiki and Tajima 1992). HgS precipitation takes place mainly in environments where high pH value and low S^{2-} are. In a reducing environment where pH is high and S^{2-} in excess, HgS precipitation will turn into a soluble mercury sulfide compound such as

HgS_2^{2-}. Organic matter can increase the solubility of mercury sulfide and transfer mercury from the solid into the liquid phase (Ravichandran et al. 1998).

In autumn, DHg in pore water in HF ranges from 10 to 21 ng L^{-1} (with an average value of 14 ng L^{-1}) at Daba, and from 7.4 to 25 ng L^{-1} (with an average value of 13 ng L^{-1}) at Houwu (see Fig. 7.26). At Daba, there are no significant change in trend from the surface to the bottom, except two peaks occurring at the depths of 4 and 17 cm. While at Houwu, within the first 5 cm depth, DHg concentration in pore water is significantly higher than at the bottom and shows a rising tendency at the depth of 18 cm before it starts to decline at the depth of 23 cm (Fig. 7.26).

In winter, DHg in pore water ranges from 10 to 25 ng L^{-1} (with an average value of 15 ng L^{-1}) at Daba, while it ranges between 9.8 and 18 ng L^{-1} (with an average value of 13 ng L^{-1}) at Houwu. As can be seen, there is no obvious seasonal variations observed between autumn and winter. At Daba, DHg is relatively higher within the first 5 cm and reaches a maximum at the depth of 5 cm (Fig. 7.27). DHg is lower below the depth of 5 cm, and it reaches yet another peak of 18 ng L^{-1} at the depth of 13 cm. At Houwu, the concentration of DHg is lower within the first 12 cm.

In spring, DHg in pore water at Daba shows a rising tendency, ranging from 11 to 27 ng L^{-1} (with a mean value of 16 ng L^{-1}). With the exception of a peak at the depth of 2 cm, concentrations do not vary significantly with depth (Fig. 7.28). While at Houwu DHg concentration varies between 16 and 28 ng L^{-1} (with an average value of 21 ng L^{-1}), we can see that concentrations are much higher than at Daba and show a declining tendency from the surface to the bottom. In summer, DHg concentration ranges from 11 to 48 ng L^{-1} (with an average value of

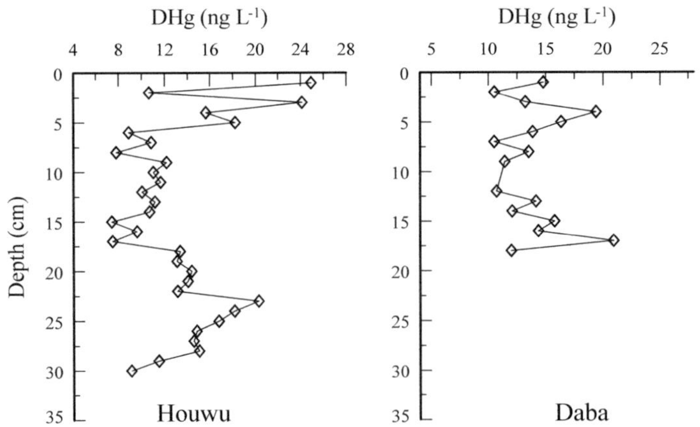

Fig. 7.26 Concentration and distribution of DHg in pore water at Daba and Houwu in Hongfeng Resevoir in autumn (redrawed from He et al. 2008d, with permission from Chinese Journal of Geochemistry)

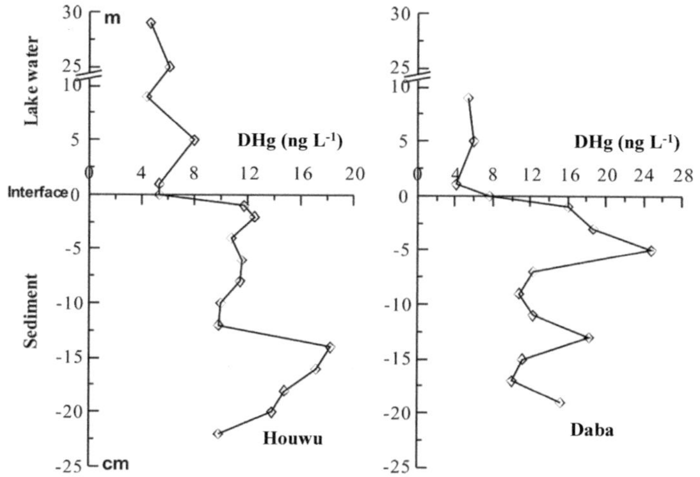

Fig. 7.27 Concentration and distribution of DHg in pore water and water columns at Houwu and Daba in Hongfeng Reservoir in winter (redrawed from He et al. 2008b, with permission from Environmental science)

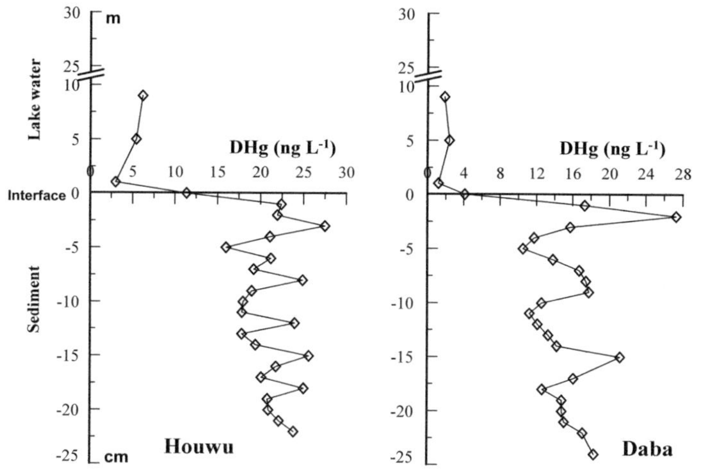

Fig. 7.28 Concentration and distribution of DHg in pore water at Daba and Houwu in Hongfeng Reservoir in spring (redrawed from He et al. 2008b, with permission from Environmental science)

24 ng L^{-1}) at Daba and from 13 to 33 ng L^{-1} (with an average value of 23 ng L^{-1}) at Houwu. It is clear that DHg concentration increases in pore water of HF in summer until reaching its annual maximum. Especially at Daba, it reaches an exceptional peak of 48 ng L^{-1} in 6–9-cm-depth pore water (Fig. 7.29).

As the equilibrium of mercury adsorption and desorption between solid/liquid phase is controlled by many factors, therefore, DHg in pore water fluctuates with

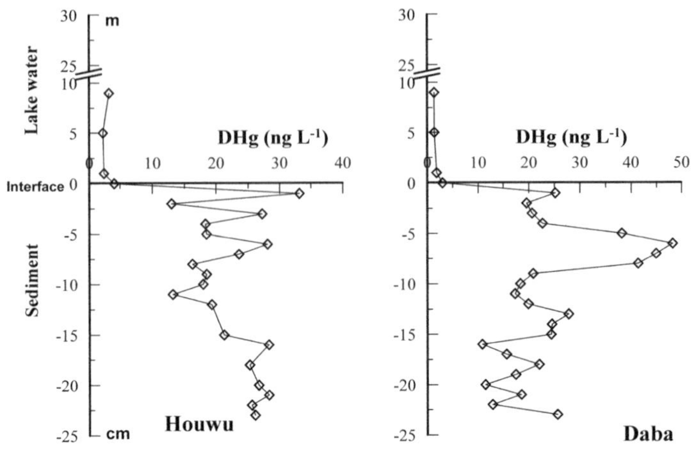

Fig. 7.29 Concentration and distribution of DHg in pore water at Daba and Houwu in Hongfeng Reservoir in summer (redrawed from He et al. 2008b, with permission from Environmental science)

depth within a relatively wide range and shows no regular pattern of variation. Peaks mostly appear within the 0–10 cm sediment depth. However, some smaller peaks may also be observed at the bottom, which may be related to the S^{2-} concentration in sediments. The mercury precipitation is enhanced by a small amount of S^{2-} appearing at the surface. Meantime, dissolution of Fe–Mn oxides under reducing condition release mercury associated with these oxides. As the depth increases and reducing condition prevails, a large amount of S^{2-} is produced, leading to more mercury precipitation as unsoluble HgS. However, the large amount of accumulated S^{2-}, HgS precipitation may be redissolved as HgS_2^{2-} in bottom water.

In summer, DHg concentrations in pore water are much higher than in autumn and winter, suggesting that seasonal changes in temperature and redox condition are also important factors controlling DHg concentration and distribution. In autumn and winter, however, sediment in solid phase and DHg concentration in pore water shows only a poor correlation ($r = 0.29$; $p = 0.035$), indicating that the mercury concentration in solid phase has little effect on that in the pore water. Average partition coefficients of the THg between solid/liquid phases at the two sampling sites (Daba and Houwu) are 3.2×10^4 and 3.0×10^4 L kg^{-1}, respectively in winter, while in summer, the coefficients, respectively, decrease to 1.7×10^4 and 1.8×10^4 L kg^{-1}. This further proves that temperature plays a very important role in equilibrium between mercury adsorption and desorption. Partition coefficients in HF are significantly higher than those observed by Jiang (2005) in WJD and DF, this discrepancy may be related to the relatively high organic matter content in HF.

(2) Concentrations and seasonal variation of MeHg in pore water

Seasonal characteristics of DMeHg concentration and distribution in pore water along the sediment column at Houwu and Daba are illustrated in Figs. 7.30 and 7.31. In autumn, DMeHg concentration varies between 0.23 and 0.82 ng L^{-1} at Daba (with an average value of 0.43 ng L^{-1}) and between 0.16 and 0.60 ng L^{-1} at Houwu (with an average of 0.32 ng L^{-1}). DMeHg concentration at Daba shows no gradient variation from the surface to the bottom with the exception of a peak at the depth of 2 cm below the surface. The DMeHg concentrations at Houwu are comparatively lower than at Daba and show an obvious declining trend from the surface to the bottom.

In spring, DMeHg concentration in pore water ranges from 0.78 to 4.2 ng L^{-1} at Daba and from 0.22 to 3.0 ng L^{-1} at Houwu (with average values of 2.1 and 1.3 ng L^{-1}, respectively). Compared with other seasons, DMeHg concentration is significantly higher in spring, reaching its annual maximum. At Daba, the DMeHg concentrations begin to decline after a constant increase from 1 to 4 cm depth, where it reaches its maximum value. It yet again reaches a smaller peak at the depth of 7 cm before showing a gradual decrease with the increase of depth. At Houwu, the DMeHg shows a relatively significant variation within 0–10 cm sediment depth. As can be seen from the figure, DMeHg concentration is significantly higher at the surface than at the bottom (Fig 7.32).

In summer, MeHg concentration in pore water varies between 0.90 and 1.7 ng L^{-1} at Daba and between 0.61 and 1.3 ng L^{-1} at Houwu (with mean values of 1.2 and 0.90 ng L^{-1}, respectively). As observed in winter, the highest values occur at the surface at neither of the two sampling sites. At Daba, there is no significant concentration gradient between the surface and the bottom, expect a

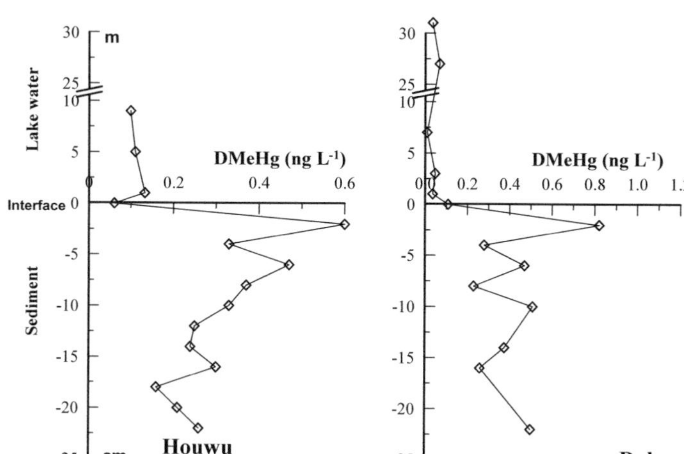

Fig. 7.30 Concentrations and seasonal variation of DMeHg in sediment pore water at Daba and Houw in Hongfeng Reservoir in autumn (redrawed from He et al. 2008b, with permission from Environmental science)

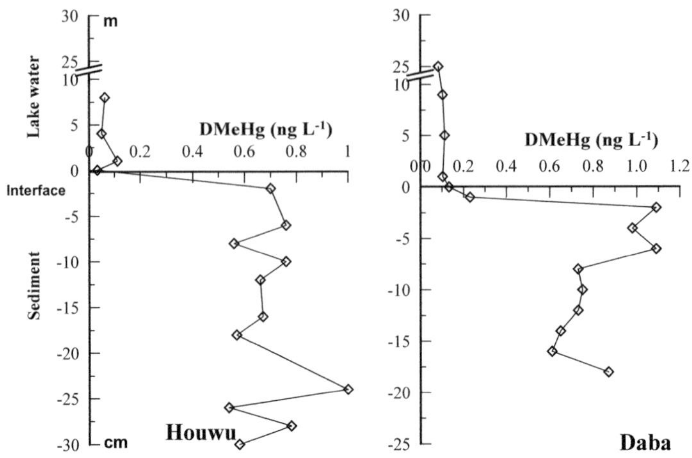

Fig. 7.31 Concentrations and seasonal variation of DMeHg in sediment pore water at Daba and Houw in Hongfeng Reservoir during winter (redrawed from He et al. 2008b, with permission from Environmental science)

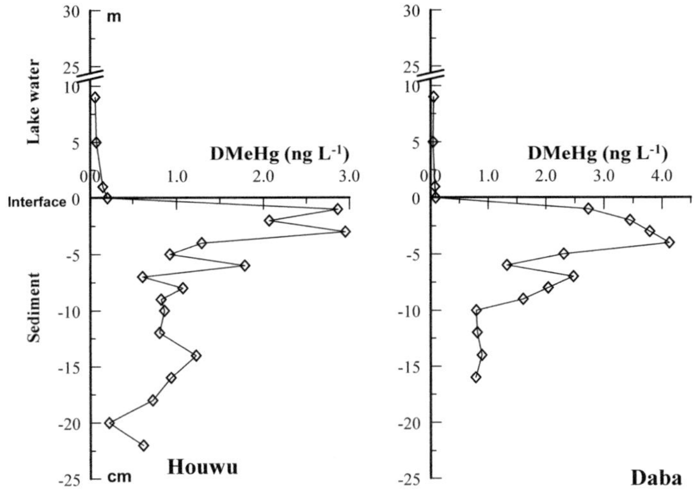

Fig. 7.32 Concentrations and seasonal variation of DMeHg in sediment pore water at Daba and Houw in Hongfeng Reservoir during spring (redrawed from He et al. 2008b, with permission from Environmental science)

peak at the depth of 8 cm. While at Houwu, DMeHg concentration shows a slight decrease within the first 10 cm, but it starts to rise again at the depth of 10 cm and reaches its maximum at the depth of 16 cm (Fig 7.33).

Similar to the DMeHg distribution in sediments in the solid phase, both the maximum and maximum average values of DMeHg occur in spring. In addition,

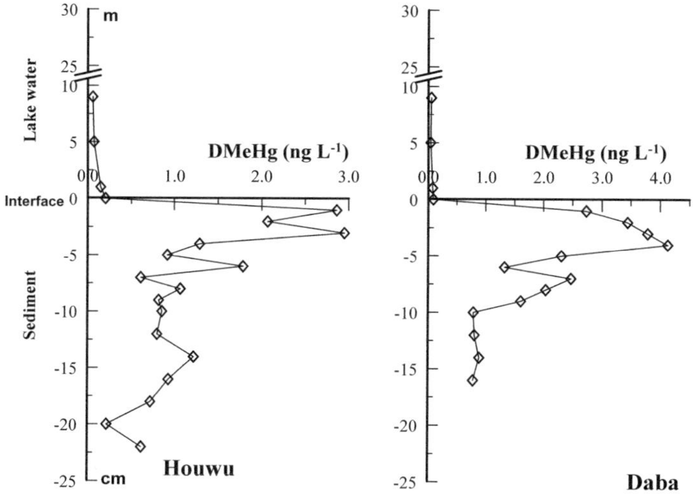

Fig. 7.33 Concentrations and seasonal variation of DMeHg in sediment pore water at Daba and Houwu in Hongfeng Reservoir during summer (redrawed from He et al. 2008b, with permission from Environmental science)

vertical distribution of DMeHg follows a similar pattern. However, contrary to that observed for DHg, there is a significant correlation ($r = 0.70$, $p < 0.001$) between pore water DMeHg and sediment MeHg. This indicates that MeHg concentration distribution in the solid phase and in pore water are closely related, both may be subject to the same mechanisms, such as the methylation process. Distribution of pore water DMeHg has its own features, it is quite low in summer, but slightly higher than in autumn and winter. Like the MeHg profile distribution in the solid phase, higher values of pore water DMeHg are also distributed within the first 8 cm at the surface, though not exactly at the same place. This proves that the pore water concentration is not only influenced by many factors in the solid/liquid distribution mechanism, but may also be related to the methylation. Especially in summer, peak values in pore water occur at the depth of 8 and 16 cm, which suggests that DMeHg may be being produced at the bottom. The distribution of sediment MeHg indicates that no intense methylation happens at the bottom of sediments, but the Hg^2 activity available is present in the liquid phase; therefore, mercury methylation and demethylation in the liquid phase are supposed to be more sensitive. Wang (2001) pointed out that pore water chemistry is an indicator of early sediment diagenesis. Many changes in sediments which are difficult to observe can lead to significant differences in chemical behavior of solutes in pore water. In HF, due to the small proportion of DMeHg in sediments and its intense demethylation, methylation that can be reflected in pore water may not necessarily be able to be reflected timely in the solid phase. According to Wang (2003), there is an intense reduction of iron at the bottom of sediments in HF, which may have created conditions for sediment methylation at the bottom. Hines et al. (2004) found that the peak value of DMeHg concentration in pore water in Spring Lake in September 2001 occurred from the depth of 20 to 31 cm at the bottom rather than at the surface. They assumed that higher sediments methylation rates at the bottom were caused by the extension of bacteria sulfate reduction activity to a deeper area under the higher temperature in summer.

The average partition coefficients of the solid/liquid MeHg at Daba in all seasons are 6.1×10^3 L kg^{-1} in autumn, 4.8×10^3 L kg^{-1} in winter, 1.9×10^3 L kg^{-1} in spring, and 3.0×10^3 L kg^{-1} in summer. While at Houwu, the average partition coefficients are 6.9×10^3, 3.2×10^3, 2.5×10^3, and 2.7×10^3 L kg^{-1}, respectively. Both the two sampling sites present the following order: autumn > winter > summer > spring. In summer, the average concentration of solid MeHg remains at a similar level as in autumn and winter, but the solid/liquid partition coefficient is significantly lower than in autumn and winter. This indicates that summer is in favor of the release MeHg from the solid phase into the liquid phase, showing a consistency with the distribution of solid/liquid THg. MeHg partition coefficient between solid phase and pore water in HF ranges between 6.5×10^2 and 1.1×10^5 L kg^{-1}, which is similar to the marine sediment partition coefficient reported by Lyon et al. (1997) and which is higher than that (3.1×10^2–3.3×10^3 L kg^{-1}) reported by Jiang (2005) at WJD and DF. This is probably due to the relatively higher OM% and stronger solid adsorption of MeHg. A significant correlation between the solid/liquid partition coefficient of MeHg and organic matter is also observed in HF during summer ($r = 0.6$, $p < 0.01$), suggesting that organic matter is also an important factor affecting the solid/liquid partition in sediment profiles.

The above analysis shows that the concentration of MeHg in solid phase and pore water are closely related, apart from being influenced by the solid/liquid partition coefficients, both are controlled by the production process of MeHg. In addition, factors such as temperature and organic matter also play important roles in the solid/liquid partition of sediment MeHg.

(3) Diffusion flux of inorganic mercury and MeHg to overlying water

Sediment–water interface is not only an important physical interface in a water body, but also an important chemical interface. Sediments and the overlying water usually have very different physical and chemical characteristics, so intense exchanges of matter and energy occur at the sediment–water interface. The upward migration and diffusion of dissolved substances in pore water is an important factor affecting chemical characteristics of the overlying water. It is also one of the major ways how contaminants buried in sediments re-release the overlying water and cause secondary pollution. Therefore, physical and chemical behaviors of substances in the sediment/interface have received the attention from researchers.

The diffusion and migration of substances in sediments occur mainly through four different ways: molecular diffusion, dispersion, bioturbation, and hydrodynamic disturbance, among which molecules diffusion is the most important way how solutes in sediment are transported through pore water (Krom and Berner

1980). Generally, the diffusion flux of dissolved components can be estimated by two approaches; one is to estimate the diffusion static flux using concentration gradient and diffusion mathematical model, and the other one is to measure directly the diffusion flux of solutes using a flux chamber. Gill et al. (1999) observed that the flux measured is higher than the one estimated mathematically and suggested that this discrepancy is due to interference caused by the activity of living organisms close to the sediment–water interface. As a seasonal hypoxia reservoir, HF has little activity of benthic fauna (Zhang 1999), so we estimated the diffusion flux of the DHg and DMeHg at the sediment/water interface by measuring concentration gradient and using Fick's First Law. When calculating the diffusion flux of dissolved substances in pore water, we excluded effects of physical advection, etc. (see Chap. 3 for the detailed formulas).

First, we calculated concentration differences using mercury concentration in pore water at both the surface and the bottom. Then we took the average concentration of DHg and DMeHg at the depths of 1 and 2 cm below the surface as parameters. Based on the above parameters and concentration differences, we worked out the diffusion flux of DHg and DMeHg at the sediment/water interface (Table 7.7). The diffusion flux of mercury in pore water in HF shows significant seasonal variations (summer > spring > winter). The seasonal diffusion flux of the DMeHg is largely consistent with the seasonal mercury concentration in pore water, and its highest value occurs in spring and shows no obvious differences in other seasons. As for DMeHg, concentration in pore water is slightly higher in summer than in winter and autumn; yet, its diffusion flux decreases as concentration rises in the overlying water. The diffusion flux of DMeHg at the sediment–water interface

Table 7.7 Diffusion flux of mercury in pore water and its contribution rate the overlying water at Houwu and Daba in Hongfeng Reservoir

Date of sampling		2003.11		2004.2		2004.5		2004.9	
Sampling site		Houwu	Daba	Houwu	Daba	Houwu	Daba	Houwu	Daba
THg	Concentration gradient (pg/cm^4)	–	–	6.8	9.8	11	18	19	22
	Flux of diffusion ($ng\ m^{-2}\ day^{-1}$)	–	–	45	65	72	120	120	140
	Contribution rate	–	–	0.05	0.05	0.09	0.37	0.30	0.56
MeHg	Concentration gradient (pg/cm^4)	0.54	0.71	0.67	0.53	2.3	3.0	0.70	0.58
	Flux of diffusion ($ng\ m^2\ day^{-1}$)	4.9	6.0	6.0	4.7	20	27	6.4	5.2
	Contribution rate	0.43	0.50	0.65	0.18	1.4	3.7	0.10	0.30

In HF is in the same range (5–35 ng m^2 day^{-1}) to that of sediment mercury in Haibin reported by Hammerschmidt et al. (2004). According to Gill et al. (1999), MeHg diffusion flux in Lavaca Bay varies within a wider range (0.2–1500 ng m^{-2} day^{-1}), reaching its maximum in the early spring. However, Covelli et al. (1999) observed that the maximum flux in the Gulf of Trieste occurred in autumn and winter.

The diffusion flux of mercury calculated using Fick's First Law is just a static one, while the actual diffusion flux of sediment is influenced by many factors such as the "dual effect", adsorption, and ion exchange reaction. All these factors can reduce sediment diffusion flux. On the other hand, factors such as sediment bioturbation and vertical exchange of pore water can increase mercury release. In addition, environmental parameters of the sediment/water interface, like redox conditions, pH, etc., will affect the diffusion flux of mercury. Therefore, the calculated diffusion flux reflects a migration trend driven by the concentration gradient. Some studies suggest that the aerobic layer of surficial sediments is a barrier preventing the upward diffusion of sediment mercury (Gagnon et al. 1996), whereas some studies show that the mercury diffusion flux increases with the decrease of pH, DO (e.g., Ullrich et al. 2001). In HF, however, the maximum emission flux of sediment MeHg does not occur in summer; this is related to the fact that HF is a hypoxia reservoir, where methylation is mainly affected by the seasonal migration of the redox zone. By assuming that molecular diffusion is the only way for the upward migration of sediment mercury and the lake water is well mixed, we can work out to what level the overlying water is affected by the upward diffusion of dissolved mercury in sediments. See Sect. 3 of Chap. 3 for detailed calculation methods and formulas.

Calculation results show that, in HF, the overlying water is most affected by the upward diffusion of DHg in summer and slight in winter (summer > spring > winter). While DMeHg in sediments has the greatest impact on the overlying water in spring and the least in summer, it affects the overlying water to the similar extent in autumn and winter. We can also see that DMeHg has greater impact at Daba than at Houwu, with the exception of winter.

7.1.2 Biogeochemical Processes of Mercury in the Baihua Reservoir

7.1.2.1 Sampling Sites Descriptions and Sample Collection

The sampling was conducted at the upstream (Sites: HQ, XMC, YJZ), midstream (Site: MT), and downstream (Sites: DB, BF) of the BH, and numerous inflow and outflow rivers (Fig. 7.34). The sampling periods include November 2002 (Autumn), March 2003 (Spring), August 2003 (Summer), March 2004 (Spring), and August 2004 (Summer). Different samples such as water columns, sediments,

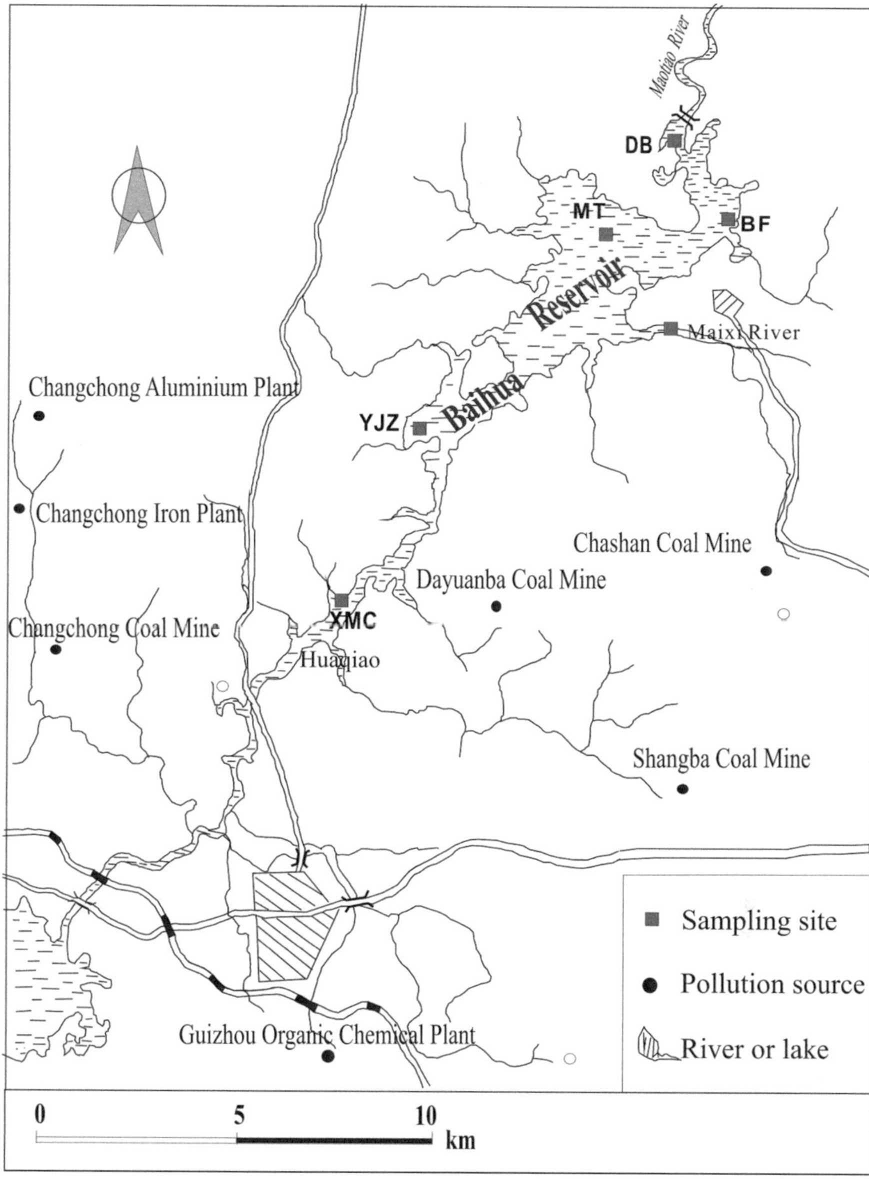

Fig. 7.34 Sampling sites of the Baihua Reservoir

pore waters, and inflow and outflow surface water were collected. Water samples in the reservoir were taken from different depths throughout the entire water columns, while only surface water was sampled in the inflows and outflow of the reservoir.

The sediment cores at sampling sites DB, BF, and YJZ were collected in November 2002, and the sediment cores at XMC and MT were collected in March 2003 for the THg analysis. Sample preparation and analysis methods are shown in Sect. 7.1.1.

Quality control for the THg determinations was addressed with method blanks, blank spikes, matrix spikes, certified reference materials (GBW07305, IAEA405), and blind duplicates. The average THg concentration of the geological standard of GBW07305 was 0.10 ± 0.01 mg kg^{-1} ($N = 6$), which is comparable with the certified value of 0.10 ± 0.02 mg kg^{-1}. The percentage of recoveries on spiked samples ranged from 91 to 112% for THg in sediment samples. The relative percentage difference was <0.6% for THg analysis in sediment samples. The average MeHg concentration of 5.45 ± 0.50 ng g^{-1} ($n = 6$) was obtained from IAEA405 with the certified value of 5.49 ± 0.53 ng g^{-1}. Limit of detection was 0.003 ng g^{-1} for MeHg in sediment sample. The average percentage of recovery of spiked samples was 97.9% for MeHg in sediment. The relative percentage difference was <4.5% for total Hg in water and pore water samples.

7.1.2.2 Mercury Species in the Water Columns

1. General water quality characteristics

The Baihua Reservoir is located in the subtropical area of southwestern China. Water columns were sampled in spring (March, dry season) and summer (August, wet season), and the sampling sites were dependent on the local conditions. In the dry season, due to less rainfall and huge agricultural water consumption, the water level was low; the rainfall increased significantly in the wet season; however, due to more the industrial and agricultural water consumption at the same time, the water level only increased slightly. The main water supply for the BH came from the outflow of the HF (which located at the upstream of the BH), tributaries of the BH, and direct precipitation. The outflow of the BH was the Maotiao River, downstream of the dam. River runoff, the amount of pollutants discharged from pollution sources, temperature, water parameters (pH, DO, COD, salinity, water temperature), and other factors vary seasonally and spatially. These factors may affect the concentration, distribution, and transformation of mercury.

Monthly variations of air temperature and rainfall in the BH basin are shown in Fig. 7.35. In March of each year, the average temperature was approximately 10 °C and the rainfall was less than 50 mm. The BH received very little river runoff during this period, and the average water depth was less than 10 m according to the actual measurement. The water might not have thermal stratification during this time due to shallow depth and lower air temperature. The highest annual temperature of BH occurs from July to August, and the rise of the water temperature resulted in obvious thermal stratification, which would promote the transformation and migration of mercury. From April to July, abundant rainfall causes increased

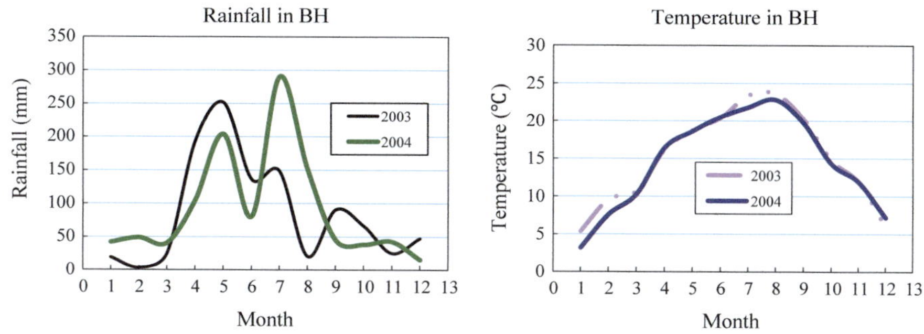

Fig. 7.35 Variation of rainfall and air temperature of the BH during 2003–2004 (data provided by Guizhou Meteorological Bureau)

water supply to the reservoir, but also increased contribution of PHg through river runoff.

Main factors influencing seasonal variations of water mercury include the seasonal thermal stratification of water body, and the changes of Hg input through changing runoff. Seasonal thermal stratification of the water body directly led to the change of seasonal hypoxia in the bottom water, and to the distribution of other substances (e.g., biological material and microorganisms) in the water column. Variations of water parameters for the BH are shown in Figs. 7.36 and 7.37. No significant stratification of pH, dissolved oxygen (DO), and conductivity was observed in the spring and autumn of 2003 and 2004. When the air temperature rose in a certain period of time, it may have resulted in a short-term thermal stratification. For instance, the surface temperature was significantly higher than that of the bottom water at the DB site in March 2003. The temperature difference was up to 6 °C, but this situation disappeared as the temperature dropped.

In summer, due to the continuous high air temperature, water in the BH has a relatively stable thermal stratification. The surface water temperature was significantly higher than that of the bottom water. The pH, DO, and electrical conductivity were also significantly stratified, especially dissolved oxygen showed anaerobic conditions in the lake bottom (Fig. 7.38). As a freshwater lake, the water salinity of the BH changed in a narrow range, i.e., 0.1–0.5‰ with an average value of 0.2‰. Affected by atmospheric acid deposition, the sulfate content was significantly elevated, and the average value was approximately 100 mg L^{-1} in 2002.

2. Species and seasonal variation of mercury in water

The seasonal variation of air temperature, rainfall, and other climatic parameters largely control the stratification and mixing of water in the BH, the temperature, dissolved oxygen, and other physicochemical characteristics of the water column. These changes also have a great influence on the concentration, distribution, migration, and transformation of mercury in the BH.

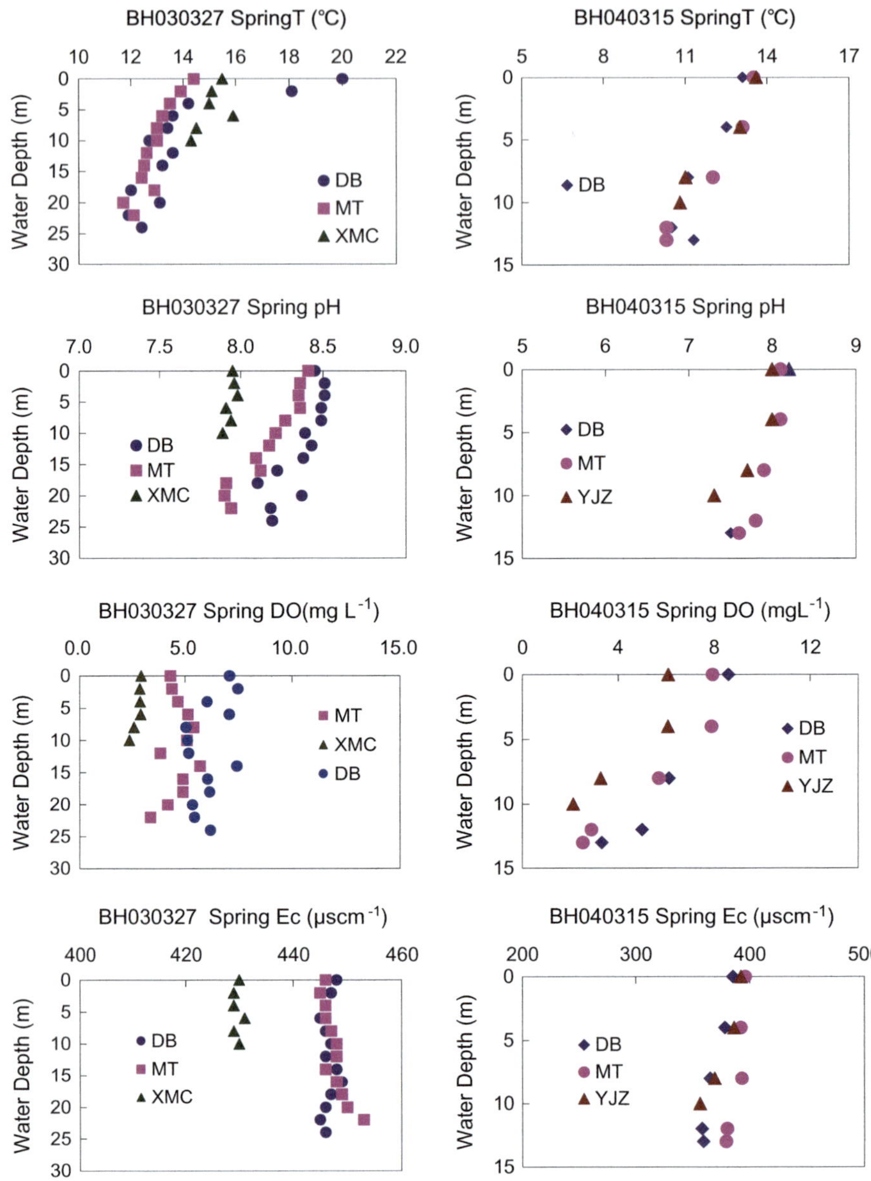

Fig. 7.36 Variation of water parameters at different sampling points of the Baihua Reservoir (*Note* Daba—DB, Matou—MT, Yanjiaozhai—YJZ, Xiemeichang—XMC)

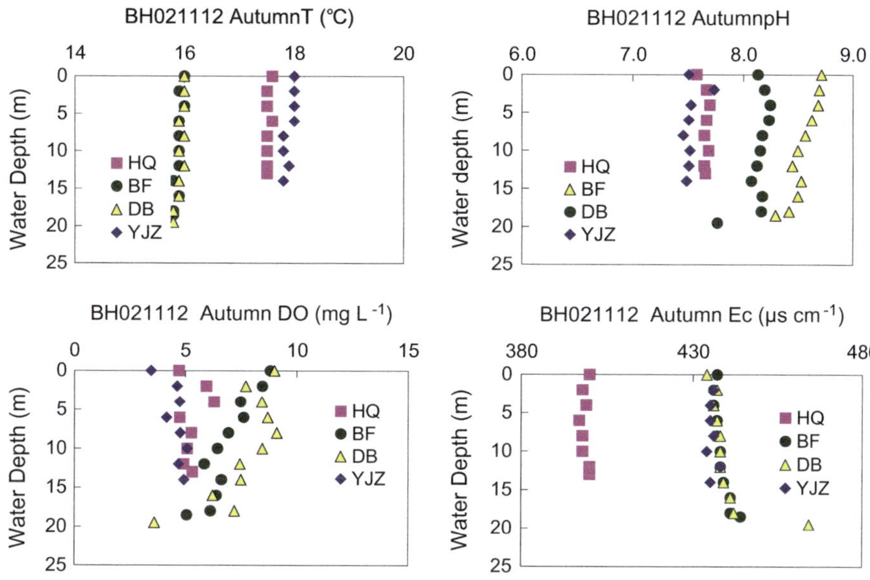

Fig. 7.37 Variation of water parameters at different sampling points of the Baihua Reservoir in autumn 2003–2004 (*Note* Daba—DB, Yanjiaozhai—YJZ, Bengfang—BF, Huaqiao—HQ)

(1) THg, DHg, and PHg in the water column

The main water supply of the BH is the HF and several tributaries around the Reservoir (Fig. 7.34). The main mercury contributor to the BH is the Beimen River, which is the main sewage drain of the Guizhou Organic Chemical Plant (GOCP) discharging mercury-containing sewage. This river also accepts the municipal wastewater and sewage from the Qingzhen City. Mercury in the Beimen River directly affects the mercury content of the BH. In the dry season, the rainfall and runoff from the surrounding tributaries (such as the Xiaohekou River, the Dayuanba River, and the Xiangpi River) was small. In March 2004, there were no water flows in rivers near the HQ. This completely cut the mercury release off the GOCP, and therefore THg in the BH waters was decreased by approximately 70%. In the wet season, abundant rainfall caused the increase of runoff from the tributaries, and therefore increased input of mercury (especially the PHg). Meanwhile, the exchange cycle of lake water was shortened in the wet season, resulting in a greater diffusion dilution of mercury and a stronger ability for self-purification. These two aspects together result in the similarity of water THg contents between wet and dry seasons, but the proportion of PHg was significantly higher than in the wet season (Table 7.8). From upstream to downstream of the BH, the proportion of PHg in the THg gradually decreased, because of the settling of PHg.

In the dry season, when the air temperature was low, convection between the surface and bottom water easily occurred, and therefore the water was mixed well, and the temperature difference between the surface and bottom waters was only

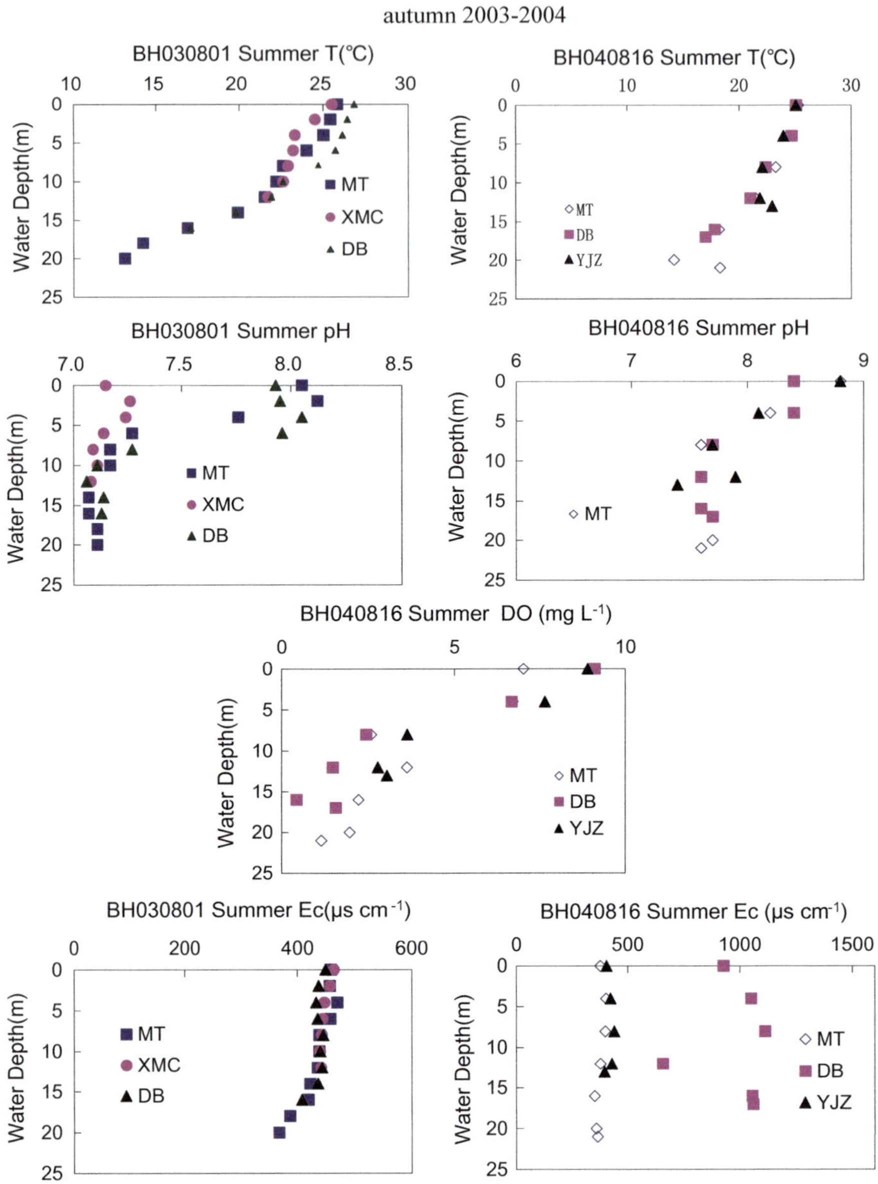

Fig. 7.38 Variation of water parameters at different sampling sites of Baihua Reservoir in summer 2003–2004

approximately 1–2 °C. In the wet season, however, the water temperature between the surface and bottom layers was apparently different and the mercury content also exhibited large vertical differences (Table 7.9).

Table 7.8 Comparison of main mercury species at different sampling points of Baihua Reservoir in wet and dry seasons

Year	Season	Upstream		Midstream		Downstream	
		THg (ng L^{-1})	PHg/THg (%)	THg (ng $L^{-1}L$)	PHg/THg (%)	THg (ng L^{-1})	PHg/THg (%)
2003	Wet season	50	91	26	60	23	61
2003	Dry season	42	72	35	42	28	46
2004	Wet season	12	60	17	44	8.6	39
2004	Dry season	11	42	12	25	9.5	25

Table 7.9 Seasonal variation of mercury species at different sampling sites of the Baihua Reservoir in the wet and dry seasons (ng L^{-1})

Year	Season		Upstream			Midstream			Downstream		
			THg	DHg	PHg	THg	DHg	PHg	THg	DHg	PHg
2003.3	Dry season	Mean	42	15	30	34.8	17.5	14.7	28.5	15.3	13.2
		Max	63	21	42	75.1	34.0	47.8	38.4	27.2	26.4
		Min	31	12	17	22.8	11.5	4.8	20.2	10.7	7.6
2003.8	Wet season	Mean	51	18	46	25.6	10.2	15.3	22.7	9.0	13.8
		Max	150	20	133	53.6	18.6	35.0	44.6	17.5	27.1
		Min	32	14	17	7.0	1.5	5.5	15.1	5.4	7.2
2004.3	Dry season	Mean	11	6.3	4.6	36.8	15.1	21.7	9.5	7.1	2.4
		Max	15	8.2	8.6	75.1	34.0	47.8	11.8	9.5	2.8
		Min	8.4	2.7	0.90	7.0	1.5	4.8	7.8	5.6	1.8
2004.8	Wet season	Mean	12	5.0	7.5	16.7	9.4	7.3	8.6	4.6	3.4
		Max	110	6.9	105.7	56.2	17.6	45.4	86.3	8.5	77.8
		Min	6.9	3.6	1.7	10.0	2.9	3.0	6.7	3.2	1.7

In the dry season (March), PHg in the lake water increased slightly with increased depth, with the highest mercury at the sediment–water interface. In the wet season, PHg was lower (Fig. 7.39), which might be caused by the algae that bloom and enrich the surface water in mercury in the spring. In the wet season (August) when the temperature was the highest, the water in the reservoir has formed a stable vertical stratification, and the biomass in different temperature layers is different. These organisms accumulated mercury in their bodies, resulting in obvious peaks at depths of 6–10 meters and at the sediment–water interface in the vertical profile. The stratification prevents vertical water flow, impeding the diffusion of mercury released from the pore water to the overlying water. Mercury concentration in the water increased with depth, and the highest value was also located at the sediment–water interface.

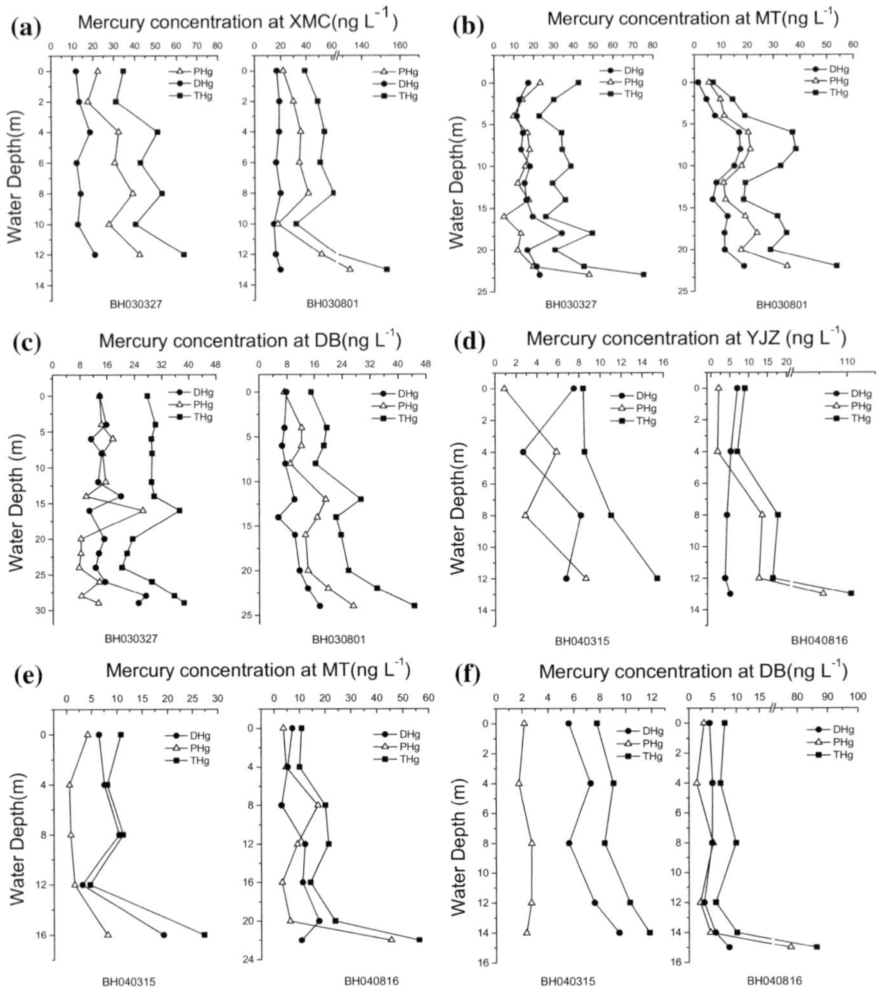

Fig. 7.39 Vertical variation of mercury species in the water column at different sampling points of the Baihua Reservoir in different years and seasons. **a**, **d**—Upstream; **b**, **e**—midstream; **c**, **f**—downstream

The main Hg source for the BH was sewage from the GOCP located upstream as mentioned before, and the mercury concentration decreased from upstream to downstream, whereas the proportion of PHg to THg decreased gradually (Fig. 7.40). Mercury is higher in the middle and bottom layers, which mainly consists of PHg. In the dry season, PHg and dissolved mercury contents were roughly equal, whereas in the wet season, mercury specifications were influenced by the water flow of the upstream.

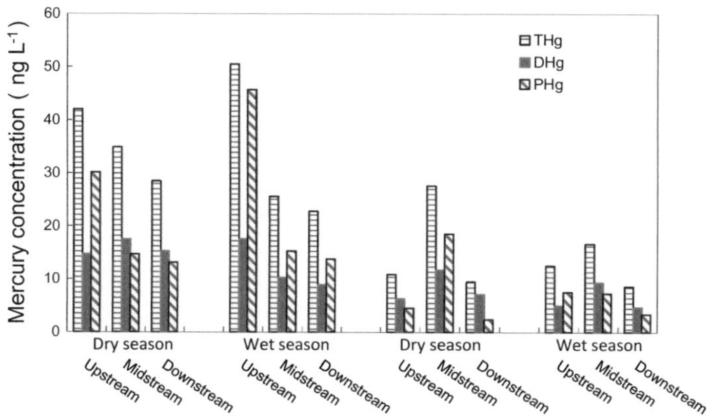

Fig. 7.40 Changes of mercury concentration at different sampling points of the Baihua Reservoir with distance from the mercury sources in different years

Mercury in the water was variable at different locations. For the upstream sites (e.g., HQ, XMC, and YJZ) that mainly receive water from the HF, Changchong River, Dayuanba River, and industrial wastewater and domestic sewage from the GOCP and other enterprises in Qingzhen City, the mercury concentration in waters was high. The highest value was 127 ng L^{-1}, especially at HQ, and the average mercury concentration was 70 ng L^{-1}. At midstream sites, such as MT, the mercury gradually settled down and was purified through the narrow watercourse. However, this site was open to visitors, so the disturbance intensity is larger than other sampling points. At midstream of the Maixi River and the Shangmaixi River, the rivers received wastewaters from fertilizer plants and coal mines that contain high mercury content, so the mercury level was still high at midstream but lower than the upstream. DB at the downstream was located in front of the BH dam, and mercury content was the lowest. Although it received sewage from the Nanmen River through the pump house, it would not have significant impact on the mercury concentration because the Nanmen River was small and was frequently dried up in the dry season.

(2) RHg and DGM

Dissolved Hg^0 usually contributes approximately 97% of dissolved gaseous mercury (DGM) in water (Vandal et al. 1991). Hg^0 is produced during various processes that reduce Hg^{2+} compounds in waters (Nelson et al. 1973; Steven et al. 2002) or humus (Alberts et al. 1974; Miller 1975; Allard and Arsenie 1991; Mason and Fitzgerald 1993; Nriagu 1994; Sullivan and Mason 1999; Tseng et al. 2004). The average reactive mercury (RHg) in water (including the sediment interface water) of the BH was 3.8 ng L^{-1} (range: 0.30–12 ng L^{-1}). The average DGM was 0.11 ng L^{-1} (range: 0.01–0.30 ng L^{-1}). In the nonpolluted lakes, RHg and DGM were generally low. For instance, the RGM is lower than 1 ng L^{-1} (Meuleman et al. 1995), and DGM ranges from 0.01 to 0.30 ng L^{-1} (Kotnik et al. 1991;

Sullivan and Mason 1998). Compared to previous results, the RHg in the BH is slightly higher, whereas DGM was in the background level.

Although the contents of RHg and DGM in the water are very low in the lake system, these two Hg species play a very important role in the Hg geochemical cycle through the mutual transformation process. RHg can be transformed to Hg^0 or MeHg in water. RHg methylated into MeHg increases the degree of mercury toxicity on the ecosystem, but when RHg is transformed into Hg^0 under the facilitation of light and biological processes, its potential harm to the ecosystem is reduced. The transformation of RHg and Hg^0 is affected by many factors, such as the pH value, light (Amyot et al. 1997; Lanzillotta et al. 2002; Tseng et al. 2004), and microorganisms (e.g., Vandal et al. 1991; Steven et al. 2002), thereby causing a seasonal variation of the concentration of these two mercury species.

RHg in the BH in the dry season was generally lower than in wet season (Fig. 7.41). The average RHg content in the dry season (excluding the interface water) was 2.2 ng L^{-1} for the upstream, 1.7 ng L^{-1} for the midstream, and 1.6 ng L^{-1} for the downstream. The average RHg content in the wet season was 4.7 ng L^{-1} for the upstream, 3.0 ng L^{-1} for the midstream, and 1.9 ng L^{-1} for the

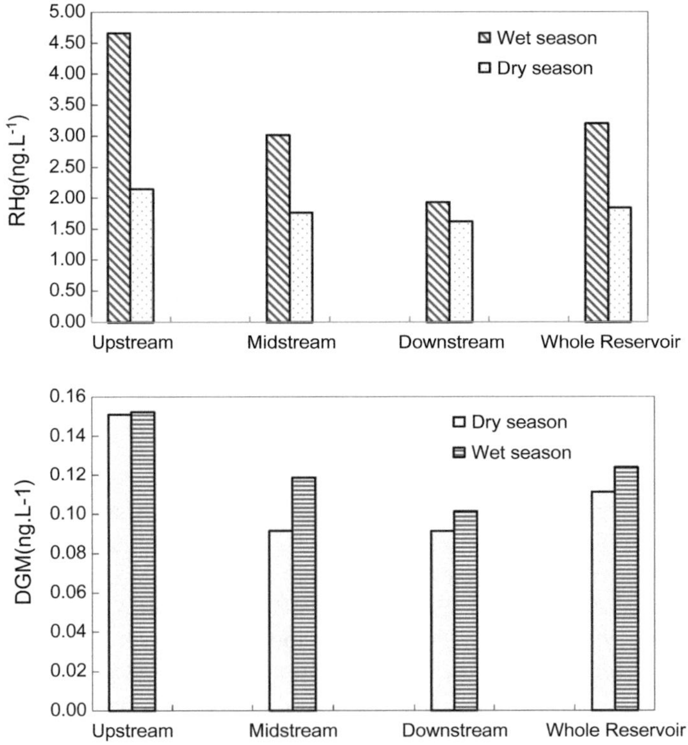

Fig. 7.41 Seasonal variation of DGM and RHg in the water of the Baihua Reservoir with distance to the pollution source

downstream. The annual average of RHg in the water was 2.5 ng L^{-1}. The average DGM in the BH in the dry season (excluding the interface water) was 0.15 ng L^{-1} for the upstream, 0.09 ng L^{-1} for the midstream, and 0.09 ng L^{-1} for the downstream. DGM in the wet season was 0.15 ng L^{-1} for the upstream, 0.12 ng L^{-1} for the midstream, and 0.10 ng L^{-1} for the downstream. The annual average of RHg content in the water was 0.12 ng L^{-1}. The seasonal variation of RHg and DGM reflected the changes of temperature and microbial activity that trigger the transformation of different mercury species in water. The DGM in the dry season was generally lower than in the wet season, and the DGM in the upstream was higher than in the midstream and downstream. This showed that the production of DGM in the wet season is higher with higher sunlight and that the DGM was higher when close to the mercury pollution source. This phenomenon was consistent with the distribution of RHg.

RHg in the upstream was significantly higher than the midstream and downstream, and it increased with the increase of water depth. The upstream was close to the pollution source, and hence all mercury species were higher and the potential for other mercury species to convert to RHg was great. In the bottom layer, the RHg was higher, both due to the conversion of other forms of the mercury that was discharged from the pollution sources to RHg and due to the emission from sediment pore water to the overlying water. The DGM in the vertical profile from the surface to the bottom decreased gradually in BH, but there also appeared higher peaks at some individual points of the bottom water, indicating that the formation and release of elemental mercury was the result of both biological and abiotic processes. In general, the DGM was in a supersaturation state. The flux of Hg^0 into the atmosphere can be affected by light intensity, temperature, pH, the speciation and number of microorganisms in water, and water quality. In the BH, the sun's radiation was obviously variable in different sampling periods, whereas other parameters might not vary too much. The average DGM content of three sampling points gradually decreased from the upstream to downstream.

In the water column of the BH, RHg increased with water depth and RHg in the sediment–water interface was generally higher in summer. This might be related to the release of RHg from the sediment pore water. RHg in the sediment surface pore water could reach up to tens to hundreds ng L^{-1}. DGM generally decreased with the increase of water depth, but occasionally some individual sampling points appeared as peaks at the middle and bottom layers (Fig. 7.42). The DGM in surface water was in the supersaturation state under the effect of photoreduction, and it was released to the atmosphere at the water/gas interface (Feng et al. 2002).

(3) MeHg in the water column

The results for TMeHg and DMeHg in water samples of the BH in March 2004 (representing the cold and dry season) and August 2004 (representing the warm and wet season) are shown in Fig. 7.43. There were significant differences in the TMeHg, PMeHg, and DMeHg in the water among different seasons. The average TMeHg (excluding the sediment–water interface) was 0.14 ng L^{-1} in the dry

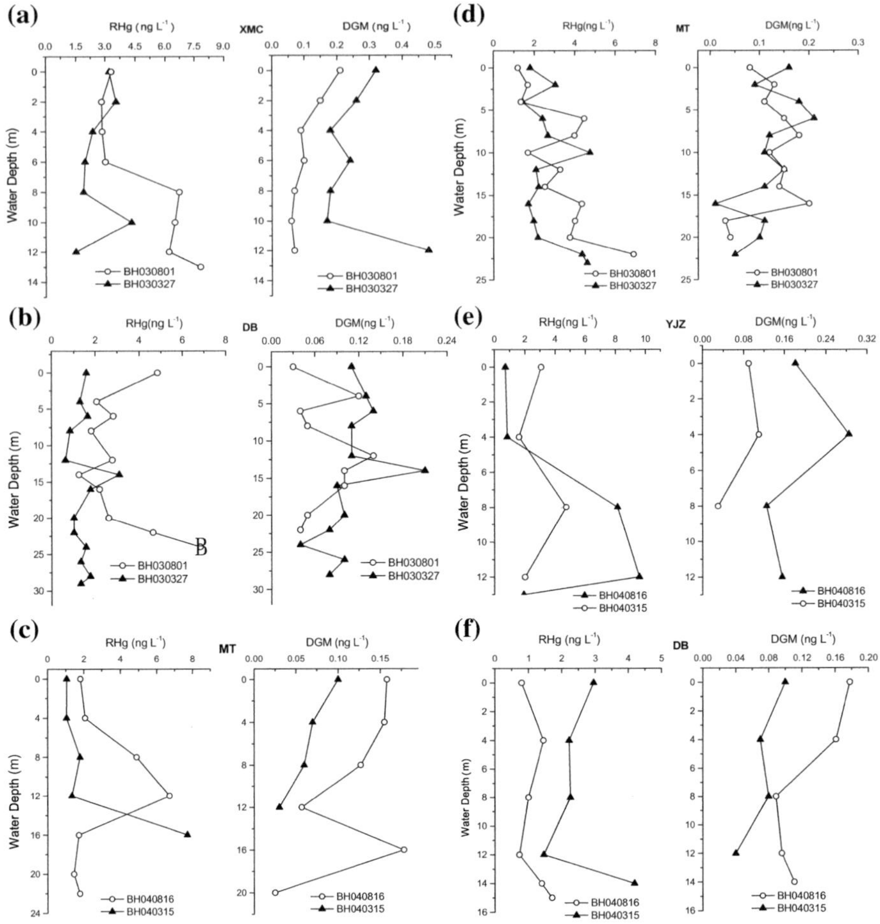

Fig. 7.42 Trend of Hg concentration in the water column of the Baihua Reservoir in different years and seasons. **a**, **d**—Upstream; **b**, **e**—midstream; **c**, **f**—downstream

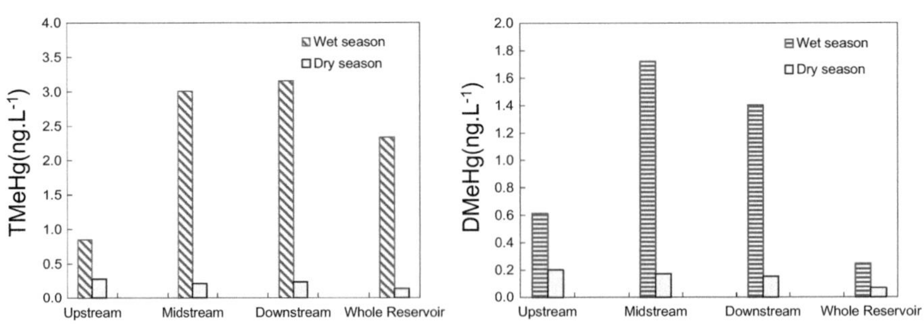

Fig. 7.43 Seasonal variation of methylmercury in the Baihua Reservoir

season and 2.3 ng L^{-1} in the wet season. The annual average was 1.3 ng L^{-1} in the whole reservoir. In general, the background value of MeHg in the water was approximately 0.08–0.42 ng L^{-1} (Vandal et al. 1998; Lyons et al. 1999). It can be seen that in the cold season, MeHg in the BH was in the range of background, but in the warm season higher than the background value.

The rainfall and river runoff increase in summer, resulting in the washing of mercury-polluted sediment into the sewage ditches and into the BH thereafter. An important source of inorganic mercury and MeHg in the reservoir was the sewage from the GOCP. The GOCP had used an activated carbon tower (FT) as a mercury removal device. The wastewater discharged from the production workshop was compared before and after the FT treatment. It showed that the MeHg content decreased by approximately 10 times when the FT was used. However, due to poor management, the raw wastewater leaked into the sewage ditch from the storage tank with wall cracks, therefore, exacerbating the mercury pollution in the sewage ditch sediment. Therefore, not only inorganic mercury but also a small amount of MeHg in the sewage ditch will finally end up in the aquatic ecosystem of the BH, resulting in a higher MeHg level in the wet season. Another reason for the higher MeHg in the wet season was that the microbial methylation in water and sediment was stronger than in the cold season. Therefore, seasonal variation of MeHg in the water was observed. In the warm season, the temperature stratification in the water generally occurred at a depth of 8 m in the BH, and dissolved oxygen decreased rapidly at this layer. The drop of dissolved oxygen led to the distribution of water organisms and microorganisms in this layer, and this is conducive to the formation and enrichment of MeHg (Vandal et al. 1998). In summer, the increase of MeHg at the sediment–water interface was ascribed to the reason that the seasonal temperature stratification caused the oxygen deficit at the water/sediment interface, favoring the methylation of inorganic mercury by anaerobic bacteria, and therefore increased MeHg levels. This was consistent with the vertical distribution of MeHg in other lakes (Vandal et al. 1998).

Many studies showed that the ratio of TMeHg to THg in freshwater systems was higher than in the estuarine water environment, and the percentage of TMeHg to THg was generally less than 5% (Coquery et al. 1997; Mason and Sullivan 1999). However, some data indicated that this ratio can be as high as 30% in freshwater lakes and rivers, and can even reach 37% in the anaerobic bottom water of natural lakes with seasonal stratification (Kudo et al. 1982; Meili 1997; Leermakers et al. 1996). The mean percentage of TMeHg to THg in water of the BH was 1% in the cold season and 12% in the warm season. Obviously, as an artificial freshwater, the BH has been built for 36 years, and the characteristic of the BH was between the natural and a new reservoir. THg in the water and sediment of the BH with a long-term mercury pollution history was far higher than the uncontaminated lakes. Compared to other lakes, the methylation ratio of mercury in the water of the BH was relatively low. The formation of MeHg was mainly related to water temperature, and the higher temperature in summer was more favorable for biological methylation of mercury.

The distribution of MeHg in the lake system showed obvious spatial variations. Due to the physical and chemical characteristics of water at each sampling site, and the distance to the pollution sources, the distribution of MeHg had evident spatial differences (Fig. 7.44). Mercury-containing wastewater from the GOCP that was located in the upper reaches of the BH met with the BH through the Beimen River, becoming an important source of MeHg to the BH. During the sampling period in March 2004 (cold and dry season), there was no wastewater flow into due to the river being dry, so MeHg in the water of the BH was low. In addition, the MeHg level in the BH (except the sediment–water interface) was comparable to the HF, which is not affected by mercury pollution and is located upstream of the BH. MeHg in the water column of the BH in March 2004 was uniformly distributed, with the only significant increase at the interface water. The reasons may include the following: (1) the water was mixed well in spring, meaning that the material distribution in the water column was homogeneous, and mercury concentration changed little with the water depth (except the sediment–water interface water); and (2) the increase of MeHg in the interface water was mainly ascribed to the diffusion of MeHg from the sediment pore water to overlying water.

The horizontal distribution of MeHg in the BH in the cold season was relatively even. With the drying up of the tributary in winter, the input of mercury through the sewage was cut off. At the same time, the frequency of water exchange in the reservoir dramatically decreased, and the disturbance from the input of the tributary was greatly reduced. Eventually, the BH was relatively static, and the disturbance from sediment was very small. Therefore, the MeHg difference between each sampling point was small, especially for MT and DB, representing the midstream and the downstream, respectively. For YJZ, where THg was the highest, the corresponding MeHg was also slightly higher than that in MT and DB. However, in the warm and wet season, the input of mercury increased. Meanwhile, the exchange frequency of reservoir water was also enhanced by the accelerated river runoff. There was no decreasing trend from the upstream to the downstream, even at YJZ, where the water surface is narrow. MeHg there was slightly lower than in the middle and the downstream reaches the BH. However, at MT and DB, where the water surface is open and fish culture activity is thriving, the hydraulic retention time was slightly longer and more abundant organic matter accumulated in the surface sediment. Under suitable water temperature, the rate of methylation increased so MeHg was higher. At these three sampling points, MeHg peaked at an 8-m-deep layer and at the sediment–water interface, reflecting the microbial genesis of MeHg (Vandal et al. 1998).

The diffusion flux of mercury from the sediment pore water to the overlying water was determined by the concentration gradient, but the seasonal variation of THg in sediment was small, so the diffusion flux was mainly determined by the mercury concentration in the interface water. According to the experimental data, the diffusion flux of mercury from the sediment pore water to the overlying water in the wet season was higher than in the dry season, with 1.13 and 0.59 g a^{-1}, respectively. Obviously, the contribution of MeHg from the sediment pore water to overlying water was higher in the warm season than in the cold season.

Fig. 7.44 Vertical distribution of MeHg in the water column at each sampling point of the Baihua Reservoir in different seasons. **a**—Upstream; **b**—middle stream; **c**—downstream

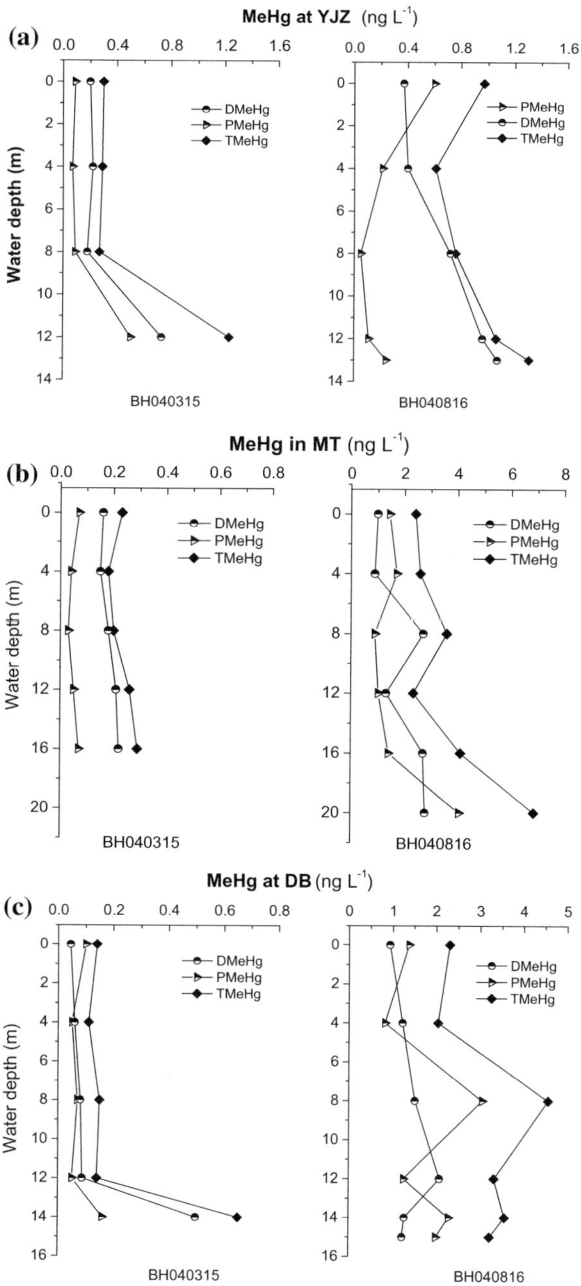

(4) Effect of water parameters on the distribution of mercury in water

The concentration, speciation, distribution, and transformation of mercury are influenced by many factors, such as water temperature, DO, pH, DOC, etc. Bruce (1997) determined THg and TMeHg in 12 low-alkalinity lakes in northeast Minnesota in the spring, summer, and autumn of 1992–1994. He found that DOC and MeHg were strongly positively correlated. The correlation of dissolved mercury concentration and DOC in the water of the BH was negatively correlated ($r = -0.466$, $p < 0.01$, $n = 47$) in spring, but positively correlated ($r = 0.443$, $p < 0.01$, $n = 41$) in summer. This indicated that the DOC in spring was mainly endogenous organic matter and these organisms could absorb part of the mercury, thereby reducing the dissolved mercury in the water, whereas in summer, the DOC was partly from the surface runoff and rivers. As a result, the increase in DOC also increased the input of mercury. Therefore, the two parameters were positively correlated (Fig. 7.45).

According to the correlation analysis, water temperature and DGM had a significant positive correlation (Spring: $r = 0.530$, $p < 0.01$, $n = 41$; Summer: $r = 0.311$, $p < 0.05$, $n = 46$). The reason was that the increase of water temperature is mainly caused by the increase of sunlight, and Hg^{2+} was transformed into elemental mercury by photoreduction, causing increased DGM in the water (Fig. 7.46).

The correlation analysis between water parameters and DMeHg is shown in Fig. 7.47. DMeHg had a significant negative correlation with DO (Spring: $r = -0.749$, $p < 0.01$, $n = 14$; Summer: $r = -0.643$, $p < 0.01$, $n = 18$), a strong negative correlation with pH (Spring: $r = -0.606$, $p < 0.05$, $n = 14$; Summer: $r = -0.549$, $p < 0.05$, $n = 18$), and a significant positive correlation with temperature ($r = 0.440$, $p < 0.05$, $n = 32$). The increase of temperature, decrease of pH, and the anaerobic water environment may facilitate mercury methylation. The

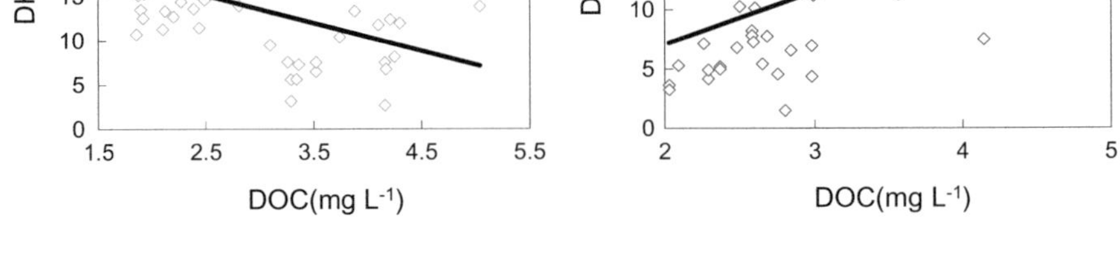

Fig. 7.45 Correlation analysis of dissolved mercury and DOC in the water of the Baihua Reservoir in different seasons

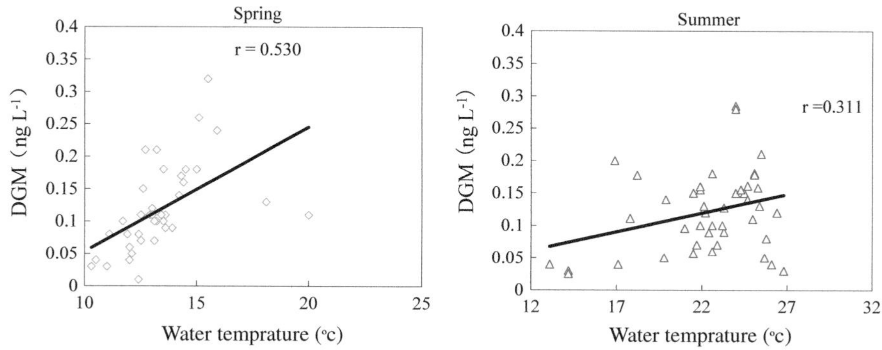

Fig. 7.46 Correlation analysis of water temperature and DGM in the water column of the Baihua Reservoir in different seasons

negative correlation between DMeHg and pH may be explained by the fact that the pH gradually decreased from the surface to bottom water of the BH. The bottom water was mostly anoxic and conducive to the formation of MeHg.

Compared with other lakes, the BH had lower Cl^{-}, higher SO_4^{2-}, and DOC, but these parameters had no significant correlation with the proportion of MeHg (Table 7.10). The BH had a direct source of mercury, and its mercury pollution mainly came from the GOCP, and the natural Hg sources accounted for only a very small amount of mercury into the reservoir. THg and the physical and chemical parameters in the water were affected by different sources, so there was no significant correlation between them. However, the change of water chemistry still had a certain effect on the transformation and migration of mercury, especially the seasonal variation of mercury methylation. In addition, compared to the HF and other lakes from North Europe and North America, the THg and TMeHg in the BH were significantly elevated, indicating serious mercury pollution in the BH.

- Distribution of mercury species in the inflow and outflow rivers of the Baihua Reservoir

Higher mercury was found in the inflow rivers or sampling sections of the Beimenqiao, the Sewage Ditch, and the HQ. The wastewater from the Beimenqiao and the Sewage Ditch mixed with the discharged water of the HF into the BH at HQ. The sewage ditch of the GOCP had been cleaned, and THg in the water of the sewage ditch was obviously low. However, the Beimenqiao had not been cleaned up, and it had accumulated a large amount of mercury from the GOCP, and sewage and wastewater from ferroalloy plants, a coking plant, and other industrial enterprises. Waters from the Beimenqiao River had THg of up to 200 ng L^{-1}. Based on the literature, field investigations, and experimental results, the main inputs of mercury into the BH were the HQ estuary, the Nanmen River, and the Maixi River. The only output was the Maotiao River. The flow rate of most rivers was small (often drying up), and the mercury content was low, so they had a slight effect on

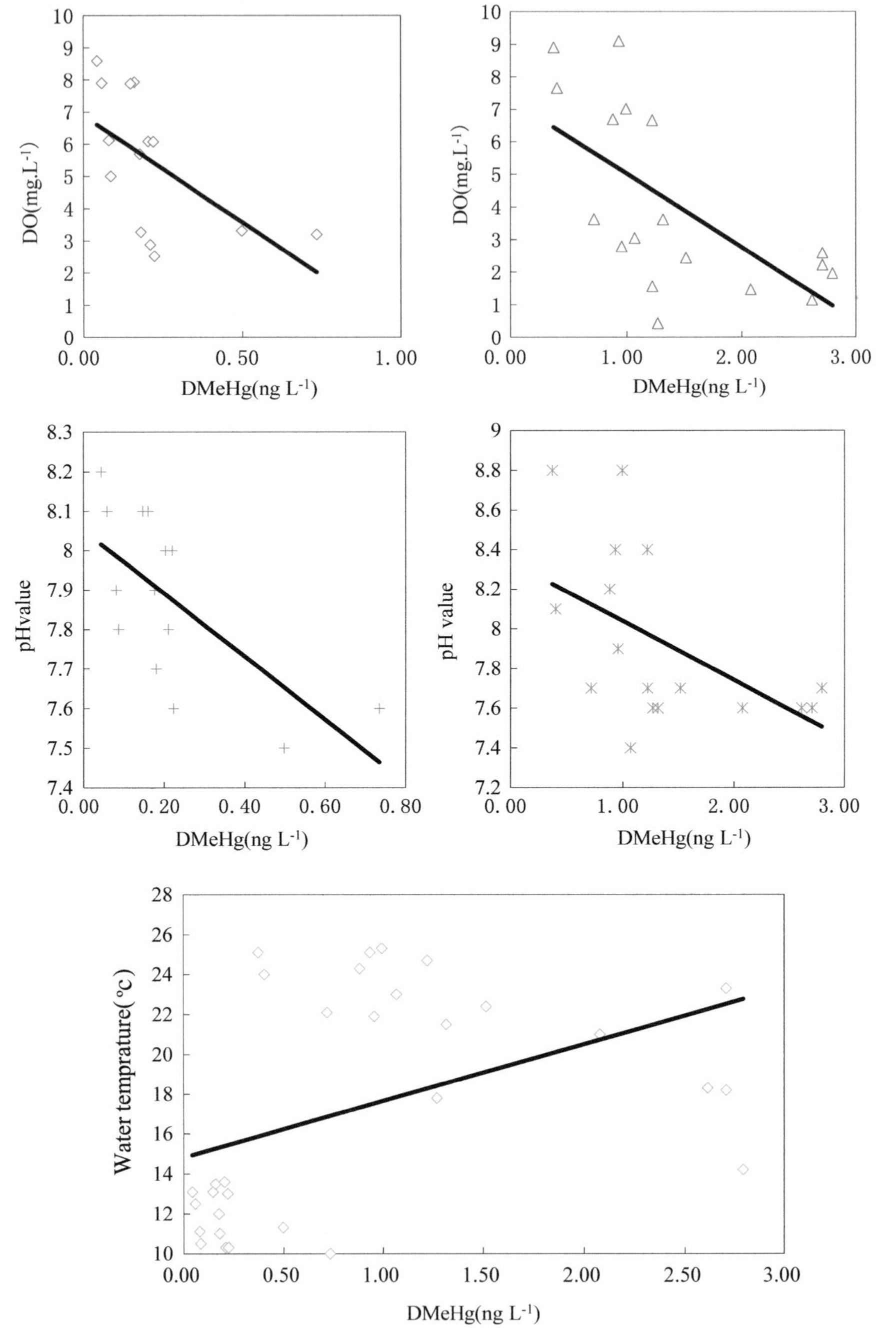

Fig. 7.47 Correlation of water quality parameters and DMeHg in the water column of the Baihua Reservoir

Table 7.10 Comparison of mercury and physical and chemical water parameters in the Baihua Reservoir with other nonpolluted lakes

National	Lake	Water physical and chemical parameters					Hg in water		
		pH	Cl^-	SO_4^{2-}	TOC	DOC	THg	MeHg	MeHg/THg
U.S.A	Allequash[a]	7.6	0.2	3.9	–	4.1	0.8	0.13	–
	Rock[b]	7.2	0.8	1.6	–	9.1	2.1	0.16	–
	Helen[c]	6.2	0.9	2.0	–	19	3.1	0.40	–
	Minnesota 12 lakes[d]	6.6	–	48	–	6.4	1.1	0.10	11
	Piney Creek[e]	7.3	20–24	7	–	5.7–6.6	0.96	–	–
	Deep Creek[f]	6.9	6.2	14	–	2.5–2.6	0.56	–	–
	Lake Habee[g]	6.8	3.0	8.8–11	–	3.4–3.5	0.40	–	–
Finland	Sargent[h]	7.7	–	–	–	6.7	<2.7	<0.3	–
Canada	60 lakes[i]	5.3	0.1	1.9	11	–	3.6	–	–
China	HF[j]	7.9	5.8	61	–		81	0.18	–
	BH[k]	7.9	6.2	98	–	27	22	1.3	1–20

Note [a–c]Watras et al. 1998, [h]Gorski et al. 2003; [d]Bruce 1997, [i]Vaidya et al. 2000, [e–g] Castro et al. 2002, [j]He 2007; [K]this study; unit for water standard parameter is mg/L, unit for mercury concentration is ng L^{-1}, for MeHg/THg is %

the mercury input/output balance of the BH. Therefore, in this work, we selected the following four main tributaries, the HQ estuary, the Nanmen River, the Maixi River, and the Maotiao River, as the research objects.

3. Mercury in inflows and outflow
(1) Water chemicals characters

Different rivers received sewage from different sources that contained different pollutants. In addition, river runoff changed greatly with seasonal change, which affected the extent of sewage dilution of wastewater that discharged into the river (Table 7.11). The change of water temperature was mainly affected by the change of the air temperature in different seasons. In spring and autumn, the sampling time of different rivers might be in the morning, noon, and afternoon. The water temperature at different sampling points may vary widely due to the diurnal variation of air temperature, but this change in temperature had little effect on the mercury concentration and transformation of each river. The main factors that influence the concentration and migration of mercury in rivers are the runoff and flow rate. The higher runoff not only dilutes the mercury concentration but also increases the amount of particulate matter in the rivers, thus increasing the PHg content.

Table 7.11 Annual variation of water parameters in the main tributaries of the BH

Sampling site	Sampling date	T (°C)	DO (mg L^{-1})	EC (μs cm^{-1})	TDS	pH	Sal (‰)	DOC (mg L^{-1})
HQ	2003.03	22.4	1.7	616	–	8.06	–	3.25
Nanmen River		24.3	6.5	293	–	8.14	–	1.62
Maixi River		22.3	5.8	563	–	7.97	–	1.93
Maotiao River		13.1	6.7	427	–	8.57	–	2.02
HQ	2003.08	22.8	–	440	218	7.41	0.2	3.7
Nanmen River		23.4	–	603	302	7.67	0.3	1.8
Maixi River		22.4	–	1167	605	7.41	0.6	1.2
Maotiao River		23	–	451	223	7.35	0.2	2.9
HQ	2004.03	–	–	–	–	–	–	–
Nanmen River		24.5	4.4	293	–	8.14	0.2	–
Maixi River		22.3	5.8	563	–	7.97	0.4	–
Maotiao River		31.1	6.7	206	–	8.61	0.2	–
HQ	2004.08	23.1	3.91	449	227	7.6	0.2	2.8
Nanmen River		22.5	8.79	418	212	8.4	0.2	0.3
Maixi River		21.6	8.98	654	341	7.8	0.3	1.5
Maotiao River		22.9	4.22	389	196	7.8	0.2	1.6

(2) Concentration and distribution of mercury in the main rivers

The distribution and seasonal variation of mercury species in different sampling points are shown in Table 7.12. The HQ was the largest water inlet of the BH, and its THg was higher in the dry season than in the wet season. The main reason is the water discharged from the HF decreased in the dry season, and the dilution of mercury-containing wastewater from the GOCP and other pollution sources decreased, whereas for other rivers, the THg in the dry season was slightly lower than in the wet season, since mercury in these rivers mainly depends on the content of particulate matter, and the water flow increased during the wet season. The reservoir accepted more mercury from surface runoff; therefore, the opposite effect of the increase of dilution and increase of particulate matter resulted in a higher THg in the wet season than in the dry season. THg in inflow water was higher than that of outflow water, indicating that the majority of mercury that enters the

Table 7.12 Distribution of different mercury species in the different seasons of 2003–2004 in the inflows and outflows of the Baihua Reservoir (ng L^{-1})

Date	Site	THg	DHg	PHg	RHg	TMeHg	DMeHg	PMeHg
Dry season	HQ	67.59	16.51	47.33	3.75	0.45	0.26	0.19
	Nanmen River	20.99	8.98	8.37	3.88	0.26	0.24	0.02
	Maixi River	18.66	6.82	11.11	0.93	0.25	0.20	0.05
	Maotiao River	13.50	7.15	5.36	1.45	0.22	0.18	0.04
Wet season	HQ	55.25	18.29	36.96	8.71	1.05	0.43	0.63
	Nanmen River	24.81	11.07	13.74	2.12	0.97	0.14	0.83
	Maixi River	19.56	9.87	9.69	1.50	1.43	0.19	1.24
	Maotiao River	15.63	7.67	7.96	1.35	0.50	0.27	0.23

reservoir settles down, especially the PHg (Table 7.12). There was obvious seasonal variation of MeHg in the rivers. MeHg in the dry season was low, and this might be affected by the temperature, since the water temperature in March was low and the low temperature was not conducive to the biological methylation of mercury. However, in the field, we did not find a clear difference in water temperature between the dry and wet seasons. This is because the temperature of the surface water had risen in the short term with strong sunlight during the sampling time. Moreover, PMeHg in rivers in the wet season was significantly higher than in the dry season. Also, the percentage of PMeHg to TMeHg was higher in the wet season. This might be related to the increasing number of particles in the water, especially the biologic and microbial particles.

7.1.2.3 Biological Geochemistry of Mercury in Sediment Cores

Sediment in lake systems was not only a sink but also a potential source of the overlying water (Covelli et al. 1999), and once contaminated, it might be a threat to the aquatic life for many years (Kudo 1992). The resuspension of sediment and the diffusion of mercury from pore water to overlying water were considered as important reasons for mercury pollution in the overlying water (Bloesch 1995). Mercury emission fluxes from sediment to the overlying water could be estimated through the concentration difference between the sediment pore water and the overlying water. Its contribution to the Hg budget of the entire reservoir was also assessed (Kotnik 2000). In addition, there were a large number of benthic aquatic organisms and fish living at the bottom of the reservoir, and their food was mainly the organic debris in the surface sediment, so mercury content in surface sediment would influence the degree of damage to the organisms. For those lakes historically contaminated by mercury, we can reconstruct the pollution degree and pollution history through the sediment profiles.

The seasonal variation of THg in the sediment and pore water was not obvious. In this study, we collected five sediment cores from the upstream to the downstream (Fig. 7.34) in autumn of November 2002 and spring of March 2003, and determined the THg in sediment and pore water. In Guizhou Province, the water temperature in spring (March) and autumn (November) was not different and the average water temperature was approximately 16 °C, with the difference between the surface and bottom layer being only 0.2 °C. In March, there was less rainfall and the reservoir was drained for irrigation, so the average water depth was only approximately 10 m. However, in November, the rainfall was greater than in March, and the reservoir was in the storage period; therefore, the water depth increased significantly.

1. THg in sediment profiles

When the bottom water of the reservoir is disturbed, the loose surface sediment enters the overlying water, becoming a potential source of mercury. Moreover, the mercury concentration difference between the sediment pore water and the overlying water determines the mercury flux direction. It can be seen that mercury concentrations in surface sediment and pore water were the key points for the research of mercury mass balance in the reservoir ecosystems. To understand the history and the present situation of mercury pollution in the sediment of the BH, we focused not only on the surface sediment, but also on the sediment profiles (including the pore water) throughout the reservoir, namely, in the upstream, the midstream and the downstream.

Five sediment cores in the BH were sampled, including DB (BH0211-1), BF (BH0211-2), YJZ (BH0211-3), MT (BH0303-4), and XMC (BH0303-5). Although the distributions of mercury in different sediment cores were not completely consistent, the general trends were same, especially for cores of 1, 2, and 4. Although the length of these cores varied, the Hg distribution trends were very similar. THg in these cores increased gradually from the bottom to the surface and peaked in the middle. Mercury decreased from the middle to the surface, suggesting that the BH had a history of serious mercury pollution and that it was greatly reduced after the removal of mercury at the GOCP. Overall, THg in the surface 1–3 cm of sediments of the five cores remained at the roughly same level, suggesting that there was no more mercury discharged from the GOCP in recent years, and the level of mercury in river sediments did not increase (Fig. 7.48).

The sediment of the BH had been seriously contaminated with mercury. The critical point for evaluating whether sediment was polluted by mercury or not was choosing the background criteria. Background values of mercury in sediments were determined based on the following different three aspects: (A) take the average content in shale as the global standard; (B) take ancient sediment that formed before the industrialization as the background value; and (C) take the values of pristine rivers and lakes that are seldom affected by contamination as the regional background. In the Second Songhua River of China, a mercury content of 0.036 mg kg^{-1} in silty clay sediment was taken as the background, and we take 8 times this value, 0.3 mg kg^{-1}, as the critical value for evaluation. In Poyang Lake, the mercury content in the soils of nonpolluted cultivated land was taken as the background value to evaluate the mercury pollution in sediment. The Geography Research

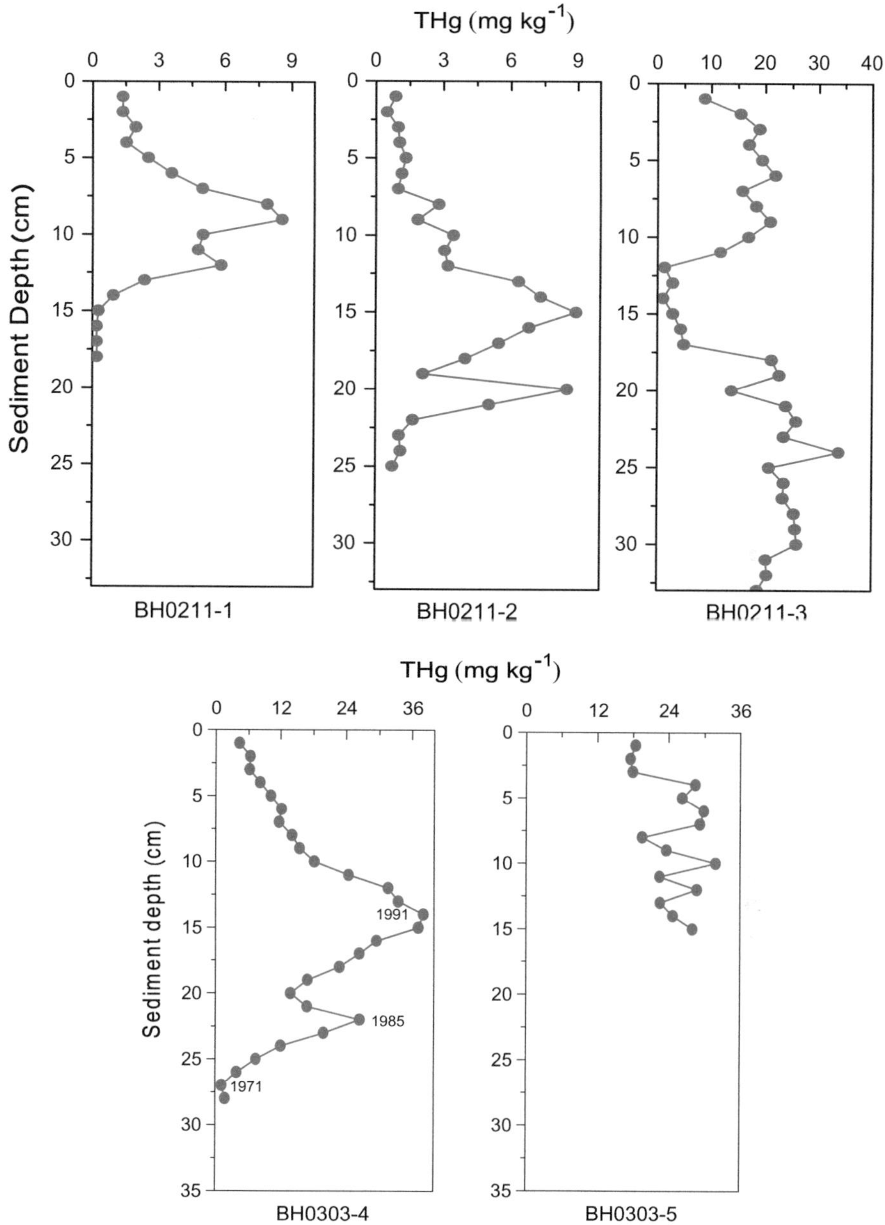

Fig. 7.48 Distribution of THg in the sediment cores of the Baihua Reservoir (redrawed from Yan et al. 2008, with permission from Elsevier)

Institute of Jiangsu Province and Nanjing University gave the standard of 0.3 mg kg^{-1} for evaluation of mercury pollution in the sediments of Poyang Lake, based on the background investigation of different types of soil and Quaternary sediments in southern Jiangsu Province and the basic situation of the lake. Guizhou is composed of many yellow soils, and the average mercury content of yellow soil was 0.262 mg kg^{-1} according to the Guizhou Provincial Environmental Protection Research Institute. Therefore, we take 0.26 mg kg^{-1} as the evaluation standard for the BH (Wang et al. 1992). The THg in the sediment of the BH was in the range of 0.26–38.9 mg kg^{-1} with a mean value of 6.5 mg kg^{-1}. The average THg in the sediment of XMC (BH0303-5) in the upstream was 24.6 mg kg^{-1}, which was considerably higher than the lakes without contamination (Anirudh et al. 2003). Despite the GOCP stopping the discharge of Hg-containing sewage at present, the large amount of sediment deposited in the rivers was still the main source of mercury pollution in the BH.

As shown in Fig. 7.49, the average THg was gradually reduced along the flow direction from the upstream to the downstream (BH0303-5, BH0211-3, BH0303-4, BH0211-2, BH0211-1, representing XMC, YJZ, MT, BF, and DB, respectively). The THg in sediment decreased approximately 60–87%. This was because most mercury in water is bound to particles. With the deposition of particulate matter, mercury was also preferentially deposited in the upstream area, making the upstream pollution more serious than the other sections.

2. DHg in the pore water

As shown in Fig. 7.50, THg in the pore water of the whole reservoir averaged 815.1 ng L^{-1} (range: 6.1–5863.9 ng L^{-1}). Mercury in the pore water of DB and MT manifested a significant peak in the vertical direction, whereas THg in the pore water of the other three cores had no obvious trend in the vertical direction, which may be related to the content of organic matter in the sediments. Compared with the overlying water, THg in the pore water was much higher, indicating Hg in

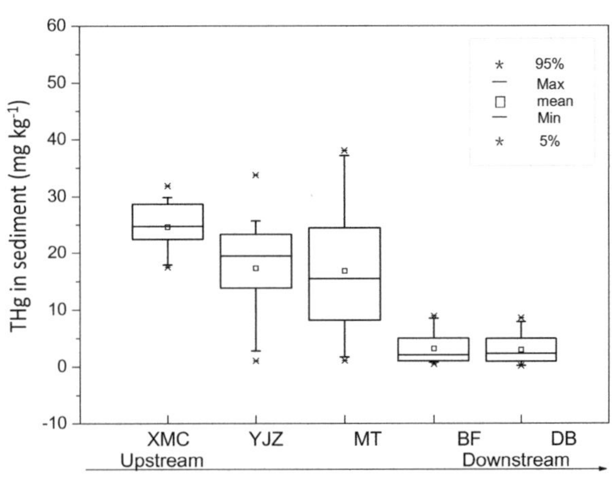

Fig. 7.49 Distribution of mercury in sediments from the downstream to the upstream of the Baihua Reservoir (redrawed from Yan et al. 2008, with permission from Elsevier)

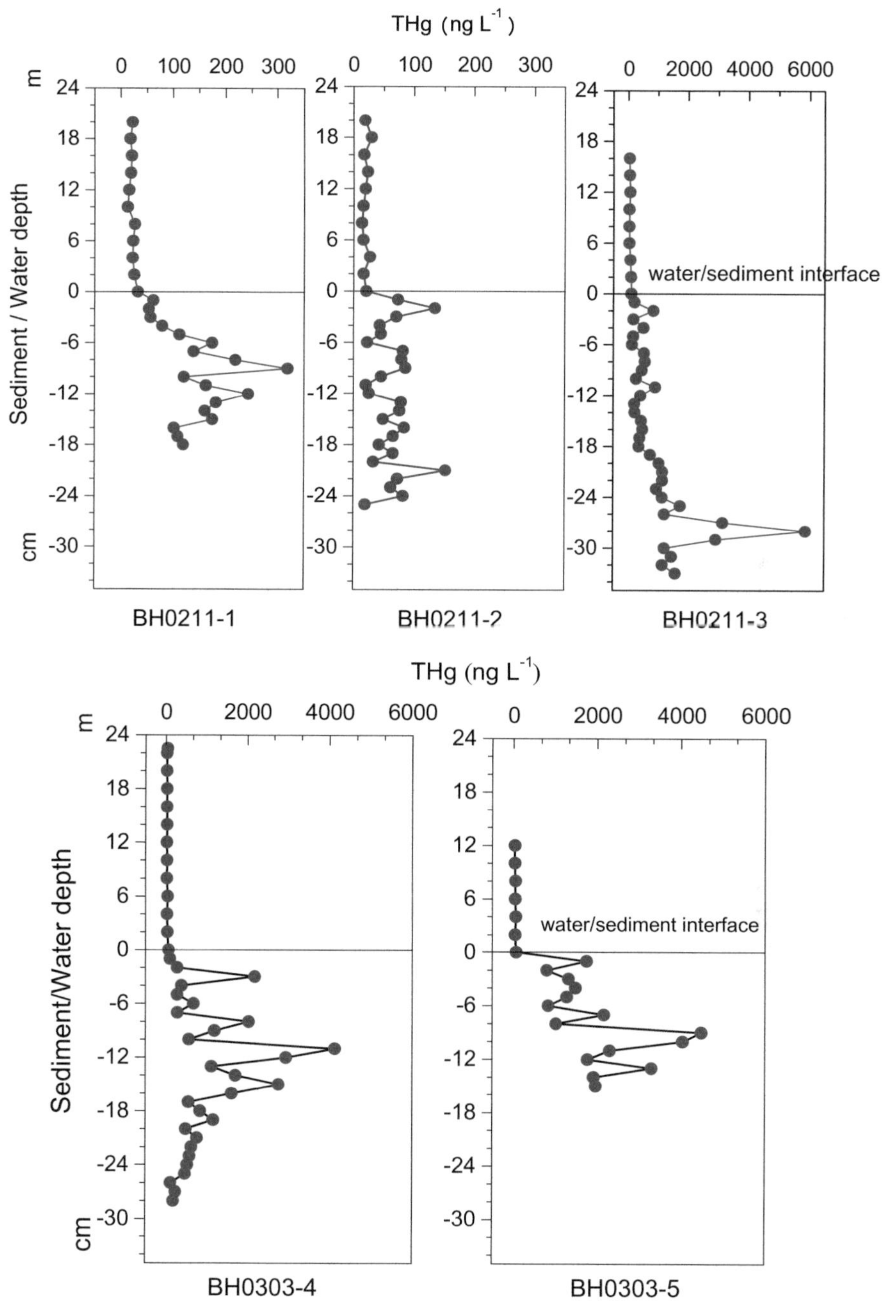

Fig. 7.50 Mercury distribution in the sediment pore water of the Baihua Reservoir (redrawed from Yan et al. 2008, with permission from Elsevier)

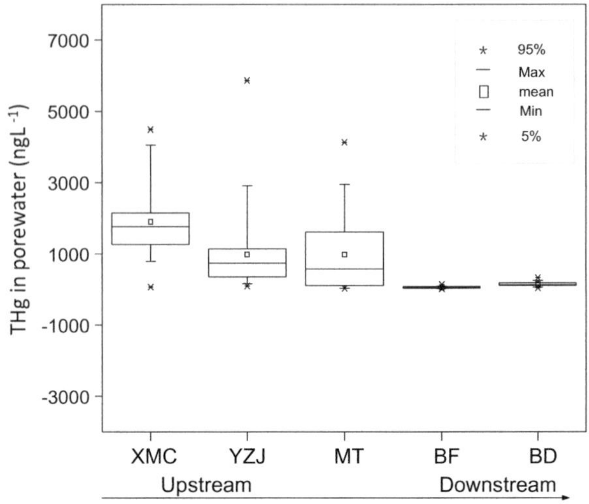

Fig. 7.51 Horizontal distribution of THg in pore water of the Baihua Reservoir (redrawed from Yan et al. 2008, with permission from Elsevier)

sediments that are likely to enter the pore water and further spread to the overlying water, and hence causing secondary pollution in the reservoir. In sediments with higher THg, if THg in pore water was also higher, it would then increase the dissolved mercury in the reservoir water. On the contrary, if sediment has high mercury while the pore water has low mercury, this indicates that most mercury in the pore water was absorbed into the solid phase of the sediment. When the sediment is disturbed, it increases the PHg in the overlying water. However, if the sediment was not disturbed or disturbed to a less extent, mercury in the overlying water would not increase too much.

The average THg in the pore water decreased from the upstream to the downstream, as shown in Fig. 7.51. The trend was the same as that of THg in the sediment. Mercury in pore water of BF and DB was equivalent in the downstream, with DB slightly higher than BF, probably due to the lower organic matter content at DB resulting in slightly higher dissolved mercury in the pore water. XMC, located approximately 1.5 km from HQ where different tributaries enter into the BH, was the first place that mercury enters the reservoir from the GOCP. Although the GOCP stopped discharging mercury-containing sewage, sediment contaminated previously by mercury in the riverbed is washed out into BH. Therefore, mercury in the sediment at XMC was the highest, and where the porosity of sediment was high, so the mercury was readily desorbed from the solid phase into the liquid phase, making mercury the highest in pore water.

7.1.2.4 Mercury Exchange Flux Over the Sediment–Water Interface

In general, for lake or reservoir systems, a large amount of particulate matter and mercury carried by the Inflow Rivers will settle down in the broad lake areas and become the sediment (Wang 1993). When accumulated to a certain degree in the sediment, mercury will transfer from the high concentration medium to the low concentration medium by the driving force of the concentration gradient, commonly diffused from the sediment pore water to the overlying water.

Generally, the THg concentration in lake sediments was approximately 0.01–0.3 mg kg^{-1} (Anirudh et al. 2003), but in the BH, the average THg in sediment was 6.5 mg kg^{-1}, with the highest value being 38.9 mg kg^{-1}; therefore, the average THg in the BH was more than 20 times higher than that of lake sediments without contamination and even 150 times higher during the stage with most serious pollution (Nguyen et al. 2005). It was seen that the BH had been severely polluted by mercury. There was strong sediment resuspension in the BH, and the mercury in water was mainly in particulate forms (Wang 1993). Moreover, mercury in surface sediments could be transported into the overlying water by human or biological disturbance, as well as the water current, wind, and other natural factors. Ultimately, secondary pollution will be formed. High mercury in pore water was also an important source of mercury in the overlying water. Therefore, the mercury exchange flux between the sediment and the overlying water was mainly divided into two parts: first, the deposition of mercury in water to the sediment and the release of mercury in the pore water to the overlying water; second, the resuspension of sediment to the overlying water, especially to the interface water. However, the exact contribution of resuspension to the overlying water needs to be accurately estimated due to the natural and anthropogenic disturbance in the surface sediments.

1. Estimation of the annual deposition of mercury from water to sediment

The BH was built in 1966, and the sediment sampling was performed in 2002. The sediment accumulation process includes not only the deposition process but also hydraulic abatement; therefore, the sediment thickness at different sites was different. The thickness of sediment in the BH was in the range of 15–33 cm with an average of 23.8 cm. The accumulation rate was 0.92 cm a^{-1} according to the thickest sediment, which was similar to the HF (0.93 cm a^{-1}) estimated by Wan et al. (2000). Based on the following parameters, namely, the deposition rate of surface sediment (0.06 g cm^{-2} a^{-1}), average mercury concentration in the first 1 cm layer of sediment (6330 ng g^{-1}) and total sediment area (1.45×10^{11} cm), the annual deposition of THg into the sediment of the BH was calculated to be 55,071 g through the formula as follows:

$$\mathrm{M}_{\mathrm{Hg-sed}} = A_{\mathrm{sed}} \times V_{\mathrm{sed}} \times C_{\mathrm{sed}} \tag{7.1}$$

$\mathrm{M}_{\mathrm{Hg-sed}}$—Amount of THg that enters into sediment per year, g a^{-1}
A_{sed}—Total deposition area, cm^2
V_{sed}—Annual sedimentation rate, g cm^2 a^{-1}
C_{sed}—Mercury concentration in surface sediments, ng g^{-1}.

2. Estimation of the annual release of mercury from pore water to the overlying water

Mercury diffusion from the pore water to the overlying water could be calculated by Fick's law. The mercury exchange flux between the pore water and the overlying water depended on the mercury concentration gradient. When mercury in the pore water was higher than in the overlying water, mercury would be released into the overlying water, and vice versa. This process could be expressed by the following formulas (Kotnik 2000):

$$F_{\text{sed-wat}} D_0 \times [(C_{\text{sed}} C_{\text{wat}})/1000] \tag{7.2}$$

$$D_0 \left[D_{\text{wat}}^{-1} + D_{\text{wat}}^{-3/4}\right] \tag{7.3}$$

$$M_{\text{Hg-sed-wat}} A_{\text{sed}} \times 365 \times F_{\text{sed-wat}} \times 10^{-9} \tag{7.4}$$

$F_{\text{sed-wat}}$—Mercury exchange flux between the overlying water and the pore water (ng m^{-2} day^{-1})
D_0—The mass transfer coefficient of mercury in the ideal solution (m day^{-1})
C_{sed}—Mercury concentration in sediment pore water (ng m^{-3})
C_{wat}—Mercury concentration in overlying water (ng m^{-3})
D_{wat}—Mass transfer coefficient of mercury in water (m day^{-1})
$M_{\text{Hg-sed-wat}}$—Annual release of mercury from pore water to overlying water (g a^{-1})
A_{sed}—Interface area between the sediment and overlying water (m^2).

The values of D_0 and D_{wat} depend on the molecular weight of mercury and its compounds. Higher values of D_0 and D_{wat} correspond to smaller molecular weight. The molecular weights of Hg^0, Hg^{2+} and MeHg compounds are between 200 and 1500. Based on this, the diffusion rates are in the range of 10^{-6}–$10^{-5}$$cm^2$ day^{-1}, the D_{wat} value should be 0.04–0.2 m day^{-1}, and the D_0 value should be 0.01–0.08 m day^{-1}. Since the proportion of MeHg to THg in the pore water and the overlying water of the BH was small, the molecular weight of inorganic mercury was relatively small, so the diffusion rate was relatively large. Hence, we chose the maximum value, 0.08 m day^{-1}, for D_0. Through formulas 7.2, 7.3, and 7.4 and the concentration gradient values listed in Table 7.13, we calculated the total amount of mercury released from the sediment pore water to the overlying water annually,

Table 7.13 Mercury concentration between in the pore water and overlying water of Baihua Reservoir

Mercury species	THg	RHg	DMeHg
Hg in pore water (ng L^{-1})	297.1	48.3	5.2
Hg in overlying water (ng L^{-1})	46.2	5.9	1.1
Hg Concentration gradient (ng L^{-1})	250.9	2.4	4.1
Annual release (g a^{-1})	106.2	17.9	1.7

which was approximately 106 g a^{-1}, and accounted for only 0.4% of THg input of the BH. Since the proportions of RHg and DMeHg were small, the high mercury concentration in the interface water was not caused by the diffusion of mercury in the pore water, but the resuspension of surface sediments.

3. Contribution of sediment resuspension to mercury in the water
   Mercury in sediments was relatively stable, but mercury in the loose sediment could be suspended in the water when the water flow or biological disturbance was strong. This phenomenon is known as mercury resuspension. Mercury resuspension is strongly influenced by the water flow, wind speed, and human disturbance. The resuspension process occurred simultaneously with the deposition of particles into the sediment, and these two processes were opposite. Therefore, we calculated the flux of resuspension according to the sinks of adsorbed mercury on the suspended particles, that is, the settlement rate of sediment each year (Table 7.13).

7.1.3 Biogeochemical Cycling of Mercury in the Aha Reservoir

7.1.3.1 Sampling Sites Description and Sample Collection

The sampling sites included the center area (site A), the upstream area (site B), and all of the inflow and outflow rivers of the AH (Fig. 7.52). Samples were collected in March 2005 (low flow season—LF) and August 2005 (high flow season—HL) and included depth-divided water, pore water, and core sediments of the reservoir as well as the surface water of the inflow and outflows using an acid-cleaned, Teflon lined, 10-L Nisiki sampler from a wooden boat. Dissolved gaseous mercury (DGM), reactive mercury (RHg), dissolved and particulate mercury (DHg, PHg), total mercury (THg), dissolved and particulate methylation (DMeHg, PMeHg), DHg and DMeHg in pore water; and THg, MeHg, and total organic carbon (OM%) in core sediment were analyzed. Parameters, such as water temperature, pH, and dissolved O_2 (DO), were measured using a portable multimeter (Henna, Italy) immediately after sampling. Undisturbed sediment cores were collected using an SWB-1, which is a custom-designed sediment core sampler. Quality control for Hg and MeHg determinations was addressed with method blanks, blank spikes, matrix spikes, certified reference materials of sediment (GBW07405; CRM580), and blind duplicates. MeHg could be detected at concentrations above 0.01 ng L^{-1} at a blank level of 0.045 ng L^{-1} in water samples. The detection limit for THg in water samples was 0.2 ng L^{-1} at a blank level of 0.3 ng L^{-1}. Limits of determination were 0.01 ng g^{-1} for total Hg and 0.003 ng g^{-1} for MeHg in sediment samples, respectively. The average total Hg concentration of the geological reference material GBW07405 was 0.30 ± 0.01 ng g^{-1} (n = 5), which is comparable with

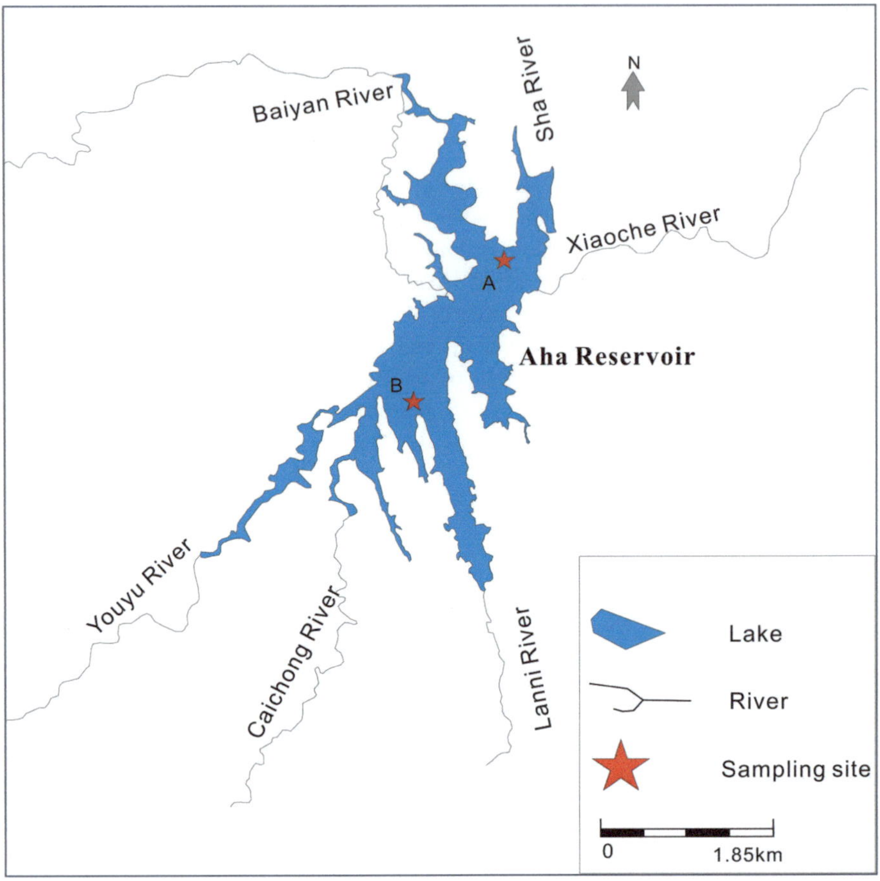

Fig. 7.52 Sampling map and sites

the certified value of 0.29 ± 0.04 ng g^{-1}. An average MeHg concentration of 70.6 ± 0.6 ng g^{-1} (n = 7) was obtained from CRM580 with a certified value of 70.2 ± 3 ng g^{-1}. Recoveries on matrix spikes of MeHg in water samples were in the range of 88.2–108.4%. The relative percentage difference was <8.5% for total Hg in sediment and water samples.

7.1.3.2 Mercury Species in the Water Columns

1. Physical and chemical characteristics of the water in the Aha Reservoir

The temperature stratification of water affects the physical, chemical, and biological processes in the aquatic environment of a reservoir. The occurrence and intensity of water stratification in different seasons are mainly affected by the geography and size of the reservoir, climate, wind force, and river discharge. The average water depth of the AH is 13.2 m, and the maximum is 24 m. The balance of heat in the water column of the AH is different in different seasons because of changes in solar radiation and the atmosphere's temperature.

The water temperature of the AH varied little in the vertical direction in the dry season, with variation between 8.9 and 10.7 °C from the surface to the bottom. In the summer, the surface water temperature rises due to stronger solar radiation and increased irradiation time. This results in larger temperature variance in the water column. In the center of the AH, at a depth of approximately 16 m, we noted a thermocline where the water temperature was 26.9 °C at the surface and 13.4 °C at the bottom. Water stratification restricts an exchange between the upper and lower layers of water and can result in significantly different physical and chemical water properties and ecosystem structure from surface to bottom (Fig. 7.53)

The pH of the AH water ranged from 7.5 to 8.5, showing alkaline. The pH varied slightly in the water column (Fig. 7.54).

The dissolved oxygen (DO) concentrations were relatively high in the AH water, with a range of 8.2–10 mg L^{-1} in the low flow season, but they decreased from the surface (8.7 mg L^{-1}) to bottom water (3.6 mg L^{-1}) in the high flow season, and the vertical pattern of dissolved oxygen was similar to the water temperature. For example, the DO dropped rapidly from surface (8.7 mg L^{-1}) to sub-bottom water over an approximate distance of 12 m (3.5 mg L^{-1}), while it varied little at depths greater than 12 m (Fig. 7.55).

In general, the phytoplankton photosynthesis is the primary process of oxygen production in a reservoir and enriches the content of dissolved oxygen in water, along with the air–water exchange and river input. Organic matter metabolism is the primary process of oxygen consumption. Therefore, dissolved oxygen appears supersaturated in the upper layer of water since phytoplankton, like algae, photosynthesizes in the euphotic zone in the summer. The dissolved oxygen content measures at low levels because of respiration and the relatively fast degradation of

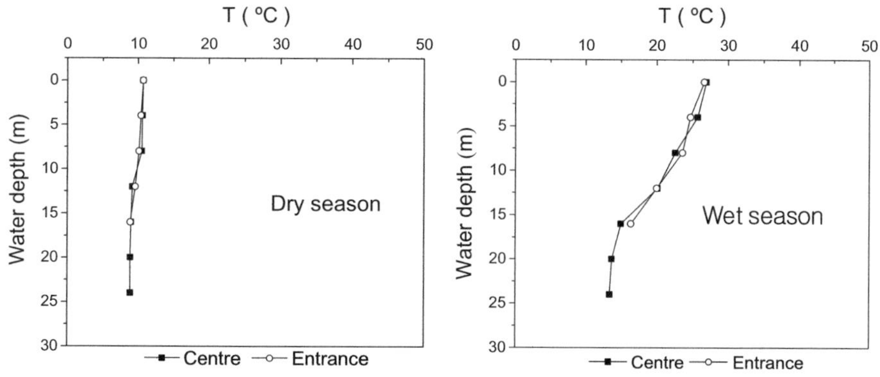

Fig. 7.53 Vertical water temperature of the Aha Reservoir (redrawed from Feng et al. 2011, with permission from Elsevier)

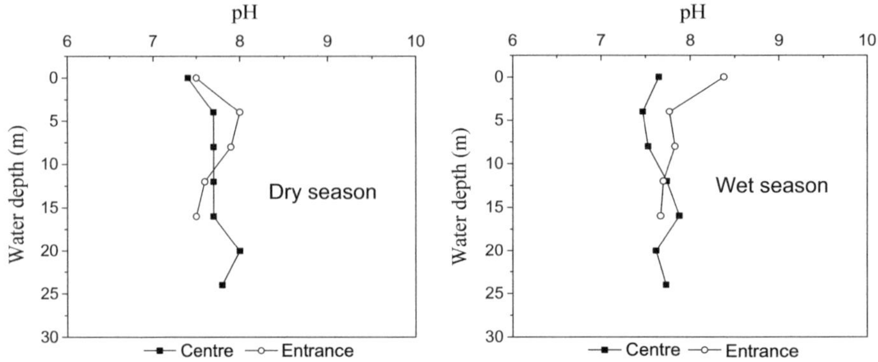

Fig. 7.54 The vertical pH values of the Aha Reservoir water (redrawed from Feng et al. 2011, with permission from Elsevier)

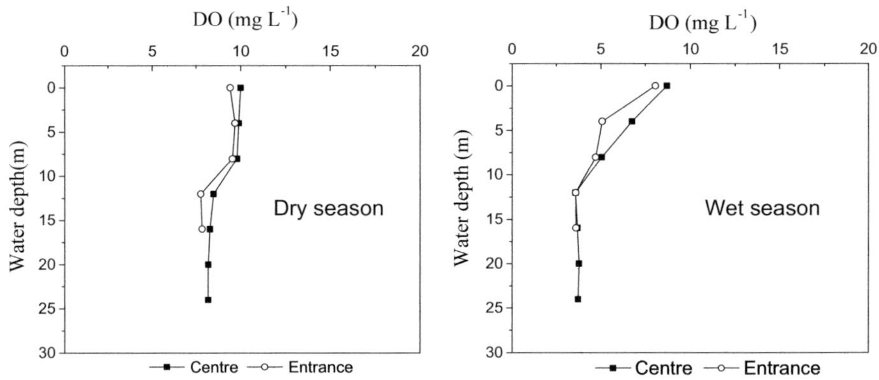

Fig. 7.55 The vertical DO concentrations of the Aha Reservoir water (redrawed from Feng et al. 2011, with permission from Elsevier)

organic matter that consumes oxygen. Water stratification restricts an exchange between the upper and lower layers of water, which results in the gradual stratification of dissolved oxygen. Therefore, the redox boundary of the sediment and water interface changes as seasons change (Luo 2001).

2. Seasonal variation and distributions of mercury species in water columns

(1) THg in water columns

THg concentrations in the AH water in the low flow season ranged from 2.1 to 20.0 ng L^{-1}, with a mean value of 6.2 ± 4.7 ng L^{-1} (Fig. 7.56). These concentrations ranged from 2.1 to 20.0 ng L^{-1}, with a mean value of 7.0 ± 5.3 ng L^{-1}, in the center of the reservoir (site B) and from 2.4 to 13.0 ng L^{-1}, with a mean value of 5.1 ± 4.1 ng L^{-1}, in the upstream area (site A). In the center, THg concentrations showed a trend of decline–rise–decline–rise as water depth increased. THg

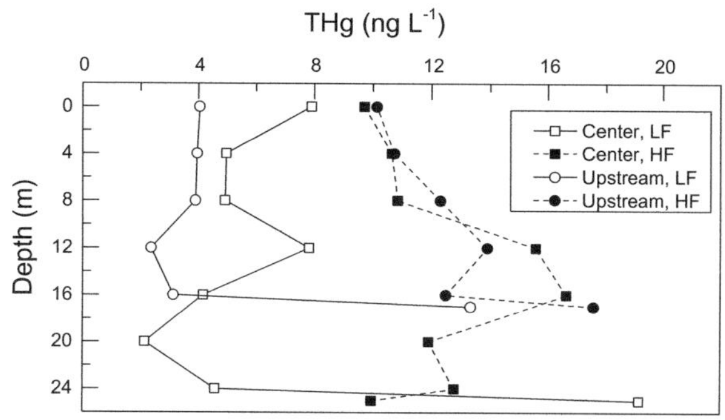

Fig. 7.56 Seasonal variations of THg in Aha Reservoir water (redrawed from Bai et al. 2006, with permission from Acta Scientiae Circumstantiae)

concentrations were high in the middle layer and low in the surface and sub-bottom layers, with a minimum value recorded at 20 m. In contrast, the THg concentrations varied little in the upstream area, except for a maximum value (13.0 ng L^{-1}) recorded in the bottom layer.

In the high flow season, THg concentrations ranged from 9.7 to 17.6 ng L^{-1}, with an average 12.5 ± 2.5 ng L^{-1}, which is nearly 2 times higher than the range recorded in the low flow season (Fig. 7.56). The THg concentrations in the high flow season ranged from 9.7 to 16.6 ng L^{-1}, with a mean value of 12.3 ± 2.6 ng L^{-1}, in the center of the reservoir (site B) and from 10.1 to 17.6 ng L^{-1}, with a mean value of 12.9 ± 2.7 ng L^{-1}, in the upstream area (site A). In the center area, THg concentrations showed distinct variance within the water column: high in the middle layer and low in the surface and bottom layers, with a maximum value at a depth of 16 m. In the upstream area, the THg concentrations increased from the surface to the bottom, except in the sub-bottom layer, where we recorded a relatively low value.

(2) DHg in water columns

The concentrations of dissolved mercury in the AH water ranged from 1.7 to 8.4 ng L^{-1}, with an average of 3.1 ± 1.7 ng L^{-1}, in the low flow season and from 3.8 to 9.9 ng L^{-1}, with an average of 5.1 ± 1.7 ng L^{-1}, in the high flow season (Fig. 7.57). In the center of the reservoir, the concentrations of dissolved mercury gradually decreased as depth increased. In the storage area of the reservoir, concentrations of dissolved mercury increased with depth and peaked at 8 m. There were no obvious differences between upper and lower layers. The water dissolved mercury concentration of the AH was significantly higher in the wet season, ranging from 3.8 to 9.9 ng L^{-1}, with an average of 5.1 ng L^{-1}. The concentrations of dissolved mercury in the center of the reservoir gradually increased as depth increased. Due to the frequent exchange between surface runoff and water input in the storage area, the upper and lower layers changed little, while the concentrations at the bottom increased significantly. This occurred because decomposition organic matter at the surface produced bubbles, causing the settlement suspension to refloat, in turn causing the release of soluble ionic mercury from the overlying water (Xu et al. 1999).

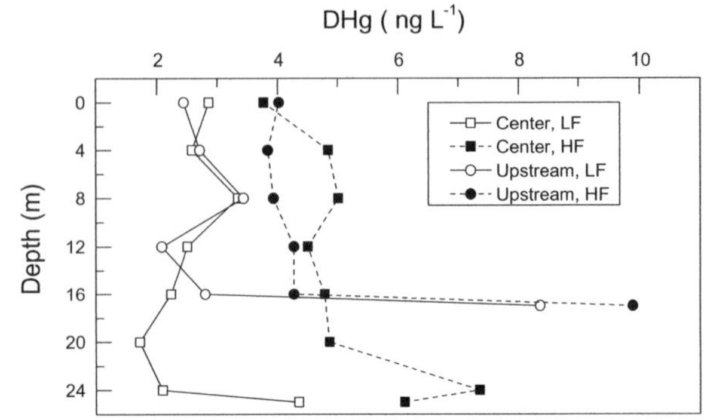

Fig. 7.57 Seasonal variations of dissolved mercury in Aha Reservoir water (redrawed from Bai et al. 2006, with permission from Acta Scientiae Circumstantiae)

The proportions of dissolved mercury to the THg in the AH water were 48.8 ± 19.0 and 43.5 ± 11.9% in the center area in low and high flow seasons, respectively, and 76.2 ± 13.8 and 38.1 ± 9.4% in the upstream area in low and high flow seasons, respectively. There were significant positive correlations between dissolved mercury and THg in both the center ($r = 0.68$, $p < 0.01$, $n = 14$) and upstream area ($r = 0.79$, $p < 0.01$, $n = 10$) of the reservoir (Fig. 7.58).

Overall, precipitation increased significantly and the frequency of water exchange increased during the wet season, making the concentrations of dissolved mercury lower than those in the dry season. Studies on the WJD, DF, and BH of the Wujiang River Basin have confirmed this observation (Jiang 2005; Yan 2005). However, in the AH, concentrations of dissolved mercury in the wet season are higher than those in dry season. During the wet season, the bottom of the reservoir experiences anaerobic conditions, which spur the migration of iron and manganese to the overlying water and the redox boundary layer moves up into the water column (Wang 2003). This causes the adsorbed mercury in the sediment and interstitial water to release into the water, resulting in greater concentrations of THg than those caused by the dilution of rainfall and surface runoff.

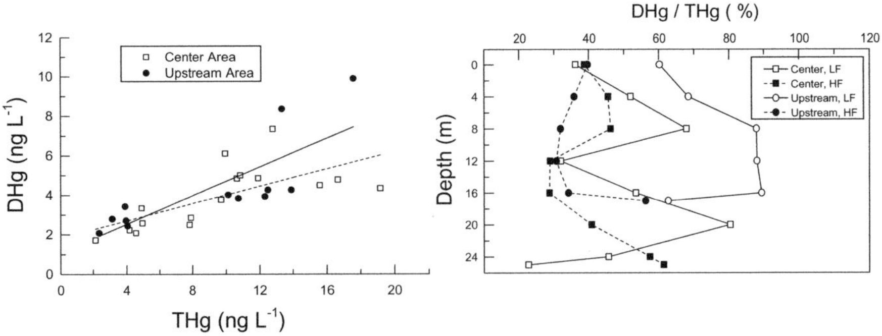

Fig. 7.58 Correlation between dissolved mercury (DHg) and total mercury (THg) in Aha Reservoir water

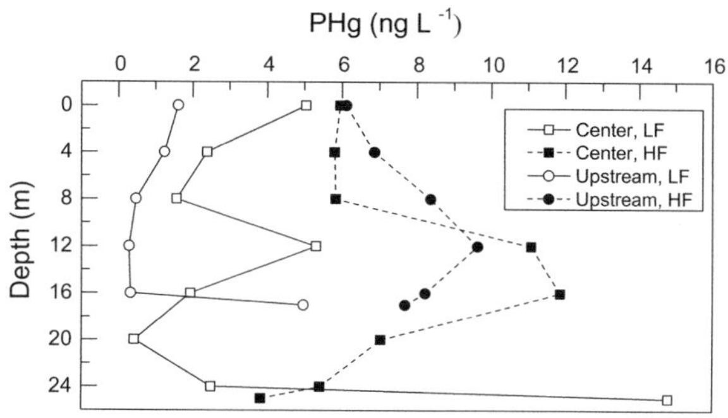

Fig. 7.59 Seasonal variations of particulate mercury in Aha Reservoir water (redrawed from Bai et al. 2006, with permission from Acta Scientiae Circumstantiae)

(3) PHg in water columns

The concentrations of particulate mercury in the AH in the dry season ranged from 0.3 to 14.8 ng L^{-1}, with an average of 3.1 ± 3.8 ng L^{-1} (Fig. 7.59). In the center (site B), particulate mercury showed a trend of decline–rise–decline–rise as water depth increased, like THg, which was high in the middle layer, low in the surface and sub-bottom layers, and highest at the bottom. The high peak at 12 m may be attributed to the adsorption of algae, since biological processes are active in this layer. However, in the upstream area, particulate mercury varied little as water depth increased. In both sites, the highest concentrations of particulate mercury may be attributed to the resuspension of sediment by bioturbation that releases mercury from sediment (Xu et al. 1999) and the accumulation of sinking algae cells that tend to adsorb mercury (Jiang et al. 2004).

The concentrations of particulate mercury ranged from 3.8 to 11.9 ng L^{-1}, with an average of 7.4 ± 2.3 ng L^{-1} in the wet season, which were significantly higher than the concentrations recorded in the dry season. In the center (site B), particulate mercury varied from 0 to 8 m yet dramatically increased from 8 to 16 m (5.8–12 ng L^{-1}) where there were high concentrations of suspended particulate matter but low concentrations of dissolved mercury. This may be attributed to the adsorption of mercury onto the algae. However, there was no significant correlation between particulate mercury and suspended particulate matter in the whole water column (r = −0.04). This was mainly due to calcareous additions to the Youyu and Baiyan Rivers by humans, which resulted in high concentrations of suspended particulate matter in the reservoir's surface water during the wet season.

Proportions of particulate mercury to THg in AH water were 51.2 ± 19.0 and 56.5 ± 11.9% in the center area in the low and high flow seasons, respectively, and 23.7 ± 13.8 and 61.9 ± 9.4% in the upstream area in the low and high flow seasons, respectively. There were strong positive correlations between particulate mercury and THg in both the center ($r = 0.96$, $p < 0.01$, $n = 14$) and upstream areas ($r = 0.92$, $p < 0.01$, $n = 10$) of the reservoir (Fig. 7.60).

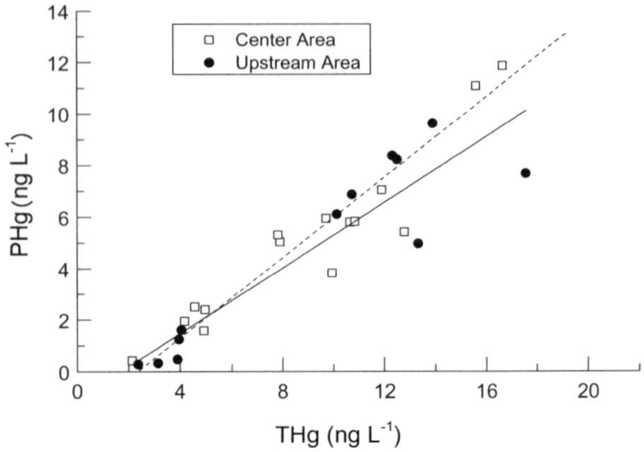

Fig. 7.60 Correlation between particulate mercury (PHg) and total mercury (THg) in water of Aha Reservoir

(4) DGM in water columns

The concentrations of dissolved gaseous mercury in AH water ranged from 0.04 to 0.09 ng L^{-1}, with an average of 0.06 ± 0.02 ng L^{-1} in the low flow season, and ranges of 0.04–0.09 ng L^{-1} in the center and 0.04–0.06 ng L^{-1} in the upstream area (Fig. 7.61).

In the summer, solar radiation increases, temperature rises, and algae blooms. We observed that concentrations of DGM in the AH in the wet season were much higher those observed in the dry season, with a range of 0.23–0.31 ng L^{-1} and an average of 0.27 ng L^{-1} in the whole reservoir, 0.24–0.31 ng L^{-1} in the center area, and 0.23–0.32 ng L^{-1} in the upstream area. Solar radiation plays an important role in forming dissolved gaseous mercury (Feng et al. 2002, 2003). The photoreduction of Hg^{2+} to Hg(0) by solar radiation was the main source of dissolved gaseous mercury in AH water in the wet season. This photoreduction reaction was mainly driven by UV-A. The reaction mechanism may involve Hg^{2+} reduction by an enzyme on the cell surface of algae. Previous studies have shown that UV-A (UV-B) can contribute up to 25% of the generation of dissolved gaseous mercury (Zhang 1996).

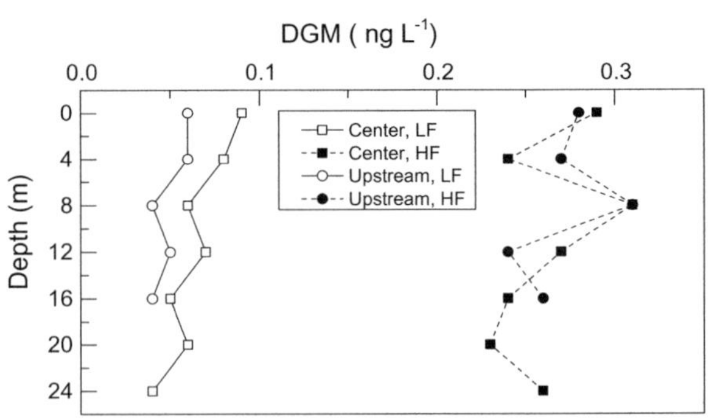

Fig. 7.61 Seasonal variances of dissolved gaseous mercury (DGM) in Aha Reservoir water (redrawed from Bai et al. 2006, with permission from Acta Scientiae Circumstantiae)

(5) RHg in water columns

The concentrations of active mercury in AH water ranged from 0.1 to 1.1 ng L^{-1}, with an average of 0.4 ng L^{-1} in the dry season, and varied little from the surface to the bottom, but with a maximum recorded at the bottom (Fig. 7.62). The density current from river to reservoir (the salinity of river water was 4 to 6‰, while the salinity of reservoir water was 2 to 4‰), caused by adding calcareous lime, accelerated mixing and exchanging across the sediment–water interface, so the reactive mercury concentration increased in water at the bottom.

The concentrations of reactive mercury in AH water ranged from 2.1 to 4.5 ng L^{-1}, with an average of 3.1 ng L^{-1} in the wet season, much higher than those recorded in the dry season. Concentrations of reactive mercury decreased from the surface to the bottom yet reached a maximum of 4.5 ng L^{-1} at the bottom in the center. In the upstream area, concentrations of reactive mercury varied little from the surface to a depth of 8 m, decreased from 2.8 to 2.1 ng L^{-1} as depth increased, and suddenly rose to 3.9 ng L^{-1} at the bottom.

The ratios of active mercury to THg were 13 and 45% in dry and wet seasons, respectively. A higher concentration of active mercury in the bottom water in the wet season was mainly attributed to the dissolution of iron and manganese oxides along with river input associated with density current. The iron and manganese oxides, which act as good adsorbents due to their large surface areas, were reduced to soluble iron and manganese ions in the oxygen deficit conditions at the bottom of the reservoir, thus releasing activated mercury from sediment into the water. Therefore, the precipitation and dissolution of Fe–Mn oxides played an important role in the migration and transformation of mercury at the water–sediment interface.

In contrast, the concentrations of active mercury were lower in the wet season than in the dry season in other reservoirs in southwest China, including the DF and WJD in the Wujiang drainage area and BH. This is mainly due to the dilution effect (Yan 2005; Jiang 2005). Two reasons explain why concentrations of activity mercury were higher in the wet season rather than the dry season. On one hand, the

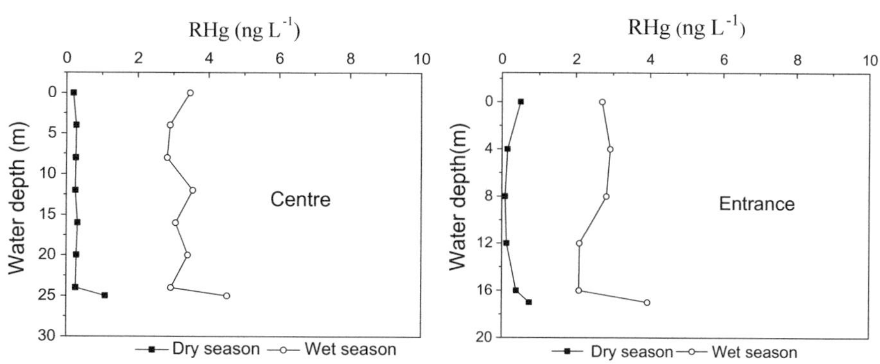

Fig. 7.62 Seasonal variances of RHg in Aha Reservoir water (redrawed from Bai et al. 2006, with permission from Acta Scientiae Circumstantiae)

AH is oxygen deficient in the wet season, which increases dissolved mercury and methylmercury. On the other hand, particulate matter properties differed from wet to dry seasons, which affected the distribution of dissolved mercury and methylmercury between the particulate phase and the water phase.

(6) MeHg in water columns

MeHg concentrations of AH water ranged from 0.03 to 0.43 ng L^{-1}, with an average of 0.25 ng L^{-1}, in the whole reservoir, ranged from 0.06 to 0.40 ng L^{-1} in the center area, and ranged from 0.03 to 0.43 ng L^{-1} in the upstream area in the dry season. Little variation was observed across the surface to bottom water. In contrast, MeHg concentrations were much higher in the wet season, ranging from 0.26 to 2.05 ng L^{-1}, with an average of 0.66 ng L^{-1} in the whole reservoir. MeHg concentrations ranged from 0.26 to 2.05 ng L^{-1} in the center area, and gradually decreased from the surface to a depth of 8 m, and then increased as depth increased, reaching 2.05 ng L^{-1} (approximately 21% of the THg) at the bottom. The concentration of methylmercury was 0.30–1.02 ng L^{-1} in the upstream area and had similar vertical profile compared to that observed in the center area. It gradually decreased from the surface to a depth of 8 m and then increased to 1.02 ng L^{-1} at the bottom (Fig. 7.63). In the summer, the concentrations of dissolved oxygen in the center and upstream areas were 4.52 and 3.61 mg L^{-1}, respectively, suggesting that the bottom water (12 m or deeper) was oxygen deficit. In sum, methylmercury in AH water mainly resulted from the release of MeHg in the sediment pore water.

The average ratios of MeHg to THg in the AH were 4 and 6% in the dry and wet seasons in the center area, respectively, and 8 and 5% in the upstream area, respectively. MeHg in the AH water did not have a significant correlation with THg in the wet ($r = -0.13$, $p > 0.05$, $n = 14$) or dry season ($r = 0.362$, $p > 0.05$, $n = 14$). This means that the generation and translocation of MeHg was not strongly correlated with MeHg in the reservoir water (Jiang 2005). In contrast, the MeHg had a significant positive correlation with RHg ($r = 0.638$, $p < 0.01$, $n = 28$) and

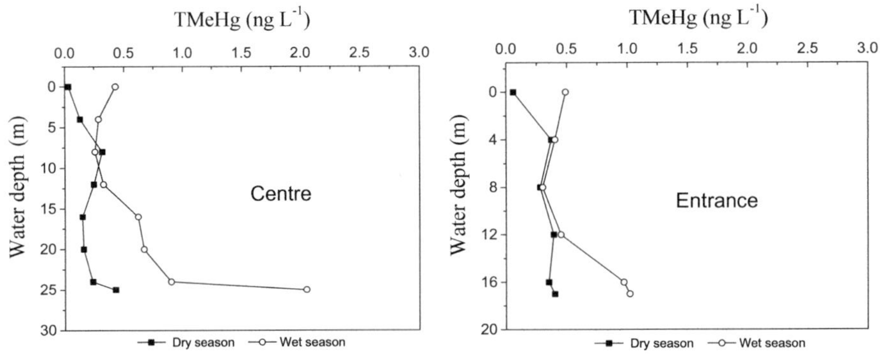

Fig. 7.63 Seasonal variation of methylmercury (MeHg) in the Aha reservoir water (redrawed from Bai et al. 2006, with permission from Acta Scientiae Circumstantiae)

DHg ($r = 0.581$, $p < 0.01$, $n = 28$). Compeau and Bartha (1987) noted that MeHg did not have the same proportion to MeHg in different samples and found lower proportions of MeHg in higher MeHg concentrations of sediments.

(7) DMeHg in water columns

The concentration of DMeHg in AH water in the dry season ranged from 0.01 to 0.30 ng L^{-1}, with an average of 0.14 ng L^{-1}, and was slightly higher at the bottom than at the surface (Fig. 7.64). The water was mixed well in the vertical direction in the dry season, as determined by the vertical profiles of water temperature and salinity. Therefore, little vertical variation of DMeHg was observed. The concentration of dissolved methylmercury in the center area was 0.13 ng L^{-1} on average, slightly increased as depth increased, and reached a maximum of 0.23 ng L^{-1} at the bottom. It was 0.18 ng L^{-1} on average in the upstream area with the highest concentration of 0.39 ng L^{-1} at 12 m. The AH experienced hypoxia at the bottom in the summer, and then the redox interface gradually extended to the upper water where methylation occurred. In addition, iron and manganese oxides in the sediment were reduced to soluble iron and manganese ions, which released the mercury absorbed before. Therefore, DMeHg concentrations of AH water increased significantly in the wet season, ranging from 0.22 to 1.25 ng L^{-1}, with an average of 0.49 ng L^{-1}. DMeHg concentrations decreased slightly at first and then increased gradually as water depth increased, finally peaking at 1.25 ng L^{-1} at the bottom in the center area. DMeHg concentrations showed similar vertical trends in the upstream area, decreasing slightly from the surface to a depth of 8 m, and then increasing and finally reaching 0.76 ng L^{-1} at the bottom. It is worth noting that methylmercury and DMeHg concentrations increased from a depth of 8 m to the bottom.

The proportions of DMeHg to the TMeHg were 63.7 and 74.2% in the dry and wet seasons, respectively. There was a significant positive correlation between DMeHg and TMeHg ($r = 0.979$, $p < 0.01$, $n = 28$). There was not a significant correlation between DMeHg and DOC ($r = 0.226$, $p > 0.05$, $n = 28$).

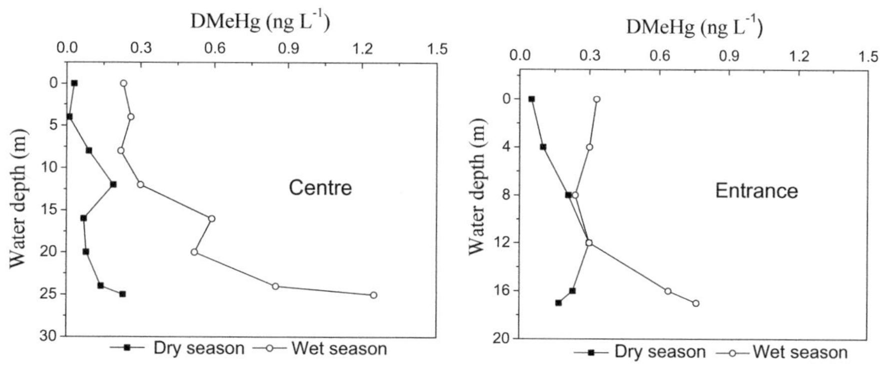

Fig. 7.64 Seasonal variation of dissolved methylmercury (DMeHg) in the Aha reservoir (redrawed from Bai et al. 2006, with permission from Acta Scientiae Circumstantiae)

(8) PMeHg in water columns

The concentrations of PMeHg in the AH water were 0.01–0.27 ng L^{-1}, with an average of 0.12 ng L^{-1}. These concentrations first increased as depth increased, reaching the highest concentrations of 0.23 ng L^{-1} and 0.27 ng L^{-1} at 8 and 4 m in the center area and upstream areas, respectively, and then decreased but increased at the bottom. The adsorption of PMeHg by algae in the biologically active layer at depths of 5–10 m resulted in high concentrations of particulate MeHg. Concentrations of particulate mercury increasing again at the bottom of the reservoir may be attributed to the resuspension of sediment caused by hydrodynamic and biological disturbance. Particulate mercury was composed of inorganic particulate mercury, organic particulate mercury, and particulate mercury derived from organisms such as bacteria, algae, and plankton. Particulate mercury was mainly associated with mineral particles and debris, but PMeHg was closely related to organisms. In a freshwater lake, the distribution of mercury and methylmercury is mainly controlled by the adsorption or desorption of particulate matter and the redox conditions at the water–sediment interface (Lawson et al. 2001).

The concentration of PMeHg of AH water ranged from 0.02 to 0.79 ng L^{-1}, with an average of 0.17 ng L^{-1} in the wet season, much higher than that recorded in the dry season (Fig. 7.65). The concentration of PMeHg had little variation in the water column, except for the bottom layer water that had the highest value (0.79 ng L^{-1}) in the center area. In contrast, the concentration of PMeHg (0.26 ng L^{-1}) did not increase in the upstream area.

The average proportions of PMeHg to TMeHg in AH water were 54.2 and 25.7% in the dry and wet seasons, respectively. The Pearson correlation coefficients of PMeHg and THg were 0.356 (n = 14) and −0.24 (n = 14) in the dry and wet seasons, which suggests that there was no significant correlation between them. Similarly, Pearson correlation coefficient between particle methylmercury and

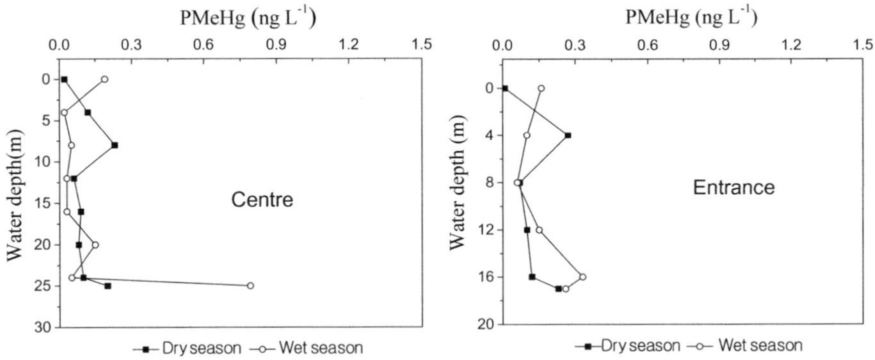

Fig. 7.65 Seasonal variation of PMeHg in Aha reservoir (redrawed from Bai et al. 2006, with permission from Acta Scientiae Circumstantiae)

suspended particulate matter was 0.205 (n = 28), suggesting no significant correlation. In conclusion, the concentration of PMeHg in AH water was not controlled by the main river input and the suspended particles in the reservoir.

7.1.3.3 Mercury Species in Inflows and Outflow

Five main rivers enter the AH: the Youyu River, Caichong River, Lanni River, Shahe River, and Baiyan River. The total annual water flux was 102 million m^3. The outlet of the dam is the only path for the discharge of water from the AH, and Xiaoche River is located downstream from the dam.

There are 216 small coal mines in the drainage areas of the Youyu, Cai Chong, and Baiyan Rivers that enter the AH. 80% of coal mines are located in the small mountain watershed of the Youyu River. Overall, privately owned coal mines were small scale and mobile with untreated and irregular wastewater discharge. Furthermore, abandoned gangue was leached and flushed by rainwater and surface runoff. This was the diffuse source of pollution in the upper reaches of the AH. In the summers of 1985 and 1990, concentrations of iron and manganese in the reservoir water exceeded standards and caused the color of the water in the whole reservoir to change to yellow, which seriously affected the water supply for residents of Guiyang. Therefore, a series of measures for controlling pollution in the AH were taken beginning in 1995, including the Xiao Gezhai (12,000 m^2) and Yangmeichong (17,000 m^2) collecting pools and the Liangshuijing, Changzha, Xuechang, and Xiazhai Dams, built according to the wastewater status and the surrounding environment of the Youyu River drainage area. Meanwhile, neutralization and precipitation from adding lime and an aeration system helped the water entering the AH finally meet the level III standard for surface water.

1. River water quality characters

The physical and chemical parameters of river water measured in the dry and wet seasons are listed in Table 7.14. The pH of the river water was above 7.5, mainly due to the alkaline environment caused the limestone bedrock of AH basin and the lime added to the river.

The dissolved oxygen concentration of the Lannigou River was very low in wet season, mainly due to mining and factories in the Lannigou River Basin. More than 10,000 people live in the Lannigou River catchment and the sewage of domestic, industrial, and agricultural discharge directly flows untreated into the AH. It is estimated that more than 10 million tons of wastewater discharge into the AH via the Lannigou River every year.

Youyu River runs through the village of Anxiang, Mai Ping in the Huaxi District, where an intensive coal mine is located. There were more than 300 mines years ago, but this number has reduced to 29 mines, which belong to the Lindong Mine Company. Now a series of water treatments, such as adding lime to the river,

Table 7.14 Inflows /Outflow water quality parameters of Aha Reservoir (redrawed from Bai et al. 2007, with permission from Acta Mineralogica Sinica)

Date	Site	T (°C)	pH	DO (mg L^{-1})	TDS	ES (μs cm^{-1})	Sal (‰)
Dry season	Xiaochehe river	9.8	7.8	9.31	375	545	0.4
	Lannigou	14.3	7.6	5.97	332	543	0.3
	Youyu river	15.9	7.9	7.75	618	1026	0.6
	Caichong river	15.7	7.7	5.26	427	715	0.4
	Sha river	15.3	8.2	9.87	388	645	0.4
	Baiyan river	13.9	8.1	8.89	383	614	0.4
Wet season	Xiaochehe river	16.7	7.5	6.96	341	567	0.3
	Lannigou	23	7.8	0.68	309	610	0.3
	Youyu river	24	8	6.66	573	1101	0.6
	Caichong river	21.8	7.3	3.68	443	844	0.4
	Sha river	26.1	8.5	3.72	300	623	0.3
	Baiyan river	24.5	8.1	2.46	383	614	0.4

have been initiated. The quality of Youyu River's water has improved; the concentration of DO was 6.7 mg L^{-1}, the conductivity was 1101 s cm^{-1}, and the salinity was 0.6% in the wet season.

2. Concentrations and distributions of mercury species

The concentration of THg in the river water entering the AH ranged from 7.1 to 47.9 ng L^{-1}, with an average of 19.6 ng L^{-1}, with 16.8 ng L^{-1} observed in the dry season, and 19.9 ng L^{-1} observed in the wet season. The average concentration of particulate mercury ranged from 0.7 to 40.3 ng L^{-1}, with an average of 15 ng L^{-1}, with 12.4 ng L^{-1} observed in the dry season (66% of the THg), and 15.4 ng L^{-1} observed in the wet season (Table 7.15). Particulate mercury in the river was positively correlated with the THg ($r = 0.979$, $p < 0.01$, $n = 10$) (Fig. 7.66).

The RHg concentrations of river water entering the AH ranged from 0.3 to 2.4 ng L^{-1}, with an average of 1.3 ng L^{-1}, with 0.5 ng L^{-1} observed in the dry season (4.4% of THg), and 2 ng L^{-1} observed in the wet season (16.8% of THg), much higher than that in the dry season. A large amount of mercury that adsorbed on suspended particles became RHg when the DHg decreased in summer, so the concentration of RHg in the wet season was higher than that in the dry season (Fig. 7.67).

The concentrations of methylmercury in rivers entering the AH ranged from 0.16 to 2.50 ng L^{-1}, with an average of 1.03 ng L^{-1}, with 0.87 ng L^{-1} observed in the dry season, and 1.34 ng L^{-1} observed in the wet season. High temperatures in the wet season accelerated the methylation of mercury. Lannigou River had the highest concentrations of mercury and methylmercury in the dry season, and Shahe River had the highest concentration of methylmercury in the wet season. Because of the

Table 7.15 Concentrations of different forms of mercury in Aha Reservoir input /output stream (ng L^{-1}) (redrawed from Bai et al. 2007, with permission from Acta Mineralogica Sinica)

Season	Hg species	River					
		Xiaochehe river	Lannigou	Youyu river	Caichong river	Sha river	Baiyan river
Dry season	THg	4.7	39.9	7.1	17.1	22.9	9.0
	DHg	3.5	10.0	2.5	2.3	4.7	3.2
	PHg	1.2	29.9	4.5	14.8	18.2	5.8
	MeHg	0.21	2.11	0.41	0.47	0.80	0.56
	DMeHg	0.10	0.48	0.27	0.16	0.70	0.04
	PMeHg	0.11	1.63	0.14	0.31	0.09	0.52
	RHg	0.12	0.25	1.02	0.30	0.57	0.25
Wet season	THg	6.2	33.2	11.6	47.92	10.8	9.9
	DHg	5.8	3.8	8.1	7.6	10.1	6.9
	PHg	0.4	29.4	3.4	40.3	0.7	3.0
	MeHg	1.04	2.36	1.44	0.25	2.50	0.16
	DMeHg	0.35	1.22	0.17	0.16	0.40	0.08
	PMeHg	0.70	1.14	1.28	0.09	2.10	0.08
	RHg	1.92	1.23	2.19	2.25	2.43	1.96

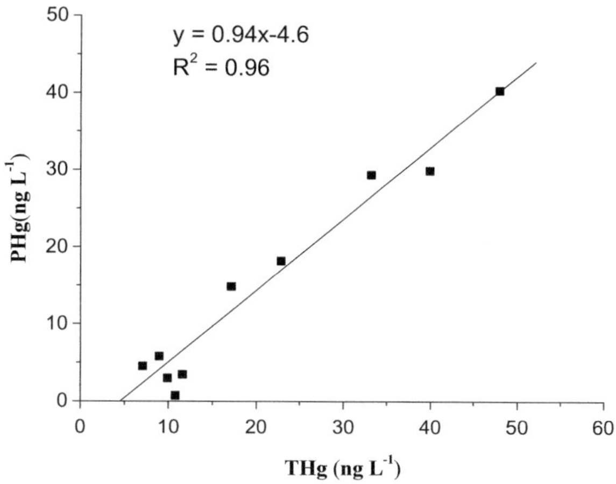

Fig. 7.66 Correlation between total mercury (THg) and particulate mercury (PHg) in the Aha reservoir

sedimentation and interception effects of the reservoir, THg, particulate mercury, active mercury, methyl mercury, and particulate mercury in the Xiaoche River—the only outlet river for the AH—were significantly lower than those concentrations observed in other rivers that enter the reservoir.

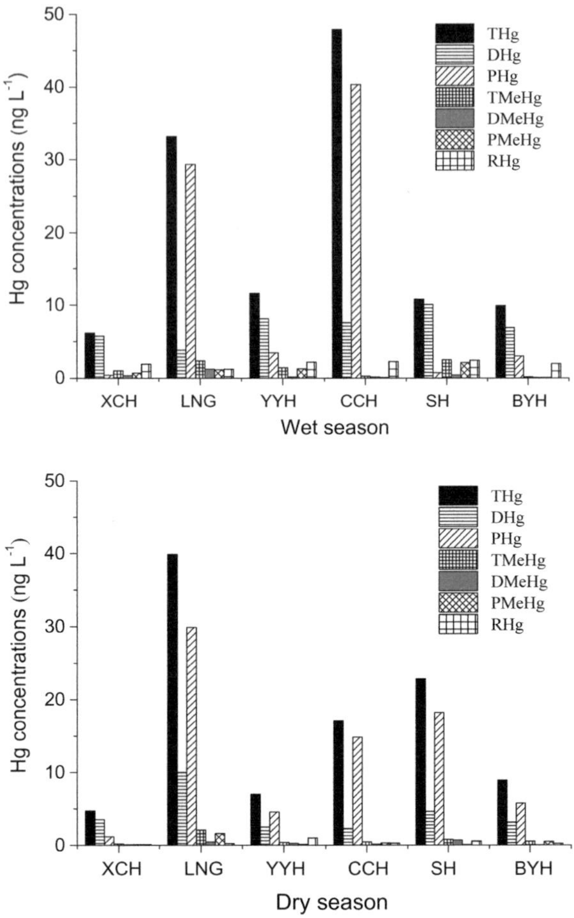

Fig. 7.67 Distribution of different mercury in the input of Aha reservoir. *Note* XCH—Xiaochehe River, LNG—Lannigou, YYH—Youyu River, CCH—Caichong River, SH—Sha River, BYH—Baiyan River (redrawed from Bai et al. 2007, with permission from Acta Mineralogica Sinica)

7.1.3.4 Mercury Species in Sediment Cores

1. Organic matter in sediment cores

SWB-1 portable sample apparatus is used to collect sediment cores as it causes no disturbance to sediments in lakes, thus ensuring that interface water is clear. In the dry season, the sediment–water interface is covered with a layer of light brown Fe–Mn nodule membrane with a thickness of about 3 cm. The surface water is clear and the color of the sediment column turns from dark brown to yellow. The dark brown part of the sediment column accounts for a length about 19 cm, and the sediment porosity of the yellow part gets higher with a texture similar to yellow soil. After entering the wet season, Fe–Mn nodule membrane at the sediment–water interface disappears, and the interface water becomes slightly turbid. Other properties of the sediment column remain similar to the dry season.

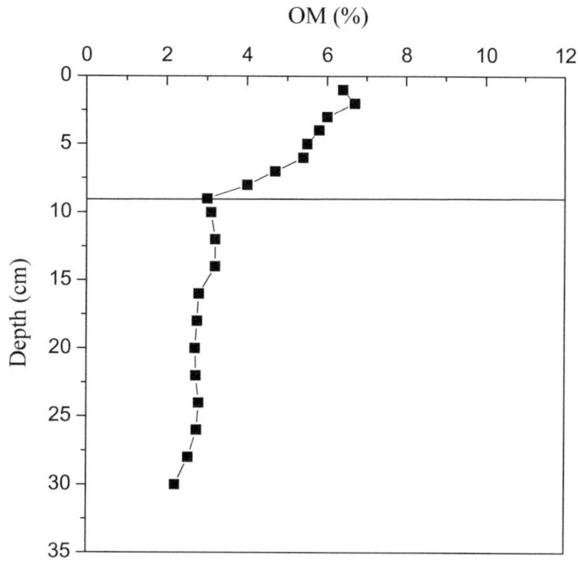

Fig. 7.68 Concentration of organic matter in sediments collected from Aha reservoir (redrawed from Feng et al. 2011, with permission from Elsevier)

The organic matter concentrations and distribution in sediments are shown in Fig. 7.68. Our data showed that the highest organic matter concentrations were present in the surface sediment layer, and decreased rapidly within the first 7 cm below the sediment surface.

2. Distribution of mercury species in sediment cores

(1) THg in sediment profiles

Sediments at the bottom of the lake are not only products of gravitational sedimentation of suspended matter in the water, but also collectors of pollutants in the water. Further, sediments are reliable records of the sedimentary history in that they have a strong adsorption to dissolved substances in the water. Therefore, the sediment is the best research object when studying the pollution history and contamination status of lakes.

The mean concentration of total Hg in sediments in AH was 210 ng g^{-1} (ranging from 160 to 252 ng g^{-1}). In the dry season, the mean concentration of total Hg in sediments in AH was 207 ng g^{-1} (160–232 ng g^{-1}), and it gets gradually higher as the depth increases and then remains stable below the depth of 18 cm. In the wet season, the average THg concentration in sediments was 211 ng g^{-1} (ranging from 166 to 252 ng g^{-1}), displaying a gradually increasing trend with depth.

THg concentration value fluctuated within the first 10 cm at the sediment surface and remained almost stable below the depth of 20 cm in the central area of the reservoir; however, sediment total Hg content remained stable below the depth of 18 cm in the inlet area. Results showed that THg concentration distribution in

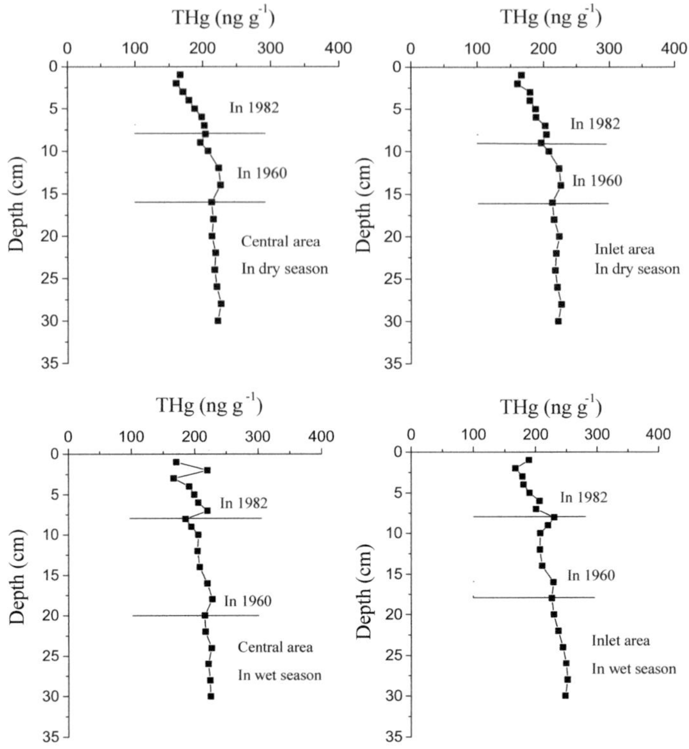

Fig. 7.69 Concentration of total mercury (THg) in sediments collected from Aha reservoir (redrawed from Feng et al. 2011, with permission from Elsevier; redrawed from Bai et al. 2011, with permission from Chinese Journal of Ecology)

sediment profiles in the dry season was similar to that in the wet season. Thus, it implied that there was no seasonal variation in the distribution of THg in sediment profiles (Fig. 7.69).

The vertical profile distribution of sediment THg could reflect the history of mercury contamination in an area (Gobeil and Cossa 1993; Canario et al. 2003). The concentrations and distribution patterns of THg in sediment collected from AH are illustrated in Fig. 7.69. THg concentrations increased gradually with depth.

The construction of AH started in 1958 and was completed and used for water storage in 1960. It began to serve as a drinking water source after its water storage capacity was expanded in 1982. We calculated the age of the sediment depth based on the sedimentation rate of AH (0.1449 g cm^{-2} a^{-1}) (Wang 2003), the section from 8 to 9 cm of the sediment column reflected the period of storage capacity expansion in 1982, and the section with a depth of 20 cm basically reflected the period when Aha Lake was transformed from a lake into a reservoir in 1958. Therefore, the history of the sediment column in AH could be divided into three sections (Fig. 7.69).

In the first section, it covered the depth of 0 cm to 9 cm of the sediment column, which reflected the period from the expansion of storage capacity till now. THg concentration in this section ranged from 160 to 230 ng g^{-1} (with an average value of 192 ng g^{-1}). The expansion of storage capacity of AH resulted in a significant rise in its water level, submerging some coal mines located close to its inflow rivers such as the Baiyan River and Youyu River. As a consequence, cinders soaked in water thus caused an influx of leaching solution and wastewater into the reservoir. Since 1985, excessively high levels of iron and manganese in the inflow and outflow water, from the waterworks located in the southern suburbs of Guiyang, had been observed, and the color of water appeared yellowish. Therefore, total Hg content at the depth of 8 cm was much higher than that at the sediment surface. Aiming to improve the water quality, relevant establishments built a number of lime delivery stations and barrages on the upper, middle, and lower reaches of the Baiyan River and Youyu River. This has greatly enhanced the pH value of water and sediments, and slowed down the velocity of the water inflow, leading to the sedimentation of heavy metals (such as iron and manganese) after their oxidation and hydroxylation, thus preventing them entering into the reservoir. Consequently, THg concentration in the sediment column at the depth of 0 to 5 cm did not increase and showed a narrow scale variation.

In the second section, it covered the depth from 9 to 20 cm of the sediment column, which reflected the period of time from the reservoir construction to storage capacity expansion. THg concentrations in this section ranged from 196 to 232 ng g^{-1} (with an average of 217 ng g^{-1}). At the time when AH was completed and filled with water in 1960, the whole water area was small and was susceptible to watershed erosion, and then the sediment porosity was lower than that below this section. In addition, the pH value of the reservoir was greatly affected by the acidic wastewater from the coal mines and the leaching solution from the coal cinders during the construction of the reservoir. As shown in Fig. 7.70 (Wang 2003), the pH value in the middle section of the sediment column was acidic, resulting in a higher THg concentration in sediments.

In the third section, it covered the part below the depth of 20 cm of the sediment column. THg concentration in this section ranged from 214 to 227 ng g^{-1} (with an average value of 219 ng g^{-1}). The color of sediment cores from this section was observed clearly, and it found that the color is light yellow with a thick and compact texture, which was completely different from the second section that is brownish black. As shown in Fig. 7.70 (Wang 2003), the carbon-to-nitrogen ratio (C/N ratio) in this sediment column was significantly higher than 24, which was also significantly higher than that in the second section. In general, the C/N ratio of terrestrial vascular plants is higher than 20 and the C/N ratio of lower aquatic plants ranges between 4 and 10. Therefore, we speculate that these sediments in this section were original deposits, in which organic matter mainly came from terrestrial ecosystems. As the wastewater influx from mines continued to accumulate in sediments, THg concentration in this section was the highest in the whole sediment column. Apart from the acidity input of the coal mines, the associated mercury in mines located in the river basin was attributed to the increase of Hg concentration in the environment.

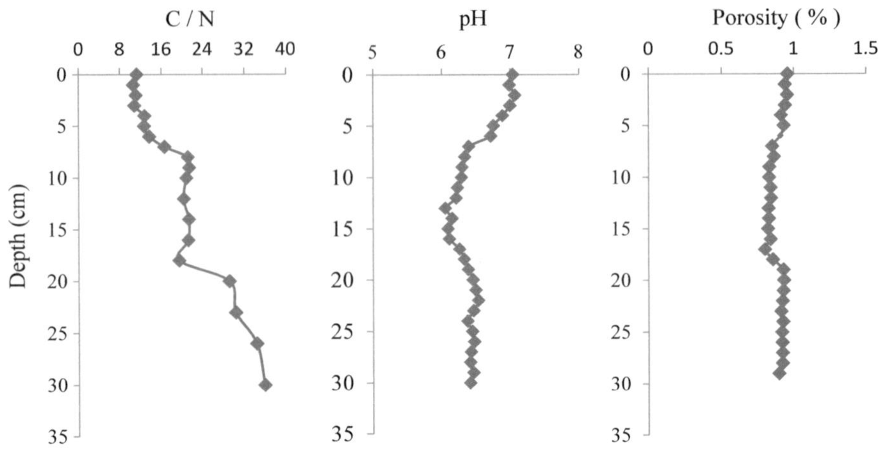

Fig. 7.70 C/N, pH, and porosity (%) in the sediment profile collected from Aha reservoir (compiled by Wang 2003)

(2) MeHg in sediment profiles

Mercury methylation may be caused by biological or nonbiological factors, or both. For instance, microbial mercury methylation was accompanied by the process which organic tin converted inorganic mercury into methylmercury (Domagalski et al. 2004). More and more researchers were becoming aware that mercury methylation is mainly caused by microbial bacteria (Lawson et al. 2001; Leermakers et al. 2001; Lindqvist et al. 1991). Compeau and Bartha (1987) confirmed that methylmercury produced by biomethylation in sediments was approximately one order of magnitude higher than methylmercury produced by abiotic methylation in an anaerobic environment.

The methylmercury concentrations of sediments in AH ranged from 0.2 to 7.2 ng g^{-1} (with an average of 1.8 ng g^{-1}). In the dry season, methylmercury concentration varied from 0.2 to 3.0 ng g^{-1} (with an average of 1.1 ng g^{-1}), and the concentrations of dissolved oxygen in the bottom waters at the both sampling sites of AH were 8.5 mg L^{-1} and 7.9 mg L^{-1}, respectively. Demethylation rate under aerobic conditions was much higher than that under hypoxic conditions (Steffan and Korthals 1994). As shown in Fig. 7.71, we observed that the concentration of methylmercury in sediments was generally highest in the surface soil (0–2 cm) at the both sampling sites of the AH, and decreased with depth.

In the wet season, the concentrations of methylmercury in sediments ranged from 0.50 to 7.25 ng g^{-1} (with an average of 2.45 ng g^{-1}). Furthermore, the highest concentration of methylmercury in sediments was observed at the surface. During the summer, the bottom of the reservoir was under an anaerobic condition (3.78 mg L^{-1} in the inlet area and 3.61 mg L^{-1} in the central area), and so, the Hg

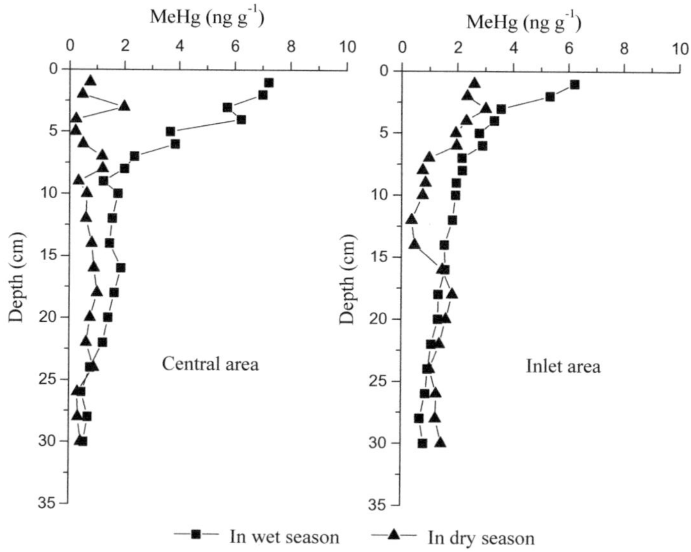

Fig. 7.71 Concentration of methylmercury (MeHg) in sediments collected from Aha reservoir (redrawed from Feng et al. 2011, with permission from Elsevier; redrawed from Bai et al. 2011, with permission from Chinese Journal of Ecology)

methylation rate was much higher than the demethylation rate under anaerobic conditions. Besides, a favorable environment for methylation production sediment was created by the rising temperature at the bottom of the reservoir (16.4 °C in the wet season, 8.9 °C in the dry season), the organic matter from dead algae and other plankton that sunk down to the reservoir bottom, and the large amount of organic matter brought by the inflow rivers. Hence, methylmercury concentration in the wet season was higher than in the dry season.

The average ratios of methylmercury to total mercury (MeHg/THg) in sediments of the AH were 0.85% (ranging from 0.12 to 4.22%) and 0.91% (ranging from 0.15 to 3.28%) at the central area and the inlet area. This is consistent with the observations that the average ratio of MeHg to THg was 1–1.5% or even less than 0.5% in rivers and seas (Steffan and Korthals 1994; Xun et al. 1987; Gilmour and Henry 1991).

Factors like the types of land using and the organic matter concentrations play important roles in the migration and conversion of Hg and MeHg. Wetlands and peatlands are considered as sensitive ecosystems of MeHg production and important MeHg sources in the freshwater (Baldi and Parati 1995). The C/N ratio in the entire sediments profile ranged between the values of organic matter of aquatic and terrestrial sources, while the terrestrial sources were characterized with high C/N ratio. The submerged soils before the construction of AH were mainly yellow soil under open forest and vegetation and paddy soil. The degradation of organic matter of AH was mainly occurred in the upper 4 cm of the sediments, while the degradation of organic nitrogen is mainly occurred in the upper 8 cm of the sediments, resulting in low C/N ratio in the surface layer. The C/N ratio in the sediments below the depth of 8 cm turns to the value of organic matter from terrestrial source, since there was no degradation of organic nitrogen blow this depth. The organic matter in AH reflected the features of terrestrial sources. The variation of MeHg concentrations below the depth of 18–20 cm revealed the type of land using of the submerged soils before flooding, which was confirmed to be paddy soil.

(3) Influence of SRB on mercury methylation

Microorganisms play a key role in the evolutionary process of the Hg biogeochemical cycle in the aquatic ecosystems. For example, microorganisms play a vital role in the conversion of Hg^{2+} into MeHg and DeMeHg and the reduction of Hg^{2+} to Hg^0 (Caldwell et al. 2000; Ikingura and Akagi 1999). Mercury compounds are toxic to most microorganisms in the freshwater, but many studies have found that many bacteria are adaptable to Hg (Tremblay et al. 1998). A significant positive correlation was observed between microbes in sediments and the distribution of Hg as well as Hg compounds (Akagi et al. 1995). Many studies have shown that SRB are major Hg-methylating microorganisms in sediments in freshwater and estuaries (Compeau and Bartha 1987). SRB includes many genera, among which desulfurization intestinal bacteria, desulfurization leaf bacteria, desulfurization bacteria, and desulfurization bacteria are common types. However, not all SRB are involved in Hg methylation, instead, some SRB are involved in Hg demethylation (Gilmour and Henry 1991). The sulfate concentration in sediments of AH is relatively high and large amounts of SRB can significantly affect Hg methylation. Because the abundance of SRB ranged with depth of the sediment profile, the sediment profiles were divided into two sections at the depth of 10 cm to study Hg methylation.

The distribution of SRB in the sediment–water interface in spring was reported in Table 7.16. The results showed that the abundance of SRB reached the peak at the depth of 2 cm below the water interface. With the increase of depth, the abundance of SRB and the sulfate reduction rate decreased. Owing to the distribution of SRB, sulfate reduction at the sediment–water interface in AH mainly occurred at the

Table 7.16 Distribution of sulfate-reducing bacteria in sediment–water interface of Aha reservoir (Wang 2003)

Depth (cm)	SRB (10^4 cell g^{-1})	SO_4^{2-} (mg L^{-1})
0	2.2	–
1	22	1200
2	103	–
4	5	1100
6	8	891
8	<1	470
10	<1	516
12	<1	402
14	4	451
16	<1	500

surface (a few centimeters) of the sediment (Wang 2003). We found that MeHg concentration in the sediments in the central area in the dry season was the highest (3.98 ng g^{-1}) at the depth of 3 cm below the surface, where the highest concentration of DMeHg (3.01 ng g^{-1}) in the pore water was also observed at this depth. As the redox interface in AH during spring located at the surface (a few centimeters) of the sediments, different types of SRB may methylate inorganic Hg (IHg) under moderate anaerobic conditions (Benoit et al. 1999). Since the peak of MeHg concentrations in the sediment coincides with the peak of SRB, we can conclude that SRB may be the major contributor to MeHg in sediments.

(4) Effects of FeRB on mercury methylation

From the distribution of MeHg concentrations in the sediments (Fig. 7.71), it can be seen that the peak value of MeHg concentrations in the sediments occurs within the first 10 cm below the sediment surface and MeHg concentration below the depth of 10 cm did not decrease but remained stable. Besides SRB, are there any other methylator of Hg?

According to Wang's study (2003) on the distribution of iron and manganese in the pore water of the sediment profile collected in front of the dam in the same season, we could see that the concentration of iron ion within the first 8 cm below the surface remained stable. However, the concentration of iron ions in the pore water displayed a rapid increase at the depth of 18 to 30 cm in the sediment profile, with a sustained concentration around 20 mg L^{-1} (Fig. 7.72). The distribution of abundance of FeRB shows a similar pattern. With the increase of depth, the abundance of Fe-reducing bacteria in the pore water increased from 10^5 cell g^{-1} at the depth of 12 cm to 2×10^5 cell g^{-1} at the depth of 15 cm in the sediments. The

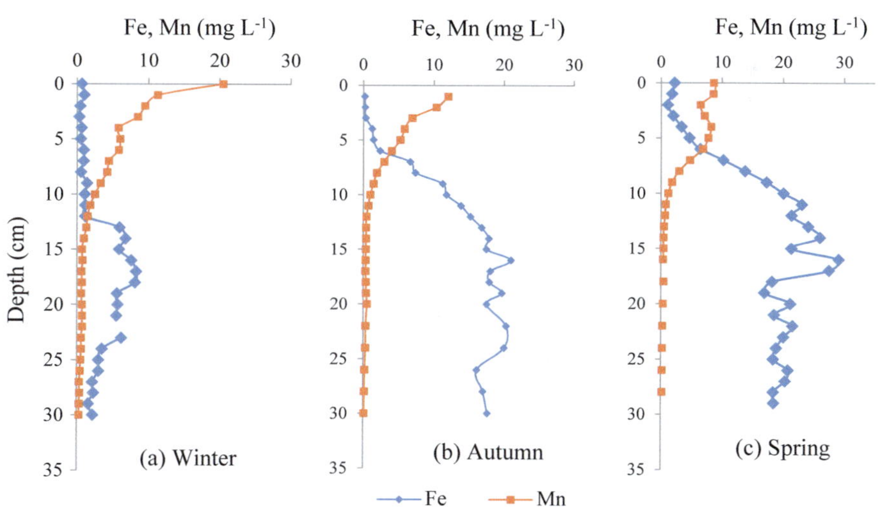

Fig. 7.72 Seasonal distribution of iron and manganese in the pore water collected from Aha reservoir (compiled by Wang 2003)

abundance of FeRB remained basically stable around 2×10^5 cell g^{-1}, indicating that FeRB mainly exists at the bottom of the sediments under the reduction condition (Warner et al. 2004). We observed the peak of MeHg concentration at the depth of 10 cm of the sediment profiles. Although MeHg concentration decreased at the depth of 20 cm, it still remained a certain concentration. This may be attributed to the contribution of FeRB on Hg methylation (Fleming et al. 2006). Fleming et al. (2006) first observed that FeRB can convert IHg into MeHg in the natural environment. In the sediments rich in iron, Hg methylation by FeRB is an important source of MeHg in the sediments. Our data also supported the view that FeRB could methylate Hg in the sediments.

3. Distribution of mercury species in pore water of sediment
(1) DHg in pore water

The release of Hg is a bidirectional process (from sediments to water and from water to sediments) characterized by high intensity and large quantity. A change in Hg concentration in the pore water is the most obvious manifestation of Hg releasing from the sediments. In the dry season, THg concentration in the pore water of the sediment collected from AH ranged from 2.7 to 19.2 ng L^{-1} (with an average of 6.3 ng L^{-1}). In the central area of the reservoir, DHg concentration in the pore water of the sediment ranged from 2.7 to 11.6 ng L^{-1} (with an average of 6.7 ng L^{-1}). The peak value of 11.6 ng L^{-1} was observed at the depth of 1 cm below the sediment surface. As the depth increasing, DHg showed a gradually decreasing trend. In the inlet area, DHg concentrations in the pore water ranged from 3.1 to 19.1 ng L^{-1} (with an average of 5.9 ng L^{-1}) and the maximum concentration of 19.1 ng L^{-1} was observed at the depth of 4 cm below the sediment surface (Fig. 7.73). In general, DHg in the pore water of the sediments was significantly higher than those in the overlying water. When there was a concentration gradient between the pore water of the sediments and the water bodies, DHg will be diffused from the pore water of the sediments into the water bodies.

In the wet season, DHg in the pore water of the sediments of AH ranged from 7.5 to 92.1 ng L^{-1} (with an average of 34.1 ng L^{-1}), which was significantly higher than those in the dry season. This was mainly attributed to the fact that temperature of the water body in the wet season was much higher than those in the dry season; in addition, the mineralization and degradation of organic matter in the sediments were the driving force of these processes. In the aerobic sediments, sedimentary organic matter was metabolized by different microbial communities, including bacteria, actinomycetes, and fungi, which could completely mineralize organic molecules and produce CO_2 directly. Under anaerobic conditions, different anaerobic bacteria communities can degrade organic matter by different steps. The mineralization and degradation of organic matter led to the changes of pH and redox potential in the sedimentary environment, which result in the reduction and dissolution of heavy metals in the solid phase. Meantime, the Hg adsorbed on iron and manganese particles could be desorbed, which could enter into the pore water of the sediments and migrate to the overlying water driven by the concentration

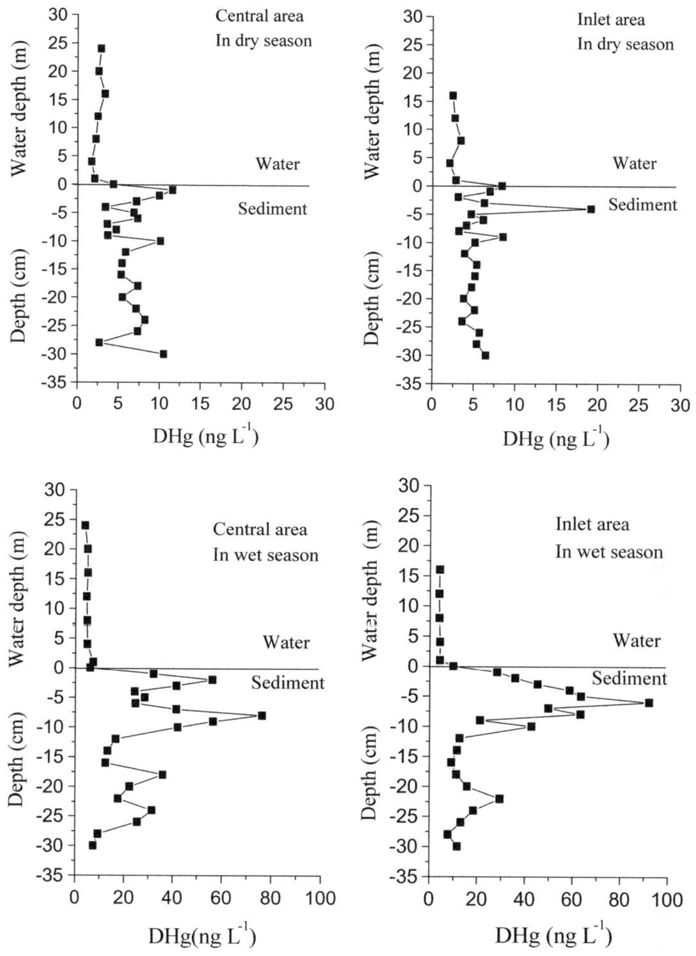

Fig. 7.73 Concentration of dissolved mercury (DHg) in pore water collected from Aha reservoir (redrawed from Feng et al. 2011, with permission from Elsevier; redrawed from Bai et al. 2006, with permission from Acta Scientiae Circumstantiae)

gradient. DHg concentrations in the pore water of the sediments in the central area varied from 7.5 to 76.2 ng L^{-1} (with an average of 30.8 ng L^{-1}), and reached a peak at the depth of 8 cm below the sediment surface. As the increase of depth, DHg concentrations gradually decreased. DHg concentrations in the pore water of the sediments in the inlet area ranged from 7.8 to 92.1 ng L^{-1} (with an average of 32.7 ng L^{-1}). The maximum value of DHg (19.1 ng L^{-1}) was observed at the depth of 6 cm below the surface.

The partition coefficient between the sediment and pore water varied significantly. IHg varied from 10^4 to 10^6 L kg^{-1} (Lyon et al. 1997). The partition coefficients of inorganic Hg between the sediment and the pore water in the dry season and wet season were 3.9×10^4 and 1.1×10^4 L kg^{-1}, respectively. As could be seen, the partition coefficient was higher in the dry season than that in the wet season. As temperature rises in summer, the bottom of the reservoir was in an anaerobic environment where anaerobic microbes like SRB and FeRB were active and abundant. It facilitated the conversion of inorganic Hg in the pore water to MeHg (Ikingura and Akagi 1999).

(2) DMeHg in pore water

DMeHg concentration in the pore water of the sediments of AH in the dry season varied between 0.06 and 1.57 ng L^{-1} (with an average of 0.62 ng L^{-1}). In the central area of the reservoir, DMeHg concentration ranged from 0.07 to 1.27 ng L^{-1} (with an average of 0.46 ng L^{-1}). In the inlet area, DMeHg concentrations varied from 0.06 to 1.57 ng L^{-1} (with an average of 0.72 ng L^{-1}). DMeHg concentrations in the pore water of the sediments were much higher than those in the overlying water; in other words, there was a concentration gradient between the sediment and the overlying water.

MeHg concentrations in the pore water of the sediments in the wet season varied from 0.33 to 4.20 ng L^{-1} (with a mean value of 1.11 ng L^{-1}), which were significant higher than those in the dry season. DHg concentrations in the pore water of the sediments in the central area of the reservoir ranged from 0.43 to 4.20 ng L^{-1} (with an average of 1.28 ng L^{-1}) and reached a peak at the depth of 3 cm below the sediment surface. DHg concentrations decreased gradually with the increase of depth of sediments. DHg concentrations in the pore water varied from 0.33 to 3.42 ng L^{-1} (with a mean value of 0.94 ng L^{-1}). The maximum value of MeHg concentration was observed at the depth of 2 cm below the sediment surface, reaching a value of 3.42 ng L^{-1}.

As shown in Fig. 7.74, MeHg concentrations in the pore water of the sediments were much higher than those in the overlying water body both in the wet and dry seasons. MeHg concentrations in the pore water were significantly correlated to MeHg concentrations in the sediments (in the dry season, $r = 0.575$, $p < 0.01$, $n = 40$; in the wet season, $r = 0.409$, $p < 0.01$, $n = 40$, Fig. 7.75). The results were similar to those observed also in reservoirs of the Wujiang River Basin by Jiang (2005). The MeHg concentrations in the pore water of the sediments were affected by the solid–liquid equilibrium between the sediments and the pore water to a certain extent.

Partition coefficient of MeHg between the sediments and the pore water varied from 10^3 to 10^5 L kg^{-1} (Gagnon et al. 1997; Fitzgerald et al. 1994). Partition coefficients of MeHg between the solid and liquid sediment interfaces in AH were as follows: 2.6×10^3 L kg^{-1} in the dry season, 2.5×10^3 L kg^{-1} in the wet season, 2.2×10^3 L kg^{-1} in the central reservoir, and 3.6×10^3 L kg^{-1} in the inlet area.

THg concentrations in the pore water of the sediments were generally higher than those in the overlying water (Canavan et al. 2000). MeHg in the pore water of the sediments also accounts for higher proportion of THg than those in the

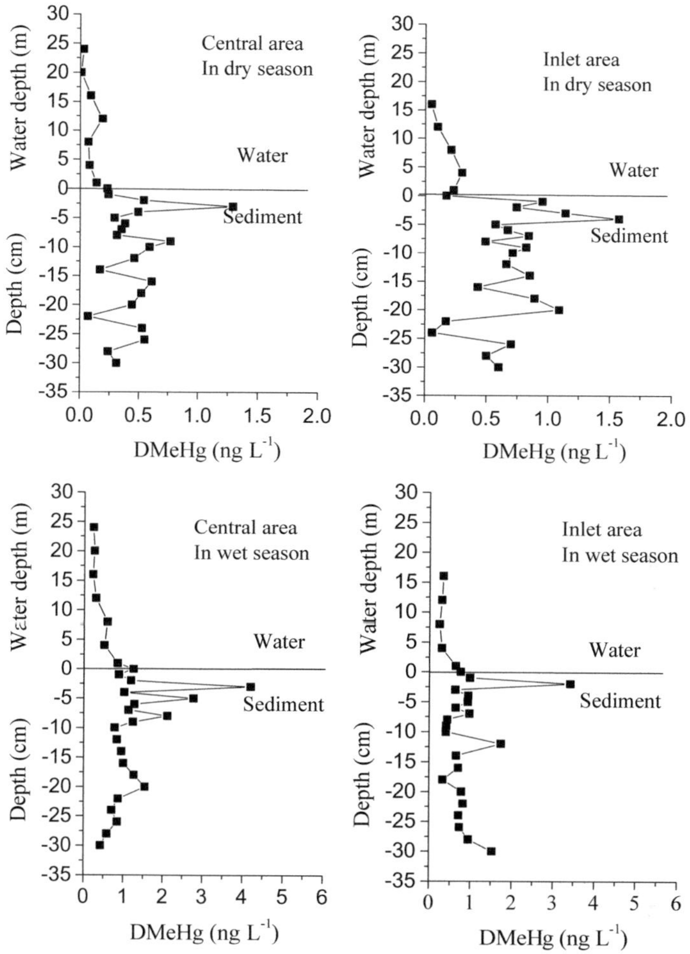

Fig. 7.74 Seasonal variation of dissolved methylmercury (DMeHg) in the pore water collected from Aha reservoir (redrawed from Feng et al. 2011, with permission from Elsevier; redrawed from Bai et al. 2006, with permission from Acta Scientiae Circumstantiae)

overlying water bodies. In this study, we found that DMeHg accounts for 1–29% of the DHg in the pore water in the dry season, averaging 11%. The ratios of DMeHg to DHg ranged from 1 to 14% with an average of 5%.

(3) Diffusion flux of inorganic mercury and methylmercury to water

The concentration of inorganic Hg (IHg) and MeHg in the pore water of the sediments was higher than those in the overlying water, showing a concentration gradient between sediments and the overlying water. For deepwater reservoirs, molecular diffusion caused by concentration gradient is a major process that facilitates the matter exchange between sediments and overlying water bodies.

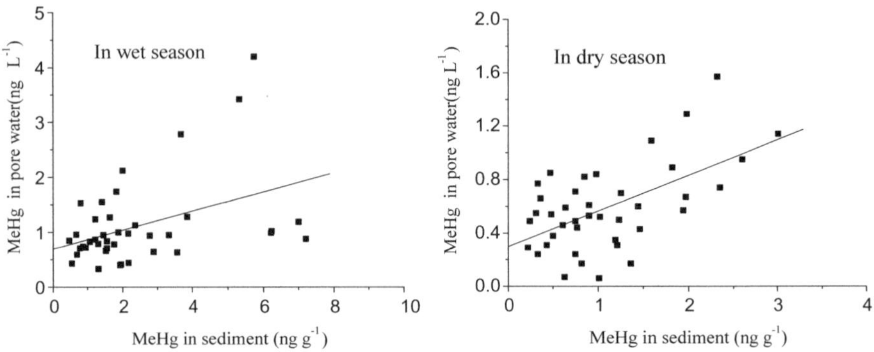

Fig. 7.75 Correlation between MeHg in sediments and MeHg in pore water collected from Aha reservoir

Table 7.17 Annual release fluxes of inorganic DHg and MeHg in Aha reservoir

	DHg	MeHg
Pore water (ng L^{-1})	92.1	4.20
Overlying water (ng L^{-1})	3.1	0.30
Concentration gradient (ng L^{-1})	89.0	4.17
Diffusion fluxes (g a^{-1})	73.7	3.45

Through the Hg concentration in the profile of the pore of the sediments, we can calculate the diffusion fluxes of DMeHg and inorganic Hg in the sediments/water interface under steady-state conditions using the First Fick's Law. It can estimate the contribution of sediment inorganic Hg and MeHg to the overlying water body. The method and formula of calculation are shown in Chap. 3 (Table 7.17).

It can be seen that the annual diffusion of DHg and MeHg from pore water of the sediments to the overlying water body reached 73.7 and 3.45 g, respectively, which accounted for 1 and 3% of the annual total amount input by river, respectively.

References

Abernathy AR, Cumbie PM (1977) Mercury accumulation by largemouth bass (*Micropterus salmoides*) in recently impounded reservoirs. Arch Environ Contam Toxicol 17:595–602

Akagi H, Malmb O, Kinjoa Y, Haradac M, Branchesb FJP, Pfeifferb WC, Katoa H (1995) Methylmercury pollution in the Amazon, Brazil. Sci Total Environ 175:85–95

Alberts JJ, Schindler JE, Miller RW (1974) Elemental mercury evolution mediated by humic acid. Science 184:895–897

Allard B, Arsenie I (1991) Abiotic reduction of mercury by humic substances in aquatic systems-an important process for the mercury cycle. Water Air Soil Pollut 56:457–464

Amyot M, Mierle G, Lean D, Mcqueen DJ (1997) Effect of solar radiation on the formation of dissolved gaseous mercury. Genchimica et Cosmochimica Acta. 61:975–987

Anirudh RMA, Rokade DV, Borole Zingde MD (2003) Mercury in sediments of Ulhas estuary. Mar Pollut Bull 46:846–857

Baeyens W, Meuleman C (1998) Behavior and speciation of mercury in the Scheldt estuary (water, sediments and benthic organisms). Trace Metals in the Westerschelde Estuary. Hydrobiologia 336:63–79

Bai WY, Feng XB, Sun L, He TR, Fu XW, Jiang HM (2006) The concentration and distribution of different mercury species in the water columns and sediment porewater of Aha Lake. Acta Scientiae Circumstantiae 26:91–98 In Chinese, with English abstract

Bai WY, Feng XB, Jin ZS, Sun L, Yan HY (2007) The influence of reivers on the transport and fate of mercury species in the Aha reservoir. Acta Mineralogica Sinica 27:218–224 In Chinese, with English abstract

Bai WY, Feng XB, He TR, Qiu GL, Yan HY (2011) Distribution patterns of total mercury and methylmercury in Aha Reservoir sediment. Chin J Ecol 30:976–980 In Chinese, with English abstract

Baldi FM, Parati F (1995) Dimethylmercury and dimethylmercury-sulfice of microbial origin in the biogeochemical cycle of Hg. Water Air Soil Pollut 80:205–220

Benoit JM, Gilmour CC, Mason RP (1999) Sulfide controls on mercury speciation and bioavailability to methylating bacteria in sediment pore waters. Environ Sci Technol 33: 951–957

Benoit JM, Gilmour CC, Heyes A, Mason RP, Miller C (2003) Geochemical and biological controls over methylmercury production and degradation in aquatic systems. In: Cai Y, Braids OC (Eds) ACS symposium series, biogeochemistry of environmentally important trace metals, vol 852. pp 262–297

Berman M, Bartha R (1986) Control of the methylation process in a mercury-polluted aquatic sediment. Environ Pollut Series B 11:41–53

Bloesch J (1995) Mechanisms, measurement and importance of sediment resuspension in lakes. Mar Freshw Res 46:295–304

Bloom NS, Moretto LM, Scopece P, Ugo P (2004) Seasonal cycling of mercury and monomethyl mercury in the Venice Lagoon (Italy). Mar Chem 91:85–99

Bruce AM (1997) Mercury cycling in low alkalinity lakes and factors influencing bioavailability. In: A thesis submitted to the faculty of the graduate school of the University of Minnesota

Caldwell CA, Canavan CM, Bloom NS (2000) Potential effects of forest fire and storm flow on total mercury and methylmercury in sediments of an arid-lands reservoir. Sci Total Environ 260:125–133

Canario J, Vale C, Caetano M, Madureira MJ (2003) Mercury in contaminated sediments and pore waters enriched in sulphate (Tagus Estuary, Portugal). Environ Pollut 126:425–433

Canavan CM, Caldwell CA, Bloom NS (2000) Discharge of methylmercury-enriched hypolimnetic water from a stratified reservoir. Sci Total Environ 260:159–170

Castro MS, McLaughlin EN, Davis SL, Morgan IIRPII (2002) Total Mercury Concentrations in Lakes and Fish of Western Maryland, USA. Arch Environ Contam Toxicol 42:454–462

Compeau GC, Bartha R (1987) Effect of salinity on mercury-methylating activity of sulfate-reducing bacteria in estuarine sediments. Appl Environ Microbiol 53:261–265

Coquery M, Cossa D, Sanjuan J (1997) Speciation and sorption of mercury in two macro-tidal estuaries. Mar Chem 58:213–227

Covelli S, Faganeli J, Horvat M, Brambati A (1999) Porewater distribution and benthic flux measurements of mercury and methylmercury in the Gulf of Trieste (Northern Adriatic Sea). Estuar Coast Shelf Sci 48:415–428

Cox JA, Carnahan J, Dinuzio J (1979) Source of mercury in fish in new impoundments. Arch Environ Contam Toxicol 23:779–783

Dalziel JA (1995) Reactive mercury in the eastern North Atlantic and southeast Atlantic. Mar Chem 49:307–314

Domagalski JL, Alpers CN, Slotton DG, Suchanek TH, Ayers SM (2004) Mercury and methylmercury concentrations and loads in the Cache Creek watershed, California. Sci Total Environ 327:215–237

Driscoll CT, Blette V, Yan C, Schofield CL, Munson R, Holsapple J (1995) The role of dissolved carbon in the chemistry and bioavailability of mercury in remote Adirondack lakes. Water Air Soil Pollut 80:499–508

Eckley CS, Watras CJ, Hintelmann H (2005) Mercury methylation in the hypolimnetic waters of lakes with and without connection to wetlands in northern Wisconsin. Can J Fish Aquat Sci 62:400–411

Feng XB, Sommar J, Gardfeldt K, Lindquist O (2002) Exchange flux of total gaseous mercury between air and natural water surfaces in summer season. Sci China series D 45:211–220

Feng XB, Tang SL, Shang LH, Yan HY, Sommar J (2003) Total gaseous mercury in the atmosphere of Guiyang, PR China. Sci Total Environ 304:61–72

Feng XB, Bai WY, Shang LH, He TR, Qiu GL, Yan HY (2011) Mercury speciation and distribution in Aha Reservoir which was contaminated by coal mining activities in Guiyang, Guizhou, China. Appl Geochem 26:213–221

Fitzgerald WF, Mason RP, Vandal GM, Dulac F (1994) Air-water cycling of mercury in lakes. In: Watras C, Huckabee J (eds) Mercury pollution: integration and synthesis. Lewis Publishers, Chelsea, pp 203–220

Fleming EJ, Mack EE, Green PG, Nelson DC (2006) Mercury methylation from unexpected sources: molybdate-inhibited freshwater sediments and an iron-reducing bacterium. Appl Environ Microbiol 72:457–464

French KJ, Scruton DA, Anderson MR (1999) Influence of physical and chemical characteristics on mercury in aquatic sediments. Water Air Soil Pollut 110:347–362

Fujiki M, Tajima S (1992) The pollution of Minamata Bay by mercury. Water Sci Technol 25:133–140

Gagnon C, Pelletier E, Mucci A, Fitzgerald WF (1996) Diagenetic behavior of methylmercury in organic-rich coastal sediments. Limnol Oceanogr 41:428–434

Gagnon C, Pelletier E, Mucci A (1997) Behaviour of anthropogenic mercury in coastal marine sediments. Mar Chem 59:159–176

Gill GA (1986) Marine biogeochemistry of mercury. Ph.D dissatation in university of Connecticut, Storrs

Gill GA, Bloom NS, Cappellino S, Driscoll CT (1999) Sediment-water fluxes of mercury in Lavaca Bay, Texas. Environ Sci Technol 33:663–669

Gilmour CC, Henry EA (1991) Mercury methylation in aquatic systems affected by acid deposition. Environ Pollut 71:131–169

Gobeil C, Cossa D (1993) Mercury in sediments and sediment pore water in the Laurentian Trough. Can J Fish Aquatic Sci 50:1794–1800

Gorski PR, Cleckner LB, Hurley JP, Sierszen ME, Armstrong DE (2003) Factors affecting enhanced mercury bioaccumulation in inland lakes of Isle Royale National Park USA. Sci Total Environ 304:327–348

Hammerschmidt CR, Fitzgerald WF, Lamborg CH, Balcom PH, Visscher PT (2004) Biogeochemistry of methylmercury in sediments of Long Island Sound. Mar Chem 90:31–52

He TR (2007) Biogeochemical cycling of mercury in Hongfeng Reservior, Guizhou, China. In: Ph. D dissertation in the Graduate School of the Chinese Academy of Sciences (GSCAS), Beijing. (In Chinese, with English abstract)

He TR, Feng XB, Guo YN, Meng B, Li ZG, Qiu GL (2008a) Distributions of reactive and dissolved gaseous mercury and controlling factors in Hongfeng Reservoir. Res Environ Sci 21:14–17 In Chinese, with English abstract

He TR, Feng XB, Guo YN, Meng B, Li ZG, Qiu GL, Liang L (2008b) Geochemical cycling of mercury in the sediment of Hongfeng Reservoir. Environ Sci 29:1768–1774 In Chinese, with English abstract

He TR, Feng XB, Guo YN, Qiu GL, Li ZG, Liang L, Lu J (2008c) The impact of eutrophication on the biogeochemical cycling of mercury species in a reservoir: a case study from Hongfeng Reservoir, Guizhou China. Environ Pollut 154:56–67

He TR, Feng XB, Guo YN, Qiu GL, Yan HY, Meng B (2008d) Distribution and speciation of mercury in the Hongfeng Reservoir, Guizhou Province, China. Chin J Geochem 27:097–103

He TR, Wu YG, Feng XB (2010) The impact of eutrophication on distribution and speciation of mercury in Hongfeng Reservoir, Guizhou Province. J Lake Sci 22:208–214 In Chinese, with English abstract

Hines NA, Brezonik PL, Engstrom DR (2004) Sediment and porewater profiles and fluxes of mercury and methylmercury in a small seepage lake in northern Minnesota. Environ Sci Technol 38:6610–6617

Hurley JP, Watras CJ, Bloom NS. (1994) Distribution and flux of particulate mercury in four stratified seepage lakes. In: Mercury Pollution-Integration and Synthesis

Ikingura JR, Akagi H (1999) Methylmercury production and distribution in aquatic Systems. Sci Total Environ 234:109–118

Jackson TA (1988) The mercury problem in recently formed reservoirs of northern Manitoba (Canada): effects of impoundment and other factors on the production of methylmercury by microorganisms in sediments. Can J Fish Aquatic Sci 45:97–121

Jiang HM (2005) Effects of hydroelectric reservoir on the biogeochemical cycle of mercury in the Wujiang River. In: Ph.D dissertation in the Graduate School of the Chinese Academy of Sciences. (In Chinese, with English abstract)

Jiang HM, Feng XB, Dai QJ, Wang YC (2004) Preliminary study on speciation and distribution of mercury in Wujiang River. Environ Chem 23:556–561 In Chinese, with English abstract

Kim JP, Fitzgerald WF (1986) Sea-air partitioning of mercury in the Equatorial Pacific Ocean. Science 231:1131–1133

Korthals ET, Winfrey MR (1987) Seasonal and spatial variation in mercury methylation and demethylation in an Oligotrophic Lake. Appl Environ Microbiol 53(10):2397–2404

Kotnik J (2000) Modeling of mercury and its compounds in Lake Velenje. In: PhD Dissertation of Polytechnic Nova Gorica School of Environmental Sciences, Slovinia

Kotnik J, Horvat M, Fajon V (2002) Mercury in small freshwater lakes: a case study: Lake Velenje, Slovenia. Water Air Soil Pollut 134:319–339

Krabbenhoft DP, Hurley J, Olson ML, Cleckner LB (1998) Diel variability of mercury phase and species distributions in the Florida Everglades. Biogeochemistry 40:311–325

Krom MD, Berner RA (1980) Adsorption of phosphate in anoxic marine sediment. Limnol Oceanoger 25:797–806

Kudo A (1992) Natural and artificial mercury decontamination-Oyyawa River and Minamata Bay (Yatsushiro Sea). Water Sci Technol 26:217–226

Kudo A, Nagase H, Ose Y (1982) Proportion of methylmercury to the total amount of mercury in river waters in Canada and Japan. Water Res 16:1011–1015

Lanzillotta E, Ceccarini C, Ferrara R (2002) Photo-induced formation of dissolved gaseous mercury in coastal and offshore seawater of the Mediterranean basin. Sci Total Environ 300:179–187

Lawson NM, Mason RP, Laporte JM (2001) The fate and transport of mercury, methylmercury, and other trace metals in Chesapeake Bay Tributaries. Water Res 35:501–515

Leermakers M, Meuleman C, Baeyens W (1996) Mercury distribution and fluxes in Lake Baikal. In: Baeyens W, Ebinghaus R, Vasiliev O (eds) Global and regional mercury cycles: sources, fluxes and mass balances. Kluwer Academic Publishers, Dordrecht, Netherlands

Leermakers M, Galletti S, De Galan S, Brion N, Baeyens W (2001) Mercury in the southern North Sea and Scheldt Estuary. Mar Chem 75:229–248

Liang XB, Zhu JM, Liu CQ, Wei ZQ, Wang FS, Wan GJ, Huang RG (2003) Enzymatic and Microbial Degradation of Organic Matter in HonFeng Lake of Guizhou Province. Quaternary Sci 23:565–572 In Chinese, with English abstract

Liang XB, Zhu JM, Liu CQ (2004) Enzymatic and microbial degradation of organic matter in Lake Hongfeng, Guizhou Province China. Chin J Geochem 23:81–88

Lindberg SE, Harriss RC (1974) Mercury-Organic matter associations in estuarine sediments and interstitial water. Environ Sci Techol 8:459–462

Lindqvist O, Johansson K, Bringmark L, Timm B, Aastrup M, Andersson A, Hovsenius G, Håkanson L, Iverfeldt Å, Meili M (1991) Mercury in the Swedish environment-recent research on cause, consequence and corrective methods. Water Air Soil Pollut 55:1–261

Lucotte M, Montgomery S, Begin M (1999) Mercury dynamics at the flooded soil–water interface in reservoirs of Northern Quebec: in situ observations. In: Lucotte M, Schetagne R, Therien N et al (eds) Mercury in the Biogeochemical Cycle, Natural Environments and Hydroelectric Reservoirs of Northern Quebec. Springer, Berlin

Luo SS (2001) The geochemical indication of Fe, Mn and S in recent lacustrine sediments on Yun-Gui Plateau. In: Ph.D. dissertation in the Graduate School of the Chinese Academy of Sciences. (In Chinese, with English abstract)

Lyon BF, Ambrose R, Rice G (1997) Calculation of soil–water and benthic sediment partition coefficients for mercury. Chemosphere 35:791–808

Lyons WB, Welch KA, Bonzongo JC (1999) Mercury in aquatic systems in Antarctica. Geophys Res Lett 26:2235–2238

Ma YJ, Wan GJ, Liu CQ, Zhou JY, Huang RG (2000) Seasonal migration of redox boundary and its effects on water quaintly in Lake Lugu Yunnan Province. Acta Sci Circumst 20(1):27–32 In Chinese, with English abstract

Mason RP, Fitzgerald WF (1993) The distribution and biogeochemical cycling of mercury in the equatorial Pacific Ocean. Deep-sea Res Part I Oceanogr Res Pap 40:1897–1924

Mason RP, Sullivan KA (1999) The distribution and speciation of mercury in the South and equatorial Atlantic. Deep-sea Res Part II Top Stud Oceanogr 46:937–956

Mason RP, Morel FMM, Hemond HF (1995) The role of microorganisms in elemental mercury formation in natural waters. Water Air Soil Pollut 80:775–787

Meili M (1997) Mercury in lakes and Rivers. In: Sigel A, Sigel H (eds) Metal Ions in Biological Systems, Mercury and its effect on environment and biology, vol 34. Marcel Dekker Inc., New York

Meuleman C, Leermakers M, Baeyens W (1995) Mercury speciation in lake Baikal. Water Air Soil Pollut 80:539–551

Miller DR (1975) The role of humic acid in the uptake and release of mercury by freshwater sediments. Verhandlungen des Internationalen Verein Limnologie 19:2082

Miskimmin BM, Rudd JWM, Kelly CA (1992) Influence of dissolved organic carbon, pH, and microbial respiration rates on mercury methylation and demethylation in lake water. Can J fish Aquatic Sci 49:17–22

Nelson JD, Blair W, Brinckma FE, Colwell RR, Iverson WP, Neala H, Patruck B, Danielr E (1973) Sediment and Porewater Profiles and Biodegradation of phenylmercuric acetate by mercury resistant bacteria. Appl Microbiol 26:321–326

Nguyen HL, Leermakers M, Kurunczi S, Bozo L, Baeyens W (2005) Mercury distribution and speciation in Lake Balaton, Hungary. Sci Total Environ 340:231–246

Nriagu JO (1994) Mechanistic steps in the photoreduction of mercury in natural waters. Science of Total Environment. 154:1–8

Porvari P (1998) Development of fish mercury concentrations in Finnish reservoirs from 1979 to 1994. Sci Total Environ 213:279–290

Ravichandran M, Aiken GR, Reddy MM, Ryan JN (1998) Enhanced dissolution of cinnabar (mercuric sulfide) by dissolved organic matter isolated from the Florida Everglades. Environ Sci Technol 32:3305–3311

Regnell O, Ewald G (1997) Factors controlling temporal variation in methylmercury levels in sediment and water in a seasonally stratified lake. Limnol Oceanogr 42:1784–1795

Regnell O, Hammar T, Helgee A, Troedsson B (2001) Effects of anoxia and sulfide on concentrations of total and methyl mercury in sediment and water in two Hg-polluted lakes. Can J Fish Aquat Sci 58:506–517

Reimers RS, Krenkel PA (1974) Kinetics of mercury adsorption and desorption in. sediments. J Water Pollut Control Fed 46:352–365

Schetagne R, Verdon R, Langlois C, Doyon JF. (1995) Mercury evolution (1978–1995) in fishes of the La Grande hydroelectric complex, Quebéc, Canada. In: Manuscript, Poster presented in Fourth International Conference on Mercury as a global pollutant. Hamburg, Germany, 4–8 August

Steffan RJ, Korthals ET (1994) Effect of acidification on mercury methylation, demethylation, and volatilization in sediments from an acid-susceptible lake. Appl Environ Microbiol 54:2003–2009

Steven DS, Nelson JOD, Lean DRS (2002) Microbial reduction and oxidation of mercury in freshwater lakes. Environ Sci Technol 36:3064–3068

Sullivan KA, Mason RP (1998) The concentration and distribution of mercury in Lake Michigan. Sci Total Environ 213:213–228

Sunderland EM, Gobas FAPC, Heyes A, Branfireun BA (2004) Speciation and bioavailability of mercury in well-mixed estuarine sediments. Mar Chem 90:91–105

Tremblay A, Lucotte M, Schetagne R (1998) Total mercury and methylmercury accumulation in zooplankton of hydroelectric reservoirs in northern Québec Canada. Sci Total Environ 213:307–315

Tseng CM, Lamborg C, Fitzgerald WF, Engstromd R (2004) Cycling of dissolved elemental mercury in Arctic Alaskan lakes. Geochimicaet Cosmochimica Acta. 68:1173–1184

Ullrich SM, Tanton TW, Abdrashitova SA (2001) Mercury in the aquatic environmental review of factors affecting methylation. Criti Rev Environ Sci Technol 31:241–293

Vaidya OC, Howell GD, Leger DA (2000) Evaluation of the distribution of mercury in lake in Nova Scotia and Newfoundland (Canada). Water Air Soil Pollut 117:353–369

Vandal GM, Mason RP, Fitzgerald WF (1991) Cycling of volatile mercury in temperate lakes. Water Air Soil Pollut 56:791–803

Vandal GM, Mason RP, McKnight D, Fitzgerald WF (1998) Mercury speciation and distribution in a polar desert lake (Lake Hoare, Antarctica) and two glacial meltwater streams. Sci Total Environ 213:229–237

Verta M, Matilainen T (1995) Methylmercury distribution and partitioning in stratified Finnish forest lakes. Water Air Soil Pollut 80:585–588

Wan GJ, Lin WZ, Huang RG, Chen ZL (2000) Dating Characteristics and Erosion Tracing of ^{137}Cs Vertical Profile in Hongfeng Lake Sediment. Chin Sci Bull 19:1487–1990 (In Chinese)

Wang ZP (1993) The mercury geochemical research of Baihua lake in Guiyang. In: Master dissertation in the Graduate School of the Chinese Academy of Sciences (GSCAS), Beijing. (In Chinese, with English abstract)

Wang YC (2001) Biogeochemical Processes of Nutrients (P, N and C) at the Sediment–Water Interface in Two Reservoirs: Hongfeng Rservoir and Baihua Reservoir, Guizhou, China. In: PhD dissertation in the Graduate School of the Chinese Academy of Sciences (GSCAS), Beijing. (In Chinese, with English abstract)

Wang FS (2003) The interface geochemical behavior of seasonal hypoxia lakes trace metal elements. In: PhD Dissertation in the Graduate School of the Chinese Academy of Sciences (GSCAS), Beijing. (In Chinese, with English abstract)

Wang PJ, Gu YH, Li DH (1992) The discussion on the state of mercury in mater body in Baihua Reservoir. J Guizhou Norm Uni 10:29–32. (In Chinese, with English abstract)

Warner KA, Roden EE, Bonzongo JC (2004) Microbial mercury transformation in anoxic freshwater sediments under iron-reducing and other electron-accepting conditions. Environ Sci Technol 38:352

Watras CJ, Bloom NS, Claas SA, Morrison KA, Gilmour CC, Craig SR (1995) Methylmercury production in the anoxic hypolimnion of a dimictic seepage lake. Water Air Soil Pollut 80:735–745

Watras CJ, Back RC, Halvorsen S, Hudson RJM, Morrison KA, Wente SP (1998) Bioaccumulation of mercury in pelagic freshwater food webs. Sci Total Environ 219:183–208

Weber JH (1993) Review of possible paths for abiotic methylation of mercury (II) in the aquatic environment. Chemosphere 26:2063–2077

Xiao HY (2002) The nitrogen biogeochemical cycle in seasonal hypoxia lakes. In: Ph.D dissertation in the Graduate School of the Chinese Academy of Sciences (GSCAS), Beijing. (In Chinese, with English abstract)

Xiao HY, Liu CQ. (2004) Discrimination between extraneous nitrogen input and interior nitrogen release in lakes. Sci China Series D Earth Sci 47:813–821

Xu YR, Xu ZJ, Xiang S (1999) Vertical distribution of Fe and Mn and optimal pumping depth in a seasonal oxygen shortage reservoir. J Environ Sci 19:147–152. (In Chinese, with English abstract)

Xun L, Campbell NER, Rudd JWM (1987) Measurement of specific rates of net methyl mercury production in the water column and surface sediments of acidified and circumneutral lakes. J Fish Aquat Sci 44:750–757

Yan HY (2005) The Methodological Development of Mercury Species in Environmental Samples and the Mass Balance of Mercury in Baihua Reservoir, Guizhou, China. In: Ph.D dissertation in the Graduate School of the Chinese Academy of Sciences (GSCAS), Beijing. (In Chinese, with English abstract)

Yan HY, Feng XB, Shang LH, Qiu GL, Dai QJ, Wang SF, Hou YM (2008) The variations of mercury in sediment profiles from a historically mercury-contaminated reservoir, Guizhou province China. Sci Total Environ 407:497–506

Zhang JX (1996) Photoinduced formation of dissolved gaseous mercury in the lake. Environ Sci Technol 2:16–22 (In Chinese, with English abstract)

Zhang W (1999) The environment characteristics and eutrophication of Hongfeng and Baihua reservoir. Guiyang: Guizhou Science and Technology Press (In Chinese, with English abstract)

Chapter 8
Mercury Mass Balance in Reservoirs with Different Ages

Abstract Reservoirs play a complicated role in the transportation of mercury (Hg) the river reservoir ecosystems. Under different environmental conditions, reservoirs can be a sink for Hg to the outflow river, a source of Hg to the inflow river, or a place for Hg to transform from inorganic Hg (IHg) to methylmercury (MeHg). The different pathways for regulation and operation in the reservoir lead to distinct water exchange frequencies and residence times as well as distinct source-sink characteristics for the reservoir. In this chapter, we (1) estimated the input/output fluxes of Hg species (THg and MeHg) in a reservoir and (2) assessed the role of different reservoir stages on the source/sink of Hg in the transportation and transformation of Hg in a river reservoir ecosystem.

Keywords Mercury · Mass balance · Reservoir · Wujiang river basin

8.1 Description of the Mass Balance Budget Calculations

Mass inputs to reservoirs originate from river inflow, direct atmospheric Hg deposition, and direct runoff from upland water and groundwater. Mass outputs from reservoirs are primarily from reservoir discharge, evaporation from the water surface into the atmosphere and other pathways, e.g., industrial and agricultural water consumption.

The Wujiang River is a typical precipitation-driven, deep-valley river. Precipitation is the primary water recharge source. The water cycling in the Wujiang River Basin is very complex, e.g., high-intensity rainfall during the rainy season, numerous runoffs, and a complicated groundwater system due to the karst geomorphology. A part of the runoff is transported into the reservoir via a river, and the other part of the runoff infiltrates the underground to recharge the groundwater. In addition, the groundwater can recharge the surface water through underground streams, debouchure, wellsprings, and so on. Given the complicated water cycle systems, we assumed that 50% of the surface runoff enters the reservoir, and we ignored the influence of the groundwater supply and surface water leakage because

X. Feng et al., *Biogeochemical Cycle of Mercury in Reservoir Systems in Wujiang River Basin, Southwest China*, https://doi.org/10.1007/978-981-10-6719-8_8

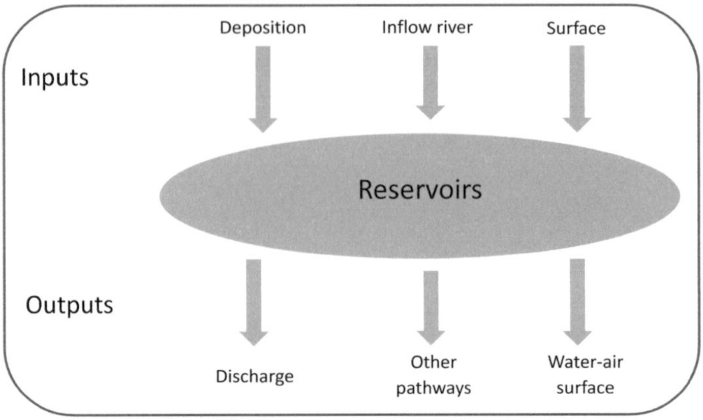

Fig. 8.1 Input–output pathways of Hg in reservoirs in Wujiang River Basin

of the mass balance for the groundwater supply and surface water leakage during one hydrologic year. Figure 8.1 shows the Hg mass balance in the reservoir, and all data in this study were obtained in 2006. Summary of input/output pathways of THg, MeHg, and total suspended solid (TSS) in reservoirs in Wujiang River Basin is shown in Table 8.1.

Input–output calculations for the total mercury (THg), methylmercury (MeHg), and particulate mercury (PHg) concentration were performed as described by St Louis et al. (1994, 2004). The basic equation used in the input–output budget calculations for each of the reservoirs was:

$$\text{Net flux} = \sum \text{output} - \sum \text{input} \tag{8.1}$$

Table 8.1 Summary of input/output pathways of THg, MeHg, and total suspended solid in reservoirs in Wujiang River Basin

Inputs and outputs pathways		Comments
Inputs	Deposition	Concentrations measured in bulk deposition; rainfall data collected from the local meteorological station
	Inflow river	Concentrations measured; water volume data collected from a hydrological station nearby
	Surface runoff	Concentrations measured; water data collected from a hydrological station nearby
Outputs	Discharge of reservoir	Concentrations measured; water data collected from a hydrological station nearby
	Evasion of Hg^0 through water–air surface	Hg^0 evasion fluxes were calculated based on DGM measurement
	Other ways	Industrial and agricultural water consumption (irrigation water use, evaporation, consume by water plant, etc.)

If the net flux <0, it means that the reservoir acts as a sink; otherwise, the reservoir is a source.

8.2 Water Balance in Reservoirs

8.2.1 Water Input from Wet Deposition

The weather in the Wujiang River Basin is controlled by the subtropical, moist monsoon climate. Atmospheric precipitation is mostly rain with very little snow. Hence, the amount of precipitation in this study refers to the rainfall amount. The annual rainfalls for the Puding Reservoir (PD), Yinzidu Reservoir (YZD), Hongjiadu Reservoir (HJD), Dongfeng Reservoir (DF), and Wujiangdu Reservoir (WJD) in the mainstreams of the Wujiang River Basin were 1203, 1069, 881, 970, and 693 mm, respectively, during 2006. The average annual rainfall was 963 ± 193 mm, which is close to the multi-year average (1100 mm). The monthly rainfall for the five selected reservoirs in the Wujiang River Basin is shown in Fig. 8.2.

The rainfall mainly occurred during the period from May to October in 2006, and this period accounted for approximately 80% (~779 mm) of the annual rainfall for each reservoir. The water input into the reservoir from rainfall (Tables 8.2, 8.3, 8.4, 8.5, 8.6 and 8.7) can be estimated using Eq. 8.2.

$$V_{\text{rainfall}} = R_{\text{rainfall}} \times A_{\text{reservoir}} \times 10^{-3} \tag{8.2}$$

where V_{rainfall} is the annual water volume added to the reservoir through rainfall (m); R_{rainfall} is the annual rainfall of the study area (mm); and $A_{\text{reservoir}}$ is the water area of the reservoir for the normal water level (m^2).

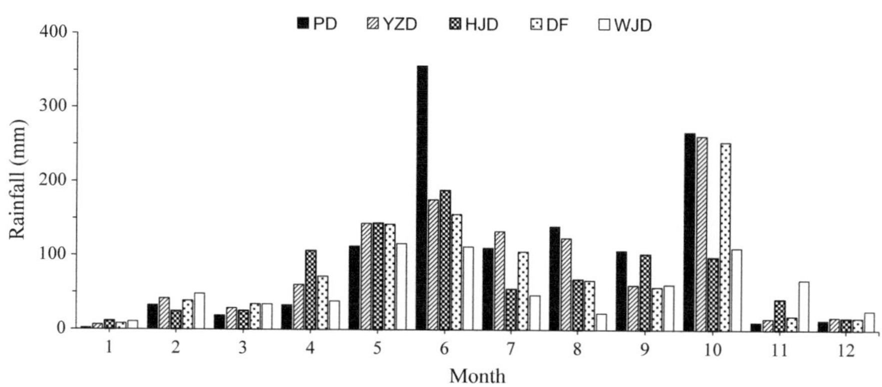

Fig. 8.2 Monthly distributions of rainfall in reservoirs in Wujiang River Basin

Table 8.2 Input–output water volume of Puding reservoir (PD) through different pathways ($\times 10^4$ m^3)

Month	Input				Output	
	Deposition	Inflow river		Surface runoff	Discharge of reservoir	Other
		Sancha river	Boyu river			
1	4.2	6106.8	100.2	69829.7	2223.1	172464.6
2	62.8	4040.1	66.3		4886.8	
3	36.6	4205.1	69.0		5383.6	
4	63.7	1840.3	30.2		5650.6	
5	216.6	3508.7	57.6		3160.5	
6	686.5	48807.4	800.9		39139.2	
7	212.3	116189.0	1906.5		44220.4	
8	269.3	26623.3	436.9		15320.4	
9	205.4	11638.1	191.0		15940.8	
10	513.6	51157.4	839.4		35756.6	
11	20.0	16251.8	266.7		23976.0	
12	25.4	7365.6	120.9		6642.4	
Total	2316.4	297733.5	4885.5	69829.7	202300.4	172464.6
	374765.0				374765.0	

Table 8.3 Input–output of water to Yinzidu reservoir (YZD) through different pathways ($\times 10^4$ m^3)

Month	Input			Output	
	Precipitation	Discharge of PD	Surface runoff	Discharge of reservoir	Other
1	9.6	2223.1	6553.594	4847.9	7185
2	62.6	4886.8		5636.7	
3	43.4	5383.6		5303.2	
4	90.6	5650.6		4225.0	
5	215.1	3160.5		3053.4	
6	263.0	39139.2		27656.6	
7	199.4	44220.4		49791.5	
8	185.0	15320.4		21105.8	
9	90.1	15940.8		14955.8	
10	391.5	35756.6		50112.9	
11	22.6	23976.0		23328.0	
12	25.6	6642.4		23489.6	
Total	1598.5	202300.4	6553.594	203267.2	7185
	210452.5			210452.5	

Table 8.4 Input–output of water to Hongjiadu reservoir (HJD) through different pathways ($\times 10^4$ m^3)

Month	Input				Output	
	Precipitation	Inflow river		Surface runoff	Discharge of reservoir	Other
		Liuchong river	Luojiao river			
1	92.6	8195.9	1790.3	117750.6	14838.3	139833.3
2	197.2	6870.5	1500.8		21022.8	
3	201.3	6669.2	1456.8		19230.9	
4	855.7	9175.7	2004.4		8864.6	
5	1160.0	8303.0	1813.7		11490.3	
6	1515.8	22109.8	4829.7		2358.7	
7	443.6	35408.4	7734.8		23328.9	
8	548.2	19900.5	4347.1		25203.7	
9	822.7	18817.9	4110.7		30430.1	
10	788.1	25846.6	5646.0		12293.9	
11	333.3	16433.3	3589.7		22550.4	
12	131.2	15882.9	3469.5		29301.7	
Total	7089.6	193613.8	42293.7	117750.6	220914.4	139833.3
	360747.7				360747.7	

Table 8.5 Input–output of water to Dongfeng reservoir (DF) through different pathways ($\times 10^4$ m^3)

Month	Input				Output	
	Precipitation	Inflow river		Surface runoff	Discharge of reservoir	Other
		Liuchong river (Yachaihe section)	Sancha river (Yachaihe section)			
1	15.4	14838.3	4847.9	21873.1	16927.5	23438.4
2	74.1	21022.8	5636.7		20853.5	
3	65.4	19230.9	7981.6		23355.6	
4	137.0	8864.6	12001.0		25764.5	
5	271.8	11490.3	3053.4		8115.6	
6	297.5	2358.7	56168.6		42301.4	
7	200.7	23328.9	49791.5		53246.6	
8	127.1	25203.7	21105.8		64495.9	
9	109.8	30430.1	14955.8		50284.8	
10	483.6	12293.9	50112.9		71191.9	
11	35.8	22550.4	23328.0		54406.1	
12	31.3	29301.7	23489.6		62728.1	
Total	1849.6	220914.4	272472.8	21873.1	493671.5	23438.4
	517109.9				517109.9	

Table 8.6 Input–output of water to Suofengying reservoir (SFY) through different pathways ($\times 10^4$ m^3)

Month	Input				Output	
	Precipitation	Inflow river		Surface runoff	Discharge of reservoir	Other
		Discharge of DF	Maotiao river			
1	4.6	16927.5	1363.3	44019.7	19445.2	62857.9
2	22.2	20853.5	3057.9		33675.3	
3	19.6	23355.6	3377.5		32783.6	
4	41.0	25764.5	4738.2		41342.4	
5	81.3	8115.6	7242.4		22605.7	
6	89.0	42301.4	10189.2		72498.2	
7	60.0	53246.6	13116.1		79816.3	
8	38.0	64495.9	2796.2		55710.7	
9	32.8	50284.8	1915.5		44582.4	
10	144.6	71191.9	8045.9		80512.7	
11	10.7	54406.1	6622.6		55287.4	
12	9.3	62728.1	2630.2		52094.9	
Total	553.1	493671.5	65094.9	44019.7	540481.2	62857.9
	603339.2				603339.2	

8.2.2 *Water Input from the Inflow River*

The tributary inflow river for each reservoir is listed in Table 5.1 (Chap. 5). The water volume added to the reservoir from the inflow river can be estimated using Eq. 8.3.

$$V_{\text{river}} = F \times T \tag{8.3}$$

where V_{river} is the water volume added to the reservoir from the inflow river (m^3); F is the flow rate of the inflow river (m^3 s^{-1}); and T is the time (s).

The largest annual water input was the inflow river of WJD through the discharge from Suofengying Reservoir (SFY). This input reached a maximum value in June and October (798,160,000 and 805,130,000 m^3, respectively). The smallest water input was from the Boyu River (BY), which is the inflow river for PD, and the minimum value was 300,000 m^3 in April. Generally, the input from the inflow river into the reservoir is controlled by seasonal variations, and the flow rates increase during the rainy season and decrease during the dry season.

Table 8.7 Input–output of water to Wujiangdu reservoir (WJD) through different pathways ($\times 10^4$ m^3)

Month	Input						Output	
	Precipitation	Inflow river				Surface runoff	Discharge of reservoir	Other
		Mainstream of Wujiang river	Yeji river	Xifeng river	Pianyan river			
1	50.2	19445.2	2115.9	745.6	2089.2	70507.6	41300.9	152137.3
2	230.4	33675.3	1741.8	633.2	991.9		30046.5	
3	163.5	32783.6	2705.2	565.8	723.2		37417.2	
4	184.0	41342.4	2488.3	782.1	1995.8		37376.6	
5	555.0	22605.7	4821.1	1561.8	4499.7		55683.9	
6	534.9	72498.2	9953.3	4102.5	4095.4		43493.8	
7	223.7	79816.3	4339.0	8908.7	3053.4		91547.7	
8	107.6	55710.7	2598.0	5327.6	1660.6		64308.4	
9	291.1	44582.4	2462.4	4421.6	1062.7		47511.4	
10	525.8	80512.7	3481.9	1330.8	2892.7		46202.4	
11	322.2	55287.4	2695.7	1200.2	2436.5		60471.4	
12	123.8	52094.9	1982.0	773.6	1901.7		55817.9	
Total	3312.1	590354.8	41384.7	30353.5	27402.6	70507.6	611178.0	152137.3
	763315.3						763315.3	

8.2.3 Water Input from the Surface Runoff

Because we did not measure the runoff flow data, the water input into the reservoirs from the surface runoff was estimated using Eq. 8.4

$$V_{\text{runoff}} = D_{\text{runoff}} \times A_{\text{runoff}}\left(m^2\right) \times 50\% \tag{8.4}$$

where V_{runoff} is the water volume added to the reservoir through the surface runoff (m^3); D_{runoff} is the multi-year average depth of the runoff (m); and A_{runoff} is the catchment area of the reservoir (m^2). We assumed 50% of the surface runoff entered the reservoir, and we ignored the influence of the groundwater supply and surface water leakage.

8.2.4 Water Output from Reservoirs

Reservoir discharge is the main pathway for water output from reservoirs. The other outputs mainly refer to industrial and agricultural water consumption, evaporation, consumption by water plants, and so on. This study also includes the water pool for the reservoir. The fluxes in the discharge water from the reservoirs were estimated using Eq. 8.3. The other output (Tables 8.2, 8.3, 8.4, 8.5, 8.6 and 8.7) was estimated using Eq. 8.5, which assumes a water balance for the input–output. Generally, the fluxes in the discharge water are controlled via the operation of the reservoir, and the maximum value was at WJD and the minimum at PD.

$$\text{Output}_{\text{other}} = \sum \text{input} - \text{Output}_{\text{discharge}} \tag{8.5}$$

where $\text{Output}_{\text{other}}$ is the output water volume through other pathways (m^3); $\sum$ *input* is the total water input volume into the reservoir (m^3); and $\text{output}_{\text{discharge}}$ is the water output volume through the reservoir discharge (m^3).

8.2.5 Input–Output Water Budgets in Reservoirs

As shown in Tables 8.2, 8.3, 8.4, 8.5, 8.6 and 8.7, the largest annual water input–output flux was at WJD (7633,153,000 m^3), which was followed by SFY (6033, 392,000 m^3), DF (5171,099,000 m^3), PD (3747,650,000 m^3), HJD (3607, 477,000 m^3), and YZD (2104,525,000 m^3). The water input came from the inflow river for each of the reservoirs in the Wujiang River Basin. The water input during the rainy season (60% of the total input, from May to October 2006) was higher than the input during the dry season. The water output from the reservoir discharge during the rainy season constituted $\geq 60\%$ of the total output for each of the reservoirs, with the exception of HJD. The water output from the reservoir discharge was mainly controlled via the generation dispatch schedule. The average annual water input to the reservoirs from the inflow rivers was 4137,660,000 ± 1960,080,000 m^3 during 2006, which accounted for 86.8 ± 11.8% (ranging from 65.4 to 96.1%) of the total input. The water input through precipitation was 70,900,000 ± 5530,000 m^3, which contributed 0.7 ± 0.7% (0.1–2%) of the total input. The water input from groundwater and direct surface runoff was 550,890,000 ± 399,070,000 m^3, which contributed 12.5 ± 11.3% (ranging from 3.1 to 18.6%) of the total input. The water input from groundwater, surface runoff, and precipitation are controlled by the areas of the Wujiang River Basin and the surface area of the reservoir. The Sancha River (SC) and Liuchong River (LC), which are located at the upper end of the Wujing River, are the main inflow rivers for PD and HJD, respectively. For the other four reservoirs (YZD, DF, SFY, and WJD), the water input mainly comes from the discharge water from the upstream reservoirs. Reservoir discharge was the predominant water output pathway for each of the reservoirs in the Wujiang River Basin and accounted for 79.5 ± 18.1% of the total output.

8.3 Input–Output Budgets for the Total Mercury and Methylmercury in Reservoirs

8.3.1 Total Mercury and Methylmercury Inputs from Precipitation

The input flux for THg (MeHg) from precipitation was estimated using the following:

$$F_{\text{precipitation}} = \sum V_{\text{month}} \times C_{\text{month}} \times 10^{-6} \tag{8.6}$$

where $F_{\text{precipitation}}$ is the THg (MeHg) flux from wet precipitation (g); V_{month} is the monthly water input to the reservoir from wet precipitation (m^3), and C_{month} is the monthly averaged THg (MeHg) concentration in the precipitation (ng L^{-1}).

The monthly inputs of THg and MeHg to the reservoirs from precipitation are shown in Tables 8.8 and 8.9. The annual THg input was 477.1 g to PD, 570.9 g to YZD, 2795.1 g to HJD, 691.1 g to DF, 207.7 g to SFY, and 1891.8 g to WJD. The annual MeHg input was 4.25 g for PD, 2.80 g for YZD, 12.62 g for HJD, 3.68 g for DF, 1.10 g for SFY, and 7.91 g for WJD. With the exception of WJD, the THg and MeHg inputs to the reservoirs from precipitation were significantly higher during the rainy season than the dry season ($p < 0.05$). The inputs during the rainy season accounted for 68.1 and 74.6% of the annual input from precipitation, respectively. However, the annual THg input from precipitation during the wet season accounted for only 44% of the annual input from precipitation, which was

Table 8.8 Input of total Hg (THg) to reservoir through precipitation (g)

Month	PD	YZD	HJD	DF	SFY[a]	WJD
1	3.5	13.5	63.2	22.2	6.6	74.8
2	30.8	47.5	106.6	67.5	20.2	300.6
3	13.7	27.1	90.9	37.6	11.2	177.1
4	44.6	57.7	719.2	69.5	20.8	174.6
5	74.8	132.0	841.6	169.3	50.6	205.9
6	98.5	72.1	306.0	73.4	22.0	253.0
7	25.7	44.8	56.8	49.2	14.7	43.6
8	55.2	22.8	113.6	22.9	6.9	23.2
9	18.9	6.7	109.9	13.0	3.9	41.5
10	95.1	119.7	218.9	131.5	39.3	265.2
11	2.4	8.5	113.7	10.0	3.0	272.6
12	13.9	18.5	54.6	24.8	7.4	59.7
Total	477.1	570.9	2795.1	691.1	206.7	1891.8

[a]Data concerning THg concentration in deposition and corresponding rainfall in SFY were obtained from nearby DF sampling station

Table 8.9 Input of methyl Hg (MeHg) to reservoir through precipitation (g)

Month	PD	YZD	HJD	DF	SFY[a]	WJD
1	0.03	0.04	0.76	0.06	0.02	0.21
2	0.20	0.15	0.76	0.19	0.06	1.06
3	0.11	0.11	0.68	0.14	0.04	0.41
4	0.09	0.16	1.12	0.28	0.08	0.47
5	0.52	0.25	2.43	0.51	0.15	1.24
6	1.18	0.41	2.09	1.00	0.30	1.72
7	0.41	0.48	1.03	0.42	0.13	0.31
8	0.38	0.34	0.63	0.13	0.04	0.21
9	0.44	0.07	1.33	0.13	0.04	0.56
10	0.83	0.67	1.18	0.68	0.20	0.61
11	0.03	0.06	0.38	0.07	0.02	0.72
12	0.04	0.06	0.24	0.07	0.02	0.39
Total	4.25	2.80	12.62	3.68	1.10	7.91

[a]Data concerning MeHg concentration in deposition and corresponding rainfall in SFY were obtained from nearby DF sampling station

probably because of the higher THg concentrations in the precipitation during the dry season.

The increase in the THg concentration in the precipitation during the dry season is attributed to Hg contamination in the regional ambient air due to anthropogenic Hg emissions, such as residential coal combustion, Hg emissions from a calcium factory in Xifeng county, and Hg emissions from an iron factory near WJD.

Table 8.10 Total Hg (THg), methyl Hg (MeHg), and total suspended solid (TSS) inputs to Puding reservoir (PD) through inflow river

Month	THg (g)		MeHg (g)		TSS (t)	
	SC[a]	BY[a]	SC	BY	SC	BY
1	191.8	4.3	6.31	0.10	76.0	2.8
2	198.4	3.5	4.73	0.07	87.1	1.4
3	131.8	3.0	4.52	0.07	86.9	1.9
4	67.5	1.5	2.28	0.04	61.4	1.0
5	167.3	3.4	5.32	0.08	132.5	2.0
6	2216.8	33.8	79.14	1.18	2413.6	40.5
7	5412.1	87.4	164.47	2.99	4627.8	93.3
8	1049.2	18.9	29.01	0.54	661.3	14.5
9	401.6	5.9	12.91	0.25	306.9	6.4
10	3320.4	91.2	73.26	1.07	2338.6	50.1
11	1059.1	25.1	21.07	0.29	664.7	8.9
12	481.9	9.6	8.10	0.12	203.7	3.1
total	14697.9	287.8	411.11	6.80	11660.6	225.9

[a]SC, Sancha river; BY, Boyu river

The correlation analysis suggested that there was not a significant correlation between Hg (THg and MeHg) inputs and Hg concentrations in the precipitation (THg: $r = 0.18, p = 0.13, n = 72$; MeHg: $r = -0.005, p = 0.97; n = 72$). However, Hg (THg and MeHg) inputs significantly correlated with the rainfall (THg: $r = 0.26$, $p = 0.03$, $n = 72$; MeHg: $r = 0.47$, $p < 0.01$, $n = 72$) and the surface areas of the reservoirs (THg: $r = 0.55, p < 0.01, n = 72$; MeHg: $r = 0.65, p < 0.01$, $n = 72$). The statistical results suggested that the Hg (THg and MeHg) inputs to the reservoirs in the Wujiang River Basin were primarily controlled by the rainfall and the surface areas of the reservoirs.

8.3.2 Total Mercury, Methylmercury, and Total Suspended Solid Inputs from the Inflow Rivers

The inputs of THg, MeHg, and TSS to reservoirs from the inflow rivers were estimated using Eq. 7.7, and the calculated data are summarized in Tables 8.10, 8.11, 8.12, 8.13 and 8.14.

$$F_{\text{river}} = \sum V_{\text{month}} \times C_{\text{month}} \times 10^{-6} \quad (8.7)$$

where F_{river} is the inputs of THg, MeHg, and TSS from the inflow rivers to the reservoirs (g); V_{month} is the monthly water input to reservoirs from the inflow rivers

Table 8.11 Total Hg (THg), methyl Hg (MeHg), and total suspended solid (TSS) inputs to Houjiangdu reservoir (HJD) through inflow river

Month	THg (g)		MeHg (g)		TSS (t)	
	LC[a]	LJ[a]	LC	LJ	LC	LJ
1	258.4	65.4	6.77	1.82	169.6	50.4
2	207.6	42.4	6.94	1.28	83.7	17.5
3	168.3	32.8	7.67	1.49	95.8	22.2
4	325.4	66.6	10.40	2.67	253.9	59.0
5	408.6	71.1	15.37	3.04	234.1	59.8
6	1186.0	224.8	34.54	6.87	1468.9	249.4
7	1967.3	357.4	66.38	13.47	1357.0	271.8
8	1003.2	146.3	22.77	7.69	501.7	61.5
9	850.2	97.9	15.81	4.47	838.2	169.9
10	2661.2	258.4	31.52	11.31	1355.6	269.1
11	600.5	196.0	12.93	3.08	540.0	139.4
12	375.3	220.2	21.66	5.03	486.8	114.9
Total	10011.9	1779.4	252.76	62.23	7385.2	1485.0

[a]LC, Liuchong river; LJ, Luojiao river

Table 8.12 Total Hg (THg), methyl Hg (MeHg), and total suspended solid (TSS) inputs to Dongfeng reservoir (DF) through inflow river (Yachi river section)

Month	THg (g)		MeHg (g)		TSS (t)	
	LC[a]	SC[a]	LC	SC	LC	SC
1	377.2	143.3	12.96	5.30	352.0	108.6
2	488.8	130.8	18.92	6.15	246.8	80.7
3	618.3	164.1	21.54	5.57	316.1	63.9
4	268.2	107.4	10.72	5.22	349.1	292.6
5	447.7	110.3	15.13	4.03	580.1	167.5
6	71.8	872.3	3.40	34.38	72.2	1854.3
7	593.0	856.1	31.87	42.34	614.0	1360.6
8	996.1	623.5	33.62	29.40	460.6	219.3
9	719.3	333.5	24.68	12.48	952.5	259.4
10	582.3	1746.7	17.39	64.66	691.3	1204.0
11	565.8	739.4	18.62	19.85	483.2	287.8
12	777.7	192.9	26.75	5.94	630.0	590.8
Total	6506.0	6020.4	235.60	235.31	5747.9	6489.4

[a]LC, Liuchong river; SC, Sancha river

(m^3), and C_{month} is the monthly averaged THg, MeHg, and TSS concentrations in the inflow rivers (ng L^{-1}).

As shown in Table 8.10, the THg, MeHg, and TSS inputs from SC to PD were 14697.9 g; 287.8 g; and 11660.6 mg in 2006, respectively, and the THg, MeHg,

Table 8.13 Total Hg (THg), methyl Hg (MeHg), and total suspended solid (TSS) inputs to Suofengying reservoir (SFY) through inflow river

Month	THg (g)		MeHg (g)		TSS (t)	
	Discharge of DF	MT[a]	Discharge of DF	MT	Discharge of DF	MT
1	367.8	36.0	21.04	1.54	363.1	29.4
2	594.3	93.6	27.31	3.36	244.2	89.1
3	493.5	111.7	29.64	4.42	420.4	83.1
4	737.4	99.3	36.34	5.46	317.6	153.8
5	209.7	108.0	9.98	9.09	91.4	224.0
6	795.4	259.8	75.69	16.10	637.4	478.8
7	1164.1	337.8	124.22	19.71	806.3	454.6
8	1170.2	48.5	159.72	4.28	600.5	51.4
9	1119.6	34.9	114.68	3.48	603.4	50.7
10	1929.8	313.4	117.28	12.07	582.5	340.4
11	1136.5	270.9	59.73	10.54	459.3	142.8
12	1547.5	112.8	61.93	2.80	488.8	68.3
Total	11265.8	1826.7	837.54	92.84	5614.9	2166.2

[a]MT, Maotiao river

and TSS inputs from the Boyu River (BY) to PD were 287.8 g, 6.80 g, and 225.9 mg in 2006, respectively. The THg, MeHg, and TSS inputs to YZD from the discharge of PD were 4966.6 g; 407.63 g; and 4871.0 mg, respectively. The THg, MeHg, and TSS inputs from LC to HJD were 10011.9 g; 252.76 g; 7852.2 mg, respectively, and the THg, MeHg, and TSS inputs from the Luojiao River (LJ) to HJD were 1779.4 g; 62.23 g; and 1485.0 mg, respectively (Table 8.11).

The LC (Yachi River (YC) section) transported 6506.0 g of Hg, 235.6 g of MeHg, and 5747.9 mg of TSS to DF, and SC (YC section) transported 6026.4 g of THg, 235.31 g of MeHg, and 6489.4 mg of TSS to DF. The THg, MeHg, and TSS inputs to SFY from the discharge of DF were 11265.8 g, 837.54 g, and 5614.9 mg; the corresponding data for the MiaoTiao River (MT, input to DF) were 1826.7 g, 92.84 g, and 2166.2 Mg for THg, MeHg, and TSS, respectively. The THg, MeHg, and TSS inputs from the mainstream of the Wujiang River (WJ) to WJD were 12812.2 g; 702.2 g; and 17615.4 mg; the corresponding data for the THg, MeHg, and TSS inputs to WJD were 1446.5 g, 67.1 g, and 1737.2 mg for the Yeji River (YJ); 2654.9 g, 75.3 g, and 1485.9 mg for the Xifeng River (XF); and 1225.8 g, 35.1 g, and 778.3 mg for the Pianyan River (PY), respectively.

The correlation analysis suggested that: (1) the THg inputs to the reservoirs from the inflow rivers did not correlate with the THg concentrations in the inflow rivers ($r = 0.03$, $p = 0.76$, $n = 156$), but the input did significantly correlate with the river flow ($r = 0.86$, $p < 0.01$, $n = 156$); (2) the MeHg input to the reservoirs from the inflow rivers did significantly correlate with the MeHg concentrations in the inflow rivers ($r = 0.23$, $p < 0.01$, $n = 156$) and the river flow ($r = 0.93$, $p < 0.01$, $n = 156$); (3) the TSS input to the reservoirs from the inflow rivers did significantly

Table 8.14 Total Hg (THg), methyl Hg (MeHg), and total suspended solid (TSS) inputs to Wujiangdu reservoir (WJD) through inflow river

Month	THg (g)				MeHg (g)				TSS (t)			
	Mainstream of WJ[a]	YJ[a]	XF[a]	PY[a]	Mainstream of WJ	YJ	XF	PY	Mainstream of WJ	YJ	XF	PY
1	528	70	39	69	21	1.7	1.2	1.5	415	28	32	45
2	790	82	34	33	23	1.6	1.3	1.1	846	37	20	15
3	1138	129	31	23	34	2.7	1.1	0.9	1289	87	18	12
4	1339	128	34	67	49	4.8	2.2	2.1	1371	140	30	53
5	979	175	100	151	55	9.7	3.8	9.0	1093	320	70	154
6	1449	464	343	192	92	23.7	9.3	4.8	4241	667	230	161
7	1707	119	834	182	145	7.0	21.5	4.1	3154	224	460	66
8	1259	56	461	102	64	2.3	11.5	2.0	735	50	188	31
9	768	90	361	49	66	2.8	11.6	1.7	746	38	209	51
10	963	50	253	181	67	6.9	9.3	4.5	2119	64	142	144
11	969	53	103	115	36	2.0	1.8	2.1	898	49	38	33
12	936	30	62	61	51	1.8	0.9	1.5	710	34	49	15
Total	12812	1447	2655	1226	702	67.1	75.3	35.1	17615	1737	1486	778

[a]WJ, Wujiang river; YJ, Yeji river; XF, Xifeng river; PY, Pianyan river

correlate with the TSS concentration in the inflow rivers ($r = 0.23$, $p < 0.01$, $n = 156$) and the river flow ($r = 0.79$, $p < 0.01$, $n = 156$). These results suggested that the THg inputs into the reservoirs from the inflow rivers are predominantly controlled by the river flow, and the MeHg and TSS inputs into the reservoirs from the inflow rivers are controlled by both the river flow and the MeHg and TSS concentrations in the inflow rivers.

Although the concentrations of THg, MeHg, and TSS in the tributaries (e.g., BY, LJ, and XF) were elevated, the water flow in these tributaries was relatively lower compared with the mainstream. Therefore, the THg, MeHg, and TSS inputs to the reservoirs from the mainstream rivers were significantly higher than the inputs from the tributaries, which implies that the tributary contributions to the THg, MeHg, and TSS inputs were less pronounced. The water input from XF to WJD accounted for only 4% of the total water input, and the THg and MeHg inputs to WJD through XF provided 11 and 7% of the total inputs, respectively. A similar phenomenon was also observed in YJ and PY, which are the tributaries of WJD.

Since particles tend to have an affinity for Hg, Hg transportation via river is affected by seasonal hydrological conditions. Therefore, Hg inputs from inflow rivers to the reservoirs change with the variations in hydrological conditions, which are impacted by the seasonal variations in the river flow. The Hg inputs in the rainy season were much higher and can be an order of magnitude greater than the inputs in the dry season. The THg, MeHg, and TSS inputs from the inflow rivers to the reservoirs in the rainy season constituted 70% (THg), 77% (MeHg), and 77% (TSS) of the total inputs, which suggested that the Hg input contributions to the reservoirs in the wet season were significantly elevated compared to the dry season.

The statistical analysis showed that: (1) the THg input to the reservoirs from the inflow rivers significantly correlated with the TSS input ($r = 0.89$, $p = 0.02$, $n = 6$) and the catchment area of the reservoirs ($r = 0.81$, $p = 0.05$, $n = 6$), but it did not correlate with the water flow ($r = 0.61$, $p = 0.20$, $n = 6$); (2) the MeHg input significantly correlated with the water flow ($r = 0.88$, $p = 0.022$, $n = 6$), but it did not correlate with the TSS input ($r = 0.47$, $p = 0.34$, $n = 6$) or the catchment area of the reservoirs ($r = 0.12$, $p = 0.82$, $n = 6$).

8.3.3 Total Mercury and Methylmercury Inputs from Surface Runoff

The Hg in soil and Hg from anthropogenic emissions are transported into reservoirs via surface runoff. The THg, MeHg, and TSS inputs to the reservoirs through surface runoff were estimated using Eq. 8.8.

$$F_{\mathrm{runoff}} = V_{\mathrm{runoff}} \times C_{\mathrm{runoff}} \times 10^{-6} \tag{8.8}$$

where F_{runoff} is the THg, MeHg, and TSS inputs from surface runoff into the reservoirs (g); V_{runoff} is the yearly water inputs to the reservoir from the surface runoff (m^3); C_{runoff} is the THg, MeHg, and TSS concentrations in water samples from the surface runoff (ng L^{-1}).

The data concerning the concentrations of THg, MeHg, and TSS in water samples from the surface runoff were unavailable for this study. Because the surface runoff is predominantly formed during the wet season (especially during the periods from May to October), the THg, MeHg, and TSS concentrations in the inflow rivers were used to represent the concentrations in the surface runoff. Our calculated data showed that the annual THg, MeHg, and TSS inputs to the reservoirs via surface runoff were 2475.9 ± 2206.6 g, 88.6 ± 63.4 g, and 2045.1 ± 1723.1 mg, respectively, for the six reservoirs (Tables 8.18, 8.19, 8.20, 8.21, 8.22 and 8.23). The potential Hg sources in the surface runoff were from soil detachment, transportation via rainfall, and Hg bonding with plant residues and detritus. The Hg inputs into the reservoirs from surface runoff were controlled by numerous factors, i.e., land use patterns, human activities, vegetation coverage, and so on. Therefore, uncertainties in the estimate of Hg input from surface runoff are inevitable in this study. Compared to the Hg inputs from precipitation, the Hg species inputs from surface runoff were slightly higher. The ratios of the specific THg, MeHg, TSS inputs from surface runoff to the total inputs were comparable to the ratios of the water input from surface runoff to the total water input for each reservoir (Tables 8.24, 8.25, 8.26, 8.27, 8.28 and 8.29).

8.3.4 Total Mercury and Methylmercury Outputs from Reservoir Discharge

The THg, MeHg, and TSS outputs from reservoirs via discharge (outflow from the reservoir) were estimated using Eq. 7.9, and the calculated data are shown in Tables 8.15, 8.16 and 8.17.

$$F_{discharge} = \sum V_{month} \times C_{month} \times 10^{-6} \quad (8.9)$$

where $F_{discharge}$ is the THg, MeHg, and TSS outputs from the reservoir via discharge (g); V_{month} is the monthly water output from the reservoir via the discharge (m^3); and C_{month} is the monthly averaged THg, MeHg, and TSS concentrations in the discharged water (ng L^{-1}).

The calculated data showed that the annual THg outputs from discharge were 16202.6 g for WJD, 11698.5 g for SFY, 11265.8 g for DF, 4966.6 g for PD, 4801.0 g for HJD, and 4401.7 g for YZD. The annual MeHg outputs from discharge were 1617.93 g for WJD, 837.31 g for DF, 771.1 g for SFY, 407.63 g for PD, 306.35 g for HJD, and 296.16 g for YZD. Similarly, the TSS outputs from reservoirs through discharge were 7986.5 mg for SFY, 7330.3 mg for WJD,

Table 8.15 Output of total Hg (THg) from reservoirs through discharge (g)

Month	PD[a]	YZD[a]	HJD[a]	DF[a]	SFY[a]	WJD[a]
1	64.6	107.4	412.0	367.8	452.3	1355.4
2	130.9	121.6	305.5	594.3	528.0	668.0
3	121.0	119.4	413.1	493.5	860.6	1214.0
4	123.5	85.5	213.2	737.4	1160.3	1090.0
5	58.9	62.3	198.5	209.7	539.1	1828.7
6	898.3	517.1	49.0	795.4	1102.4	1009.0
7	981.5	722.8	430.9	1164.1	1379.3	1935.4
8	319.7	425.7	411.5	1170.2	972.9	1317.3
9	333.6	321.1	970.3	1119.6	1146.9	1019.8
10	1136.8	1373.7	350.1	1929.8	1606.2	1450.5
11	606.5	420.5	441.4	1136.5	1130.1	1643.0
12	191.3	124.5	605.6	1547.5	820.3	1671.5
Total	4966.6	4401.7	4801.0	11265.8	11698.5	16202.6

[a]PD, Puding reservoir; YZD, Yinzidu reservoir; HJD, Hongjiadu reservoir; DF, Dongfeng reservoir; SFY, Suofengying reservoir; WJD, Wujiangdu reservoir

Table 8.16 Output of methyl Hg (MeHg) from reservoirs through discharge

Month	PD[a]	YZD[a]	HJD[a]	DF[a]	SFY[a]	WJD[a]
1	2.76	4.99	16.05	21.04	24.99	83.99
2	6.18	7.89	20.94	27.31	27.29	62.89
3	6.60	7.49	21.17	29.64	43.27	76.71
4	8.46	6.41	11.37	36.34	54.89	83.72
5	5.17	4.82	15.68	9.98	33.53	124.72
6	82.11	42.63	3.54	75.69	80.01	105.62
7	114.69	62.66	47.81	123.99	130.41	313.27
8	40.22	31.69	43.43	159.72	91.43	213.99
9	37.50	26.28	41.29	114.68	75.54	150.28
10	63.03	68.03	17.54	117.28	91.82	115.33
11	31.50	25.28	24.59	59.73	61.33	138.93
12	9.42	7.98	42.93	61.93	56.60	148.49
总计	407.63	296.16	306.35	837.31	771.10	1617.93

[a]PD, Puding reservoir; YZD, Yinzidu reservoir; HJD, Hongjiadu reservoir; DF, Dongfeng reservoir; SFY, Suofengying reservoir; WJD, Wujiangdu reservoir

5641.9 mg for DF, 4871.0 mg for PD, 3478.8 mg for YZD, and 2599.5 mg for HJD.

The statistical analysis showed that: (1) the THg output from reservoirs through discharge significantly correlated with the THg concentrations in the discharged

Table 8.17 Output of total suspended solid (TSS) from reservoirs through discharge (t)

Month	PD[a]	YZD[a]	HJD[a]	DF[a]	SFY[a]	WJD[a]
1	25.4	65.0	442.2	363.1	222.1	501.8
2	50.5	59.2	221.4	244.2	515.2	829.6
3	136.4	75.5	275.6	420.4	76.5	249.4
4	67.7	34.6	115.4	317.6	481.4	268.8
5	66.3	25.0	192.2	91.4	1659.3	399.2
6	873.9	681.9	21.8	637.4	1390.4	536.2
7	918.4	574.1	0.0	806.3	1207.6	1541.7
8	233.5	177.0	84.0	600.5	1168.1	514.5
9	446.3	658.1	202.9	603.4	891.6	665.2
10	1402.4	748.4	39.9	582.5	836.5	510.0
11	529.3	196.9	205.0	459.3	1112.9	589.0
12	120.8	183.0	799.1	488.8	338.3	724.9
Total	4871.0	3478.8	2599.5	5614.9	7986.5	7330.3

[a]PD, Puding reservoir; YZD, Yinzidu reservoir; HJD, Hongjiadu reservoir; DF, Dongfeng reservoir; SFY, Suofengying reservoir; WJD, Wujiangdu reservoir

water ($r = 0.36$, $p = 0.002$, $n = 72$) and the water flow ($r = 0.94$, $p < 0.01$, $n = 72$) for all six reservoirs; (2) the MeHg output from the reservoirs via discharge significantly correlated with the MeHg concentrations in the discharged water ($r = 0.80$, $p < 0.01$, $n = 72$) and the water flow ($r = 0.47$, $p < 0.01$, $n = 72$); (3) the TSS output from the reservoirs via discharge significantly correlated with the concentrations in the discharged water ($r = 0.74$, $p < 0.01$, $n = 72$) and the water flow ($r = 0.88$, $p < 0.01$, $n = 72$). The THg, MeHg, and TSS outputs from the reservoirs via discharge in the wet season accounted for 53, 63, and 62% of the annual total outputs via discharge, which were slightly higher than that observed in the season.

8.3.5 Total Mercury and Methylmercury Outputs from Other Pathways

The THg, MeHg, and TSS outputs from other pathways (e.g., industrial and agricultural water consumption, water irrigation use, evaporation, consumption by water plants and so on) were estimated using Eq. 8.10.

$$F_{\text{other}} = V_{\text{other}} \times C_{\text{other}} \times 10^{-6} \tag{8.10}$$

where F_{other} is the output fluxes of THg, MeHg, and TSS from other pathways (g); V_{other} is the water outputs from other pathways (m^3); C_{other} is the average annual concentrations of THg, MeHg, and TSS in water samples from reservoirs in 2006.

The calculated data for the THg, MeHg, and TSS outputs via other pathways are summarized in Tables 8.18, 8.19, 8.20, 8.21, 8.22 and 8.23. The annual THg outputs from the reservoirs via other pathways were 3052.4 g a^{-1} for PD, 93.0 g a^{-1} for YZD, 2181.4 g a^{-1} for HJD, 335.6 g a^{-1} for DF, 867.4 g a^{-1} for SFY, and 2871.4 g a^{-1} for WJD. The annual MeHg outputs from other pathways were 179.5 g a^{-1} for PPD, 7.4 g a^{-1} for YZD, 125.8 g a^{-1} for HJD, 32.0 g a^{-1} for DF, 56.6 g a^{-1} for SFY, and 265.8 g a^{-1} for WJD. The annual TSS outputs from reservoirs via other pathways were 3552.8 g a^{-1} for PD, 214.8 g a^{-1} for YZD, 1664.0 g a^{-1} for HJD, 361.0 g a^{-1} for DF, 577.0 g a^{-1} for SFY, and 2738.5 g a^{-1} for WJD.

8.3.6 Elemental Mercury Emission Over the Water–Air Surface

The Hg^{2+} in the surface water can be reduced to Hg^0 via photo reduction, which leads to Hg^0 emission into the atmosphere from the surface water. This pathway is an important process to reduce the Hg burden in the reservoir. Based on the observed data, the current study found that the atmospheric wet deposition fluxes of Hg were higher than the Hg^0 emission over the water–atmosphere surface in PD, YZD, HJD, DF, and SFY. Furthermore, the Hg^0 emission quantity over the water–atmosphere surface was 0.5–0.8 times the wet deposition fluxes of Hg (see detail in Chap. 4).

The Hg^0 emission over the water–atmosphere surface in WJD was approximately 3.3 times higher than the wet deposition fluxes of Hg, and this was attributed to the elevated dissolved organic matter (DOC) concentrations in WJD (see detail in Chap. 4). Previous studies reported that DOC, even at a low concentration level, can significantly improve the photoreduction of Hg^{2+} in water (Xiao et al. 1995; Costa and Lisss 1999; Ravichandran et al. 2000). Jiang (2005) further observed a significant correlation between DOC and dissolved gaseous Hg (DGM) concentrations in WJD. Such results are consistent with the observations in the Petit-Sanut Reservoir and Baihua Reservoir (Muresan et al. 2007; Feng et al. 2004). Furthermore, the air–water Hg^0 emission in the Petit-Sanut Reservoir can reach 4.2 g a^{-1}, which was approximately 1.5 times higher than the Hg input from precipitation (Muresan et al. 2007); the air–water Hg^0 emission in the Baihua Reservoir was 725 g a^{-1}, which was 1.2 times higher than the corresponding Hg input from precipitation (Feng et al. 2004). The annual re-emission of Hg^0 over the water–air surface in Michigan Lake was 7.8 μg m^{-2} (449 kg), which accounted for approximately 74% of the annual Hg input from wet deposition (Landis and Keeler 2002). The Hg mass balance study in Ontario Lake from 1998 to 1999 reported that the annual Hg^0 emission was 8.8 μg m^{-2} (167 kg), which accounted for 63% of the annual Hg input from wet deposition (Vijiayaraghavan et al. 2005).

Table 8.18 Annual input–output fluxes of water volume, Hg species, and total suspended solid (TSS) in Puding reservoirs (PD)

Parameters	Input				Total input	output			Total output	Net flux	Source/Sink
	Deposition	SC[a]	BY[a]	Surface runoff		Reservoir discharge	Other ways	Water–air surface			
Water volume ($\times 10^4$ m^3)	2316.4	297733.5	4885.5	69829.7	374765.0	202300.4	172464.6	–	374765.0	–	–
THg (g $year^{-1}$)	477.1	14697.9	287.8	3544.4	19007.3	4966.6	3052.4	519.8	8538.8	−10468.5	Sink
MeHg (g $year^{-1}$)	4.2	411.1	6.8	95.6	517.8	407.6	179.5	–	587.1	69.4	Source
TSS (t $year^{-1}$)	–	11660.6	225.9	2814.4	14700.9	4871.0	3552.8	–	8423.8	−6277.1	Sink

[a]SC, Sancha river; BY, Boyu river

Table 8.19 Annual input–output fluxes of water volume, Hg species, and total suspended solid (TSS) in Yinzidu reservoirs (YZD)

Parameters	Input			Total input	Output			Total output	Net flux	Source/Sink
	Deposition	Discharge of PD[a]	Surface runoff		Reservoir discharge	Other ways	Water–air surface			
Water volume ($\times 10^4$ m^3)	1598.5	202300.4	6553.6	210452.5	203267.2	7185.3	–	210452.5	–	–
THg (g $year^{-1}$)	570.9	4966.6	150.0	5687.6	4401.7	93.0	354.8	4849.5	−838.1	Sink
MeHg (g $year^{-1}$)	2.8	407.6	14.3	424.7	296.2	7.4	–	303.5	−121.2	Sink
TSS (t $year^{-1}$)	–	4871.0	160.1	5031.1	3478.8	214.8	–	3693.7	−1337.4	Sink

[a]PD, Puding reservoir

Table 8.20 Annual input–output fluxes of water volume, Hg species, and total suspended solid (TSS) in Hongjiadu reservoirs (HJD)

Parameters	Input				Total input	Output			Total output	Net flux	Source/Sink
	Deposition	LC[a]	LJ[a]	Surface runoff		Reservoir discharge	Other ways	Water–air surface			
Water volume ($\times 10^4$ m^3)	7089.6	193613.8	42293.7	117750.6	360747.7	220914.4	139833.3	–	360747.7	–	–
THg (g $year^{-1}$)	2795.1	10011.9	1779.4	5810.7	20397.1	4801.0	2181.4	1980.3	8962.7	−11434.4	Sink
MeHg (g $year^{-1}$)	12.6	252.8	62.2	178.5	506.1	306.3	125.8	–	432.2	−73.9	Sink
TSS (t $year^{-1}$)	–	7385.2	1485.0	4691.0	13561.2	2599.5	1664.0	–	4263.5	−9297.6	Sink

[a]LC, Liuchong river; LJ, Luojiao river

Table 8.21 Annual input–output fluxes of water volume, Hg species, and total suspended solid (TSS) in Dongfeng reservoirs (DF)

Parameters	Input				Total input	Output			Total output	Net flux	Source/Sink
	Deposition	LC[a]	SC[a]	Surface runoff		Reservoir discharge	Other ways	Water–air surface			
Water volume ($\times 10^4$ m^3)	1849.6	220914.4	272472.8	21873.1	517109.8	493671.5	23438.4	–	–	517109.8	–
THg (g $year^{-1}$)	691.1	6506.0	6020.4	699.6	13917.1	11265.8	335.6	495.6	12097.0	−1820.1	Sink
MeHg (g $year^{-1}$)	3.7	235.6	235.3	27.2	501.8	837.3	32.0	–	869.4	367.5	Source
TSS (t $year^{-1}$)	–	5747.9	6489.4	693.0	12930.3	5614.9	361.0	–	5975.9	−6954.4	Sink

[a]LC, Liuchong river; SC, Sancha river

Table 8.22 Annual input–output fluxes of water volume, Hg species, and total suspended solid (TSS) in Suofengying reservoirs (SFY)

Parameters	Input				Total input	Output			Total output	Net flux	Source/Sink
	Deposition	Discharge of DF[a]	MT[a]	Surface runoff		Reservoir discharge	Other ways	Water–air surface			
Water volume ($\times 10^4$ m^3)	553.1	493671.5	65094.9	44019.7	603339.2	540481.2	62857.9	–	603339.2	–	
THg (g $year^{-1}$)	206.7	11265.8	1826.7	1007.7	14306.9	11698.5	867.4	143.6	12709.5	−1597.4	Sink
MeHg (g $year^{-1}$)	1.1	837.5	92.8	76.8	1008.3	771.1	56.6	–	827.7	−180.6	Sink
TSS (t $year^{-1}$)	–	5614.9	2166.2	993.0	8774.2	7986.5	577.0	–	8563.6	−210.6	Sink

[a]DF, Dongfeng reservoir, MT, Maotiao river

Table 8.23 Annual input–output fluxes of water volume, Hg species, and total suspended solid (TSS) in Wujiangdu reservoirs (WJD)

Parameters	Input						Total input	Output			Total output	Net flux	Source/Sink
	Deposition	Mainstream of WJ[a]	YJ[a]	XF[a]	PY[a]	Surface runoff		Reservoir discharge	Other ways	Water–air surface			
Water volume ($\times 10^4$ m^3)	3312.1	590354.8	41384.7	30353.5	27402.6	70507.6	763315.3	611178.0	152137.3	–	–	–	–
THg (g $year^{-1}$)	1891.8	12812.2	1446.5	2654.9	1225.8	3643.0	23674.2	16202.6	2871.4	6204.4	25278.4	1604.2	Source
MeHg (g $year^{-1}$)	7.9	702.2	67.1	75.3	35.1	138.9	1026.5	1617.9	265.8	–	1883.7	857.2	Source
TSS (t $year^{-1}$)	–	5747.9	6489.4	693.0	12930.3	5614.9	**361.0**	–	5975.9	−6954.4	汇	−14467.3	Sink

[a]WJ, Wujiang river; YJ, Yeji river; XF, Xifeng river; PY, Pianyan river

8.4 The Relative Contribution of Different Vectors to the Mercury Input–Output Budgets

8.4.1 The Relative Contribution of Different Vectors to the Mercury Input in Reservoirs

The ratios of the different pathways for water volume, Hg species, and TSS to the total input–output fluxes in the reservoirs in the Wujiang River Basin are shown in Tables 8.24, 8.25, 8.26, 8.27, 8.28 and 8.29.

(1) Atmospheric deposition and river input

Although the Hg concentration in the precipitation in the Wujiang River Basin was highly elevated compared to the Hg concentrations in the inflow rivers, the water input via precipitation (accounting for 0.7% of the total water input) is

Table 8.24 The ratios of different pathways of water volume, Hg species, and total suspended solid (TSS) to total input–output fluxes in Puding reservoirs (PD) (%)

Parameters	Input ratio				Output ratio			Net ratio
	Deposition	SC[a]	BY[a]	Surface runoff	Reservoir discharge	Other ways	Water–air surface	
Water volume	0.6	79.4	1.3	18.6	54.0	46.0	–	–
THg	2.5	77.3	1.5	18.6	58.2	35.7	6.1	−55.1
MeHg	0.8	79.4	1.3	18.5	69.4	30.6	–	13.4
TSS	–	79.3	1.5	19.1	57.8	42.2	–	−42.7

[a]SC, Sancha river; BY, Boyu river

Table 8.25 The ratios of different pathways of water volume, Hg species, and total suspended solid (TSS) to total input–output fluxes in Yinzidu reservoirs (YZD) (%)

Parameters	Input ratio			Output ratio			Net ratio
	Deposition	Discharge of PD[a]	Surface runoff	Reservoir discharge	Other ways	Water–air surface	
Water volume	0.8	96.1	3.1	96.6	3.4	–	–
THg	10.0	87.3	2.6	90.8	1.9	7.3	−14.7
MeHg	0.7	96.0	3.4	97.6	2.4	–	−28.5
TSS	–	96.8	3.2	94.2	5.8	–	−26.6

[a]DF, Dongfeng reservoir

Table 8.26 The ratios of different pathways of water volume, Hg species, and total suspended solid (TSS) to total input–output fluxes in Hongjiadu reservoirs (HJD) (%)

Parameters	Input ratio				Output ratio			Net ratio
	Deposition	LC[a]	LJ[a]	Surface runoff	Reservoir discharge	Other ways	Water–air surface	
Water volume	2.0	53.7	11.7	32.6	61.2	38.8	–	–
THg	13.7	49.1	8.7	28.5	53.6	24.3	22.1	−56.1
MeHg	2.5	49.9	12.3	35.3	70.9	29.1	–	−14.6
TSS	–	54.5	11.0	34.6	61.0	39.0	–	−68.6

[a]LC, Liuchong river; LJ, Luojiao river

Table 8.27 The ratios of different pathways of water volume, Hg species, and total suspended solid (TSS) to total input–output fluxes in Dongfeng reservoirs (DF) (%)

Parameters	Input ratio				Output ratio			Net ratio
	Deposition	LC[a] (YC[a] section)	SC[a] (YC[a] section)	Surface runoff	Reservoir discharge	Other ways	Water–air surface	
Water volume	0.4	42.7	52.7	4.2	95.5	4.5	–	–
THg	5.0	46.7	43.3	5.0	93.1	2.8	4.1	−13.1
MeHg	0.7	46.9	46.9	5.4	96.3	3.7	–	73.2
TSS	–	44.5	50.2	5.4	94.0	6.0	–	−53.8

[a]LC, Liuchong river; YC, Yachi river; SC, Sancha river

Table 8.28 The ratios of different pathways of water volume, Hg species, and total suspended solid (TSS) to total input–output fluxes in Suofengying reservoirs (SFY) (%)

Parameters	Input ratio				Output ratio			Net ratio
	Deposition	Discharge of DF[a]	MT[a]	Surface runoff	Reservoir discharge	Other ways	Water–air surface	
Water volume	0.1	81.8	10.8	7.3	89.6	10.4	–	–
THg	1.4	78.7	12.8	7.0	92.0	6.8	1.1	−11.2
MeHg	0.1	83.1	9.2	7.6	93.2	6.8	–	−17.9
TSS	–	64.0	24.7	11.3	93.3	6.7	–	−2.4

[a]DF, Dongfeng reservoir; MT, Maotiao river

significantly less than the water input from the inflow rivers, which leads to limited Hg contributions from precipitation. For example, the water, THg, and MeHg inputs from precipitation only constituted 0.7 ± 0.7% (water input), 6.8 ± 4.7% (THg input), and 0.9 ± 0.8% (MeHg input) of the total inputs. However, the water,

Table 8.29 The ratios of different pathways of water volume, Hg species, and total suspended solid (TSS) to total input–output fluxes in Wujiangdu reservoirs (WJD) (%)

Parameters	Input ratio						Output ratio			Net ratio
	Deposition	Mainstream of WJ*	YJ[a]	XF[a]	PY[a]	Surface runoff	Reservoir discharge	Other ways	Water–air surface	
Water volume	0.4	77.3	5.4	4.0	3.6	9.2	80.1	19.9	–	–
THg	8.0	54.1	6.1	11.2	5.2	15.4	64.1	11.4	24.5	6.8
MeHg	0.8	68.4	6.5	7.3	3.4	13.5	85.9	14.1	–	83.5
TSS	–	71.8	7.1	6.1	3.2	11.9	72.8	27.2	–	−59.0

[a]WJ, Wujiang river; YJ, Yeji river; XF, Xifeng river; PY, Pianyan river

THg, MeHg, and TSS inputs from the inflow rivers, were 86.8 ± 11.9% (water input), 80.4 ± 12.6% (THg input), 85.1 ± 12.6% (MeHg input), and 85.8 ± 11.4% (TSS input). A previous study reported that the water, THg, and MeHg inputs from precipitation accounted for 7 ± 2.4% (water input), 18 ± 7% (THg input), and 7 ± 4.2% (MeHg input) of the total input at the Experimental Lakes Area (ELA), Canada (St Louis et al. 2004), and these values were higher than the values in this study.

The THg input contribution to the reservoirs in the Wujiang River Basin from precipitation was comparable to the results in the Lot-Garonne River system in France (<10%; Schafer et al. 2006). In the Superior Lake mass balance, the THg input from atmospheric deposition dominated the total THg inputs (annual THg input of 740 kg a^{-1}), which was approximately 2.6 times higher than that from the inflow river (annual THg input of 280 kg a^{-1}), and the MeHg inputs from atmospheric deposition (annual MeHg input of 4.7 kg a^{-1}) were comparable to the input from the inflow river (annual MeHg input of 3.5 kg a^{-1}) (Rolfhus et al. 2003).

In addition, the contribution of different vectors to the Hg input budgets are also related to the characteristics of the lake or reservoir (area of the watershed, total water volume, and so on). While the Superior Lake is the largest freshwater lake on earth, the watershed area is only 1.5 times larger than the surface area of the lake, which leads to a relatively higher Hg input contribution from atmospheric deposition to the total Hg input. In contrast, the Long Island Sound Lake has a watershed area 13 times larger than the surface area of the lake, and 99% of the total THg input comes from the inflow river and 1% from atmospheric deposition; meanwhile, 75% of the total MeHg input was from the inflow river and 13% from atmospheric deposition (Balcom et al. 2004). The study from Michigan Lake suggested that the atmospheric deposition (wet and dry deposition) was the primary pathway for THg input and accounted for 84% of the total input (Landis and Keeler 2002).

The ratio of the watershed area to the surface area of the lake/reservoir (W/S) is the key factor controlling the contribution of different vectors to the Hg input budgets, specifically the THg input. For example, a larger ratio means a smaller contribution to the THg input from deposition with a higher THg input contribution from the inflow river. The ratio of W/S in the Wujiang River Basin exhibited the following patterns: SFY (649) > PD (305) > WJD (124) > HJD (123) > DF (96) > YZD (37). Consequently, the lowest THg input contribution from precipitation to the total THg inputs was observed at SFY (1.4%), followed by PD (2.5%), DF (5%), WJD (8%), YZD (10%), and HJD (13.7%), which was consistent with the distribution patterns for the ratios of W/S in the Wujiang River Basin. In this study, the ratios of W/S throughout the six selected reservoirs in the Wujiang River Basin were much higher than the ratios in Superior Lake (1.5 times higher) and Long Island Sound (13 times higher). Therefore, a smaller THg input contribution from precipitation was observed in this study. The THg input from the inflow rivers to the reservoirs was the predominant pathway.

(2) Surface runoff

The Hg input contributions to the reservoir from the surface runoff are closely related to the watershed area of the reservoir. Our calculated data showed that the water, THg, MeHg, and TSS inputs from the surface runoff only constituted 12.5 ± 11.3% (water input), 12.9 ± 9.8% (THg input), 13.9 ± 11.8% (MeHg input), and 14.2 ± 11.4% (TSS input) of the total inputs.

For the ELA Lake, the water, THg, and MeHg inputs from the surface runoff constituted 3.0 ± 0.6% (water input), 13 ± 3% (THg input), and 2 ± 0.5% (MeHg input) of the total inputs (St Louis et al. 2004), which were comparable to the results of this study. The Hg inputs from the inflow rivers represented a higher proportion of the total inputs in this study, e.g., 86.8 ± 11.9% for water input, 80.4 ± 12.6% for THg input, 85.1 ± 12.6% for MeHg inputs, and 85.8 ± 11.4% for TSS input.

A previous study further reported that the water, THg, and MeHg inputs from the surface runoff accounted for 90% (water input), 68% (THg input), and 91% (MeHg input) of the total inputs in the ELA, Canada (St Louis et al. 2004). Compared to the results in the ELA, Canada (St Louis et al. 2004), the contributions of the water and MeHg inputs from the river to the total inputs at the Wujiang River Basin were lower. Additionally, higher THg input contributions to total inputs from the surface runoff were observed in this study. Based on the Hg mass balance study in the Chesapeake Bay, the Hg input from the inflow river was comparable to the corresponding value from atmospheric deposition. The two pathways accounted for approximately 50% of the total Hg input (Cossa et al. 1996; Mason et al. 1997).

8.4.2 The Relative Contribution of Different Vectors to the Mercury Output from Reservoirs

Our calculated data showed that the water, THg, MeHg, and TSS outputs from the reservoirs via discharge constituted 79.5 ± 18.1% (water input), 75.3 ± 18.6% (THg input), 85.5 ± 12.6% (MeHg input), and 78.8 ± 17.1% (TSS input) of the total outputs.

The water, THg, MeHg, and TSS outputs from the reservoirs through other pathways (e.g., industrial and agricultural water consumption, water irrigation, evaporation, and consumption by water plants) accounted for 20.5 ± 18.1% (water output), 13.8 ± 13.5% (THg output), 14.5 ± 12.6% (MeHg output), and 21.2 ± 17.2% (TSS output) of the total outputs. The THg output from Hg^0 emission over the water–air surface supplied 10.9 ± 9.9% of the total outputs. Discharge was the predominant pathway for the Hg outputs from the reservoirs in the Wujiang River Basin.

8.5 Net Fluxes and Stocking Rates of Mercury Species in Reservoirs

8.5.1 Net Fluxes of Mercury Species in Reservoirs

Because the six selected reservoirs in the Wujiang River Basin are located in the regional background area, the Hg in the ambient air is relatively low (see detail in Chap. 4). Therefore, we assumed that the Hg input to the reservoirs from dry deposition was not an important pathway and was insignificant compared to the other pathways (e.g., surface runoff and inflow rivers). The predominant pathways for THg, MeHg, and TSS input to the reservoirs are from surface runoff and inflow rivers, and the main pathway for THg, MeHg, and TSS output from the reservoirs is from discharge (outflow river).

Our study observed that the different reservoirs in the Wujiang River Basin had different roles in the total Hg, MeHg, and particle transportation within the river reservoir ecosystem. The six selected reservoirs were net sinks for TSS annually. Moreover, the inflow rivers were the primary pathway for TSS input to the reservoirs. As shown in Table 8.30, the TSS net fluxes in the reservoirs in the Wujiang River Basin exhibited the following distribution patterns: WJD (−14467.3 mg a^{-1}) > HJD (−92797.6 mg a^{-1}) > DF (−6954.4 mg a^{-1}) > PD (−6277.1 mg a^{-1}) > YZD (−1337.4 mg a^{-1}) > SFY (−210.6 mg a^{-1}). Generally, erosion, transportation, and deposition processes occur when rivers flow. The flow rate of the river and the geologic conditions determine the function and the strength of the function. When the slope of the riverbed slows or the increasing transport mass leads to a slow flow rate, the carrying capacity of the river is weakened, and the carried materials will be deposited into the sediment. This function is more pronounced after rivers flow into the reservoir.

The reservoirs in the Wujiang River Basin, where the net sinks for THg annually with the exception of WJD. The THg net fluxes in the reservoirs in the Wujiang

Table 8.30 Net fluxes of total Hg (THg), methyl Hg (MeHg), and total suspended solid (TSS) in reservoirs in Wujiang River Basin

Reservoir	Construction time (year)	MeHg			THg	TSS	
		Net flux (g $year^{-1}$)	Storage rate (%)	Net flux (g $year^{-1}$)	Storage rate (%)	Net flux (g $year^{-1}$)	Storage rate (%)
SFY	2005	−180.6	−17.9	−1597.4	−11.2	−210.6	−2.4
HJD	2004	−73.9	−14.6	−11434.4	−56.1	−9297.6	−68.6
YZD	2003	−121.2	−28.5	−838.1	−14.7	−1337.4	−26.6
PD	1995	69.4	13.4	−10468.5	−55.1	−6277.1	−42.7
DF	1995	367.5	73.2	−1820.1	−13.1	−6954.4	−53.8
WJD	1979	857.2	83.5	1604.2	6.8	−14467.3	−59.0

[a]PD, Puding reservoir; YZD, Yinzidu reservoir; HJD, Hongjiadu reservoir; DF, Dongfeng reservoir; SFY, Suofengying reservoir; WJD, Wujiangdu reservoir

River Basin had the following distribution patterns: HJD (11434.4 g a^{-1}) > PD (−10468.5 g a^{-1}) > DF (−1820.1 g a^{-1}) > SFY (−1597.4 g a^{-1}) > YZD (−838.1 g a^{-1}). WJD was the net source of THg with net fluxes of +1604.2 g a^{-1}. In the uncontaminated river systems, the Hg in the river water tends to bind with particulate materials. The particulate-bound Hg can be deposited in sediment through sedimentation, which leads to the THg net sink for the reservoir. As shown in Chap. 4, the fluxes in the Hg^0 emission from the surface water to the atmosphere in WJD were significantly elevated, which indicated the THg source was WJD. Our results were consistent with the previous observations conducted in Petit-Saut (Muresan et al. 2008).

Our results showed that YZD, HJD, and SFY were the net sinks for MeHg annually, and the corresponding net MeHg fluxes were −121.2 g a^{-1} (YZD), −73.9 g a^{-1} (HJD), and −180.6 g a^{-1} (SFY). PD, DF, and WJD acted as the net sources for MeHg, with net MeHg fluxes of +69.4 g a^{-1} (PD), +367.5 g a^{-1} (DF), and +857.2 g a^{-1} (YZD).

8.5.2 Storage Rates for Mercury Species in Reservoirs

The inflow rivers are the primary pathways for water and Hg species input into reservoirs, and discharge dominates the output pathways. Our results showed that the average ratios of the water, THg, and TSS outputs from the reservoirs through discharge to the total outputs were 79.5 ± 18.1% (water input ratio), 59.7 ± 27.4% (THg input ratio), and 47.6 ± 27.2% (TSS input ratio) throughout the six selected reservoirs in the Wujiang River Basin. This indicated that some proportion of THg and TSS were still stored in the reservoirs.

In this study, we defined the Hg storage rate as the ratio of the net Hg flux to the total Hg input. Furthermore, the positive net MeHg flux was generally attributed to the newly formed MeHg (net Hg methylation) in the reservoirs. Subsequently, the MeHg storage rate in the reservoirs is usually recognized as the transformation rate. Our calculated data showed that the THg storage rates in the reservoirs were −55.1% in PD, −14.7% in YZD, −56.1% in HJD, −13.1% in DF, −11.2% in SFY, and 6.8% in WJD (Table 8.30). Similarly, the corresponding MeHg storage/transformation rates were 13.4% in PD, −28.5% in YZD, −14.6% in HJD, −73.2% in DF, −17.9% in SFY, and 83.5% in WJD. The TSS storage rates in the reservoirs were −42.7% in PD, −26.6% in YZD, −68.6% in HJD, −53.8% in DF, −2.4% in SFY, and −59% in WJD (Table 8.30).

The THg storage rates in PD and HJD (located in the upper end of the Wujiang River Basin) were highly elevated compared to the values in the other four reservoirs (SFY, DF, WJD, and YZD), which indicated that the reservoirs located upstream captured more Hg than the downstream reservoirs. The MeHg in the reservoirs is from the in situ IHg methylation and MeHg input from the inflow rivers. The positive net MeHg fluxes in PD, DF, and WJD suggested the net Hg methylation, which can be transferred to downstream ecosystems through reservoir discharge. Furthermore, the transformation rates of MeHg in PD, DF, and WJD reached 13.4, 73.2, and 83.5%, respectively (Table 8.30), which further suggested that a large proportion of the MeHg formed in the reservoirs via Hg methylation.

8.5.3 Possible Factors Controlling the Net Fluxes and Storage Rates of Mercury Species in Reservoirs

Spatial and seasonal differences in the physical–chemical characteristics are observed between the reservoirs and river ecosystems. Given the strong hydrodynamic conditions and frequent water exchange, thermal stratification of the water body is absent in a river system. However, thermal stratification is usually observed in a water body such as a reservoir. Generally, clear thermal stratification of the water column occurs in reservoirs with a large total water volume and water depth, especially during the summer. The dissolved oxygen in the surface water layer barely diffuses into the water column below the thermocline. Therefore, the vertical profiles of the dissolved oxygen are pronounced and correlate with the thermal stratification. Combined with the intensive bacterial decomposition of the settled degradable organic matter, a low dissolved oxygen concentration (anaerobic environment) in the hypolimnion of reservoirs is observed. Consequently, the seasonal and vertical stratifications in the water column significantly impact the transportation and transformation of Hg in reservoir systems, specifically the formation of MeHg. Earlier studies suggested that the concentrations and distributions of MeHg in the water columns of reservoirs were controlled by different factors in different seasons, i.e., SRB-induced Hg methylation during the wet season and diffusion of MeHg from the sediments to the water during the dry season (Muresan et al. 2008).

In addition, the basic characteristics of reservoirs and the geology/geomorphology of the watershed (especially the water residence time in the reservoir and organic matter content in the reservoir) are very important factors controlling the biogeochemical cycling of Hg, which impacts the function of the Hg source/sink in the reservoir. The reservoirs in the Wujiang River Basin are typically deep-valley, high-mountain gorge. The organic matter content in the flooded farmland is very poor. The organic matter in the reservoirs was primarily from internal sources. With the evolution of the reservoirs, the organic matter in the reservoirs from internal sources continually increased.

The thermal stratification of the reservoir and water residence time were estimated using Eqs. 8.11 and 8.12.

$$\alpha = F_{\text{input}} / V_{\text{reservoir}} \quad (8.11)$$

$$R = V_{\text{normal}} / F_{\text{discharge}} \quad (8.12)$$

where F_{input} is the annual water input volume of the reservoir (m^3); $V_{reservoir}$ is the reservoir storage capacity (m^3); R is the water residence time of the reservoir (day); V_{normal} is the reservoir capacity at its normal storage level; $F_{discharge}$ is the multi-year averaged flow rate of the reservoir discharge. When $\alpha < 10$, the reservoir is stably stratified, and for $\alpha > 20$, the reservoir is mixed water.

Based on the data concerning the annual water input volume, reservoir storage capacity, and reservoir capacity for the normal storage level in 2006, the water residence time and α for the six selected reservoirs in the Wujiang River Basin were calculated and are shown in Table 8.31. Our calculated data suggested that the α were all below 10 for the reservoirs in the Wujiang River Basin (HJD, YZD, WJD, DF, and PD) with the exception of SFY ($\alpha > 10$), which indicated that these reservoirs are thermally stratified. SFY has a relatively small storage capacity, short water residence time (5 days), and frequent water exchange, which make it a typical daily regulation and mixed water reservoir. However, HJD, which has the longest water residence time (555 days) and largest storage capacity, is a typical annual regulation reservoir.

As shown in Chap. 5, the six selected reservoirs in the Wujiang River Basin were classified as a primary evolutionary stage (YZD, SFY, and HJD), intermediate evolutionary stage (PD and DF), and advanced evolutionary stage (WJD) based on the age of the reservoirs. As shown in Table 8.31, PD, YZD, and DF, which had similar water residence times (PD, 45 days; DF, 36 days; YZD, 58 days), were classified as the different evolutionary stage. PD and DF intermediate evolutionary stage reservoirs were the net source of MeHg, and YZD (primary evolutionary stage) was the net sink for MeHg. Although HJD and SFY (primary evolutionary stage) had significantly different water residence times (HJD, 555 days; SFY, 5 days), they acted as the net sinks for MeHg. WJD, an advanced evolutionary stage reservoir, was a source of MeHg annually. Furthermore, the net MeHg flux in an advanced evolutionary stage reservoir (e.g., WJD) was significantly elevated compared to that in an intermediate evolutionary stage reservoir (e.g., DF and PD) (Table 8.30). These results suggested that the evolutionary stage and not the water residence time is the primary factor controlling Hg methylation in reservoirs in the Wujiang River Basin. Although PD and DF are classified as an intermediate evolutionary stage, the net MeHg flux in DF was approximately 5.3 times higher than that in PD, which indicated that there are some other factors controlling Hg

Table 8.31 Stratification and water residence time of reservoirs in Wujiang River Basin

Parameters	PD[a]	YZD[a]	HJD[a]	DF[a]	SFY[a]	WJD[a]
α	8.9	4.6	0.7	5	30	3.3
Stratification	Stratified	Stratified	Stratified	Stratified	Mixed	Stratified
water residence time (day)	45	58	555	36	5	81

[a]PD, Puding reservoir; YZD, Yinzidu reservoir; HJD, Hongjiadu reservoir; DF, Dongfeng reservoir; SFY, Suofengying reservoir; WJD, Wujiangdu reservoir

methylation in the reservoirs in addition to the evolutionary stage; this could be verified in a further study.

An elevated net MeHg flux and MeHg storage rates in reservoirs indicate an increased environmental risk to aquatic food chains and human health. Given the complicated biogeochemical processes for Hg in river reservoir ecosystems, a study on the Hg mass balance is useful to better understand the environmental effects of reservoirs on Hg cycling in aquatic ecosystems. Although uncertainties still remain, this study suggests a clear net Hg methylation in PD, DF, and WJD. The net MeHg fluxes in the reservoirs (e.g., PD, DF, and WJD) in the Wujiang River Basin, which poses a potential threat to downstream ecosystems, should be watched more closely. More importantly, we cannot deny that the reservoirs with a primary evolutionary stage could change from net MeHg sinks to net MeHg sources in the future.

References

Balcom PH, Fitzgerald WF, Vandal GM, Lamborg CH, Rolfhus KR, Langer CS, Hammerschmidt CR (2004) Mercury sources and cycling in the Connecticut River and Long Island Sound. Mar Chem 90:53–74

Cossa D, Coquery M, Gobeil C, Martin JM (1996) Mercury fluxes at the ocean margins. In: Baeyens W, Ebinghaus R, Vasiliev O (eds) Global and regional mercury cycles: sources, fluxes and mass balances. Kluwer, Netherlands, pp 229–247

Costa M, Liss P (1999) Photoreduction of mercury in sea water and its possible implication for Hg^0 air-sea fluxes. Mar Chem 68:87–95

Feng XB, Yan HY, Wang SF, Qiu GL, Tang SL, Shang LH, Dai QJ, Hou YM (2004) Seasonal variation of gaseous mercury exchange rate between air and water surface over Baihua reservoir, Guizhou. China Atmos Environ 38:4721–4732

Jiang HM (2005) Effects of hydroelectric reservoir on the biogeochemical cycle of mercury in the Wujiang River. Ph.D. dissertation, Graduate School of the Chinese Academy of Sciences, Beijing (In Chinese, with English abstract)

Landis MS, Keeler GJ (2002) Atmospheric mercury deposition to Lake Michigan during the Lake Michigan mass balance study. Environ Sci Technol 36:4518–4524

Mason RP, Lawson NM, Sullivan KA (1997) The concentration, speciation and sources of mercury in Chesapeake bay precipitation. Atmos Environ 31:3541–3550

Muresan B, Cossa D, Richard S, Dominique Y (2008) Monometylmercury sources in a tropical artificial reservoir. Appl Geochem 23:1101–1126

Muresan B, Cossa D, Richard S, Burban B (2007) Mercury speciation and exchanges at the air-water interface of a tropical artificial reservoir, French Guiana. Sci Total Environ 385:132–145

Ravichandran M, Araujo R, Zepp RG. (2000) Role of humic substances on the photochemical reduction of mercury. In: Presented at 220th American Chemical Society National Meeting, Washington, DC, 20–24 Aug 2000

Rolfhus KR, Sakanoto HE, Cleckner LB, Stoor RW, Babiarz CL, Back RC, Manolopoulos H, Hurley JP (2003) Distribution and fluxes of total and methylmercury in Lake Superior. Environ Sci Technol 37:865–872

Schafer J, Blanc G, Audry S, Cossa D, Bossy C (2006) Mercury in the Lot-Garonne River system (France): sources, fluxes and anthropogenic component. Appl Geochem 21:515–527

St Louis VL, Rudd JWM, Kelly CA, Beaty KG, Bloom NS, Flett RJ (1994) Importance of wetlands as sources of methyl mercury to boreal forest ecosystems. Can J Fish Aquat Sci 51:1065–1076

St Louis VL, Rudd JWM, Kelly CA, Bodaly RA, Paterson MJ, Beaty KG, Hesslein RH, Heyes A, Majewski AR (2004) The rise and fall of mercury methylation in an experimental reservoir. Environ Sci Technol 38:1348–1358

Vijayaraghavan K, Seigneur C, Lohman K (2005) Modeling of atmospheric mercury deposition in the Great Lakes region. In: Presented at 48th Annual IAGLR Conference, Ann Arbor, Michigan, 23–27 May 2005

Xiao ZF, Stromberg D, Lindqvist O (1995) Influence of humic substances on photolysis of divalent mercury in aqueous solution. Water Air Soil Pollut 80:789–798

Chapter 9
Bioaccumulation of Mercury in Aquatic Food Chains

Abstract Current paradigms regarding the bioaccumulation of mercury are rooted in observations that methylmercury (MeHg) biomagnifies along pelagic food chains. However, mechanisms regulating the bioaccumulation of MeHg in food chain, its initial incorporation at the base of pelagic food chains, and its subsequent trophic transfer remain controversial. Here we measured mercury (total mercury (THg) and MeHg) and stable carbon/nitrogen isotopes ($\delta^{13}C$ and $\delta^{15}N$) and use these field data from seven reservoirs to understand the transport and accumulation and influence effectors of mercury in the aquatic food chain (plankton and fish) in Wujiang River Basin, and assessed the health risk.

Keywords Mercury · Bioaccumulation · Eutrophication level · Health risk assessment

9.1 Aquatic Food Chains in Reservoirs

In ecosystems, chemical energy stored in organisms can be transmitted layer-by-layer; the sequence of various organisms linked to each other according to their trophic relationships in ecology is called a "food chain". Food chains can be divided into three types: grazing, detrital, and parasitic; the type is determined by the relationship between the organisms within the chain (Mackenzie et al. 1998; Sun et al. 2002). Studies on the transport and accumulation of mercury in the food chain have mainly focused on the main grazing food chain, namely, phytoplankton → zooplankton → herbivorous fish → omnivorous fish → predatory fish. This chapter mainly discusses the transfer, accumulation, and biomagnification of mercury by food chains in reservoirs located in the Wujiang River Basin, Guizhou province, including WJD, DF, PD, and BH.

X. Feng et al., *Biogeochemical Cycle of Mercury in Reservoir Systems in Wujiang River Basin, Southwest China*, https://doi.org/10.1007/978-981-10-6719-8_9

9.1.1 Phytoplankton

Phytoplankton refers to microphytes that float on the water, including Cyanophyta, Bacillariophyta, Chrysophyta, Xanthophyta, Pyrrophyta, Cryptophyta, Euglenophyta, and Chlorophyta, rather than bacteria and other plants. Phytoplankton are trophic level 1 of the aquatic food chain and are the most critical level. Transport and accumulation of Hg in the aquatic food chain begin with its adsorption and absorption by phytoplankton, followed by zooplankton and the next trophic levels after feeding.

1. Taxonomic composition of phytoplankton

In the flood season (July) and dry season (October) of 2007, the phytoplankton collected from WJD, DF, PD, and BH were mainly Chlorophyta, diatoms and cyanobacteria (Table 9.1), comprising approximately 30–40 species. *Microcystis aeruginosa* Kutz, *Aphanizomenon flos-aquae*(L.)Ralfs, *Melosira granulata*(Ehr.) Ralfs,), *M.varians* Ag, *Cyclotella bodanica* Eul., *C.comensis* Grun., *Attheya zachariasi* Brun, *Fragilaria capucina* Desm., *Asterionella formosa* Hust., *Synedra acus* Kutz.; *Cryptomonas erosa* Ehr., *Cr.ovata* Ehr.; *Ceratium hirundinella*(Mull.) Schr.;*Pandorina morum* (Mull.) Bory, *Eudorina elegans* Ehr., *Pediastrum simplex var.echinulatum* Wittr., *P.dulex var.clathratum* A.Brunn, *P.simplex var.duodenarium* (Bail.) Rabenh. *Coelastrum reticulatum(Dang.)* Senn, *Mougeotia parvula* Haas. *Staurastrum gracile* Ralfs ex Ralfs, and *S.manfeldtii Delp.* were the most common species found in the four reservoirs.

Figure 9.1 shows that the phytoplankton abundance was between 2.70−49.68 × 10^6 cells L^{-1}; PD and BH had the lowest and highest abundances in October, respectively. In WJD, the phytoplankton abundance was 2.88 × 10^6 cells L^{-1} in July, which was significantly higher than its abundance in October of 6.39 × 10^6 cells L^{-1}. In DF, the phytoplankton abundances were 6.89 × 10^6 cells L^{-1} in July and 5.81 × 10^6 cells L^{-1} in October. In PD, the phytoplankton abundances were 7.94 × 10^6 cells L^{-1} in July and only 2.70 × 10^6 cells L^{-1} in October, which was significantly lower than in the other reservoirs. In BH, the phytoplankton abundances were 28.08 × 10^6 cells L^{-1} in July and reached 49.68 × 10^6 cells L^{-1} in October, which was significantly higher than in the other reservoirs Fig. 9.1.

The distribution of phytoplankton abundance showed that it was mainly composed of chlorophyta and its percentage was up to 61.50% in WJD in July; the second most abundant phytoplankton was cyanobacteria, which reached 26.36% (Fig. 9.2). In October, cyanobacteria reached a percent abundance of 86.02%, ranking first. In DF in July, the abundance of phytoplankton was mainly composed of diatoms with a percentage that reached 80%. In October, diatoms were still the most abundant, reaching a percentage of 32.58%, followed by cyanobacteria and Cryptophyta with percent abundances of 24.24 and 29.00%, respectively. In PD, the abundances of Chlorophyta in July and October were 46.52 and 51.66%, respectively; the second most abundant phytoplankton were diatoms, with abundances of

Table 9.1 Phytoplankton types and community

Date	Site	Blue algae	Green algae	Diatom	Cryptophyta	Pyrroptata	Chryso-phyceae	Xantho-phyta	Total
July	WJD	5	21	3	3	1	1		34
	DF	6	12	13	3	1	1		36
	PD	6	18	15	0		1	1	41
	BH	6	19	3	2	1			31
October	WJD	4	20	11	2	1			38
	DF	6	16	10	2	1	1		36
	PD	6	18	15	0		1	1	41
	BH	5	13	13	2	1	1		35

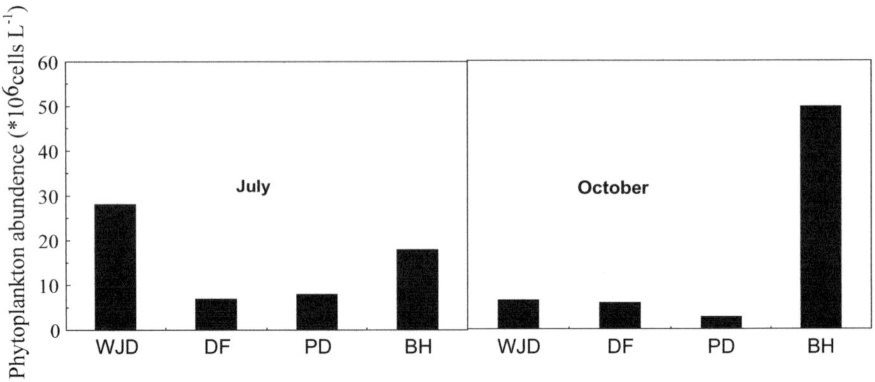

Fig. 9.1 Abundance of phytoplankton in four reservoirs

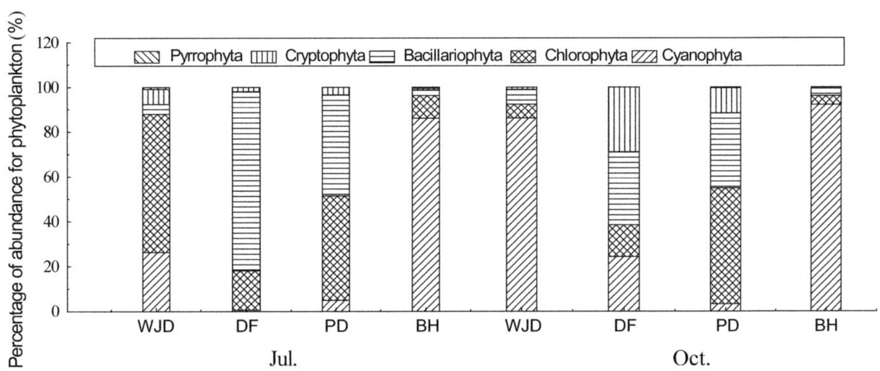

Fig. 9.2 Percentage of phytoplankton abundance

45.28 and 33.36% in July and October, respectively. In BH, the abundance of cyanobacteria reached 86.10 and 92.11% in July and October, respectively.

The distribution of phytoplankton abundance showed that it was mainly composed of green algae, and the percentage of the phytoplankton community was 61.50% in WJD in July.

2. Phytoplankton Biomass

Figure 9.3 shows that the phytoplankton biomass of the four reservoirs varied widely, and its variation was significant and largest. In WJD, the phytoplankton biomasses were 48.47 and 1.99 mg L^{-1} in July and October, respectively. In DF, the phytoplankton biomasses were 8.08 and 5.82 mg L^{-1} in July and October, respectively. In PD, the phytoplankton biomasses were 11.17 and 4.66 mg L^{-1} in July and October, respectively. In BH, the phytoplankton biomasses were 11.06 and

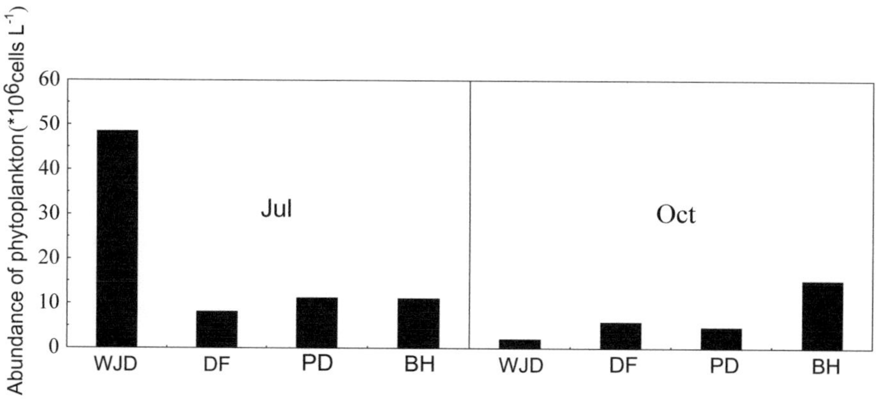

Fig. 9.3 Biomass of phytoplankton in four reservoirs

15.13 mg L^{-1} in July and October, respectively. BH was the only reservoir in which the biomass was higher in October than July.

In WJD in July, the dominant phytoplankton was chlorophyte, with a percentage that reached 71.26%; dinoflagellates were the second most abundant and reached 24.97% (Fig. 9.4). In October, there were mainly diatoms and chlorophyte, with percentages of 40.87 and 41.92%, respectively. In DF, the main phytoplankton type was diatoms, with percentages of 68.22 and 65.01% in July and October, respectively. In PD, the composition of the phytoplankton's biomass changed little between July and October; chlorophyte and diatoms had the highest proportions of 66.18 and 59.90% in July and 32.20 and 38.68% in October, respectively. In BH, the phytoplankton biomass was composed of dinoflagellates, chlorophyte, and cyanobacteria with percentages of 39.79, 33.02, and 22.98%, respectively; in October, the percentage of cyanobacteria and diatoms increased to 49.92 and

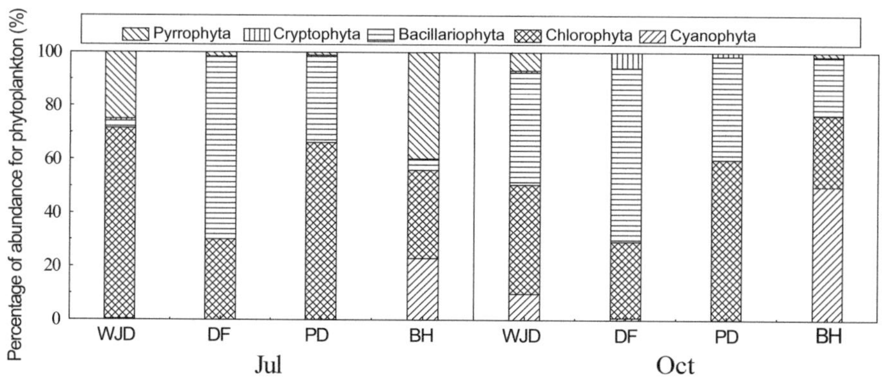

Fig. 9.4 Percentages of phytoplankton abundance

21.67%, respectively, whereas the percentage of dinoflagellates was significantly reduced by 1.44% (Wang et al. 2008, 2009).

9.1.2 Zooplankton

Zooplankton is an aquatic animal that is suspended in water and can usually only be seen through a microscope. In freshwater ecosystems, zooplankton mainly comprises protozoa, rotifers, cladocerans, and copepods, which play important roles in aquaculture, ecosystem structure and function, and studies on biological productivity.

As the primary consumer, zooplankton is the second trophic level in the classic aquatic food chain but is also key to the micro-food web in addition to the classic food web.

Zooplankton is a key level in the transfer and accumulation of mercury in the aquatic food chain, and mercury is transferred to the higher trophic level through its bioaccumulation in zooplankton.

1. Zooplankton in WJD, PD, DF and BH

Four reservoirs in Guizhou Province were surveyed in the winter of 2008. The largest group of zooplankton, 35 species, was recorded in WJD and only 14 species were found in DF. The dominant type of zooplankton was rotifers, which accounted

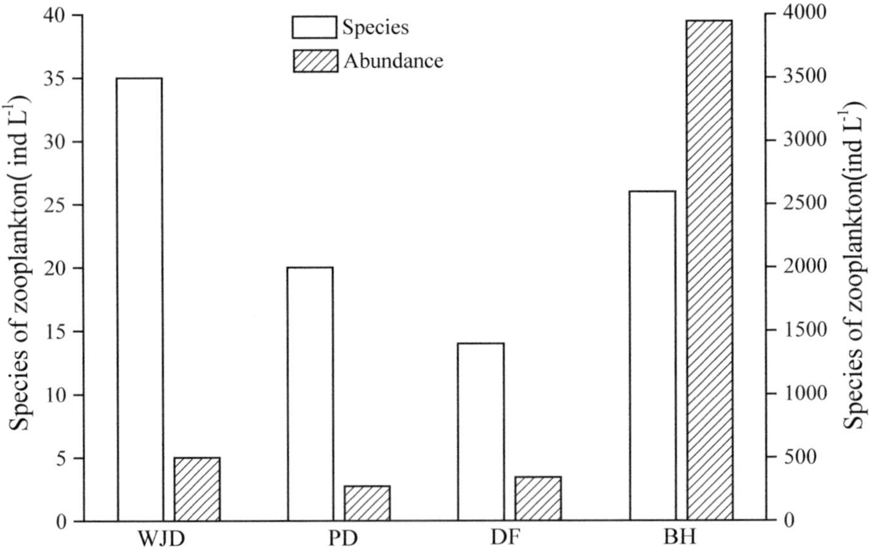

Fig. 9.5 Abundance and species of zooplankton (Sept. 2008)

for 70–89% of the zooplankton. Up to 31 species of rotifers were found in WJD, but only 11 species were found in DF (Fig. 9.5).

Of the four reservoirs, the abundance of zooplankton was highest in BH and lowest in PD at only 276 ind L^{-1} (Fig. 9.5). The dominant species of rotifers were similar in the four reservoirs, but their relative dominance was different. *Keratella rotifers* were the most dominant species in the four reservoirs (23.7–72.2%); the highest and lowest (7.8%) percentages of these rotifers were found in PD and DF, respectively (Fig. 9.6).

2. Plankton in Baihua Reservoir
(1) Plankton size

Four body sizes of plankton were investigated in BH: microplankton, microzooplankton and medium and large zooplankton. Large zooplankton (particle sizes > 610 μm) were mainly copepods (more than 70%); *N. schmackeri* were found most often, followed by *M. thermocyclopoides* and *T. taihokuensis*. Cladocera, including *D. brachyurum*, *D. cucullata*, *C. cornuta*, *B.longirostris*, and *B. deitersi*, accounted for more than 20% of the large zooplankton; the rest were nauplii, rotifers, and larger sized microalgae (less than 5%). Mesozooplankton (216–610 μm) mainly comprised the Cyclops, followed by copepodid larvae of Calanoida (approximately 54%); Cladocera was the second largest community (approximately 32%); approximately 10% of the community were nauplii and rotifers, and microalgae and other impurities accounted for less than 5%.

Small zooplankton (108–216 μm) mainly consisted of small- and medium-sized crustaceans, such as copepod larvae, *Daphnia trunks*, *textured Daphnia*, rotifers,

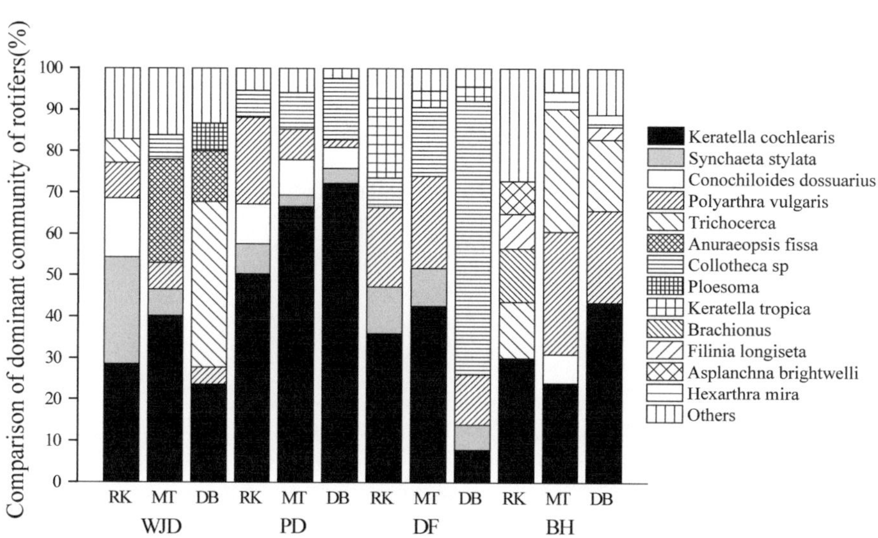

Fig. 9.6 Comparison of the dominant communities of rotifers (Sept. 2008)

and nauplii, which accounted for 73% of the community; phytoplankton accounted for approximately 27% of the community.

Approximately 90% of the microzooplankton (38–108 μm) were phytoplankton, and approximately 10% were nauplii and rotifers.

(2) Dynamics of rotifer community structure in Baihua Reservoir

The highest abundance of rotifers was found in BH in September; up to 3945 ind L^{-1} were found at the entrance of reservoir, more than the 2160 and 990 ind L^{-1} found at the wharf and dam, respectively. The abundance decreased in October to only 40 ind L^{-1} (Fig. 9.7). There was no significant difference in rotifer abundance among the three sites. The abundance of nauplii was highest in September, with abundances up to 550 and 600 ind L^{-1} at the Matou and the dam, respectively, which were higher than the inflow abundance (17 ind L^{-1}). The lowest abundance of 10 ind L^{-1} was found in October. There was no significant difference in nauplii abundance between the three sites (Fig. 9.8).

Keratella rotifers were the dominant species at the inflow of BH (means of 32.8% in 2008 and 28.8% in 2009), whereas *Polyarthra vulgaris* was the second most dominant species at both sites (means of 14.4 and 28.4%). *Polyarthra vulgaris* was the dominant species (mean of 38.9%) in the dam, and *Keratella rotifers* were the second most dominant species (mean of 29.7%). *Keratella rotifers* were the dominant species at the entrance upstream of the BH, and their numbers were generally more stable (Fig. 9.8); in the dock and dam, their numbers increased gradually (Figs. 9.9 and 9.10).

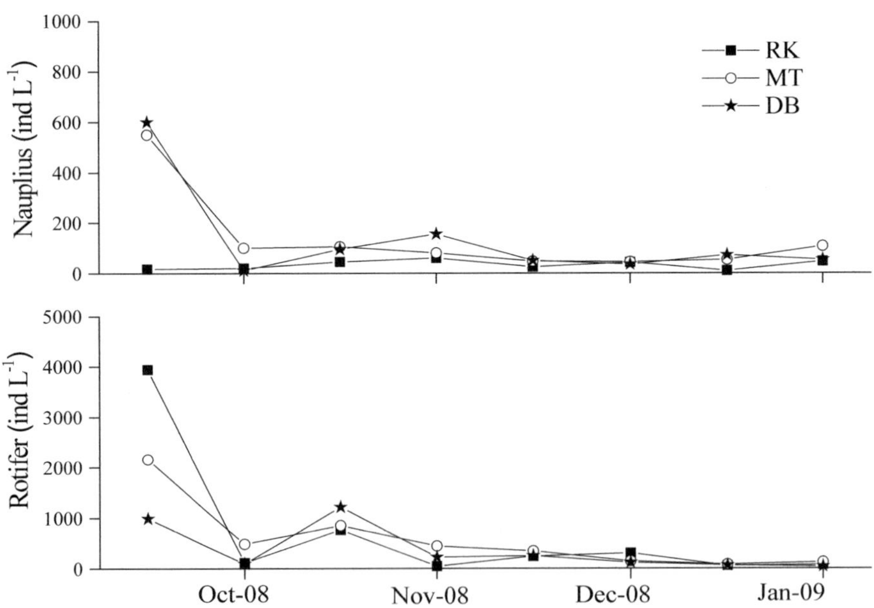

Fig. 9.7 Seasonal variations in rotifers and nauplii (Sept. 2008–Jan. 2009)

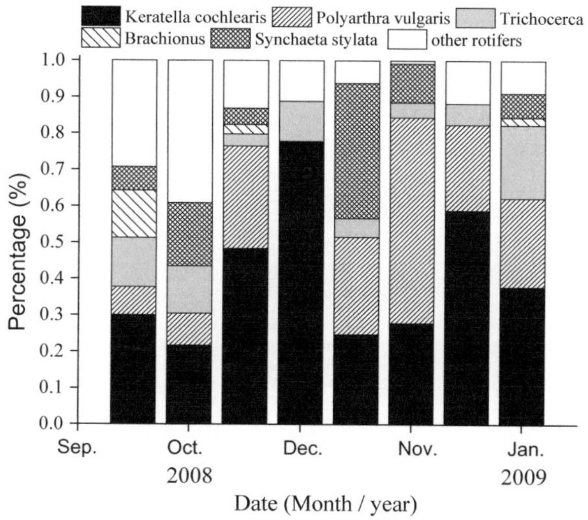

Fig. 9.8 Seasonal variation of the dominant rotifer community in the entrance of BH (Sept. 2008–Jan. 2009)

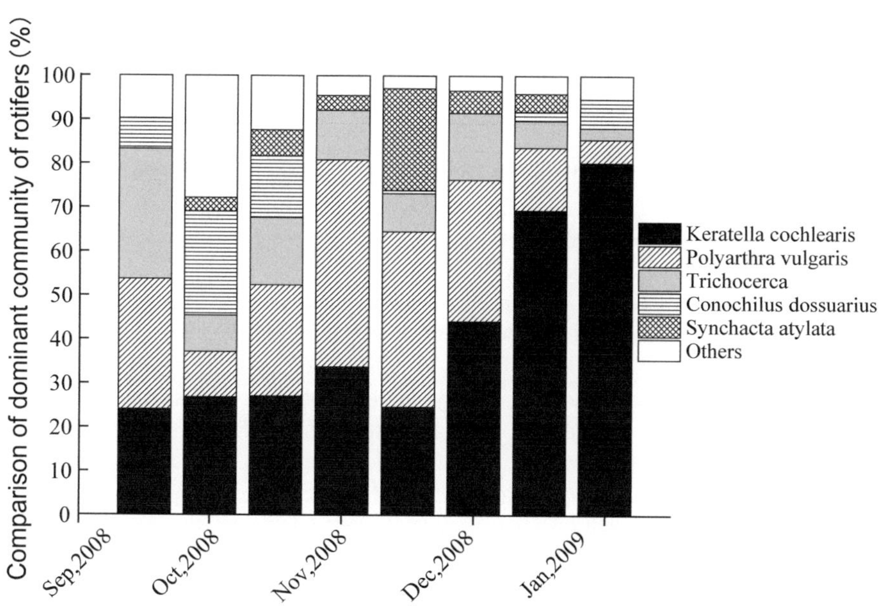

Fig. 9.9 Seasonal variation of the dominant rotifer community at the MT of BH (Sep. 2008–Jan. 2009)

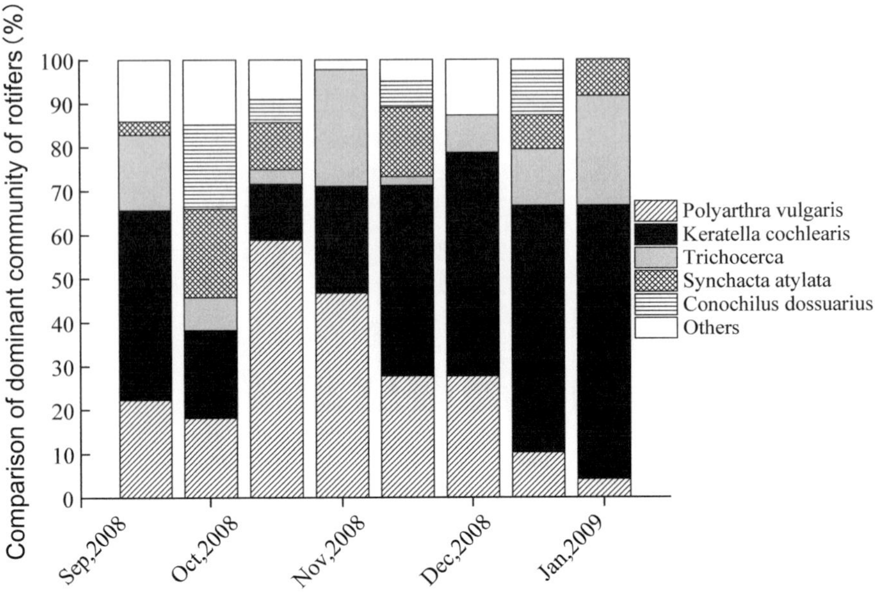

Fig. 9.10 Seasonal variation of the dominant rotifer community in DB of BH (Sep. 2008–Jan. 2009)

9.1.3 Fish

There were 133 types of fish in the Wujiang River Basin in Guizhou province. Twenty-six species belonged to endangered fish of the Upper Yangtze. Three types, Leptobotia elongata, Zong and rock carp, are included in the "China Red Data Book of Endangered Animals" (Zhang 2008).

In two field surveys conducted in 2007, 6 orders, 16 families, 62 genera, and 83 species were investigated in the Wujiang River in Guizhou. The recorded fish accounted for 63.3% of the species found. Fish of Order Cypriniformes (55 species) accounted for 66.3% of the total survey of fish species; Siluriformes were the second most abundant (17 species) and accounted for 20.5% of the species; and Perciformes (6 species) accounted for 7.2% of the species. Thirteen types of the collected fish belong to endangered species of the Upper Yangtze River, which account for 50% of all endemic fish ever recorded; two types are included in the "China Red Data Book of Endangered Animals", accounting for 66.7% of the historical record. Zones of the Wujiang River affected by the cascade hydropower stations may contain 131 species of fish, of which 29 species are endemic upstream of the Yangtze River.

Table 9.2 Fish species in HF and BH

No.	Name	Latin Name	HF	BH
1	Common carp	*Cyprinus carpio*	√	√
2	Crucian carp	*Carassius auratus*	√	√
3	Hooksnout carp	*Opsariichthys bidens*	√	√
4	Barble chub	*Spualiobarbus Curriculus*	√	
5	Black carp	*Mylopharyngodon piceus*	√	√
6	Grass carp	*Ctenopharyngodon idellus*	√	√
7	Redfinculter	*Culter alburnus*	√	
8	Topmouth culter	*Erythroculter ilishaeformis*	√	√
9	white bream	*Parabramis pekinensis*	√	
10	Richardson bream	*Magalobrame Tarminalis(Richardson)*	√	
11	Wuchang bream	*Bluntnose black bream*	√	
12	Silver carp	*Hypophthalmichthys molitrix*	√	√
13	Bighead carp	*Aristichthys nobilis*	√	√
14	Yellowcheek carp	*Hemibarbus maculates Bleeker*	√	
15	Slender topmouth gudgeon	*Pseudorasbora parva*	√	√
16	Arctic grayling	*Abbottina rivularis*	√	√
17	China-bitterling	*Acheilognathus barbatus Nichols*	√	√
18	Chinese bitterling	*Rhodeus sinensis*	√	√
19	Pond loach	*Oriental weatherfish*	√	√
20	Sinusoid	*Parasilutus asotus*	√	
21	Yellow catfish	*Pelteobagrus fulvidraco*	√	
22	Tilapia	*Oreochomis.sp*	√	
23	Red pacu	*Colossoma brachy pomum*	√	
24	Largefin longbarbel catfish	*Mystus macropterus*	√	
25	eel	*Monopterus albus*	√	
26	Small Galangal	*Pseudolaubuca sinensis*	√	
27	Sinilabeo discognathoides	*Sinilabeo rendahli*	√	
28	White semiknife carp	*Hemicculter Leuciclus (Basilewaky)*	√	√
29	Craspedacusta	*Craspedacusta sowerbyi*	√	
30	whitebait	*Hemisalanx prognathus Regan*		√

Fifty-seven species of fish were collected in field surveys conducted in 2007; nine types are endemic to the Upper Yangtze River; and the other 14 types were fish of economic importance, including S. sinensis, white snapper, petal knot fish, carp, catfish, large-mouth catfish, Yunnan light-lipped catfish, gray Schizothorax, S.kneri, flower hom fish, spring-water fish, doctor fish, grass carp, and silver carp. There are 15 types of fish in BH, including black carp, grass carp, bighead carp, crucian carp, and bream, and 29 types of fish in HF. Most of the species in BH were also found in HF Table 9.2.

9.2 Bioaccumulation and Transportation of Mercury in Food Chains

In aquatic ecosystems, mercury is transported and accumulated mainly through food chains. Generally, mercury levels are low in small and young aquatic organisms at the bottom of food chains. Mercury (mostly methylmercury) is absorbed by aquatic organisms via ingestion of food and is enriched as trophic levels increase. To understand the mechanisms of mercury and methylmercury enrichment in the food chain of the cascade reservoirs at different evolutionary stages, we collected the dominant aquatic species of different trophic levels in the reservoirs from upstream and downstream in the Wujiang River Basin.

During 2007 and 2012, a total of 454 plankton, fish and shellfish samples were collected from reservoirs of the Wujiang River Basin (Fig. 9.11). Because fishery resources and cage culture in each reservoir are different, samples collected from these reservoirs are not identical (see Table 9.3 for detailed sample information). In the laboratory, we determined fish length and body weight, and then the skinless dorsal muscle tissues were removed with clean scalpels and weighed (wet weight). After being weighed, muscle tissues stored in sealed plastic bags were freeze-dried. The dry weight of each muscle tissue was determined to calculate the wet/dry weight (W/D) ratio of each sample. They were then homogenized and kept dry until analyzed. We identified the age of fish by studying their scales. For those fishes without scales and some scaly fishes, we attempted to determine their ages by studying their otoliths (limited by equipment availability, this work is yet to be completed). Plankton was collected using plankton nets with different mesh sizes and were classified according to their particle sizes. Plankton samples were taken back to the laboratory and freeze-dried for determination of total mercury, methylmercury and stable carbon and nitrogen isotope compositions. We ground and sieved the freeze-dried plankton samples used for the determination of stable carbon and nitrogen isotopes with a 60 mesh nylon screen. We then kept the screened samples in sealed centrifuge tubes after wrapping them with aluminum foil and placed them in drying vessels prior to isotope determination.

Stable carbon and nitrogen ($\delta^{13}C$ and $\delta^{15}N$) isotopes composition was determined using a MAT-252 mass spectrometer and expressed as ‰ using $\delta^{13}C$ and $\delta^{15}N$ notations relative to the International Standard PVDB and the atmospheric nitrogen isotopic ratio. Total mercury was determined using the acid digestion-CVAFS analysis, and methylmercury was determined using the alkaline digestion-GC-CVAFS approach (Yan 2005). The total mercury concentrations in fishes of the Wujiang River Basin reservoirs was far lower than the MeHg concentration (0.5 mg kg^{-1}, wet weight) (Yan et al. 2010) specified as the pollutant safety limit of aquatic products in China, and the methylmercury concentration in the total mercury was below 50% on average (Yan et al. 2008a, b). Taking these factors into account, we only selected samples with higher total mercury for methylmercury analysis.

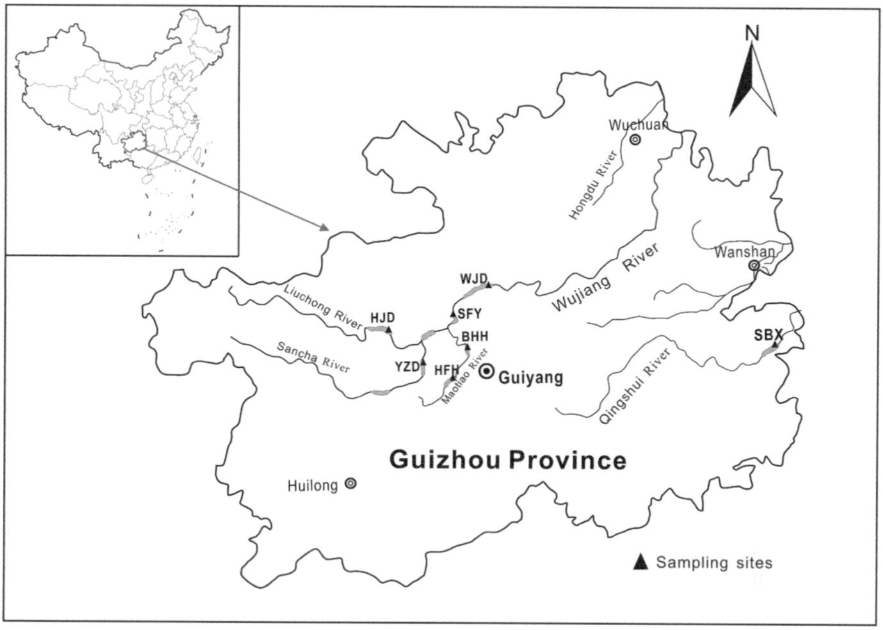

Fig. 9.11 Locations of fish sampling sites

9.2.1 Mercury Species in Plankton

Mercury transfer occurs mainly through feeding relationships in the aquatic food chain. The first trophic level includes phytoplankton and zooplankton, but the trophic level of some zooplankton is slightly higher than the trophic level of phytoplankton because they are omnivorous aquatic organisms; however, phytoplankton mainly obtains mercury from water by adsorption.

Different fish species have different feeding habits. For example, silver carp mainly feed on small-sized phytoplankton, but bighead carp mainly feed on large zooplankton; therefore, it is necessary to understand the characteristics of mercury distribution in different plankton. Plankton is too small to accurately pick out a single species and measure its total mercury and methylmercury for so many species of plankton. Therefore, based on microscopic observations, we have different mesh screening sizes to obtain different particle sizes of plankton and classify them. For particle sizes >610 μm, copepods composed more than 70% of the plankton; for particle sizes of 216–610 μm, 54% of the plankton was copepods' biological larvae; for particle sizes of 108–216 μm, the plankton was mainly small crustaceans, which accounted for the majority of the plankton at 73%; and 90% of the plankton of particle sizes 38–108 μm were phytoplankton.

The composition of phytoplankton species exhibits obvious seasonal change, and phytoplankton species composition is related to the nutrient status (Rojo 1998;

Table 9.3 Characteristics of the distribution of fish species from the reservoirs

Name	FH	Name of reservoirs						
		HJD	SFY	PD	YZD	WJD	HF	BH
		Length(cm) average ± SD(Median) Weight(kg) average ± SD(Median) Age n=						
PP	–	–	–	–	–	+	+	+
ZP	–	–	–	+	–	+	–	+
GC	HS	91.3 11500 3^+ $n = 1$	–	–	25.8 ± 3.5(25.0) 284 ± 107 (255) $2^- \sim 4^+$ $n = 14$	36.9 ± 6.9(35.8) 653 ± 391(568) $1^+ \sim 2^+$ $n = 6$	40.5 ± 16.3 (40.5) 888 ± 1008 (888) na $n = 2$	32.3 ± 2.7 (31.0) 690 ± 177 (630) $2 \sim 3^+$ $n = 5$
SF		–	–	–	–	24.8 ± 0.4(24.8) 298 ± 39(298) na $n = 2$	–	–
TP		–	–	–	–	–	19.5 ± 1.3(19.5) 145 ± 31(138) na $n = 4$	–
BB		–	–	–	–	28.9 ± 5.0(26.0) 400 ± 225(289) $1 \sim 2^+$ $n = 7$	–	–
TC	OH	17.3 ± 1.2(17.5) 100 ± 18(110)	17.3 ± 1.5(17.5) 42 ± 10(45)	–	–	–	–	13.5 ± 2.0 (13.0)

(continued)

Table 9.3 (continued)

Name	FH	Name of reservoirs						
		HJD	SFY	PD	YZD	WJD	HF	BH
		Length(cm) average ± SD(Median) Weight(kg) average ± SD(Median) Age n=						
		na n = 13	na n = 9					13 ± 3(12) $1 \sim 1^+$ n = 11
CU		18.2 ± 2.3(18.2) 210 ± 79(200) na n = 28	–	–	17.0 ± 2.5(17.5) 146 ± 67(150) $1 \sim 3^+$ n = 5	–	–	19.6 ± 2.8 (18.6) 224 ± 103 (200) $1 \sim 3$ n = 32
CC		34.1 ± 5.2(33.0) 1203 ± 614 (1000) $2^+ \sim 5^+$ n = 55	52.7 ± 14.6 (55.0) 2562 ± 1762 (2700) na n = 3	55.0 ± 7.8(60.5) 2700 ± 1096 (3475) na n = 2	25.6 ± 9.9(22.5) 391 ± 427(255) $1^+ \sim 3^+$ n = 24	27.3 ± 12.3 (22.3) 509 ± 791(155) $1^+ \sim 3^+$ n = 7	31.5 ± 9.3(31.5) 517 ± 381(455) na n = 24	737 ± 696 (450) 41.3 ± 20.6 (38.0) $1 \sim 4$ n = 26
SC	FH	–	–	–	–	50.0 ± 0.0(50.0) 1393 ± 81 (1393) 2^+ n = 2	–	39.4 ± 8.3 (40.0) 950 ± 646 (800) $1^- \sim 4^+$ n = 17
BC		–	–	–	20.7 ± 2.3(21.0) 110 ± 30(115)	48.8 ± 22.5 (39.0)	57.0 ± 35.4 (57.0)	40.4 ± 9.9 (40.3)

(continued)

Table 9.3 (continued)

Name	FH	Name of reservoirs						
		HJD	SFY	PD	YZD	WJD	HF	BH
		Length(cm) average ± SD(Median) Weight(kg) average ± SD(Median) Age n=						
					na $n = 6$	2632 ± 3235 (585) $1 \sim 1^{+}$ $n = 9$	4100 ± 5141 (4100) na $n = 2$	1317 ± 700 (1250) $2^{+} \sim 3^{+}$ $n = 17$
AB	CH	–	–	–	+	–	+	+
YHC		–	15.9 ± 1.3(16.0) 38 ± 15(40) na $n = 4$	–	–	32.0 ± 2.8(32.0) 280 ± 42(280) na $n = 2$	–	18.4 ± 5.1 (19.0) 237 ± 77(220) na $n = 6$
CF		70.5 ± 17.6 (69.5) 9128 ± 6777 (7920) na $n = 5$	77 3700 na $n = 1$	–	41.3 ± 20.6(38.0) 737 ± 696(450)na $n = 6$	48.4 ± 17.7 (443.0) 1377 ± 699 (640)na $n = 7$	34.6 ± 9.3(32.0) 301 ± 270(205) na $n = 5$	39.8 ± 16.9 (45.0) 700 ± 527 (650) na $n = 3$
VA		29.1 ± 3.7(28.8) 555 ± 120(580) na $n = 4$	–	–	–	–	–	–
BBS		30.3 ± 4.3(29.7) 473 ± 193(430)	–	–	–	28.0 ± 1.4(28.0) 275 ± 92(275)	–	–

(continued)

Table 9.3 (continued)

Name	FH	Name of reservoirs						
		HJD	SFY	PD	YZD	WJD	HF	BH
		Length(cm) average ± SD(Median) Weight(kg) average ± SD(Median) Age n=						
		na $n = 9$				na $n = 2$		
TC		15.3 ± 2.1(15.3) 48 ± 27(53) na $N = 9$	–	–	–	15.7 ± 2.8(15.5) 36 ± 21(30) na $n = 7$	19 80 na $n = 1$	–
VA		–	–	–	–	80 4620 na $n = 1$	21.8 ± 5.4(24.0) 107 ± 54(130) na $n = 9$	–
SH	BI	+	–	–	+	–	+	+
SN		+	–	–	–	–	+	+
EE		+	–	–	–	–	+	+

GC Grass carp, *CR Ctenopharyngodon idellus, CU* crucian carp, *CC* Common carp, *Cyprinus carpio*
SC Silver carp, *Hypophthalmichthys molitrix*, *BC* Bighead carp, *Aristichthys novilis*
CF catfish, *Parasilurus asotus*, *SC* Silver carp, *Carassius auratus*
Chinese hooksnout carp, *O.bidens Gunthe, SF* Sunfish, *Mola*
BB Bluntsnout bream, Megalobrama amblycephala, *TP* Tilapia, Mossabica tilapia
YHC Yellow-headed catfish Pseudobagrus *fulvidraco (Richardson)*
BL Bleeker, *Erythroculter ilishaeformis*, *BBS* Black bass, *Micropterus salmonides*
SH Shrimp, *SN* Snail, *AB* Abbottina, *VA* varicorhinus (Onychostoma) lini Wu, *TC* Topmouth culter
PP phytoplankton, *ZP*zooplankton, *BI* benthic
HJD Hongjiadu, *YZD* Yinzidu Reservoir, *SFY* Suofengying Reservoir, *PD* Puding Reservoir
HF Hongfeng Reservoir, *BH* Baihua Reservoir, *WJD* Wujiangdu Reservoir
FH feeding habits, *HS* herbivores, *CH* carnivorous, *OH* omnivorous

Table 9.4 Descriptive statistics of THg and MeHg distribution in plankton from different reservoirs

Name	Size	MeHg (ng g^{-1})		THg (ng g^{-1})	
		Range	Mean	Range	Mean
WJD	>38	10.41−16.32	12.82 ± 3.10	184.81−311.23	242.06 ± 64.05
	>150	29.28−91.18	58.62 ± 31.07	120.32−192.68	157.23 ± 36.20
PD	>38	19.40−37.09	27.01 ± 9.10	118.93−232.94	188.89 ± 61.26
	>150	23.93−25.18	24.67 ± 0.65	42.95−77.50	60.90 ± 17.32
DF	>38	13.70−44.92	33.00 ± 16.87	50.69−219.96	147.38 ± 87.18
	>150	15.21−35.69	28.01 ± 11.16	26.82−76.83	55.36 ± 25.75
BH	>38	11.79−63.45	28.98 ± 16.30	51.06−1391.57	315.62 ± 451.31
	>150	10.16−46.54	27.63 ± 13.46	50.12−334.11	114.06 ± 98.93

Reynolds et al. 2000). In this study involving the BH, WJD, DF and PD, there were more than 70 different types of phytoplankton. We ordered them in the following manner according to their abundances: *cyanobacteria, chlorophyta, bacillariophyta, pyrrophyta, euglenophyta* and *chrysophyta* species. *Cyanophyta, chlorophyta* and *bacillariophyta* accounted for more than 70% of the total.

Rotifers were the most common zooplankton species found in the WJD, DF, PD, and BH, accounting for 70–89% of the total number of planktonic animal species. WJD contained the most rotifer species (31), whereas DF contained the fewest rotifer species (11).

1. THg and MeHg in plankton from WJD, PD, DF and BH

The THg in seston (>38 μm) (254 ± 320 ng g^{-1} dry weight) was significantly higher than in zooplankton (>150 μm) (103 ± 79 ng g^{-1} dry weight) ($t = 2.4$, $p < 0.05$), and THg in the plankton was highest at the entrance to the reservoir (Fig. 9.12). The THg in the two types of plankton are significantly positively correlated ($R^2 = 0.81$, $p < 0.0001$) (Fig. 9.13).

MeHg in seston (>38 μm) was slightly lower than in plankton (>150 μm) (26.6 ± 14.6 ng g^{-1}) in all the reservoirs except WJD. MeHg in small crustacean zooplankton (>150 μm) was highest in WJD and significantly higher than in the plankton dominated by seston (>38 μm) ($t = -2.54$, $p < 0.05$) (Fig. 9.14). MeHg in small-sized plankton (>38 μm) was not related to THg, and MeHg in crustacean zooplankton (>150 μm) had a linear correlation with THg ($p < 0.05$) (Fig. 9.15), which is similar to the phenomenon observed in fish.

Correlations of physical and chemical factors with the distribution of THg and MeHg in plankton are shown in Table 9.5 the results indicated that THg in plankton of different sizes was significantly correlated and that the accumulation of all levels of inorganic mercury in plankton were similar no matter which lakes were sampled or what the conditions were. Separate analyses of pH values and THg for different

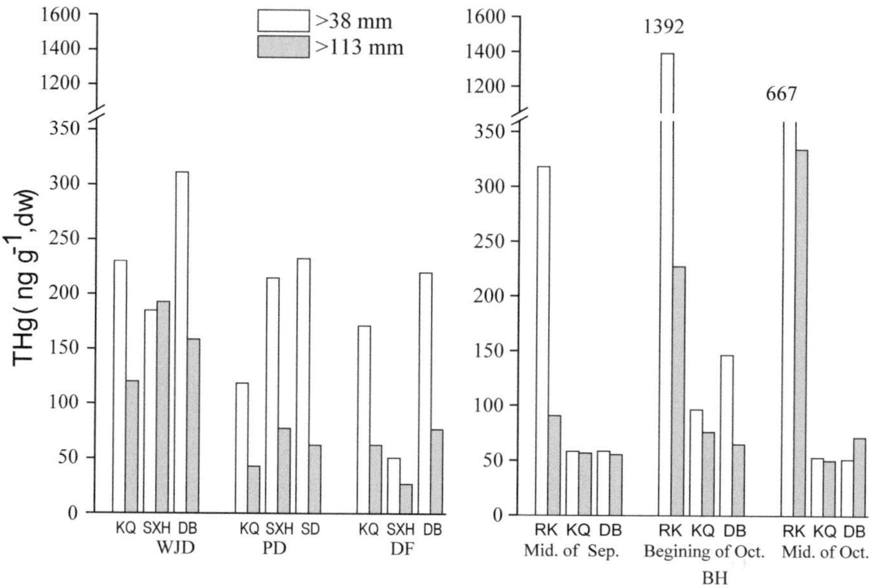

Fig. 9.12 THg in plankton from the four reservoirs

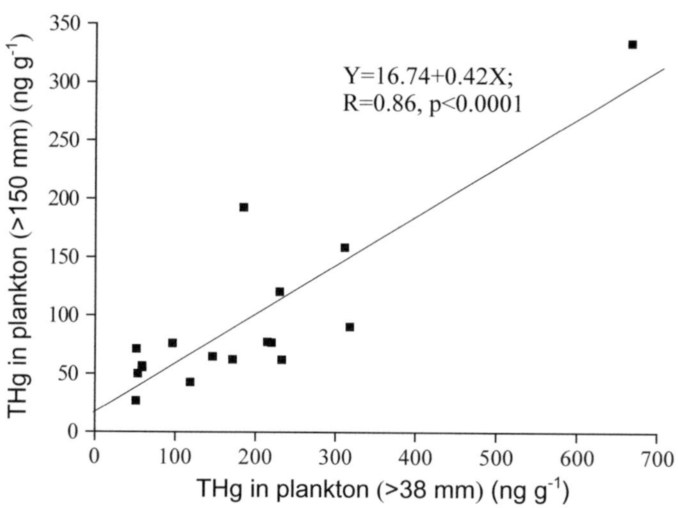

Fig. 9.13 Regression analysis of total mercury in plankton

particle sizes of plankton found that in lakes or reservoirs under natural pH conditions (pH 7.0–9.2), planktonic THg and pH are significantly negatively correlated ($p < 0.01$) and that THg decreases with increasing pH (Fig. 9.16).

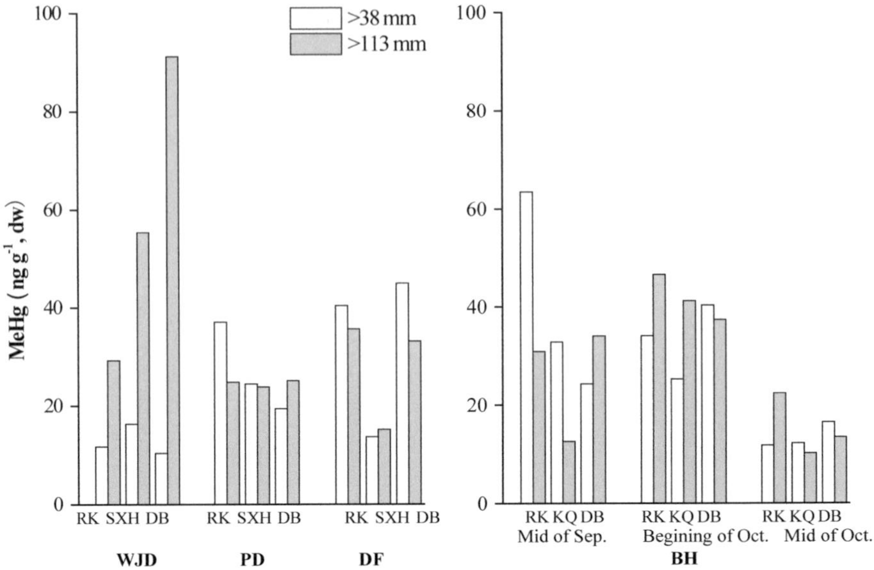

Fig. 9.14 MeHg in plankton from the four reservoirs

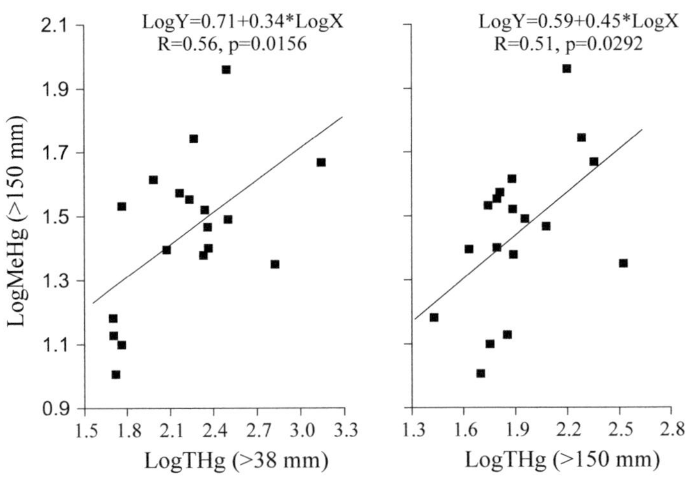

Fig. 9.15 Regression analysis of THg and MeHg in plankton

2. Seasonal variation of THg and MeHg in plankton from BH

THg and MeHg in plankton were investigated according to the four particle sizes of plankton in three sampling sites in BH (RK, DB, and MT). Their seasonal variations are shown in Figs. 9.17 and 9.18. Variance analysis showed that THg in

Table 9.5 THg and MeHg in plankton and their correlation with physicochemical factors

	T°C	pH	SD	Chl.*a*	MeHg(1)	MeHg(2)	THg(1)	THg(2)
T°C	1.00							
pH	0.65**	1.00						
SD	0.30	−0.11	1.00					
Chl*a*	−0.35	−0.04	−0.55*	1.00				
MeHg1	−0.01	0.17	−0.05	−0.05	1.00			
MeHg2	0.40	−0.49	−0.03	−0.23	−0.05	1.00		
THg1	−0.22	−0.68**	0.09	0.09	0.09	0.27	1.00	
THg2	−0.11	−0.69**	0.06	0.32	−0.25	0.36	0.72**	1.00

Note MeHg(1) and MeHg(2) represent, respectively, >38 μm and >150 μm MeHg in plankton; THg(1) and THg(2) represent, respectively, >38 μm and >150 μm THg in plankton;
*Significant correlation ($p < 0.05$)
**Extremely significant correlation ($p < 0.01$)

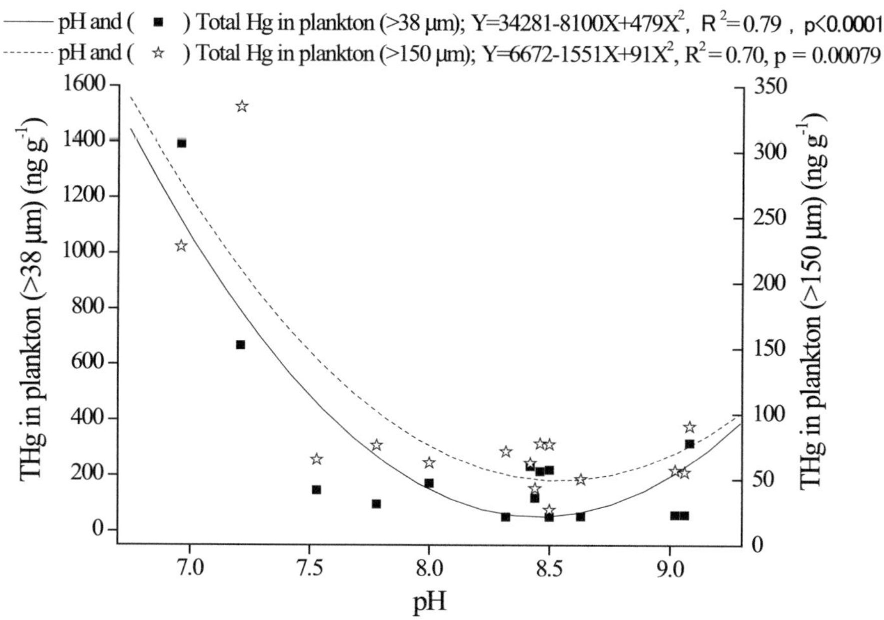

Fig. 9.16 Regression analysis of THg in plankton with pH. *Note* pH and (Black square box) THg in plankton (>38 mm); Y = 34281−8100X + 479X^2, R^2 = 0.79, $p < 0.0001$ pH and (Star) THg in plankton (>150 mm); Y = 6672−1551X + 91X^2, R^2 = 0.70, $p = 0.00079$

plankton of 38–108 μm was significantly higher than in the other three sizes of plankton ($p < 0.01$), and MeHg was significantly different at the three sites ($p < 0.05$), possibly because different hydrodynamic conditions led to significant differences in planktonic mercury levels between the three sites.

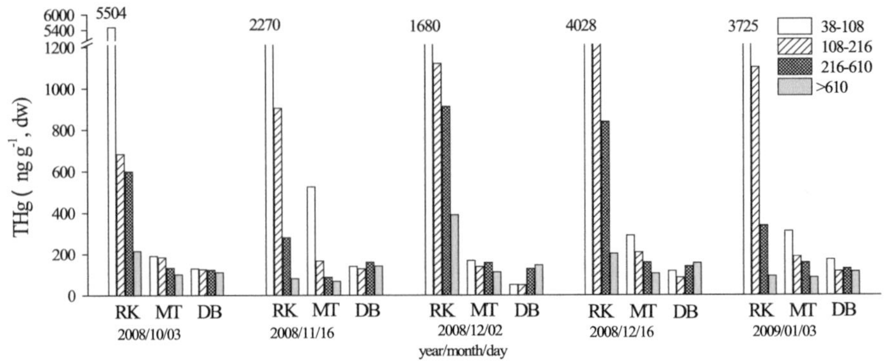

Fig. 9.17 Seasonal variation of THg in plankton from BH (redrawn from Wang et al. 2011, with permission from the alliance of crop, soil, and environmental science societies). *Note* RK–inflow; MT–middle stream; DB–downstream at dam

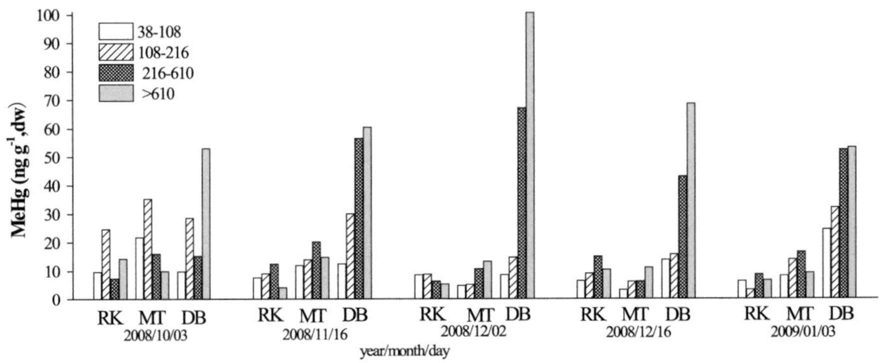

Fig. 9.18 Seasonal variation of MeHg in plankton from BH (redrawn from Wang et al. 2011, with permission from the alliance of crop, soil, and environmental science societies). *Note* RK—inflow; MT—middle stream; DB—downstream at dam

The ratios of MeHg to THg (%MeHg) for the different particle sizes of plankton differed significantly between the three sites. Generally, %MeHg was higher in large zooplankton than in the microscopic plankton and was highest at DB.

THg in plankton of 38–108 μm (1287 ± 1776 ng g^{-1}) was significantly correlated with THg in plankton of 108–216 μm (482 ± 574 ng g^{-1}) ($p < 0.01$). The former was significantly higher than the latter ($t = 2.26$, $p < 0.05$), which indicated that phytoplankton (the major component of plankton communities) had the maximum adsorption capacity for THg and that crustacean plankton mainly uptake inorganic mercury through food. Therefore, the accumulation of mercury was lower in crustacean plankton. MeHg in plankton of 108–216 μm (16.69 ± 10.66 ng g^{-1}) was significantly correlated with plankton of 38–108 μm (10.51 ± 5.86 ng g^{-1}), and the former was lower than the latter ($t = -3.5$, $p < 0.01$). THg and MeHg in

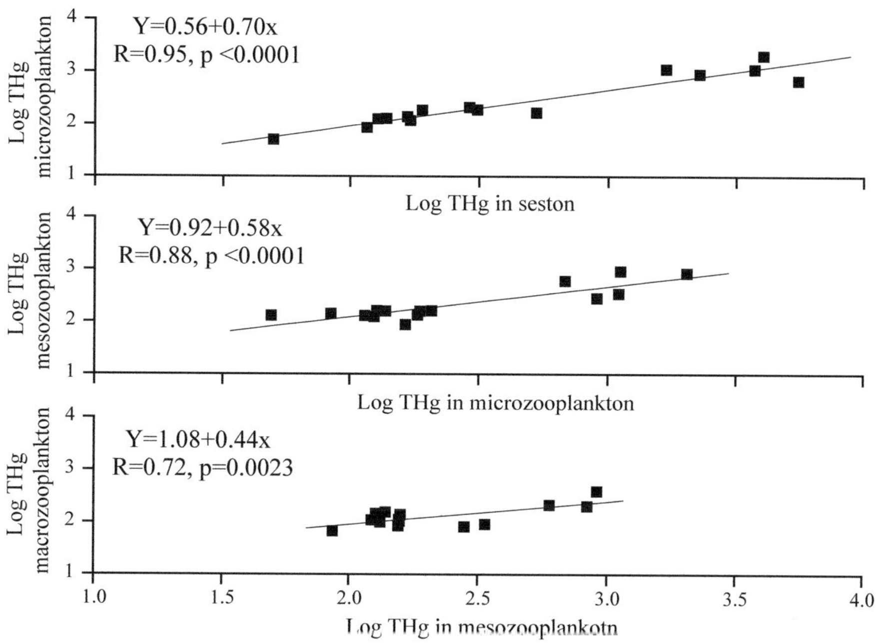

Fig. 9.19 Linear correlation of THg in plankton of different sizes

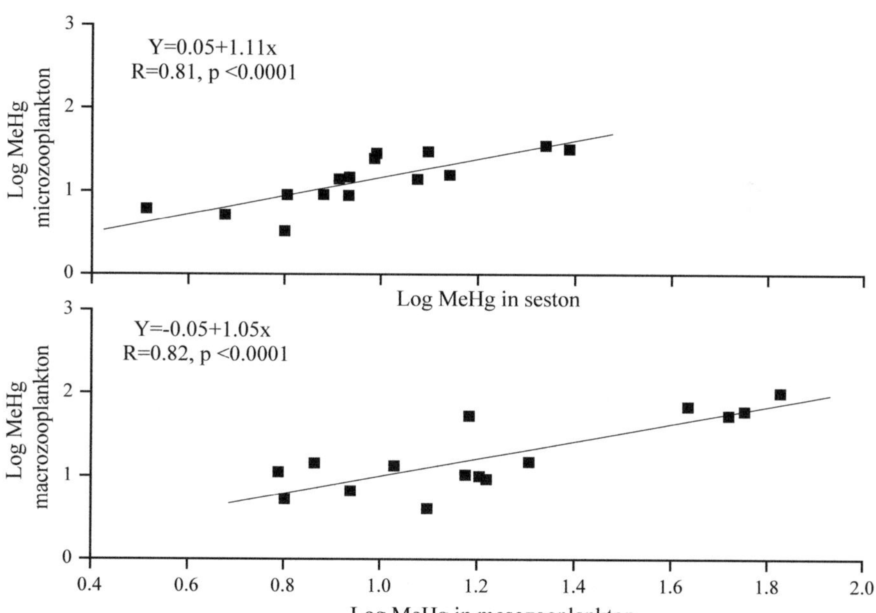

Fig. 9.20 Linear regression analysis of MeHg in low and high trophic level plankton in BH

plankton of adjacent particle sizes were significantly correlated (Figs. 9.19 and 9.20).

THg in plankton at the RK and MT sites in BH increased with decreasing particle size, but there was no obvious trend at DB (Fig. 9.21). MeHg at the RK site was not significantly different from the MeHg of the whole reservoir area but increased at DB with increasing particle size (Fig. 9.22); %MeHg increased with increasing particle size at three sites (Fig. 9.23). THg in plankton at RK and the middle of the reservoir was negatively correlated with particle size, which did not occur at DB. MeHg in plankton at DB had a positive linear correlation with particle size (Fig. 9.24). The ratio of MeHg to THg was positively correlated with particle size at three sites (Fig. 9.25).

Statistical analysis showed that the concentration of chlorophyll a was significantly negatively correlated with the total mercury content of plankton of 38–108 μm ($R = 0.55$, $P < 0.05$), indicating that eutrophication may have biologically diluted the mercury. The plankton of 38–108 μm was mainly composed of filamentous algae, alga, and microcystis species; therefore, the chlorophyll could reflect their density.

Chen and Folt (2005) found that the density of phytoplankton was negatively correlated with mercury bioaccumulation in a biota. Therefore, the higher the density, the lower the concentration of Hg, which is called the "growth dilution effect" (Sunda and Huntsman 1998). This effect also applies to zooplankton (Chen and Folt 2005). Thus, eutrophication in BH may lead to low levels of mercury in fish.

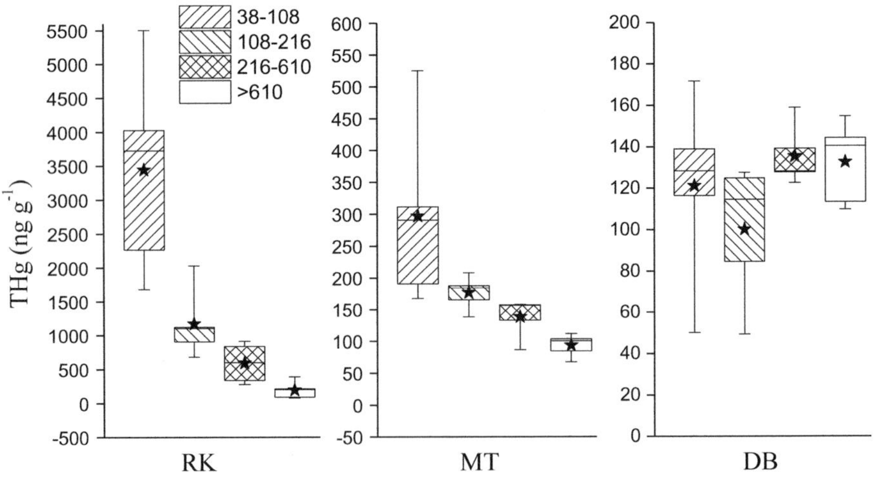

Fig. 9.21 Variation of THg in different size plankton from BH (redrawn from Wang et al. 2011, with permission from the alliance of crop, soil, and environmental science societies)

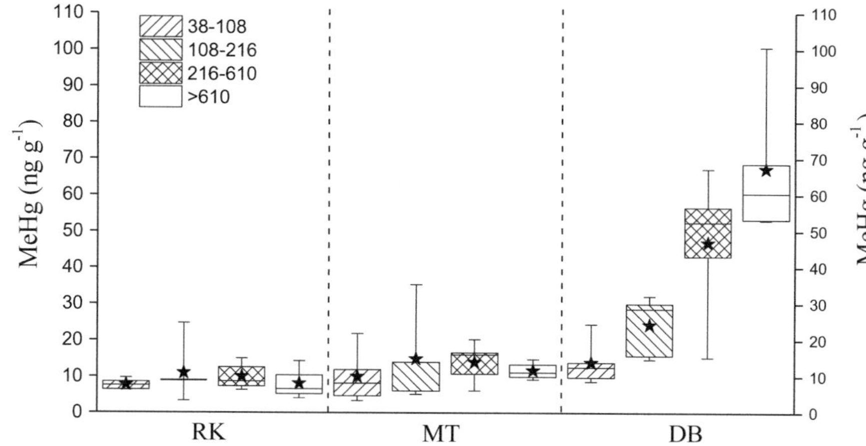

Fig. 9.22 Variation of MeHg in different size plankton from BH (redrawn from Wang et al. 2011, with permission from the alliance of crop, soil, and environmental science societies)

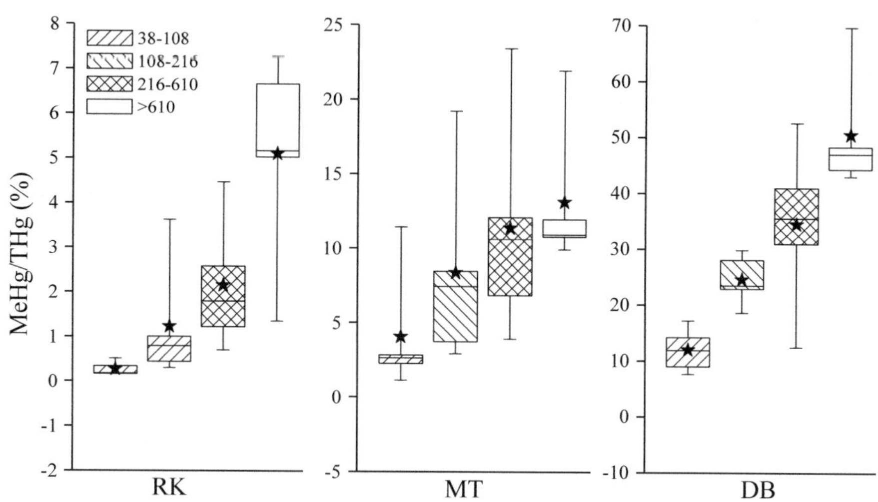

Fig. 9.23 Variation of % MeHg/THg in different size plankton from BH (redrawn from Wang et al. 2011, with permission from the alliance of crop, soil, and environmental science societies)

3. Characteristics of mercury bioaccumulation in plankton

DHg and DMeHg in the water of WJD were 6.8 ± 3.2 ng L^{-1} and 0.44 ± 0.19 ng L^{-1}, respectively (Jiang 2005). THg was 242.06 ± 64.05 ng g^{-1} for plankton >38 μm and 157.23 ± 36.20 ng g^{-1} for plankton >150 μm. MeHg was 12.82 ± 3.10 ng g^{-1} for plankton >38 μm and 58.62 ± 31.07 ng g^{-1} for plankton >150 μm. We found that THg and MeHg accumulated 3.5×10^4 and

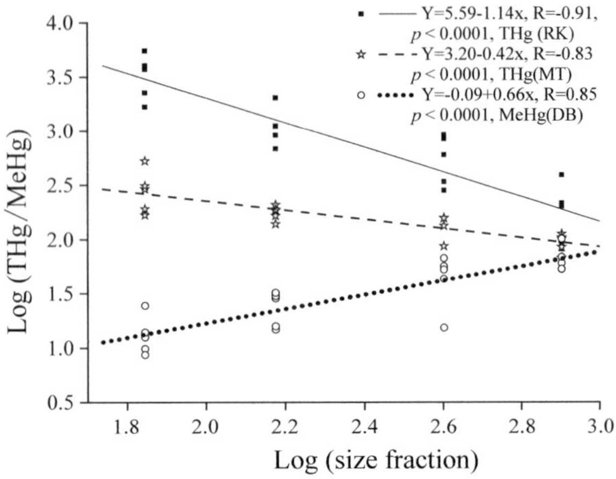

Fig. 9.24 Linear regression analysis of THg and particle size (RK and MT), MeHg and particle size (DB) in BH (redrawn from Wang et al. 2011, with permission from the alliance of crop, soil, and environmental science societies)

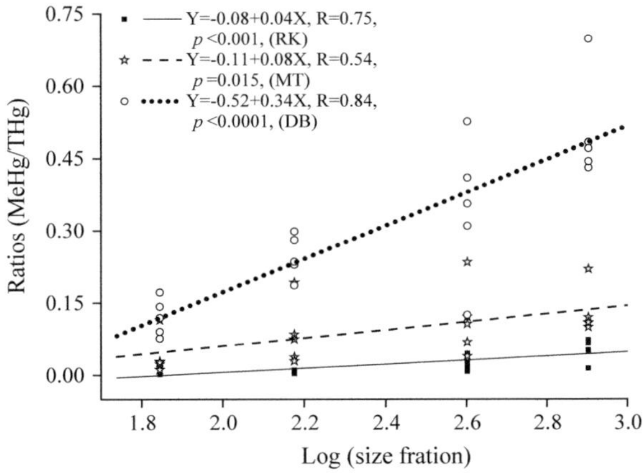

Fig. 9.25 Regression analysis of %MeHg variation with particle size in plankton in BH (redrawn from Wang et al. 2011, with permission from the alliance of crop, soil, and environmental science societies)

2.91×10^4 times, respectively, in small plankton(>38 μm) from the water. THg accumulation was not different for small (>38 μm) and larger (>150 μm) plankton, but the accumulation of MeHg was 4.57 times greater in larger plankton than in small plankton.

DHg and DMeHg in the water of DF were 5.5 ± 2.3 and 0.46 ± 0.20 ng L^{-1} respectively (Jiang 2005). THg was 147.38 ± 87.18 ng g^{-1} for plankton >38 μm and 55.36 ± 25.75 ng g^{-1} for plankton >150 μm. MeHg was 33.00 ± 16.87 ng g^{-1} for plankton >38 μm and 28.01 ± 11.16 ng g^{-1} for plankton >150 μm. We found that THg and MeHg accumulated 2.68×10^4 and 7.17×10^4 times from water, respectively, in small plankton (>38 μm). The accumulations of THg and MeHg did not differ between small plankton (>38 μm) and larger plankton (>150 μm).

DHg and DMeHg in the water of BH was 10.1 ng L^{-1} (2.0–35.1 ng L^{-1}) and 0.71 ng L^{-1} (0.04–2.79 ng L^{-1}), respectively (Yan 2005). THg was

315.62 ± 451.31 ng/g for plankton >38 μm and 114.06 ± 98.93 ng g^{-1} for plankton >150 μm. MeHg was 28.98 ± 16.30 ng g^{-1} for plankton >38 μm and 27.63 ± 13.46 ng g^{-1} for plankton >150 μm. We found that THg and MeHg accumulated 3.12×10^4 and 4.08×10^4 times from water, respectively, in small plankton (>38 μm). THg and MeHg did not accumulate to different extents between small plankton (>38 μm) and larger plankton (>150 μm).

For the BH sampling sites RK, MT, and DB, the DHg and DMeHg were 23.12 and 0.34, 10.12 and 0.71, and 7.5 and 0.22 ng L^{-1}, respectively (Yan 2005). THg and MeHg accumulated 1.49×10^4 and 2.27×10^4 times, respectively, in plankton of 38–108 μm from water at RK. THg and MeHg accumulated 1.41×10^4 and 2.94×10^4 times, respectively, in plankton (38–108 μm) from water at MT. THg and MeHg accumulated 6.29×10^4 and 1.61×10^4 times, respectively, in plankton (38–108 μm) from water at DB. However, MeHg did not accumulate from phytoplankton to zooplankton, and THg decreased in plankton with increasing plankton size at RK and MT. MeHg accumulated 1.75 times more in the plankton of 108–216 μm than in plankton of 38–108 μm, 1.94 times more in the plankton of 216–610 μm than in plankton of 108–216 μm, 1.43 times more in the plankton of >610 μm than in plankton of 216–610 μm plankton, respectively. The biomagnification factor (BMF) in the food chain of BH is shown in Fig. 9.26.

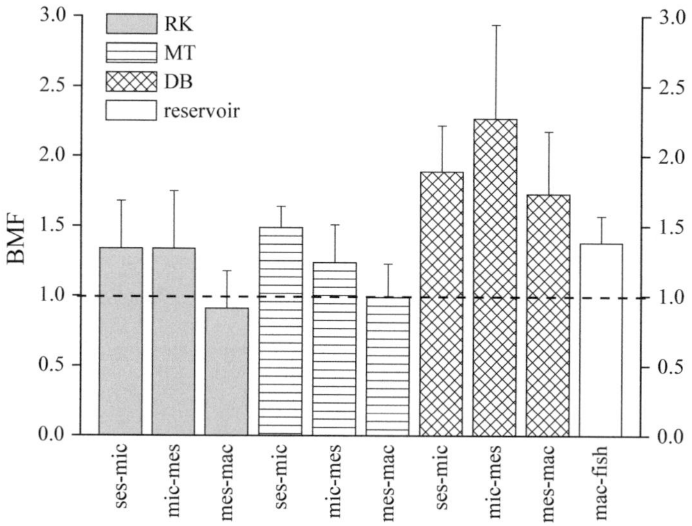

Fig. 9.26 Biomagnification factors at the different trophic levels of plankton (redrawn from Wang et al. 2011, with permission from the alliance of crop, soil, and environmental science societies). *Note* seston–ses, microzooplankton–mic, mesozooplankton–mes, macrozooplankton–mac and fish (*Aristichthys nobilis*, data from Yan et al. 2008a, b)

9.2.2 Mercury Species in Fish

1. THg and MeHg concentrations

THg levels in the 454 fish and shellfish samples collected from the seven reservoirs in the Wujiang River Basin were all below 400 ng g^{-1}, and most samples had a mercury level less than 50 ng g^{-1} (Fig. 9.27). BH was seriously contaminated with Hg from the GOCP, resulting in higher mercury concentrations in its water and sediments than in other reservoirs, but THg in fish and shellfish was not significantly higher than that in the other reservoirs. The highest THg levels (>100 ng g^{-1}) were found in fish from SFY, which was built in 2005 for water storage, followed by fish from HJD, HF, BH, WJD, PD, and YZD. This does not seem to be fully consistent with the age and evolution of the reservoirs because new reservoirs do not show higher mercury levels than the old ones. The main reason is because the sizes, ages, species, and quantities of fish collected from each reservoir were quite different.

Feeding habit is a major factor affecting mercury bioaccumulation in fish and shellfish. When ordered by feeding habit, we found that the median THg in fish was highest in carnivorous fish, followed by zooplankton-feeding fish, omnivorous fish, phytoplankton-feeding fish, benthic fish, and herbivorous fish (Fig. 9.28). However, because fish and shellfish with different feeding habits also differ in age and size, the THg in some samples do not follow this order. For example, in Wuchang bream, the largest herbivorous fish, THg is much higher than in most omnivorous, and even carnivorous, fish. In BH, zooplankton-feeding, bighead carps had the highest THg, followed by the carnivorous fish, the omnivorous fish, the benthic fish, the phytoplankton-feeding fish, and the herbivorous fish. Obviously, THg was lowest in herbivorous fish in all of the reservoirs studied, except that a relatively high level was observed in a bream from WJD (Fig. 9.30).

MeHg is the main mercury species in fish, normally accounting for 90% of THg. In this study, the ratio of MeHg to THg in fish of the Wujiang River Basin is less

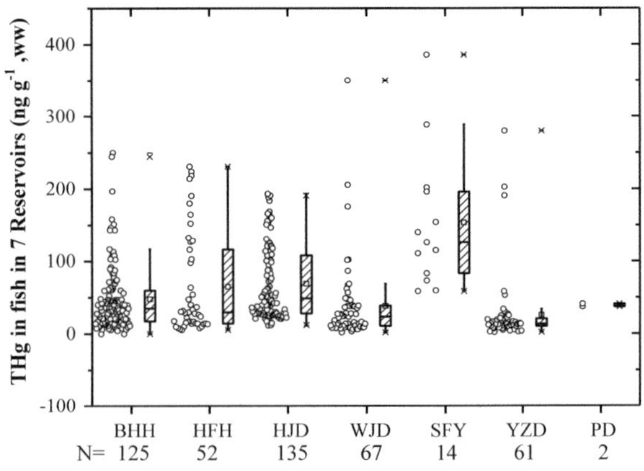

Fig. 9.27 Distribution of THg in fish and shellfish from the different reservoirs

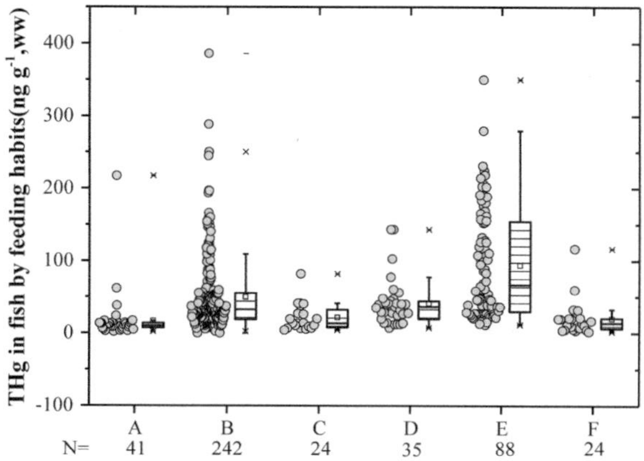

Fig. 9.28 Characteristics of THg distribution in fish and other biota. *Note* (A) Herbivores; (B) Omnivorous; (C) phytoplankton feeder; (D) zooplankton feeder; (E) carnivorous; (F) Bethic. *Note* Hongjiadu reservoir—HJD; Yinzidu reservoir—YZD; Suofengying reservoir—SFY; Puding reservoir—PD; Hongfenghu reservoir—HF; Baihuahu reservoir—BH; Wujiangdu reservoir—WJD

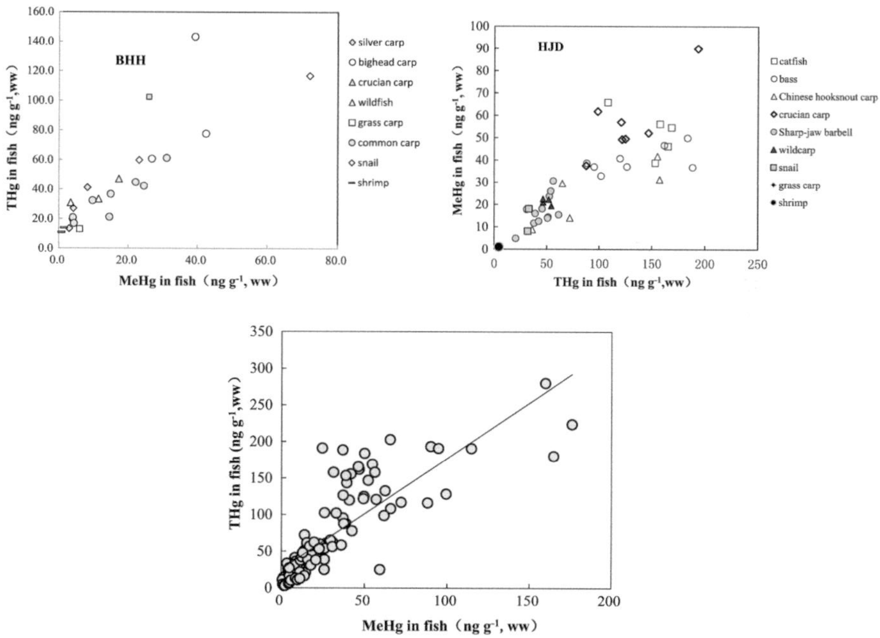

Fig. 9.29 Correlation of THg with MeHg in fish (redrawn from Yan et al. 2008a, b, with permission from Chinese Journal of Ecology). *Note* Hongjiadu reservoir—HJD; Baihua reservoir—BH

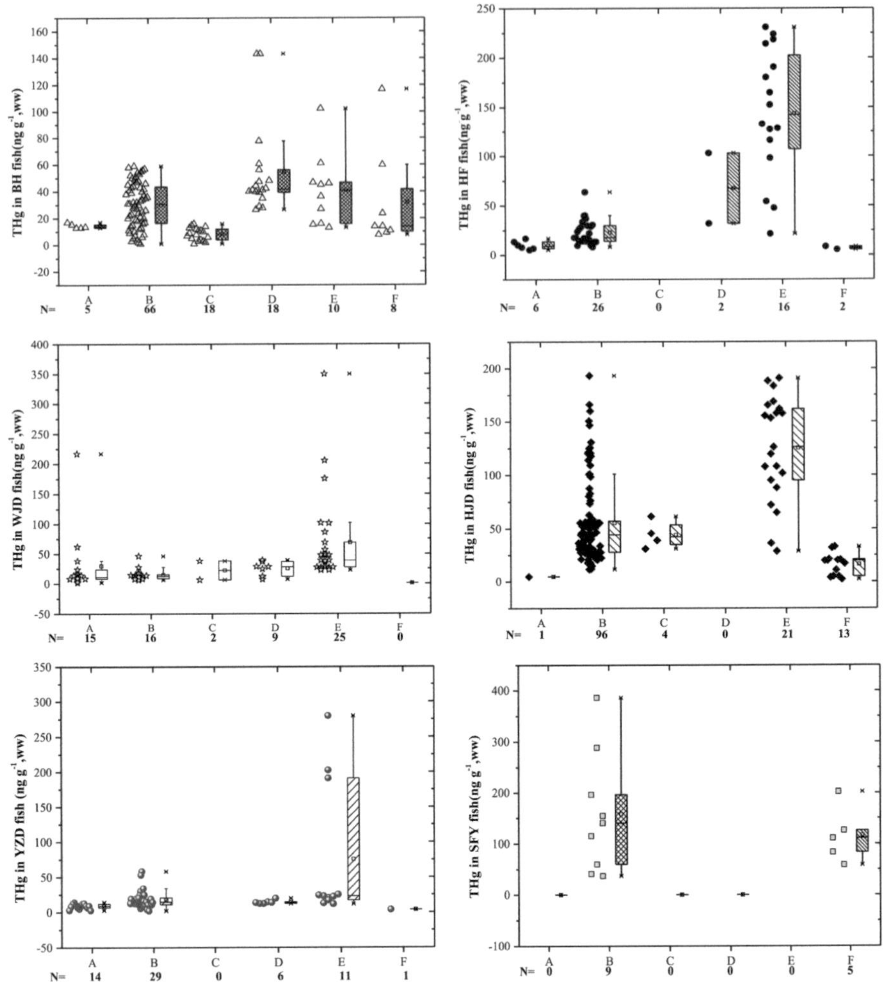

Fig. 9.30 THg concentration and distribution in fish and shellfish (redrawn from Yan et al. 2008a, b, with permission from Chinese Journal of Ecology; redrawn from Yan et al. 2010, with permission from Chinese Journal of Ecology). (A) Herbivores; (B) omnivorous; (C) phytoplankton-feeding; (D) zooplankton-feeding; (E) carnivores-feeding; (F) Benthic

than 50% in most samples. In a small proportion of carnivorous fish, %MeHg reached 60–70%. In some herbivorous fish, %MeHg is even less than 30% (Yan et al. 2008a, b; Jiang 2005; Yao 2010). The correlation between MeHg and THg in fish is always positive (Fig. 9.29) for the following reasons: first, most studies were conducted on carnivorous wild fish at high trophic positions, mostly in lakes of Northern European and North American countries, so %MeHg in fish was high; and second, the metabolism and accumulation of MeHg are slow processes in fish, but overfishing results in a shorter food chain; therefore, mercury accumulation is low in the Wujiang River Basin. There was a significant positive correlation between MeHg and THg in fish in our study (Fig. 9.29). MeHg was also higher in fish with high levels of THg, even if the ratio of MeHg to THg was low.

2. Factors influencing the bioaccumulation of mercury in fish

In the aquatic environment, mercury can be absorbed by organisms through both diffusion and active uptake mechanisms (e.g., by Na^+ and Ca^{2+}), and the degree of absorption and biomagnification depends on the mercury species.

Many types of organic matter and inorganic ligands can form compounds with mercury in natural aquatic systems. Some ligands can sequester, or prevent the passage of, mercury through biofilms, whereas others can promote the bioavailability of mercury and its absorption by organisms. In marine environments, Cl^- is an important inorganic mercury ligand. Neutral ligands of $HgCl_2$ diffuse faster than Hg^{2+} in biofilms (Gutknecht 1981), and the former is more easily absorbed in cells and tissues due to the high Cl^- concentration (Klinck et al. 2005).

Recent studies also suggest that mercury bound to small particles of thiol compounds (e.g., cysteine) can be more readily absorbed and methylated by sulfate-reducing bacteria (SRB) (Schaefer and Morel 2009). CH_3HgCl, a compound formed by methylmercury ions (CH_3Hg^+) and the two major inorganic ligands of mercury (OH and Cl^-), can be more readily absorbed by organisms than CH_3HgOH. The former prevails in environments where pH is low and Cl^- is high. In terrestrial freshwater systems, dissolved organic matter (DOM) is both a vital and complex factor that affects the bioavailability of mercury.

We use %C and dissolved organic carbon (DOC) as units of organic matter concentration in sediments and water, respectively. In systems containing relatively more mercapto- and sulfur-containing organic matter, inorganic mercury will preferentially bind to mercapto- or other sulfur-containing functional groups, leading to encapsulation and limiting mercury's bioavailability (Ravichandran 2004). However, under acidic conditions, H^+ may compete with Hg^{2+} at the binding site of DOM, thereby favoring the release of Hg^{2+}. The enables inorganic Hg to be activated, absorbed and methylated by organisms. Likewise, chemical factors, which affect the formation and availability of mercury in the aquatic environment, can have a significant impact on the speciation and concentration of mercury in organisms of the low trophic level and their consumers. According to studies conducted in China, new reservoirs have inundated large areas of forests and slowed the velocity of water flow, leading to a rapid increase in organic matter and the formation of anaerobic environments in the reservoirs. Rapid increases in methylmercury have also been observed in fish from the new reservoirs at the early stage. As the reservoirs evolve, mercury in fish will gradually return to initial levels between 10 and 30 years (Bodaly et al. 2007).

In contrast, the Wujiang River Basin is located in a karstic environment characterized by low organic matter content (C%) and high pH values. Therefore, the flooded soil during the early stages of reservoir construction is thin. The concentration of organic matter in sediments is only 1.9–4.3% and DOC is approximately 1.1–1.4 mg L^{-1}. At the beginning of their construction, no fish farming was carried out and reservoirs suffered less from industrial and agricultural pollutants. After the rivers were impounded, the velocity of water flow slowed down. Especially in those areas where the water was deep, the flow velocity was low, the water area was relatively wide and human activities were intense, the water bodies and the sediment began to evolve with time. Thick layers of sediment gradually formed at the bottom. The growth and degradation of algae led to the formation of an anaerobic environment in the water at the bottom of the reservoir, thus causing significant changes in the physical and chemical parameters and microbial communities of the entire water body. Changes in these parameters resulted in changes in mercury speciation, concentration, and distribution throughout the entire aquatic environment, thereby altering its bioaccumulation.

In particular, algal blooms caused by eutrophication provided rich food for wild fish. However, algae degradation led to elevated levels of endogenous organic matter in the water, increased the anaerobicity of the bottom water, and changed the number of biological communities; to a certain extent, it also promoted the methylation of mercury.

(1) Trophic-level status

Reservoirs are built for the purpose of power generation. In addition, aquaculture is an important function. Eleven-step cascade hydropower stations have been constructed for power generation in reservoirs of different ages in the Wujiang River Basin. The content and distribution of MeHg are obviously different in these reservoirs. We found that in older reservoirs where sediments are thick and seasonal hypoxia occurs at the bottom of the water body, MeHg levels in the sediments are higher than in the newer reservoirs (Feng et al. 2009). Therefore, in this section, we studied the relationship between reservoir evolution and MeHg levels in fish. We found that there was no significant correlation between mercury levels and reservoir age, with no obvious differences in mercury levels between reservoirs older than 30 years and the newly built reservoirs. The main reason is because the fish samples collected from each reservoir had different feeding habits and were of different sizes and quantities (Table 9.6); hence, it resulted in poor comparability. The main reason for the high mercury content in SFY is because the standing crop was limited, and the samples collected were mainly large fish. As seen in Fig. 9.15, omnivorous, carnivorous and benthic fish were available in each reservoir and were suitable for simple comparisons. Omnivorous fish mainly feed on zooplankton, algae, shrimp, and insects in water; carnivorous fish mainly feed on fish, shrimp, and insects; and benthic fish feed on sediment debris and small mollusks. THg in omnivorous fish from HJD and SFY was higher than in omnivorous fish from the other four reservoirs.

This indicates that eutrophication of reservoirs provides abundant food for fish and shortens the food chain, thereby accelerating the growth of fish. Taking this into consideration, one could conclude that the increased phytoplankton in eutrophic reservoirs led to lower MeHg concentrations in the fish. The median THg

Table 9.6 Physical and chemical parameters of reservoirs and Hg in fish

		Reservoirs						
		HF	BH	WJD	PD	YZD	HJD	SFY
THg-f	N= mean ± SD range	50 63.8 ± 69.3 5.4–230.8	125 48.8 ± 45.9 4.2–250.1	66 39.8 ± 57.4 1.8–350.2	2 39.1 ± 2.8 37.1–41.1	61 47.5 ± 47.5 4.0–279.9	135 61.7 ± 48.7 2.1–139.3	14 138.2 ± 97.3 41.1–385.9
THg-w		6.9	22.4	0.6–3.5	1.0–11.7	2.5	2.7 ± 1.3	0.9 ± 0.4
MeHg-w	Mean or	0.12–0.51	1.29	0.03–2.9	<0.05–0.5	0.03–0.16	0.12 ± 0.03	0.11 ± 0.04
THg-s	range	0.39	0.26–38.9	0.12–0.42	0.20	na	na	na
MeHg-s		2.5	8.2	0.15–8.3	na	na	na	na
C%-s		2.2–8.8	5–12	2.2–11	6.2	na	na	na
Chla		11.1–33.2	32.5–42.6	9.5–15.3	5.8–8.1	na	2.2	1.3
DOC-w		1.7–3.2	2.9–4.0	1.2	0.9–1.1	0.6	1.2	1.1
pH-w		7.0–8.8	7.5–8.2	7.2–8.8	7.2–8.3	7.2–8.3	7.1–8.1	7.4–8.0
Age		48	42	29	13	5	4	3

Note THg-f-THg in fish (ng g^{-1}, ww), THg-w-THg in water(ng L^{-1}), MeHg-w- MeHg in water (ng L^{-1}), THg-s-THg in sediment (μg g^{-1}, dw), MeHg-s MeHg in sediment (ng g^{-1}), C%-s- organic matter in sediment (%), Chla (μg L^{-1})–Chlorophyll a, DOC-w-dissolved organic matter in water, age (year) calculated till in 2008. Hongjiadu Reservoir—HJD, Yinzidu Reservoir—YZD, Suofengying Reservoir—SFY, Puding Reservoir—PD, Hongfeng Reservoir—HF, Baihua Reservoir—BH, Wujiangdu Reservoir—WJD

Table 9.7 Transfer and bioaccumulation of THg and MeHg in biota from reservoir food chains

	Name	Samples		Mercury concentrations		BAF/BCF	
			Hg	MeHg	%MeHg	For THg	For MeHg
Old Reservoirs	WJD	Water	0.86	0.11	12.9		
		Sediment	274	4.65	1.7		
		Plankton >38 μm	242.1 ± 64.1	12.8 ± 3.1	5.3	2.8E + 05	1.2E + 05
		Plankton >150 μm	157.2 ± 36.2	58.6 ± 31.1	37.3	1.8E + 05	5.3E + 05
		Fish	39.8	21.5	54.0	4.6E + 04	2.0E + 05
	HF	Water	2.1	0.2	9.5		
		Sediment	392 ± 70	3.4 ± 2.5	0.9		
		Plankton	71.2	9.1	12.8	3.4E + 04	4.6E + 04
		Fish	32.1	12.3	38.3	1.5E + 04	6.2E + 04
	BHH	Water	6.9 ± 4.5	0.5 ± 0.6	10.3 ± 9.5		
		Sediment	13700 ± 10300	6.1 ± 5.2	0.04 ± 0.3		
		Porewater (1–10 cm)	197.6 ± 67.5	0.8 ± 1.2	1.6 ± 2.7		
		Plankton < 112 μm	724.4 ± 410.4	4.0 ± 2.7	0.6 ± 0.6	1.0E + 05	8.0E + 03
		Plankton < 64 μm	862.4 ± 507.8	2.7 ± 0.4	0.5 ± 0.3	1.2E + 05	5.4E + 03
		Fish	48.8 ± 45.3	16 ± 9.4	46.5 ± 21.8	7.1E + 03	3.20E + 04
		Red worm	205.6 ± 71.6	9.46 ± 11.1	6.7 ± 4.0	1.0E + 03	1.20E + 04

(continued)

Table 9.7 (continued)

	Name	Samples		Mercury concentrations		BAF/BCF	
			Hg	MeHg	%MeHg	For THg	For MeHg
Intermediate reservoirs stage of evolution	DF	Water	0.96	0.13	13.9		
		Sediment	238	1.58	0.7		
		Plankton >38 μm	147.4 ± 87.2	33.0 ± 16.9	22.4	1.5E + 05	2.5E + 05
		Plankton >150 μm	55.4 ± 25.8	28.1 ± 11.2	50.7	5.8E + 04	2.1E + 05
		Fish	36.0	10.2	28.3	3.8E + 04	7.6E + 04
	PD	Water	1.67	0.13	7.5%		
		Sediment	195	2.50	1.3%		
		Plankton	62.80	2.9	4.6%	3.8E + 04	2.3E + 04
		Fish	39.1	na	na	2.3E + 04	na
New reservoirs	HJD	Water	0.84 ± 0.51	0.06 ± 0.01	7.1		
		Sediment	na	na	na		
		Plankton	na	na	na		
		Fish	59.4 ± 45.1	24.9 ± 21.1	39.5 ± 11.5	7.1E + 04	1.9E + 05
	SFY	Water	0.35 ± 0.12	0.04 ± 0.03	11.4		
		Sediment	na	na	na		
		Plankton	na	na	na		
		Fish	138.2	na	na	3.9E + 05	
	YZD	Water	0.5 ± 0.15	0.07 ± 0.04	14.0		
		Sediment	na	na	na		
		Plankton	na	na	na		
		Fish	40.4 ± 68.8	23.3 ± 31	57.7	8.1E + 04	3.3E + 05

Data for DF is from Jiang (2005); na-no data

concentrations in carnivorous fish from HF and HJD were the highest, whereas those in BH and YZD were relatively lower, although the maximum values were not significantly different. An exception was found in BH; the major reason is that most of the carnivorous fish in BH were Abbottina rivularis (Basilewaky) and a small number of catfish that were small in size. Another possible reason is that mercury accumulation in fish is mainly through their food intake; therefore, although the mercury concentrations in the sediment and water of BH are higher than in the other reservoirs, the mercury concentrations in fish did not seem to increase significantly.

(2) pH value

Aquatic systems with lower pHs are more conducive to mercury methylation and its bioaccumulation, whereas mercury in neutral and slightly alkaline environments tends to be more adsorbed on particulate matter, from which it is difficult to form methylmercury. Variation of water pH values can affect the mobility of inorganic mercury and microbial activity; thus it can affect the methylation of inorganic mercury and its bioaccumulation rate in the entire body of water (Kelly et al. 2003). Weakly acidic environments are especially conducive to the formation of methylmercury from inorganic mercury and its bioaccumulation. The pH values of the seven reservoirs involved in the study ranged from neutral (6.9) to slightly alkaline (9.7). Under these conditions, inorganic mercury tends to be absorbed by particles and eventually settles to the bottom. In addition, the pH of each reservoir differed little, and the accumulation of mercury in the fish was little affected by pH.

(3) Mercury concentration

Mercury speciation and concentrations in aquatic environments are important factors influencing mercury levels, especially methylmercury levels in fish. Table 9.4 shows that the MeHg concentrations in four eutrophic reservoirs were higher compared with three reservoirs with lower levels of eutrophication. This indicates that the evolution of reservoirs does increase the potential for mercury methylation. Interestingly, in the canyon-shaped deepwater reservoirs, most fish live in the upper water body where there is enough light, and oxygen is abundant. Phytoplankton is abundant within the top 10 meters of the water body, especially in eutrophic cage-culture reservoirs. Phytoplankton is an abundant food source for fish that can accelerate the growth of the fish and dilute inorganic mercury and methylmercury that they have ingested, thereby reducing their mercury accumulation. However, mercury remains high in large predatory fish, reaching values up to 350 ng g^{-1} in both types of reservoirs.

(4) Physical conditions—age, weight, and length

Mercury in fish mainly accumulates through food uptake. Mercury (especially MeHg) in food will progressively accumulate along the aquatic food chain. Because the metabolic cycle of MeHg is longer than the metabolic cycle of inorganic mercury, MeHg is more likely to accumulate and enter into organisms at higher trophic levels in the food chain. Hence, larger predatory fishes of the same species that are at the top of the aquatic food chain tend to have higher levels of mercury.

Affected by artificial bait and eutrophication, natural food chains have been disturbed in some aquatic systems, leading to a disproportion between mercury concentration and fish size. In this study, we did not find a positive correlation between mercury and body size. Fish of the same species and length collected from cage-culture reservoirs can be either wild or fish that escaped from the net cage, and the length and complexity of their food chains vary greatly. Therefore, THg concentration in fish was not always correlated with their weight and length. However, correlation between THg and length was still observed in some samples from the same batch of the same reservoir (Fig. 9.31). For example, there was a notable correlation between THg and carp weight in WJD ($r = 0.639$, $p < 0.01$, $n = 16$) and YZD ($r = 0.593$, $p < 0.01$, $n = 24$). Especially in HF, THg in samples with weights within these ranges (100–300 g, 600–900 g, and approximately 1200 g) was also significantly correlated with fish weight ($r = 0.727$, $p < 0.01$, $n = 24$).

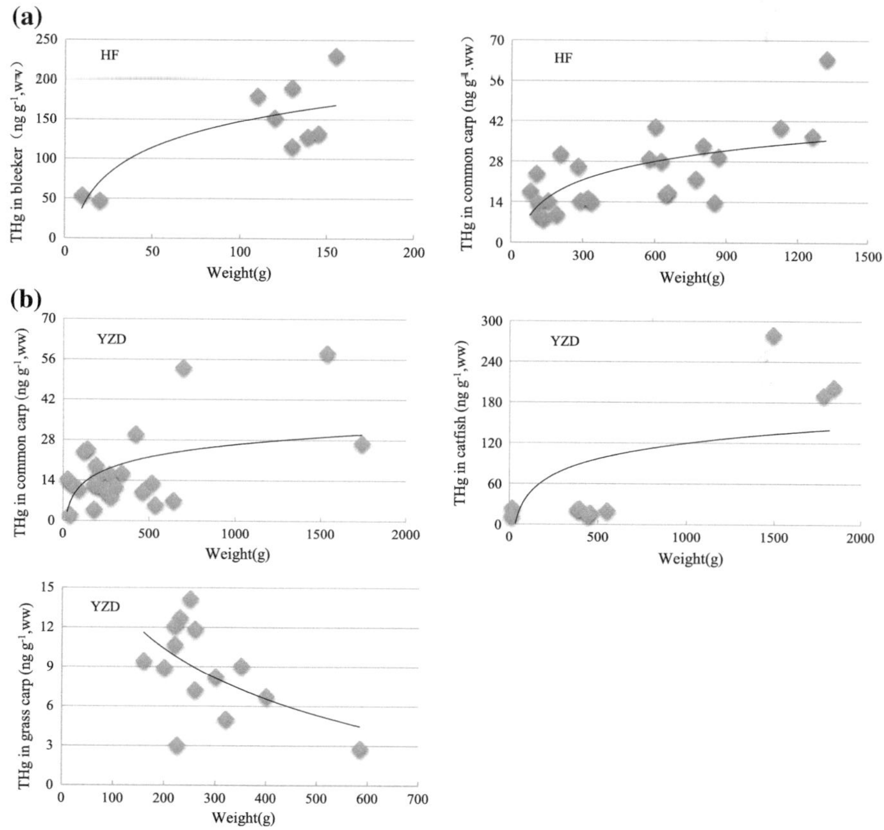

Fig. 9.31 Correlation of length and THg in fish. **a** HF; **b** YZD; **c** WJD

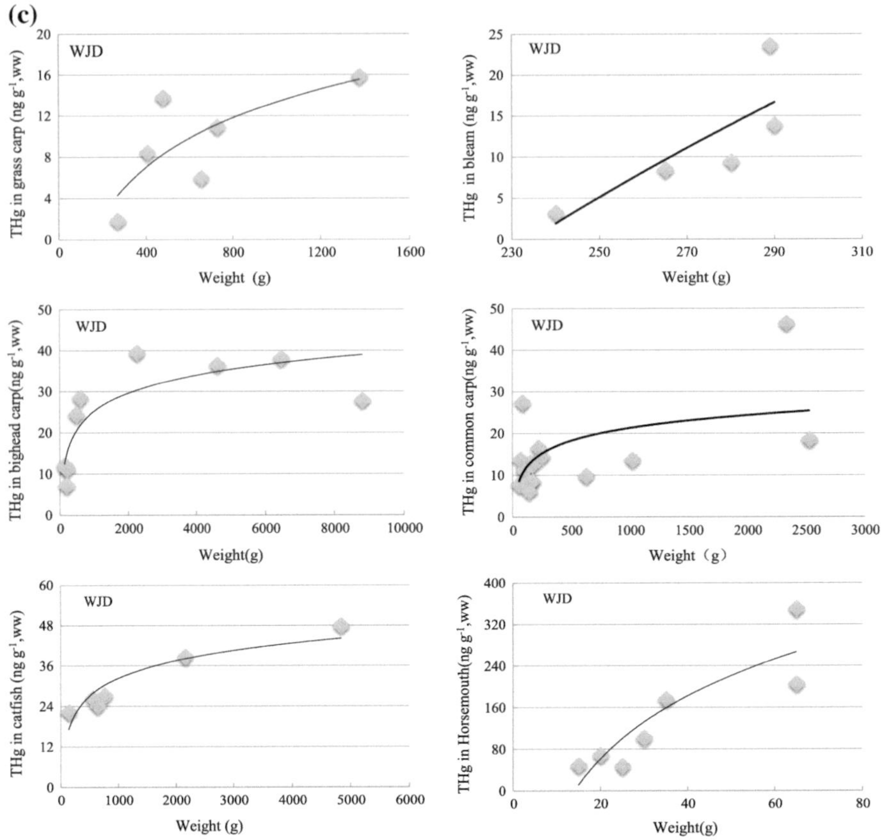

Fig. 9.31 (continued)

However, in the 135 carp samples collected from HJD in different seasons, THg was not correlated with weight.

In three reservoirs, THg concentrations in topmouth culters, hook snout carps and catfish were positively correlated with weight. In grass carps, mercury concentrations were negatively and positively correlated with fish weight in YZD and HF, respectively. A possible reason is that fish weights in YZD were not significantly different (200–600 g), whereas carp from HF displayed comparatively larger weight differences (300–1500 g) and ages; hence, there was a notable difference in mercury accumulation.

We found no correlation between mercury level and weight except in the above samples. This is due to a number of factors: when and where we caught the fish, whether the fish were all wild and how many fish samples we collected. Because most cascade reservoirs in the Wujiang River Basin are long and narrow, different parts are impacted by human activities to different degrees, and food chains in different parts of the reservoirs are different; thus, mercury contents can be different even in fish of the same species and length. Especially when the number of fish sampled is small and when wild fish may be a mixture of fish that live far away from and near net cages, where they can grow faster, the mercury accumulated in their bodies is diluted to relatively low levels. Therefore, the trophic levels, which can be determined by measuring the nitrogen isotopes of the samples, are crucial to interpreting mercury levels in fish.

3. Values of the stable isotopes $\delta^{15}N$ and $\delta^{13}C$ in biota

Generally, aquatic food chains can be divided into four levels: primary producers, consumers, predators, and decomposers. However, because organisms may be omnivorous, when based on simple deduction from general feeding habits, the classification of the consumer level, in particular, can hardly be consistent with the facts in natural environments. Thus, it is unreliable to study the trophic level of organisms by simply taking biological feeding habits as a reference. We can find out what food organisms have ingested by analyzing their stomach contents. However, because part of the food may have already been digested, and there may have been incidental food intake, there is increased uncertainty in this information. In this case, using stable carbon and nitrogen isotopes to identify the food-web structure becomes a good approach. Stable isotope analysis can be a highly efficient way for us to comprehensively understand the spatial and temporal changes in biological feeding. Stable carbon and nitrogen isotopes can be used to determine the feeding habits and trophic levels of organisms based on their specific behaviors. Heavy isotopes (e.g., ^{13}C, ^{15}N) are more apt to stay in the body of organisms than light isotopes (e.g., ^{12}C, ^{14}N); therefore, they are increased at higher levels of food chains.

The stable carbon and nitrogen isotopic compositions ($\delta^{13}C$ and $\delta^{15}N$) of organisms can be used to accurately quantify their carbon sources and the trophic level in the food web (Post 2002). Stable carbon isotopic composition ($\delta^{13}C$) is an important tool for analyzing nutritional connections in food webs (Fry et al. 1978; Peterson and Fry 1987) and can reflect the evolution of food webs and food sources over a long period (Kling et al. 1992). Organisms with similar $\delta^{13}C$ values have similar food compositions or living habits. The enrichment factor of $\delta^{13}C$ ‰ with the trophic level of the food web is approximately 1‰ (Deniro and Epstein 1981). The larger the differences in isotopic composition between organisms, the greater their differences in feeding habits. The $\delta^{13}C$ value is generally negative; it usually increases with increasing trophic levels of organisms.

The enrichment factors of the stable nitrogen isotope ratio are high (3–5‰), so it is mainly used to determine the trophic level of organisms in the food web. The higher the value, the higher the trophic level (Miyake and Wada 1967). The greater the difference in $\delta^{15}N$ values between organisms at high and low trophic levels, the greater the degree of biomagnification. Generally, $\delta^{15}N$ increases by 3–4‰ as eutrophication increases (Post 2002).

The $\delta^{13}C$ or $\delta^{15}N$ of fish samples are calculated to compare isotope ratios in organism samples using the following formula:

$$\delta^{13}\mathrm{C}\,\mathrm{or}\,\delta^{15}\mathrm{N} = \left[\left(R_{\mathrm{sample}}/R_{\mathrm{standard}}\right) - 1\right] \times 1000 \quad (9.1)$$

where R_{sample} is the $^{13}C/^{12}C$ ratio or $^{15}N/^{14}N$ ratio measured in the samples. The variation in $\delta^{13}C$, which is smaller than the variation in $\delta^{15}N$ in the food chain, can still be used to determine the food sources of organisms. For example, benthic organisms have higher $\delta^{13}C$ values than plankton. Whether there is biomagnification of mercury in organisms is assessed by the correlation between the mercury concentration and the $\delta^{15}N$ value. The correlation between mercury concentration and the $\delta^{15}N$ value is usually expressed as Log_{10} [THg] = a$\delta^{15}N$ + b. [THg] represents the THg concentration and $\delta^{15}N$ represents the nitrogen isotopic composition of the organism. The value of "a" represents the degree of mercury biomagnification in organisms.

In addition, the stable isotopes of nitrogen ($\delta^{15}N$) and carbon ($\delta^{13}C$) provide powerful tools for estimating the trophic positions of and carbon flow to consumers in food webs and determine the lengths of food chains. Using data from the literature, the mean trophic fractionations of $\delta^{15}N$ and $\delta^{13}C$ are 3.4 and 0.4‰, respectively (1 SD = 1.3‰) (Post 2002). Therefore, we used the $\delta^{15}N$ value to calculate the food chain length and TLs in all of the reservoirs. The following model is the simplest one for estimating the trophic position of a consumer:

$$\mathrm{TL}_{\mathrm{consumer}} = \left(\delta^{15}\mathrm{N}_{\mathrm{consumer}} - \delta^{15}\mathrm{N}_{\mathrm{base}}\right)/3.4‰ + \lambda \quad (9.2)$$

where $TL_{consumer}$ is the trophic level of the consumer, $\delta^{15}N_{consumer}$ is the $\delta^{15}N$ value of the consumer, $\delta^{15}N_{base}$ is the $\delta^{15}N$ value of the baseline,and λ is the trophic position of the organism used to estimate $\delta^{15}N_{base}$ (e.g., $\lambda = 1$ for a primary producer).

The trophic levels and the $\delta^{15}N$ and $\delta^{13}C$ values in food chains of the reservoirs for five different stages of evolution in the Wujiang River Basin are shown in Table 9.8. In terms of $\delta^{13}C$ ‰, the carbon isotope ratios measured in samples from YZD and HF were quite similar at approximately −22.6‰, showing restricted variations of approximately 2.0‰. This indicated that the main food sources in the two reservoirs were similar. The ratio of carbon isotopes in WJD and BH were similar at approximately −24.6‰, but the variation was greater in BH than in WJD. The reason is that aquatic samples from BH include not only fish but also phytoplankton. Because their feeding habits are different, they have different carbon sources. Of the five reservoirs, $\delta^{13}C$ ‰ was lowest and showed the greatest variation in HJD (−25.2‰). It was closer to the degree of enrichment of carbon isotopes (3.0‰) in organisms in natural aquatic ecosystems. This indicates that the reservoir experienced little anthropogenic carbon pollution during the initial stage of its construction. However, we found that the $\delta^{15}N$ value was higher than in most unpolluted aquatic systems, and the trophic level in the food chains in these reservoirs was significantly high. It suggests that complicated nitrogen input increased the $\delta^{15}N$ values and led to abnormal trophic levels, even as many as 6.

Table 9.8 Stable isotope characteristics and trophic levels of food chains in reservoirs

	YZD		WJD		BH		HF		HJD	
	$\delta^{13}C$	$\delta^{15}N$	$\delta^{13}C$	$\delta^{15}N$	$\delta^{13}C$	$\delta^{15}N$	$\delta^{13}C$	$\delta^{15}N$	$\delta^{13}C$	$\delta^{15}N$
Mean	−22.7	9.3	−24.5	10.9	−24.7	15.8	−22.3	18.6	−25.2	13.9
SD	2.1	1.6	2.7	2.7	3.3	3.9	2.0		3.0	3.8
TL	3.76		5.03		4.13		3.96		6.29	

Figure 9.32 the $\delta^{15}N$ and $\delta^{13}C$ values were positively correlated in BH and HF, whereas in the other three reservoirs, the lower the $\delta^{13}C$ value was, the higher the $\delta^{15}N$ value was. Meanwhile, according to the sample species and the absolute values of their carbon and nitrogen isotope ratios, we can conclude that BH and HF have been affected by carbon and nitrogen input due to human activity Fig. 9.32. Therefore, Log_{10} [TH_g] and $\delta^{15}N$ showed a positive correlation, but no significant relevance with $\delta^{13}C$ Fig. (9.33).

This can be manifested by higher than normal nitrogen isotope ratios in organisms at low trophic levels that are similar to nitrogen isotope ratios at the fourth trophic level in other food chains. We analyzed the slopes of the linear regressions between the carbon and nitrogen isotope ratios of samples collected in the different reservoirs and found that the slope was steepest for HJD. Of the species collected from HJD, grass carp and carnivorous fish belonged to the lowest and highest trophic levels, respectively. Carbon and nitrogen isotope ratios measured in samples from BH were similar to those from HJD. Samples collected from BH included phytoplankton, zooplankton, herbivorous fish, omnivorous fish, carnivorous fish and benthic shrimp and snails. This shows that in HJD, the food sources for organisms at different trophic levels in the food chain vary greatly and exhibit a higher degree of bioconcentration.

4. Influence of the reservoir evolution process on the bioaccumulation of mercury in food chains

In aquatic environments, mercury is mainly absorbed by organisms (such as plants, algae, and other primary producers) from the water body and then passed on along the food chain until it ultimately accumulates and magnifies in organisms at the top of the food chain. This accumulation and magnification are influenced by the types

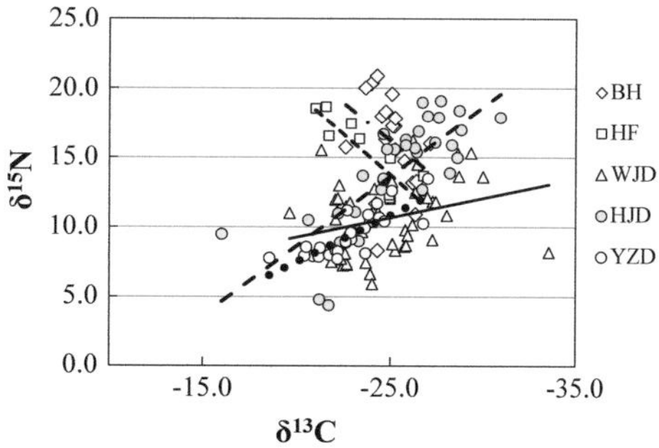

Fig. 9.32 Correlation between nitrogen and carbon isotopes

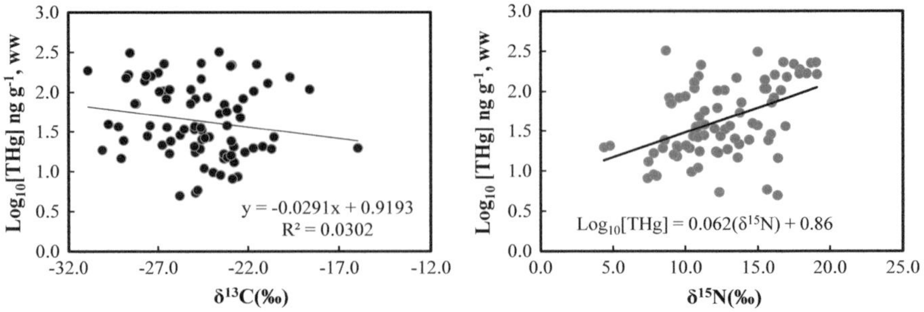

Fig. 9.33 Correlation between mercury and $\delta^{15}N/\delta^{13}C$

of mercury species and their bioavailability. With the existence of DOC, sulfide, Cl^- and OH^- and part of mercury ligand compounds, Hg^{2+} and CH_3Hg^+ are more likely to penetrate membrane tissue and be absorbed. That implies that Hg^{2+} and CH_3Hg^+ have greater bioavailability. When the concentrations of Hg^{2+} and CH_3Hg^+ are higher in organisms than in water, it is called bioconcentration. After reaching a steady state, the ratio of mercury in plankton to water is called the "bioconcentration factor" (BCF), which is usually determined in the laboratory under controlled conditions. The average BCFs of Hg^{2+} and CH_3Hg^+ in aquatic organisms are approximately 5000–9000, respectively (McGeer et al. 2003).

Tertiary consumers in the food chain are exposed to inorganic mercury and methylmercury mainly through food and water, and mercury exposure via the food chain increases at higher trophic levels. Analogous with BCF, levels of mercury exposure in tertiary consumers are expressed as bioaccumulation factors (BAFs), which are the ratios of mercury in organisms to mercury in the water body. The difference between BCFs and BAFs is that mercury sources of tertiary consumers are mainly from primary producers or organisms at high trophic levels in the water and food chain. Generally, BAFs are determined by calculations on samples collected from the field. The results show that BAFs are relatively low (10^5–10^6) in primary consumers of shorter life spans and smaller sizes (>107%) but high in organisms of longer lifespans and larger sizes (e.g., fish) and are closely related to DOC in the water (Watras et al. 1998).

Mercury in the form of methylmercury is one of the rare metals that can be enriched and magnified along the food chain. Biomagnification in the aquatic food chain, usually expressed as BMFs (Biomagnification Factors), denotes higher Hg concentrations in prey and consumers at higher trophic levels, i.e., the ratio of predator Hg to predator–prey Hg is approximately 2–10.

BH, HF, and WJD are all at later stages of evolution and are all seriously eutrophicated. The sediment depth in these reservoirs is approximately 30 cm, and the surface is rich in organic matter. Especially in the WJD aquaculture area, cage culture has led to high levels of organic matter, with more than 30% in the surficial sediments (Feng 2012). The 11-year-old PD and DF reservoirs are both at intermediate evolutionary stages. Sediments are the thickest with an OM% as high as

10% close to the dam where there is less the cage culture due to drawing off. The HJD, SFY and YZD reservoirs were built more recently, within the past 3–5 years; therefore, there are thin and mainly coarse sediment grains that have been inundated, and nutrient levels in the water bodies are rather low, which posed difficulties for plankton collection. Therefore, reservoirs were divided into three categories in Table 9.7 for analysis.

5. Variations in mercury bioaccumulation in the food chains of different reservoirs

Differences between reservoirs are mainly dependent on their age and evolutionary process and pollution sources. Reservoirs of different evolutionary stages have their own characteristics of MeHg and THg distribution and different eutrophication levels. Plankton species and abundance vary greatly, which in turn results in food chain discrepancies. Here, pollution sources refer mainly to anthropogenic mercury pollutant emissions.

According to the first few chapters, we can see that eutrophication levels are low in reservoirs at their initial stages of evolution. In three reservoirs (WJD, HF, and BH) at later evolutionary stages, plankton (especially algae) was particularly abundant, and the reservoirs were eutrophic almost year round. Therefore, plankton samples were mainly collected in these reservoirs. Generally, plankton is small in size, which makes it difficult to clearly distinguish phytoplankton (also known as microalgae) from zooplankton. Therefore, we adopted a standard approach to sample collection. The size of phytoplankton is generally assumed to be smaller than 38 or 64 μm, whereas the size of zooplankton ranges from 64 to 150 μm or from 64 to 112 μm.

Table 9.7 shows that there was no obvious correlation between THg in plankton and the evolutionary stage of the reservoirs. MeHg showed no notable pattern of variation among these reservoirs. However, the THg concentration was higher in phytoplankton than in zooplankton, whereas the MeHg concentration was higher in zooplankton. Mercury concentrations were highest in the phytoplankton of BH, where the natural background of mercury is particularly high. THg concentrations were approximately 724 ng g^{-1} in plankton smaller than 112 μm and approximately 862.4 ng g^{-1} in phytoplankton smaller than 64 μm. However, the MeHg concentration was only 2.7–4.0 ng g^{-1}, accounting for only 0.5–0.6% of the THg (Table 9.9). This showed that the extracellular polymer of microalgae had a strong adsorption and scavenging effect on inorganic mercury, primarily on divalent mercury ions (Zhang et al. 2010). The ratio of MeHg to THg (%MeHg) in benthic

Table 9.9 The Probable Daily Intake (PDI) and Hazard Quotients (HQ) of MeHg exposure via fish consumption for Guizhou adults

	Fish consumption (g day^{-1})	Body weight (kg)	PDI_{50} (ng kg^{-1} day^{-1})	PDI_{90} (ng kg^{-1} day^{-1})	RfD (ng kg^{-1} day^{-1})	HQ_{50}	HQ_{50}
Urban	9.15	60	3.16	11	100	0.03	0.11
Rural	0.93	60	0.32	1.12	100	0.003	0.01

chironomids was 6.7%, which was lower than the 26% observed in chlor-alkali-contaminated areas (Becker and Bigham 1995). Because chironomid larvae mainly live in shallow water, we did not collect chironomid larvae and benthic organism samples in the other reservoirs.

Overall, the THg enrichment factor in plankton (relative to water) was 10^4–10^5, whereas the enrichment factor of MeHg was 10^3–10^5 (Table 9.7). This shows that plankton had a slightly stronger adsorption affinity for inorganic mercury than for MeHg. Studies by Zhang et al. (2010) also showed that the extracellular secretions of planktonic microalgae had strong adsorption affinities for inorganic mercury in the water.

The enrichment factors (BAFs) of the total mercury and methylmercury in fishes were quite similar, with respective values of 10^3–10^4 and 10^4–10^5. MeHg in fish accounted for less than 50% of THg. In general, the enrichment factor of inorganic mercury in fish is much lower than that of MeHg. However, %MeHg in fish usually accounts for nearly 90% or more of the accumulated mercury. Apparently, what we found in this study is not consistent with the results of previous studies, mainly due to the small proportion of large carnivorous fish samples. MeHg is transported and accumulated mainly through the food chain, so a complete food chain is critical for MeHg accumulation. The simple food chain structure of the reservoirs in the Wujiang River Basin cannot boost the high enrichment of MeHg in organisms. In fact, this has been confirmed in previous studies. In 2003, the average THg concentration in fish collected from net cages in BH was 28 ng g^{-1}, whereas THg concentrations increased to 55.3 ng g^{-1} in samples collected in 2009 after cage culture was prohibited.

9.3 Health Risk Assessment for Mercury Exposure

Methylmercury is a potent toxicant (NRC 2000). It can be efficiently absorbed from the gastrointestinal tract. Approximately 95% of MeHg in fish ingested by volunteers was absorbed from the gastrointestinal tract. MeHg can pass through the blood–brain barrier and damage the central nervous system in the human body (Clarkson and Magos 2006). The absorption of IHg from the gastrointestinal tract has been estimated at approximately 7% in human body and it can cause renal toxicity (WHO 1990). MeHg constitutes the major portion of THg in fish tissue, generally, fish and shellfish, especially for marine fish, more than 90% of THg is in the form of MeHg. Human exposure to MeHg occurs mainly through consuming fish (Mergler et al. 2007). MeHg contamination in fish is a worldwide environmental concern, because fish contain high-quality protein and other essential nutrients with known benefits to human health (Mergler et al. 2007). Fish is an excellent source of omega-3 fatty acids such as DHA (docosahexaenoic acid) and EPA (eicosapentaenoic acid) (Meyer et al. 2003), and balancing the risks and benefits have become an increasingly important goal of fish consumption advisories (Ginsberg and Toal 2009).

The original epidemiologic report of MeHg poisoning occurred in Minamata, Japan, in 1950s. The disease is caused by a large number of local fishermen consumed fish with high MeHg level contaminated by wastewater discharged from a chemical plant (Irukayama and Irukayama 1966). The main symptom of Minamata disease is the nervous disorder. From 1956 the first case of Minamata disease reported, so far there were 2200 people diagnosed with Minamata disease. The emission of wastewater containing MeHg from the chemical plant continued to 1968. The event of Minamata disease caused serious Hg pollution in this region. Hair THg concentrations in fisherman living in this region were very high, with the average of 338.4 mg kg^{-1}, and the range was from 96.8 to 705 mg kg^{-1}. Generally, hair THg concentrations were less than 1 mg kg^{-1} for the general population. In Minamata area, many of the fetus suffered from similar cerebral palsy disease through maternal placental, but the mothers themselves had no or only mild poisoning symptoms (Harada 1978). In the 1970s, MeHg poisoning occurred in Iraq as "seeds event", because the farmers consumed the seeds adding with MeHg to prevent insect. In the 1980s, high fish Hg concentrations (>0.5 mg kg^{-1}) were found in remote lakes in Europe and North America, but there was no obvious anthropogenic pollution nearby. About 90–98% of THg were in the form of MeHg in fish tissues, which posed high health risks to the seafood consumers. After 1985–1990, international organizations and different governments, especially Europe and the United States, developed a series of relevant policies to reduce health risks of MeHg exposure from fish consumption. The United States set fish Hg consumption advices in many waters in order to prevent and reduce MeHg exposure risks in the general population (Anderson et al. 2004; Watras et al. 1998; Katner et al. 2011).

9.3.1 Toxicity and Metabolism of Mercury Species

Mercury (Hg) and its compounds are recognized as potentially hazardous materials and are rated in the top category of environmental pollutants. Mercury can significantly adverse effects on human health. The toxicity of Hg depends on its chemical form.

The inorganic forms of Hg include liquid metallic Hg and its vapor, compounds of mercurous and mercuric Hg. The ingestion of liquid metallic Hg or "quicksilver" does not appear to be toxic in itself. Health hazards from quicksilver are due to its potential to release Hg vapor. Inhaled Hg vapor can cause damage to the central nervous system due to its ability to cross the blood–brain barrier. IHg may cause a variety of adverse effects. Neurological effects, renal effects, cancer, respiratory effects, cardiovascular effects, gastrointestinal and hepatic effects, effects on the thyroid gland, effects on the immune system, effects on the skin, reproductive and developmental effects, and genotoxicity have been observed following exposure to IHg (WHO 1991; USEPA 1997; ATSDR 1999; UNEP 2002). The specific symptoms are found in the central nervous system and the kidney. Urine and feces are the principal routes of Hg elimination and the urinary route dominates when exposure is high (WHO 1991). The half-time for Hg in urine is about 2 months. Urine Hg (UHg) measurements are widely used for assessment of IHg (mainly mercury vapor) exposure in humans because UHg is thought to be indicating most closely the Hg levels present in the kidneys (Clarkson et al. 1988; Clarkson and Magos 2006; Barregard 1993).

Compared with IHg, the toxicity of MeHg is much higher. It is present worldwide in fish and marine mammals consumed by humans. MeHg is formed naturally (from anthropogenic and naturally released Hg) by biological activity in aquatic environments, and it is bio-magnified in the food chain, resulting in much higher concentrations in higher predatory fish and mammals than in water and lower organisms. Most of the THg concentrations in fish are in the form of MeHg (close to 100% for older fish). Consumption of contaminated fish and marine mammals is the most important source of human exposure to MeHg (WHO 1990; US EPA 1997). MeHg is highly toxic, and the nervous system is its principal target tissue. In adults, the earliest effects are non-specific symptoms such as paresthesia, malaise, and blurred vision; with increasing exposure, signs appear such as concentric constriction of the visual field, deafness, dysarthria, ataxia, and ultimately coma and death (Harada 1995). The developing central nervous system is more sensitive to MeHg than the adult. In infants exposed to high levels of MeHg during pregnancy, the clinical picture may be indistinguishable from cerebral palsy caused by other factors, the main pattern being microcephaly, hyperreflexia, and gross motor and mental impairment, sometimes associated with blindness or deafness (Harada 1995; Takeuchi 1982). Hair and blood Hg concentrations are both accepted as valid biomarkers of MeHg exposure, although each provides a somewhat different reflection of exposure (NRC 2000). Blood gives an estimate of exposure over the most recent one to two half-lives, with the half-life of MeHg in blood being 50–70 days, whereas hair reflects the average exposure over the growth period of the segment (NRC 2000).

9.3.2 Criteria of Risk Assessment of Mercury Exposure via Fish Consumption

The maximum level of MeHg in fish recommended by the Joint FAO/WHO Food Standards Programme CODEX Committee on Contaminants in Foods is 1.0 μg g^{-1} for predatory fish and 0.5 μg g^{-1} for other fishery products (JFSPCC 2011). This guideline has been adopted by most countries.

On the basis of the studies conducted in the Faroes Islands (cord blood Hg concentration of 58 μg L^{-1}), USEPA set the limit of 0.1 μg kg^{-1} day^{-1} as the reference dose (RfD) for MeHg (USEPA 1997). According to the calculations made in the Faroe Islands study and in another from the Seychelles (12 μg g^{-1} Hg concentration in maternal hair), the Joint FAO/WHO Expert Committee on Food Additives (JECFA) established a provisional tolerable weekly intake (PTWI) for

MeHg at 1.6 μg kg^{-1} $week^{-1}$ (JECFA 2003). The quantification of these intake values is important because the effect of low-level MeHg exposure on neurological development has been quantified. A loss of 0.18 intelligence quotient (IQ) points is thought to be associated for each part per million increase of maternal hair Hg (Axelrad et al. 2007; Bellinger 2011).

9.3.3 *Study State of Risk Assessment of Mercury Exposure*

Fish and shellfish have significant health risks because of its bioaccumulation of environmental pollutants, but fish also can provide trace elements, minerals, proteins, which have known health benefits to human body. Especially for EPA and DHA, they can reduce the risks of cardiovascular disease. Therefore, in addition to simple ingestion limits and risk index studies, more attentions are paid to the integrated analysis of health risks resulted from aquatic products consumption in the current research. The main aspects include the following: (1) comprehensive evaluation between the adverse effects of intake of contaminants (such as POPs, MeHg, Cd, PCB, etc.) and beneficial effects of nutrient ingestion (Hursky and Pietrock 2012; Vieira et al. 2011); (2) consumption advices of aquatic products on the behalf of the distribution of Hg and MeHg in aquatic food chain (Bonsignore et al. 2013; Burger 2009); (3) the suggestion of health risks of fish and shellfish consumption at different areas based on the study of Hg and Se concentrations in aquatic products and its molar ratios (Burger and Gochfeld 2011, 2013).

The results showed that: (1) consumption of small amount (about 80–100 g day^{-1}) of marine fish, which can not only ingest enough unsaturated fatty acids, but also prevent exposure to toxic heavy metals; (2) avoid eating huge quantities of fish, especially wild fish, predatory fish, and large fish (such as fish with body length more than 50 cm, tuna, and sharks), these fish can bioaccumulate more MeHg because of their high trophic level; the Se: Hg molar ratios in fish tissues decreased with increasing of fish body length (Burger and Gochfeld 2013) and the detoxification mechanism may also decrease; (3) prevent eating the aquatic products from the Hg polluted water body, especially for these bentonic organisms living close to the pollution source (Bonsignore et al. 2013).

9.3.4 *Health Risk Assessment of Methylmercury Exposure via Fish Consumption*

To estimate MeHg intake from seafood consumption, the probable daily intake (PDI) for different age groups was calculated according to the following formula:

$$\mathrm{PDI} = \frac{C \times \mathrm{IR}}{\mathrm{BW}} \tag{9.3}$$

where PDI is given in nanogram per kilogram of body weight per day (ng kg^{-1} day^{-1}); BW is body weight; C is MeHg concentration in the seafood (ng g^{-1}); and IR is the daily intake rate of the seafood (g day^{-1}). The median of MeHg concentration was 20.8 ng g^{-1} for 153 fish and shellfish samples collected from Wujiangdu, Hongfeng Reservoir, Hongjiadu, Yinzidu, and Baihua reservoir, and the 90th percentile value of MeHg concentration was 72.0 ng g^{-1}. The average daily intake of seafood for urban residents was 9.15 g d^{-1} in Guizhou in 2011, and that for rural population was 0.93 g d^{-1} according to Guizhou Statistical Yearbook (GBS 2012). The average body weight of 60 kg for adult population was according to the second National Physique Monitoring Bulletin. The PDI_{50} and PDI_{90} values were calculated using the 50th and 90th percentile values of fish MeHg concentration. The hazard quotient (HQ) is simply the ratio of the estimated exposure to an effective concentration. If the value of the HQ was less than 1, it was considered to indicate acceptable risk. In the present study, HQ was calculated by dividing the PDIs by the guideline levels of RfD (0.1 μg kg^{-1} day^{-1}) set by USEPA.

For adults in Guizhou Province, the PDI and HQ values were listed in (Table 9.9). The HQ in all age groups were less than 1, which indicates that MeHg exposure through fish consumption may not lead to adverse health effects. The PDIs of MeHg via fish consumption were relatively low, because of the low daily fish consumption for both urban and rural residents (9.15–0.93 g day^{-1}, respectively). In conclusion, the general populations in Guizhou Province had a low risk of MeHg exposure through fish consumption.

References

Anderson HA, Hanrahan LP, Smith A, Draheim L, Kanarek M, Olsen J (2004) The role of sport-fish consumption advisories in mercury risk communication: a 1998–1999 12-state survey of women age 18-45. Environ Res 95:315–324

ATSDR (1999) Toxicological profile for mercury; agency for toxic substances and disease registry. Atlanta, GA, USA

Axelrad AD, Bellinger CD, Ryan ML, Woodruff JT (2007) Dose-response relationship of prenatal mercury exposure and iq: an integrative analysis of epidemiologic data. Environ Health Perspect 115:609–615

Barregard L (1993) Biological monitoring of exposure to mercury vapor. Scand J Work Environ Health 19:45–49

Becker DS, Bigham GN (1995) Distribution of mercury in the aquatic food web of Onondaga Lake, New York. Water Air Soil Pollut 80:563–571

Bellinger DCA (2011) strategy for comparing the contributions of environmental chemicals and other risk factors to neurodevelopment of children. Environ Health Perspect 120:501–507

Bodaly RA, Jansen WA, Majewski AR, Fudge RJP, Strange NE, Derksen AJ, Green DJ (2007) Postimpoundment time course of increased mercury concentrations in fish in hydroelectric reservoirs of northern manitoba, Canada. Arch Environ Contam Toxicology 53:379–389

Bonsignore M, Salvagio MD, Oliveri E, Sprovieri M, Basilone G, Bonanno A, Falco F, Traina A, Mazzola S (2013) Mercury in fishes from Augusta Bay (southern Italy): risk assessment and health implication. Food Chem Toxicol 56:184–194

Burger J (2009) Risk to consumers from mercury in bluefish (Pomatomus saltatrix) from New Jersey: size, season and geographical effects. Environ Res 109:803–811

Burger J, Gochfeld M (2011) Mercury and selenium levels in 19 species of saltwater fish from New Jersey as a function of species, size, and season. Sci Total Environ 409:1418–1429

Burger J, Gochfeld M (2013) Selenium and mercury molar ratios in commercial fish from New Jersey and Illinois: variation within species and relevance to risk communication. Food Chem Toxicol 57:235–245

Chen CY, Folt C (2005) High plankton densities reduce mercury biomagnification. Environ Sci Technol 39:115–121

Clarkson TW, Magos L (2006) The toxicology of mercury and its chemical compounds. Crit Rev Toxicol 36:609–662

Clarkson TW, Hursch JB, Sager PR, Syversen TLM (1988) Mercury. In: Clarkson TW, Friberg L, Nordberg GF, Sager PR (eds) Biological monitoring of toxic metals. Plenum, New York, pp 199–246

Deniro MJ, Epstein S (1981) Influence of diet on the distribution of nitrogen in animals. Geochim Cosmochim Acta 45:341–351

Feng CY (2012) The influence of cage culture on the geochemical cycling of methylmercury in reservoir. Master Dissertation in the Graduate School of the Chinese Academy of Sciences (GSCAS). Beijing. (In Chinese, with English abstract)

Feng XB, Jiang HM, Qiu GL, Yan HY, Li GH, Li ZG (2009) Mercury mass balance study in Wujiangdu and Dongfeng reservoirs, Guizhou, China. Environ Pollut 157:2594–2603

Fry B, Jeng JW, Scalan RS, Parker PL, Baccus J (1978) ^{13}C food web analysis of a Texas sand dune community. Geochim Cosmochim Acta 42:1299–1302

GBS. Guizhou statistical yearbook-2012, Guizhou Bureau of Statistics. Beijing: China Statistics Press; 2012

Ginsberg G, Toal B (2009) Quantitative approach for incorporating methylmercury risks and omega-3 fatty acid benefits in developing species-specific fish consumption advice. Environ Health Perspect 117:267–275

Gutknecht J (1981) Inorganic mercury (Hg^{2+}) transport through lipid bilayer membranes. J Membr Biol 61:61–66

Harada M (1978) Congenital Minamata disease: intrauterine methylmercury poisoning. Teratology 18:285–288

Harada M (1995) Minamata disease: methylmercury poisoning in Japan caused by environmental pollution. CRC Crit Rev Toxicol 25:1–24

Hursky O, Pietrock M (2012) Chemical contaminants and parasites: assessment of human health risks associated with consumption of whitefish (Coregonus clupeaformis) from two boreal lakes in northern Saskatchewan, Canada. Sci Total Environ 424:97–103

Irukayama K, Irukayama TK (1966) Studies on the organomercury compound in the fish and shellfish from Minamata Bay and its origin. VII. Synthesis of methylmercury sulfate and its chemical properties. Nippon Eiseigaku Zasshi 21:342–343

JECFA (2003) Summary and Conclusions of the Sixty-First Meeting of the Joint FAO/WHO Expert Committee on Food Additives. Italy, Rome

JFSPCC (Joint FAO/WHO Food Standards Programme Codex Committee on Contaminants in Foods) (2011) Working Document for Information and Use in Discussions Related to Contaminants and Toxins in the GSCTFF, Fifth Session. Hague, Netherlands

Jiang HM (2005) Effects of hydroelectric reservoir on the biogeochemical cycle of mercury in the Wujiang River. Ph.D Dissertation in the Graduate School of the Chinese Academy of Sciences (GSCAS). Beijing. (In Chinese, with English abstract)

Katner A, Ogunyinka E, Sun MH, Soileau S, Lavergne D, Dugas D, Suffet M (2011) Fishing, fish consumption and advisory awareness among Louisiana's recreational fishers. Environ Res 111:1037–1045
Kelly CA, Rudd JWM, Holoka MH (2003) Effect of pH on mercury uptake by an aquatic bacterium: implications for Hg cycling. Environ Sci Technol 37:2941–2946
Klinck J, Dunbar M, Brown S, Nichols J, Winter A, Hughes C, Playle RC (2005) Influence of water chemistry and natural organic matter on active and passive uptake of inorganic mercury by gills of rainbow trout (Oncorhynchus mykiss). Aquat Toxicol 72:161–175
Kling GW, Fry B, O'Brien WJ (1992) Stable isotopes and planktonic trophic structure in arctic lakes. Ecology 73:561–566
Mackenzie A, Ball AS, Virdee SR (1998) Instant notes in ecology. Bios Scientific Publishers, Oxford
McGeer JC, Brix KV, Skeaff JM, DeForest DK, Brigham SI, Adams WJ, Green A (2003) Inverse relationship between bioconcentration factor and exposure concentration for metals: implications for hazard assessment of metals in the aquatic environment. Environ Toxicol Chem 22:1017–1037
Mergler D, Anderson AH, Chan HM, Mahaffey RK, Murray M, Sakamoto M, Stern HA (2007) Methylmercury exposure and health effects in humans: A worldwide concern. AMBIO-a J Human Environ 36:3–11
Meyer B, Mann N, Lewis J, Milligan G, Sinclair A, Howe P (2003) Dietary intakes and food sources of omega-6 and omega-3 polyunsaturated fatty acids. Lipids 38:391–398
Miyake Y, Wada E (1967) The abundance ratio of $^{15}N/^{14}N$ in marine environments. Rec Oceanogr Works Japan 9:32–59
NRC (U.S. National Research Council) (2000) Toxicological Effects of Methylmercury. National Academy Press, Washington, DC, p 344
Peterson BJ, Fry B (1987) Stable isotopes in ecosystem studies. Annu Rev Ecol Syst 18:293–320
Post DM (2002) Using stable isotopes to estimate trophic position: models, methods, and assumptions. Ecology 83:703–718
Ravichandran M (2004) Interactions between mercury and dissolved organic matter-a review. Chemosphere 55:319–331
Reynolds C, Dokulil M, Padisák J (2000) Understanding the assembly of phytoplankton in relation to the trophic spectrum: where are we now? Hydrobiologia 424:147–152
Rojo C (1998) Differential attributes of phytoplankton across the trophic gradient: a conceptual landscape with gaps. Hydrobiologia 369–370:1–9
Schaefer JK, Morel FMM (2009) High methylation rates of mercury bound to cysteine by Geobacter sulfurreducens. Nat Geosci 2:123–126
Sun RY, Li QF, Niu CJ (2002) Basic ecology. Higher education press
Sunda WG, Huntsman SA (1998) Processes regulating cellular metal accumulation and physiological effects: Phytoplankton as model systems. Sci Total Environ 219:165–181
Takeuchi T (1982) Pathology of Minamata disease. With special reference to its pathogenesis. Acta Pathol Jpn 32:73–99
UNEP (2002) Global mercury assessment. United Nations Environment Programme, Geneva
USEPA (1997) Mercury study report to the congress, EPA 452/R-97-0003. U.S. Environmental Protection Agency, Washington
Vieira C, Morais S, Ramos S, Delerue-Matos C, Oliveira MBPP (2011) Mercury, cadmium, lead and arsenic levels in three pelagic fish species from the Atlantic Ocean: Intra- and inter-specific variability and human health risks for consumption. Food Chem Toxicol 49:923–932
Wang BL, Liu CQ, Wang FS, Yu YX, Zhang LH (2008) The distributions of autumn picoplankton in relation to environmental factors in the reservoirs along the Wujiang River in Guizhou Province, SW China. Hydrobiologia 598:35–45
Wang BL, Liu CQ, Wang FS, Yu YX, Zhang LH (2009) Flow cytometric observation of picophytoplankton community structure in the cascade reservoirs along the Wujiang River, SW China. J Limnol 68:53–63

Wang Q, Feng XB, Yang YF, Yan HY (2011) Spatial and temporal variations of total and methylmercury concentrations in plankton from a mercury-contaminated and eutrophic reservoir in Guizhou Province, China. Environ Toxicol Chem 30:2739–2747

Watras CJ, Back RC, Halvorsen S, Hudson RJM, Morrison KA, Wente SP (1998) Bioaccumulation of mercury in pelagic freshwater food webs. Sci Total Environ 219:183–208

WHO (1990) Environmental health criteria 101-Methylmercury. World Health Organ, Geneva

WHO (1991) Environmental health criteria 118-inorganic mercury. World Health Organ, Geneva

Yan HY (2005) The Methodological Development of mercury species in environmental samples and the mass balance of mercury in Baihua Reservoir, Guizhou, China. Ph.D. Dissertation in the Graduate School of the Chinese Academy of Sciences (GSCAS). Beijing. (In Chinese, with English abstract)

Yan HY, Feng XB, Shang LH, Qiu GL, Dai QJ, Wang SF, Hou YM (2008a) The variations of mercury in sediment profiles from a historically mercury-contaminated reservoir, Guizhou province, China. Sci Total Environ 407:497–506

Yan HY, Feng XB, Liu T, Shang LH, Li ZG, Li GH (2008b) Present situation of fish mercury pollution in heavily mercury-contaminated Baihua reservoir in Guizhou. Chin J Ecol 27: 1357–1361

Yan HY, Rustadbakken A, Yao H, Larssen T, Feng XB, Liu T, Shang LH, Haugen TO (2010) Total mercury in wild fish in Guizhou reservoirs, China. J Environ Sci 22:1129–1136

Yao H (2010) Biogeochemical cycling of mercury in newly built reservoir, Wujiang River. Ph.D Dissertation in the Graduate School of the Chinese Academy of Sciences (GSCAS). Beijing. (In Chinese, with English abstract)

Zhang HC (2008) Study on Environmental Impact of Hydropower Development in Mainstream of Wujiang River in Guizhou Province. Postdoctoral Outbound Report, Institute of Geochemistry, Chinese Academy of Sciences. (In Chinese, with English abstract)

Zhang DY, Pan XL, Mostofa KMG, Chen X, Mu GJ, Wu FC, Liu J, Song WJ, Yang JY, Liu YL, Fu QL (2010) Complexation between Hg(II) and biofilm extracellular polymeric substances: an application of fluorescence spectroscopy. J Hazard Mater 175:359–365

Chapter 10
Primary Factors Controlling Hg Methylation in Reservoirs

Abstract In this book, we conducted an intensive study on the mercury (Hg) biogeochemical processes in a river-reservoir ecosystem in the Wujiang River Basin, Southwest China. The concentrations and distributions of Hg species in the reservoirs (water column, sediment, and sediment pore water), the inflow/outflow rivers of the reservoirs, wet deposition, and food chains within the Wujiang River Basin were systematically investigated (Chaps. 3, 5, 6, 7, and 9). Measurements of the water/air exchange flux of gaseous elemental mercury (GEM) were also conducted in the reservoirs (Chap. 4). A detailed mass balance of the total mercury (THg) and methylmercury (MeHg) in the selected six reservoirs was developed (Chap. 8). This chapter reveals the primary factors that controlling Hg methylation in reservoirs in Wujiang River Basin, Southwest China. To better understand the biogeochemical processes of Hg in Wujiang River Basin, a conceptual model of the Hg in the river-reservoir system was further developed in this chapter.

Keywords Mercury · Methylation · Factor · Wujiang river

10.1 Age of the Reservoir

10.1.1 Introduction to the Evolutionary Stage of Reservoirs

The properties and evolution of a reservoir ecosystem are controlled by the input load and accumulation of biogenic elements. After the impounding and operation, the biological, physical, and chemical processes in the reservoirs changed constantly. These physical and chemical changes and the caused various complex processes will significantly change the reservoir ecosystem. Several reservoirs were built in a small area of the Wujiang River Basin, and these reservoirs have different ages, nutrition levels, degrees of environmental pollution, and pollution sources. Through the study of the reservoirs with different evolutionary stages in the Wujiang River Basin, it can provide a scientific basis for environmental management and environmental risk assessment of high-dam and deep-canyon-type

X. Feng et al., *Biogeochemical Cycle of Mercury in Reservoir Systems in Wujiang River Basin, Southwest China*, https://doi.org/10.1007/978-981-10-6719-8_10

reservoirs in southwestern China, and provides scientific foundations for understanding and exploring the impact of construction of large water conservancy facilities on the reservoir ecosystem.

Table 10.1 summarizes the conditions of the nine reservoirs along the mainstream and its tributaries of Wujiang River. The nine reservoirs are divided into four types based on their impoundment time and pollution status: The Hongjiadu Reservoir (HJD), Suofengying Reservoir (SFY), and Yingzidu Reservoir (YZD) reservoirs are level I reservoirs (initial impoundment period, no pollution); the Puding Reservoir (PD) and Dongfeng Reservoir (DF) are level II reservoirs (intermediate stage of evolution, low pollution); the Wujiangdu Reservoir (WJD) is a level III reservoir (relatively mature stage of evolution, substantial pollution), and the Hongfeng Reservoir (HF), Baihua Reservoir (BH) and Aha Reservoir (AH) are level IV reservoirs (mature stage of evolution, industrial level of pollution). Through the study of the temporal and spatial distributions, migration and transformation of the various Hg species in the environmental media of the reservoirs at these four evolutionary stages, it can better understand the biogeochemical processes and environmental impacts of mercury (Hg) in the reservoirs at various evolutionary stages in the Wujiang River Basin. Based on this information, we also can explore the mobilization characteristics of Hg in the reservoirs with various evolutionary stages, and compare with the studies conducted in the reservoirs in North America and Europe and possible reasons and mechanism.

10.1.2 Impact of Evolutionary Stage of Reservoir on the Distribution of Mercury

Table 10.2 summarizes relevant data presented in previous chapters. The total mercury (THg) and methylmercury (MeHg) concentrations in the water and sediments in the reservoir at different evolutionary stages increase with the age. Through the calculation of the input and output fluxes of the reservoirs, it indicates that: the YZD, HJD and SFY at the initial evolutionary stage are sinks for MeHg; the PD and DF at the intermediate evolution stage, the WJD, at the relatively mature stage, and the HF, BH and AH at the mature stage, are sources of MeHg. These results indicate that the methylation capability of Hg is much stronger in mature reservoirs than those in the initial stage of evolution. With the evolution, the reservoir may convert from the sink of MeHg to the source of MeHg. In addition, the net flux and conversion rate of MeHg in the WJD (relatively mature stage) were significantly higher than those in the PD and DF (intermediate stage), which indicated that the methylation of Hg increased with the reservoir age.

The comparison of the THg and MeHg concentrations in the nine reservoirs in the Wujiang River Basin with those in North America and Europe reservoirs (Table 10.3) indicates that the MeHg concentrations in the reservoirs in the initial evolution stage are much lower than those in the newly built reservoirs in North

Table 10.1 The storage period and pollution status of the nine reservoirs in the mainstream and tributaries in the Wujiang River Basin

Reservoir	Geographical location and type in the Wujiang River	Storage period (year)	Pollution status
SFY	Midstream in the main stream,a typical deep-valley, high-mountain gorge	2005	No pollution
HJD	Upstream in the main stream,a typical deep-valley, high-mountain gorge	2004	No pollution
YZD	Midstream in the main stream,a typical deep-valley, high-mountain gorge	2003	No pollution
DF	Midstream in the main stream,a typical deep-valley, high-mountain gorge	1994	Less pollution,it is being affected by agricultural activity, resulting in the input of many nitrogen and phosphorus, so the water is in a mesotrophic state
PD	Upstream in the main stream,a typical deep-valley, high-mountain gorge	1993	Less pollution,it is being affected by agricultural activity, resulting in the input of many nitrogen and phosphorus, so the water is in a mesotrophic state
WJD	Downstream in the main stream,a typical deep-valley, high-mountain gorge	1979	Pollution,it is being affected by cage aquaculture act so the water is in a eutrophic state. ivity, and there have many chemical industries in the upstream of Xifeng river, so the water is in a eutrophic state
BH	Tributary	1966	Pollution,it is adjacent to towns and thus is being affected by Chemical Industry, Coal mining factory and domestic wastewater, so the water is in a eutrophic state
HF	Tributary	1960	Pollution,it is adjacent to towns and thus is being affected by domestic wastewater and cage aquaculture activity, so the water is in a hyper-eutrophic state
AH	Tributary	1960	Pollution,it is being affected coal mine and domestic wastewater, and an important sources of drinking water in Guiyang

America and Europe but are comparable with those in the old reservoirs there. However, the MeHg concentrations in the reservoirs in the intermediate stage are slightly lower than or close to those in the newly built reservoirs in North America and Europe but significantly higher than those in the old reservoirs there. These

Table 10.2 Summary of the nine reservoirs THg and MeHg concentrations of their water and sediment, the net flux of THg and MeHg in the mainstream and tributaries of Wujiang River

Reservoir	THg in water column (ng L^{-1})	TMeHg in water column (ng L^{-1})	THg in sediment (ng g^{-1})	MeHg in sediment (ng g^{-1})	Net flux of THg (g $year^{-1}$)	Net flux of MeHg (g $year^{-1}$)	Storage ratio (%)	Source/sink	Organic matter in sediment/flooded soil (%)
SFY	0.42–4.9	0.030–0.22	141–372	0.7–2.0	−1597	−181	−17.9	To THg, these reservoirs are a sink for THg, except WJD is a source of THg; To MeHg, YZD, HJD and SFY are a sink for TMeHg, PD, DF and WJD are a source of TMeHg	2.6–6.9
HJD	0.32–6.6	0.05–0.17	20–137	0.4–1.9	−11434	−74	−14.6		0.4–3.9
YZD	0.40–1.9	0.0028–0.44	–	–	−838	−121	−28.5		
DF	0.68–3.9	0.05–0.50	130–330	0.23–2.8	−1820	368	73.2		1.4–3.98
PD	1.0–12	0.05–0.51	148–298	0.34–6.6	−10469	69	13.4		3.0–9.0
WJD	0.60–3.5	0.033–2.9	108–1376	0.15–8.3	1604	857	83.5		2.2–11.0
BH	6.7–153	0.11–6.9	665–8073	3.3–15	–	–	–	HF and AH are a sink for THg in river, and a source of TMeHg	10–17
HF	2.0–14	0.053–2.2	197–793	0.19–8.4	–	–	–		2.2–8.8
AH	2.1–20	0.03–2.1	160–252	0.20–7.3	–	–	–		2.2–6.7

findings imply that the methylation capability of the new reservoirs in North America and Europe is stronger than that of the old reservoirs. In contrast, in the Wujiang River Basin, the methylation capability of the reservoirs in the mature stage is stronger than that of the reservoirs in the initial stage, which indicated that the methylation capability of inorganic Hg increases with the reservoir age.

However, these findings contradict the pattern among the reservoirs in North America and Europe, which revealed that the intensity of methylation of Hg decreased with increasing of the reservoir age. This marked difference could be attributed to the differences in the concentrations and sources of organic matter (OM) in the submerged soils and sediments. In North America and Europe, newly constructed reservoirs typically submerge large areas of boreal forest or wetland, where OM concentrations are as high as 30–50% (Verdon et al. 1991; Tremblay et al. 1996; Lucotte et al. 1999). Lucotte et al. (1999) found that inorganic Hg in the submerged soil can converse into MeHg and the conversion process is very active, and releasing of MeHg from the submerged soil to the overlaying water mainly derived from Hg methylation. The OM play an important role: oxygen is consumed and CO_2 is released during the process of degradation of OM, which resulted in the anaerobic and low-pH environment (favorable conditions for Hg methylation) in the bottom water (Fearnside 2001). It can speed the dissolution of inorganic Hg from the soil and vegetation and thus provides enough inorganic Hg for methylation (Gagnon and Fisher 1997; Cossa and Gobeil 2000); In addition, large amount of nutrients released from the degradation of OM can provide rich nutrient sources for microorganisms (mainly methylating bacteria, including sulfate- and iron-reducing bacteria), which enhance the activity of microorganism and promote conversion of inorganic Hg into MeHg (Winfrey and Rudd 1990; Rogers et al. 1995; Bodaly et al.

Table 10.3 THg and MeHg concentrations in reservoir water in North America and Europe (ng L^{-1})

Sampling site	Periods (year)	THg	MeHg	Data source
Experimental reservoir, Canada	0.04	0.98–6.95	0.05–3.2	Kelly et al. (1997)
Quebec reservoir, Canada	3	<5	0.01–2	Lucotte et al. (1999)
Caniapiscau reservoir, Canada	17	1.19–1.69	0.06–0.09	Schetagne et al. (2000)
Flood-control imploundments, USA	2–4	0.74–6.97	0.06–6.6	Brigham et al. (2002)
Maryland reservoir, USA	12–133	0.4–6.8	0.048–0.38	Mason and Sveinsdottir (2003)
Experimental reservoir, Canada	3	1.1–6.0	0.1–2.1	Hall et al. (2005)
Narraguinnep reservoir, USA	16	0.47–1.06	0.010–0.043	Gray et al. (2005)

1997; Heyes et al. 1998). St. Louis et al. (2004) designed a flooding experiment in a wetland and observed significant difference of MeHg production rate before and after flooding: in the first year of flooding, the MeHg production rate reached 70 mg ha^{-1}, which was 40-fold of that of before flooding (1.6–1.9 mg ha^{-1}, St. Louis et al. 2004).

Therefore, the Hg methylation in the reservoir ecosystems in Northern Europe and America summarized as follows: the degradation of OM in flooded soil can result in the environments in favor of Hg methylation; These environments can produce large amounts of nutrients, which stimulate microbiological activity and promote the conversion of HgII into MeHg; which finally increased MeHg production rate in water columns. However, as the age of reservoir increasing, the OM in the submerged soil is gradually degraded and exhausted, which significantly reduced the MeHg production rate in flooded soil, as a result, the methylation of Hg is weakened.

The Wujiang River Basin is dominated by typical karst topography, which characterized with thin soil layer, low soil fertility, and steep slopes, and dryland is the main farming mode. Because of the limited land resources and intensive farming, the organic matter content in the submerged soil the reservoir in the initial evolutionary stage is generally low (Table 10.2). As a result, the nutrient levels in the reservoirs are poor to intermediate (the primary productivity is low) (Sect. 4 of Chap. 6). The endogenous organic matter is limited and the pH of the water body is high, which is not favorable for the methylation of Hg in the reservoirs. However, with the gradual evolution of the reservoirs in the Wujiang River Basin, the primary productivity of the reservoirs gradually increased and the yield of the endogenous organic matter in the reservoirs increased. In addition, aquaculture and industrial pollution in the reservoirs have resulted in severe eutrophication in the reservoirs in mature and mature stages (see Chap. 6). The lost feed and fish excrement from aquaculture and plankton, such as algae, provide a significant source of OM (Table 10.2), and degradation of OM by microorganisms can provide ideal environments for the methylation of inorganic Hg (anaerobic and weakly acidic condition in the bottom of the water bodies).

In conclusion, the Hg methylation rate of the reservoir ecosystems in the Wujiang River Basin increases with increasing of the reservoir age. These findings are completely different from the findings obtained in Northern Europe and North America, which methylation rate decreases with increasing of the reservoir age. The submerged soils in North America and Europe were originated from boreal forest or wetland, therefore OM concentrations in newly created reservoirs were relatively high and high levels of OM can produce more favorable conditions for Hg methylation. However, the submerged soils in the newly built reservoirs in Wujiang River Basin were originated from farmland or uncovered rock of valley and revealed low levels of OM. Therefore, these newly built reservoirs were not active sites for Hg methylation. However, during the long period of evolution, anthropogenic pollution (such as cage aquaculture and discharges of industrial wastewater) will cause MeHg pollution problems in these reservoirs. It's important for policy decision for local government. To prevent MeHg pollution after the building of reservoirs, the selection of sites for the construction of reservoirs were critical: (1) choosing the landscape with low OM concentrations in soil; (2) removing plants and rubbish of the area to avoid environmental pollution prior to flooding; (3) controlling the input of extraneous organic materials or nutrients, e.g., cage aquaculture and waste discharge from chemical plant. These methods may help to mitigate Hg methylation and MeHg pollution in the newly built reservoirs in China.

10.2 Eutrophication

Eutrophication is a serious environmental problem of lakes and reservoirs all over the world. Under natural conditions, eutrophication of lakes and reservoirs occurs over thousands of years. Since the Industrial Revolution, human activities have accelerated the process of eutrophication. With the rapid development of industry, agriculture, and urbanization, discharges of industrial and domestic wastewater increased significantly. Other harmful human activities such as land reclamation from lakes, deforestation, and cage aquaculture have accelerated eutrophication of reservoirs (lakes). Eutrophication has actually pervaded in most reservoirs (lakes) in China, and many large reservoirs (lakes) have already reached eutrophic or hypertrophic states.

The methylation of inorganic Hg and demethylation of MeHg are important steps in the biogeochemical processes of Hg in the reservoirs. MeHg concentrations in the environment depend on the net methylation rate, which revealed the equilibrium between methylation and demethylation. The net methylation rate of Hg is affected by many factors, which included the activity of microorganisms, the amount of Hg that can be used by microorganisms, temperature, pH, redox potential, sulfate and sulfur concentrations, organic matter concentrations, and salinity (Ullrich et al. 2001). Therefore, eutrophication will unavoidably change the physical conditions (such as temperature and light transmittance), chemical conditions (such as redox potential and pH), and biological conditions (such as bacteria and algae populations) in the water column and sediments. These changes will inevitably affect the distribution, migration, and conversion of Hg in reservoirs (lakes), which will finally impact the health of humans and other higher organisms. However, there are very limited studies considering the impacts of eutrophication on the Hg methylation in the reservoirs (lakes), and its mechanism is not clearly understood. The reservoirs (lakes) in the Wujiang River Basin with different nutrient levels have provided natural experimental sites for these studies. Through the comparative studies of the biogeochemical processes of Hg in these reservoirs (lakes) with different nutrient levels in this basin, it can provide the theoretical basis for the prediction of environmental risks of Hg in the eutrophized reservoirs (lakes) and the reasonable suggestions for remediation t of Hg-contaminated and entropized reservoirs (lakes).

10.2.1 Evaluation of Reservoir Eutrophication in the Wujiang River Basin

1. The method of eutrophication evaluation

At present, there exist diverse methods for evaluating water eutrophication, but none of them are universally accepted. Generally, researchers refer to the evaluation method specifically used to assess the eutrophication of shallow lakes. Initially, it adopted to Carlson's Trophic State Index (TSI) method and then evolved into the revised Carlson's TSI method and Weighted Comprehensive TSI method (see, Table 10.4) (Jin and Tu 1990; Lin et al. 2001).

The trophic state of waters depends mainly on the comprehensive effect of a variety of correlated factors, such as nutritive salt content (including nitrogen (N) and phosphorus (P)) and transparency (SD). The trophic state of various reservoirs can be evaluated using the method of weighted comprehensive TSI based on the data of chlorophyll-a (Chl.a), total nitrogen (TN), total phosphorus (TP), and transparency. Based on the empirical formula for the reservoirs in China (lnChl.a = 1.912 − 1.349ln(SD), R^2 = 0.7; lnChl.a = 3.497 + 0.654ln(TP), R^2 = 0.66; lnChl.a = 1.597 + 0.972ln(TN), R^2 = 0.60), the TSI can be calculated with the formulas as follows:

$$\mathrm{TSI(Chl.a)} = 10\left(2.46 + \frac{\ln(\mathrm{Chl.a})}{\ln 2.5}\right) \tag{10.1}$$

$$\mathrm{TSI(SD)} = 10\left(2.46 + \frac{1.912 - 1.349\ln(\mathrm{SD})}{\ln 2.5}\right) \tag{10.2}$$

$$\mathrm{TSI(TP)} = 10\left(2.46 + \frac{3.497 + 0.654\ln(\mathrm{TP})}{\ln 2.5}\right) \tag{10.3}$$

$$\mathrm{TSI(TN)} = 10\left(2.46 + \frac{1.597 + 0.972\ln(\mathrm{TN})}{\ln 2.5}\right) \tag{10.4}$$

The weighted comprehensive TSI can be calculated with the formula as follows:

$$\mathrm{TSI}\left(\sum\right) = \sum_{j=1}^{m} W_j \bullet \mathrm{TSI}(j) \tag{10.5}$$

Table 10.4 Classification criteria of the correlation weighted composite trophic state index

Eutrophication index	Trophic state
<30	Oligotrophic
30–50	Mesotrophic
>50	Eutrophic

Where TSI (∑) refers to the comprehensive TSI and TSI (*j*) refers to the TSI of the *j*th type of parameters. Here, W_j refers to the weight of the TSI of the *j*th type of parameter, as formulated below:

$$W_j = \frac{r_{1j}^2}{\sum_{j=1}^{m} r_{1j}^2} \tag{10.6}$$

Where r_{1j}^2 refers to the coefficient of correlation between the *j*th parameter and Chl-a, and m refer to the number of selected main parameters, including chlorophyll, transparency, total phosphorus, and total nitrogen.

(1) Nitrogen and phosphorus criteria of trophic condition

According to the classification criteria of total nitrogen (TN) developed by Vollenweider and Kerekes (1982), trophic condition of waters can be classified as oligotrophic if the TN is below 0.2 mg L^{-1}, mesotrophic if the TN is in the range of 0.2–0.5 mg L^{-1}, eutrophic if the TN is in the range of 0.6–1.5 mg L^{-1}, and hyper-eutrophic if the TN is higher than 1.5 mg L^{-1} (see Table 10.5).

According to the classification criteria of total phosphorus (TP) developed by the trophic condition of waters, it can be classified as oligotrophic if the TP is below 0.01 mg L^{-1}, mesotrophic if the TP is in the range of 0.01–0.03 mg L^{-1}, and eutrophic if the TP is higher than 0.03 mg L^{-1}.

(2) Criteria of OECD

According to the classification criteria of the Organisation for Economic Cooperation and Development (OECD) and Gu and Shu (1988), trophic condition of waters can be classified as hyper-eutrophic if the Chl-a is higher than 78 μg L^{-1}, eutrophic if the Chl-a is in the range of 11–78 μg L^{-1}, mesotrophic if the Chl-a is in the range of 4–11 μg L^{-1}, and oligotrophic if the Chl-a is lower than 4 μg L^{-1} (see Table 10.6).

According to the classification criteria of the OECD based on phytoplankton abundance, trophic condition of waters can be classified as hyper-eutrophic if the Chl-a is higher than 10^7 cells L^{-1}, eutrophic if the Chl-a is in the range of 5×10^6–10^7 cells L^{-1}, mesotrophic if the Chl-a is in the range of 5×10^5–5×10^6 cells L^{-1}, and oligotrophic if the Chl-a is lower than 5×10^5 cells L^{-1} (Jin and Tu 1990).

Table 10.5 Nitrogen and phosphorus criteria of trophic condition

TN (mg L^{-1})	TP (mg L^{-1})	Trophic state
<0.2	<0.01	Oligotrophic
0.2–0.5	0.01–0.03	Mesotrophic
0.5–1.5	>0.03	Eutrophic
>1.5		Hyper-eutrophic

Table 10.6 Classification criteria of OECD

Chl.a ($\mu g\ L^{-1}$)	Phytoplankton abundance (cells L^{-1})	Trophic state
<4	$<5 \times 10^5$	Oligotrophic
4–11	5×10^5–5×10^6	Mesotrophic
11–78	5×10^6–10^7	Eutrophic
>78	$>10^7$	Hyper-eutrophic

2. Eutrophication of reservoirs in the Wujiang River

(1) Wujiangdu Reservoir

As described in Table 10.7, in 2005, the TSI of WJD reached the peak (meso-eutrophic or eutrophic) in the wet season, and were relatively low (mesotrophic or meso-eutrophic) in the dry season. Across the whole year, the TSI were highest in front of the dam in the wet season, and were lowest in the upstream of Pianyan River in the normal water period (Yan, 2006). During the years from 2006 to 2007, phytoplankton abundance in WJD was 9.8×10^6 cells L^{-1}, indicating a eutrophic state (see Table 10.8). In the dry season of 2007, the phytoplankton abundance was 0.51×10^6 cells L^{-1}, and the TSI was 51, indicating a meso-eutrophic state (see Table 10.9). In the wet season of 2007, the phytoplankton abundance was 1.80×10^6 cells L^{-1}, and the TSI was 50, indicating a meso-eutrophic state. Overall, during the years 2005–2007, WJD remained in the eutrophic state, and the water quality gradually deteriorated from upstream (Xuantang) to downstream (in front of the dam), as indicated by the increasingly high TSI values (see Table 10.7).

(2) Suofengying Reservoir

As described in Table 10.10, in 2007, the TSI of SFY indicated a mesotrophic state in both wet and dry seasons. Based on the classification criteria of the OECD, the phytoplankton abundance was below 5×10^5 cells L^{-1} in both wet and dry seasons, indicating an oligotrophic state. Overall, SFY remained in an oligo-mesotrophic state in 2007.

Table 10.7 Evaluation results based on weighted composite trophic state index in WJD

Sampling time	Sampling site	Dry season		Wet season		Normal period	
		TSI	Evaluate result	TSI	Evaluate result	TSI	Evaluate result
2005	Dam	40	Meso-eutrophic	50	Eutrophic	35	Mesotrophic
	Pianyan River	40	Meso-eutrophic	47	Meso-eutrophic	34	Mesotrophic
	Xifeng River	40	Meso-eutrophic	49	Meso-eutrophic	41	Meso-eutrophic
	Xuantang River	38	Mesotrophic	45	Meso-eutrophic	41	Meso-eutrophic

Data in the table from Yan (2006)

Table 10.8 Using OECD to evaluate the trophic state of WJD

Sampling time	Chl.a (μg L^{-1})	Evaluate result	Phytoplankton abundance (× 10^6 cells L^{-1})	Evaluate result
2006–2007	4.4	Mesotrophic	9.8	Eutrophic

Data in the table from Wang et al. (2008, 2009)

Table 10.9 Using TSI and OECD to evaluate the trophic state of WJD

Sampling time		TSI	Evaluate result	Phytoplankton abundance (× 10^6cells L^{-1})	Evaluate result
2007	Wet season	51	Meso-eutrophic	0.51	Meso-eutrophic
	Dry season	50	Meso-eutrophic	1.8	Meso-eutrophic

Data in the table from Wei et al. (2010)

(3) Dongfeng Reservoir

During the years 2006–2007, the phytoplankton abundance was 5.28 × 10^5 cells L^{-1} (indicating a eutrophic state) and the chlorophyll-a concentration was 2.40 μg L^{-1} (indicating an oligotrophic state) in front of the dam of DF (see Table 10.11). As shown in Table 10.12, in 2007, the TSI of the DF indicated a mesotrophic state in both the wet and dry seasons. The phytoplankton abundance was lower than 5 × 10^5 cells L^{-1} in both the wet and dry seasons, indicating an oligotrophic state. Overall, DF remained in a mesotrophic state during the years 2006–2007.

(4) Hongjiadu Reservoir

As described in Table 10.13 in 2007, the TSI of HJD indicated an oligotrophic state in both the wet and dry seasons. Based on the classification criteria of OECD, the phytoplankton abundance was lower than 5 × 10^5 cells L^{-1} (indicating an oligotrophic state) in the dry season, and was slightly higher than 5 × 10^5 cells L^{-1} (indicating a mesotrophic state) in the wet season. In 2007, HJD remained in an oligotrophic state during most periods, and remained in a mesotrophic state (in terms of phytoplankton abundance) only in summer. Overall, HJD Reservoir remained in an oligotrophic state in 2007.

Table 10.10 Using TSI and OECD to evaluate the trophic state of SFY

Sampling time		TSI	Evaluate result	Phytoplankton abundance (×10^5cells L^{-1})	Evaluate result
2007	Wet season	46	Mesotrophic	4.0	Oligotrophic
	Dry season	44	Mesotrophic	3.9	Oligotrophic

Data in the table from Wei (2010)

Table 10.11 Using TSI to evaluate the trophic state of DF between 2006 and 2007

Sampling time	Sampling site	Phytoplankton abundance (10^6 cells L^{-1})	Evaluate result	Chll.a ($\mu g\ L^{-1}$)	Evaluate result
2006–2007	Dam	5.3	Eutrophic	2.4	Oligotrophic

Data from Wang et al. (2008, 2009)

Table 10.12 Using TSI and OECD to evaluate the trophic state of DF between 2006 and 2007

Sampling time		TSI	Evaluate result	Phytoplankton abundance ($\times\ 10^5$cells L^{-1})	Evaluate result
2007	Wet season	35	Mesotrophic	3.7	Oligotrophic
	Dry season	32	Mesotrophic	4.6	Oligotrophic

Data from Wei (2010)

(5) Yinzidu Reservoir

As described in Table 10.14, during the years 2006–2007, the phytoplankton abundance of YZD Reservoir was 5.23×10^6 cells L^{-1} (slightly higher than 5.0×10^6 cells L^{-1}), indicating a eutrophic state; whereas, the chlorophyll-a concentration of YZD was as low as 2.6 $\mu g\ L^{-1}$, indicating an oligotrophic state. Overall, YZD Reservoir remained in a mesotrophic state during the years 2006–2007.

(6) Puding Reservoir

As described in Table 10.15, during the years 2006–2007, the phytoplankton abundance of PD was as high as 23.7×10^6 cells L^{-1}, indicating a hyper-eutrophic state; the chlorophyll-a concentration of PD was 7.3 $\mu g\ L^{-1}$, indicating a mesotrophic state. Overall, PD remained in a eutrophic state during the years 2006–2007.

Table 10.13 Using TSI and OECD to evaluate the trophic state of HJD

Sampling time		TSI	Evaluate result	Phytoplankton abundance ($\times\ 10^5$ cells L^{-1})	Evaluate result
2007	Wet season	27	Oligotrophic	2.3	Oligotrophic
	Dry season	24	Oligotrophic	7.7	Mesotrophic

Data from Wei (2010)

Table 10.14 Using OECD to evaluate the trophic state of YZD

Sampling time	Chl.a (μg L^{-1})	Evaluate result	Phytoplankton abundance (× 10^6 cells L^{-1})	Evaluate result
2006–2007	2.6	Oligotrophic	5.2	Eutrophic

Data from Wang et al. (2008, 2009)

Table 10.15 Using OECD to evaluate the trophic state of PD

Sampling time	Chl.a (μg L^{-1})	Evaluate result	Phytoplankton abundance (× 10^6 cells L^{-1})	Evaluate result
2006–2007	7.3	Mesotrophic	23.7	Hyper-eutrophic

Data in the table from references Wang et al. (2008, 2009)

(7) Hongfeng Reservoir

In 1996, a large-scale algal bloom broke out in HF. More specifically, the comprehensive TSI was higher than 50, indicating a eutrophic state. The degrees of eutrophication in the southern parts were more severe than those in the northern parts. In the southern parts of the reservoir, the trophic condition of waters near the Xiaojia Dam (the water supply point for Liangjiao Water Plant) was eutrophic. During the years 1997–1998, algal blooms broke out continuously at large scale. As described in Table 10.16, the trophic condition of waters in different areas in HF is still eutrophic, and the degree of eutrophication in the southern parts is still higher than that in the northern parts. Table 10.17 describes the trophic state of HF during the years 2008–2010. It can be seen that the degree of eutrophication was reduced to a mesotrophic state. In recent years, local government has paid attention to the

Table 10.16 Using weighted composite TSI to evaluate the trophic state of HF

1996			1997–1998		
Sampling site	TSI	Evaluate result	Sampling site	TSI	Evaluate result
Near the Dayang Dao	56	Eutrophic	Zhangguan	51	Eutrophic
HF administrative office	55	Eutrophic	Daba	52	Eutrophic
Huayu Dong	57	Eutrophic	Aoli	56	Eutrophic
The water intake point for Xijiao Water Plant	68	Eutrophic	Xiaojia Dam	55	Eutrophic
Yujialongtan	54	Eutrophic	Jiangjun Wan	55	Eutrophic
HF	56	Eutrophic	Liangcha Dong	56	Eutrophic

Data from Department of environmental protection of Guizhou Province (2003)

Table 10.17 Using weighted composite TSI to evaluate the trophic state of HF (2008–2010)

Sampling site		2008/11		2009/10		2010/01	
		TSI	Evaluate result	TSI	Evaluate result	TSI	Evaluate result
2008–2010	Yaodong	44	Mesotrophic	48	Mesotrophic	42	Mesotrophic
	Jiangjun	44	Mesotrophic	46	Mesotrophic	42	Mesotrophic
	Houwu	43	Mesotrophic	46	Mesotrophic	42	Mesotrophic
	Daba	43	Mesotrophic	44	Mesotrophic	46	Mesotrophic

Data from Ren et al. (2010) and Feng et al. (2011)

Table 10.18 The evaluation results based on weighted composite trophic state index in BH

Sampling time	Sampling site	TSI	Trophic state
1997–1998	Meituwan	46	Meso-eutrophic
	Yuchang dam	52	Eutrophic

Data in the table from Reference Department of environmental protection of Guizhou Province (2003)

water quality of HF and has implemented a few measures to improve the water quality and mitigate the eutrophication in the waters.

(8) Baihua Reservoir

As described in Table 10.18, BH (downstream of HF) also remained highly eutrophic during the period of the outbreak of algal blooms in HF. Table 10.19 describes the trophic state of BH during the years 2009–2010. Evidently, BH remained in a mesotrophic state in winter, and in a mesotrophic or eutrophic state in other seasons. There was no big difference of degree of eutrophication between the period of 2009–2010 and the period of 1997–1998.

(9) Aha Reservoir

As described in Table 10.20, in 2005, the total nitrogen concentration of AH ranged from 0.88 to 0.89 mg L^{-1}, indicating a eutrophic state, and the total phosphorus concentration of AH ranged from 0.063 to 0.065 mg L^{-1}, indicating a eutrophic state. Table 10.21 describes the monthly variation of the trophic state of AH in 2010. Throughout the year, AH remained in a meso-eutrophic state in October and November, and remained in a eutrophic state in other months, especially from March to May. From Tables 10.20 and 10.21, it can be concluded that AH remained in a eutrophic state in recent years.

Table 10.19 The evaluation results based on weighted composite trophic state index in BH

	26/12/2009		21/03/2010		12/06/2010		12/09/2010	
Sampling site	TSI	Trophic state	TSI	Trophic state	TSI	Trophic state	TSI	Trophic state
Yanjiao Village	39	Mesotrophic	51	Eutrophic	52	Eutrophic	52	Eutrophic
Wharf	37	Mesotrophic	51	Eutrophic	50	Mesotrophic	50	Eutrophic
Dam	41	Mesotrophic	50	Mesotrophic	50	Eutrophic	47	Mesotrophic

Data in the table from references Xia et al. (2011) and Yan et al. (2013)

Table 10.20 Using TN and TP to evaluate the trophic state of AH

Sampling time		TN (mg L^{-1})	Trophic State	TP (mg L^{-1})	Trophic state
2005	Dry season	0.90	Eutrophic	0.065	Eutrophic
	Normal period	0.88	Eutrophic	0.064	Eutrophic
	Wet season	0.88	Eutrophic	0.063	Eutrophic

Data in the table from Reference Yang et al. (2007)

Table 10.21 Using weighted composite TSI to evaluate the trophic state of AH

Month	TSI	Trophic state	Month	TSI	Trophic state
1	56	Eutrophic	7	53	Eutrophic
2	56	Eutrophic	8	53	Eutrophic
3	57	Eutrophic	9	56	Eutrophic
4	58	Eutrophic	10	46	Meso-eutrophic
5	58	Eutrophic	11	47	Meso-eutrophic
6	56	Eutrophic	12	54	Eutrophic

Data in the table from Reference Xue et al. (2011)

10.2.2 *Effects of Eutrophication on the Mercury Distribution in Reservoirs*

1. Main physical and chemical parameters of waters and sediments (flooded soils)

The objects of this study were six reservoirs (SFY, HJD, YZD, PD, DF, and WJD) along the mainstream of the Wujiang River, and three reservoirs (BH, HF, and AH) along its tributary streams. Table 10.22 gives a summary of their trophic states, the main physical and chemical parameters of their waters and sediments, THg and MeHg concentrations, and the net flux of THg and MeHg. It should be noted that in Chap. 4, the effects of water eutrophication on the Hg flux between water and atmosphere are described. In Chap. 9, the pattern of Hg enrichment (biomagnification) in the food chain is described. Therefore, this section is mainly focused on the effects of eutrophication on the distribution, transportation, and transformation of Hg in the water column and sediment in reservoirs.

As described in Table 10.22, during the sampling period, SFY and HJD (along the mainstream of the Wujiang River) had the lowest degree of water eutrophication (indicating an oligo-mesotrophic state). YZD and DF had a little higher degree of water eutrophication (indicating a mesotrophic state) than the two reservoirs above (SFY and HJD). Influenced by cage fish culture there, PD and WJD (along the mainstream of the Wujiang River) had an apparently higher degree of water eutrophication (indicating a eutrophic state) than other reservoirs along the mainstream of the Wujiang River. HF, BH, and AH (along the tributary streams of the

Table 10.22 Summary of their trophic state, the main physical and chemical parameters and THg and MeHg concentrations of their waters and sediments, the net flux of THg and MeHg in reservoirs of Wujiang River Basin

Reservoir	Trophic state	Water column				Sediment/flooded soil			Net THg flux (g y^{-1}), source/sink	Net MeHg flux (g y^{-1}), source/sink
		DO (mg L^{-1})	pH	THg (ng L^{-1})	MeHg (ng L^{-1})	OM (%)	THg (ng g^{-1})	MeHg (ng g^{-1})		
SFY	Oligo-mesotrophic	4.8–9.8	7.2–8.2	0.42–4.9	0.03–0.22	2.6–6.9	141–372	0.7–2.0	−1597/sink	−181/sink
HJD	Oligotrophic	1.6–12	6.8–8.7	0.32–6.6	0.05–0.17	0.4–3.9	20–137	0.40–1.9	−11434/sink	−74/sink
YZD	Mesotrophic	4.6–13	6.6–8.4	0.40–1.9	0.03–0.44	–	–	–	−838/sink	−121/sink
DF	Mesotrophic	4.4–11	7.2–8.3	0.68–3.9	0.05–0.50	1.4–4.0	130–330	0.23–2.8	−1820/sink	368/source
PD	Eutrophic	4.1–14	6.5–8.5	1.0–12	0.05–0.51	3.0–9.0	148–298	0.34–6.6	−10469/sink	69/source
WJD	Eutrophic	1.2–12	6.7–9.7	0.60–3.5	0.033–2.9	2.2–11	108–1376	0.15–8 Re.3	1604/source	857/source
BH	Eutrophic	0.03–14	7.4–8.9	6.7–153	0.11–6.9	10–17	665–8073	3.3–15	–	–
HF	Eutrophic	1.4–9.6	7.2–8.7	2.0–14	0.05–2.2	2.2–8.8	197–793	0.19–8.4	–	–
AH	Eutrophic	3.6–10	7.5–8.5	2.1–20	0.03–2.1	2.2–6.7	160–252	0.20–7.3	–	–

Wujiang River) are adjacent to towns and thus are being affected by industrial, agricultural and domestic wastewater, so their waters are all in a eutrophic state.

Under the influence of a special geological background (karstic environment), the waters of the reservoirs in the Wujiang River Basin are slightly alkaline and the pH of the waters varies within a very narrow range. Eutrophication has no apparent connection with the pH of waters, indicating that eutrophication has little effect on the pH of such waters. In the Wujiang River Basin, the distribution pattern of DO in the waters is quite different from that of the pH. With the exception of HJD, the DO in the reservoirs in an oligo-mesotrophic state, overall, remained high and the waters in such reservoirs remained in a well-oxygenated state in each season. However, the DO in the eutrophic reservoirs (e.g. WJD and HF) was obviously lower than that in other reservoirs, especially for waters in the bottom layer in summer. For WJD, BH and HF, the DO concentration in their waters reaches the annual minimum (1.2, 0.03, and 1.4 mg L^{-1}, respectively) representing an extremely anaerobic condition.

As described in Table 10.22, the OM in sediments/flooded soil is universally low in the oligo-mesotrophic reservoirs, for example, (2.6–6.9)% for SFY, (0.4–3.9)% for HJD, and (1.4–4.0)% in DF. In contrast, the OM in sediments is universally high in the eutrophic reservoirs, for example, 9.0, 11 and 17%, respectively, in PD, WJD and BH. For the severely eutrophic WJD, the highest organic content was found in the superficial sediments, and the organic content in superficial sediments rose gradually with increase in the degree of water eutrophication from the upstream to downstream (dam), as mentioned in Chap. 6. It is generally believed that in a eutrophic lake or reservoir, the substantial increase in inorganic nutrients (including nitrogen and phosphorus) in the waters may give rise to abnormal reproduction of autotrophic phytoplankton or aquatic macrophytes, and that the remains of dead aquatic organisms sink to the lake or reservoir bottom, thus quickly increasing the sedimentation rate at the lake or reservoir bottom. Therefore, the most direct effect of eutrophication on the sediments is that dead algae enter into the sediments where they sharply raise the organic content (Jin and Tu 1990). On one hand, the newly deposited organic matter is easily decomposed, thus becoming a source of food for anaerobes. This means that the increased organic content in the sediments arising from eutrophication, can improve the activity of anaerobes. On the other hand, a large amount of oxygen is consumed during decomposition of the organic matter by microbes, thus causing a severe depletion of oxygen, even an extremely anaerobic state, near the water-sediment interface (bottom water and superficial sediments). To conclude, the oxygen dissolution in the waters and organic matter content (or distribution) in the sediments in the reservoirs of the Wujiang River Basin are largely determined by the degree of eutrophication in those bodies of water.

2. Distribution, transportation, and transformation of mercury in water and sediments (flooded soils)

Under the influence of human activities (mainly including the discharge of industrial and municipal wastewater described in Table 10.1, the THg concentrations in the waters and/or sediments of the reservoirs (including BH, HF, and AH) along the tributary streams of the Wujiang River is universally higher than that in the reservoirs along the mainstream of the Wujiang River (as described in Table 10.22). The exception occurs in the upstream of WJD (for details about the reason, see Chap. 6). For the reservoirs in the Wujiang River Basin, however, the mechanisms that determine the concentration and distribution of MeHg in both waters and sediments (submerged soils) are quite different from those of THg, specifically:

(1) When reservoirs (including SFY and HJD) remain in an oligo-mesotrophic state, the MeHg concentration in both waters and submerged soils remains at a low level and the MeHg distribution in the profile does not exhibit any obvious regular change. This indicates that no obvious Hg methylation occurs in the waters and submerged soils (for details, see Chap. 6).
(2) When the YZD and DF remain in a mesotrophic state, MeHg concentration in both waters and sediments is a little higher than that in SFY and HJD, indicating a certain rate of Hg methylation.
(3) For the eutrophic reservoirs (including WJD, HF, and BH) in the Wujiang River Basin, the MeHg concentration in both waters and sediments is far higher than that in other reservoirs; the highest MeHg concentration arose in the anaerobic water in the bottom layer in summer. The highest MeHg concentration in the sediment was found in the superficial layer, and the positions with the highest MeHg concentration in both waters and sediments were consistent with the locations with the highest organic matter content. This indicates that there exists an active methylation process in the anaerobic waters and organic-enriched superficial sediments (such as in WJD and HF; for details, see Chaps. 6 and 7). As described in Table 10.7, while the degree of eutrophication gradually increases from upstream to downstream of the WJD, the MeHg concentration in the corresponding waters (especially the anaerobic bottom waters) and superficial sediments also gradually increased from upstream to downstream. This phenomenon agrees well with the spatial distribution pattern of the eutrophication. Therefore, it can be concluded that eutrophication can indeed promote the transformation of inorganic mercury in waters and sediments into MeHg (for details, see Chap. 6).

It is generally believed that the anaerobes controlling the methylation/demethylation of mercury include sulfate-reducing bacteria and iron-reducing bacteria. Research data shows that sulfate-reducing bacteria can raise the methylation rate of Hg and that iron-reducing bacteria can increase the net yield of MeHg indirectly by suppressing the demethylation rate (Avramescu et al. 2011). In a highly eutrophic aquatic environment, the superficial sediments contain abundant active organic matter. On one hand, the active organic matter provides a sufficient source of food for methylation bacteria. On the other hand, a large amount of oxygen is consumed during the decomposition of organic matter and thus an anaerobic environment is generated at the sediment–water interface. This facilitates the metabolic activities of sulfate-reducing bacteria. During the process of eutrophication, many algae actively absorb Hg from the water column and then enter the superficial sediments after death via the sedimentation process (Coelho et al. 2005). Then, in the process of algal decomposition, the Hg is released again into the water column, thus providing a source of active inorganic Hg sufficient for methylation. Evidently, for the highly eutrophic reservoirs in the Wujiang River Basin, the anaerobic environments of the bottom layer, and the abundant, active organic matter in the superficial sediments; provide the major driving force for the methylation of inorganic Hg while eutrophication is the essential cause of this phenomenon. With the exception of WJD (this reservoir provides a source of THg; for details about the reason, see Chap. 8), THg is sequestered in other reservoirs along the mainstream of the Wujiang River (including YZD, HJD, SFY R, DF, and PD). MeHg is sequestered in the oligo-mesotrophic reservoirs (including SFY, YZD, and HJD). However, such eutrophication of water bodies facilitates the transformation of inorganic Hg contained in waters and sediments into MeHg; so the meso-eutrophic reservoirs (such as, WJD and PD) provide sources of MeHg.

10.3 Biogeochemical Model of Mercury in Reservoir

Based on the measured data, we developed a conceptual model of the Hg biogeochemical cycling in the river-reservoir system in the Wujiang River Basin. This conceptual model can help better understand the Hg biogeochemical characteristics in reservoirs, which could be used to predict the biogeochemical evolutionary tread of Hg in river-reservoir ecosystems, such as the three gorge reservoirs of the Yangtze River.

10.3.1 Primary Evolutionary Stage of Reservoirs

An important environmental consequence of constructing reservoirs is MeHg contamination in the food web of the aquatic ecosystems (St. Louis et al. 2004; Hall et al. 2005; Lucotte et al. 1999). Elevated levels of MeHg were observed in fish in newly constructed reservoirs in North America and northern Europe in the late 1970s and early 1980s (Lodenius et al. 1983; Hecky 1991; Abernathy and Cumbie 1997; Bodaly 1997). Active MeHg production in newly built reservoirs following impoundment may persist for up to 10 years (Hall et al. 2005; St. Louis et al. 2004; Hall et al. 2005). Subsequently, the increased MeHg levels in fish from the reservoirs may last for up to 30 years after impoundment (Mailman et al. 2006; St. Louis et al. 2004; Hall et al. 2005; Lucotte et al. 1999). The methylation of inorganic Hg

(IHg) in water and sediments constitutes a key role in the cycling of Hg in aquatic systems (Fitzgerald and Mason 1997).

The decomposition of flooded vegetation and organic matter in soils may stimulate the microbial methylation of IHg to MeHg (Furutani and Rudd 1980; Lucotte et al. 1999). Given the decomposition of organic matter in submerged soil, the net Hg methylation rate decreases with an increase in the reservoir age (St. Louis et al. 2004; Hall et al. 2005; Lucotte et al. 1999). Studies in North America showed that the Hg methylation rates decreased as the age of the reservoir increased because of the decomposition of organic carbon in the flooded soils (St. Louis et al. 2004; Hall et al. 2005; Lucotte et al. 1999). Our current study showed that the MeHg production in reservoirs located in southwestern China cannot be predicted using previous observations in North America and Europe because the Hg methylation process in the Wujiang River Basin is driven by a completely different biogeochemical dynamic (Larssen 2010). Consequently, the observations obtained from the reservoirs in the Wujiang River tell a different story.

Given the karstic environment of the Wujiang River Basin, the organic matter contents in the submerged soil (range: 0.4–6.9%, see details in Chap. 6) were much lower than the organic matter concentrations (varied from 30 to 50%) in submerged soil from the boreal forest or wetlands in North America and Europe (St. Louis et al. 2004; Hall et al. 2005; Lucotte et al. 1999). The primary productivity in newly constructed reservoirs is currently represented as oligotrophic–mesotrophic due to the absence of cage culture fishing in the newly constructed reservoirs (Fig. 10.1). The current study also indicated that the major source of organic matter in the newly constructed reservoirs (e.g., YZD, SFY, and HJD) was mainly derived from the watershed input with little autochthonous contribution because of the low primary productivity (Jiang 2007). The low organic matter content in the submerged upland soil of the Wujiang River Basin may inhibit methylating microorganisms from colonizing the newly constructed reservoirs or decrease their rate of metabolism (Fig. 10.1). Furthermore, river erosion and surface runoff were the main sources of MeHg in the newly constructed reservoirs in the Wujiang River Basin. Therefore, the concentrations of MeHg in the water column and fish samples from the newly constructed reservoirs (e.g., YZD, SFY, and HJD) were significantly lower than those from DF, PD, and WJD in the Wujiang River Basin and from newly constructed reservoirs in North America and northern Europe (see details in Chaps. 6 and 8). Consequently, the newly constructed reservoirs, such as SFY, HJD, and YZD in the Wujiang River, did not show a net source of MeHg in the reservoir systems and instead represented a net sink of MeHg (see detail in Chap. 7). Therefore, the newly constructed Chinese reservoirs within the Wujiang River (YZD, SFY, and HJD) were not active sites of net Hg methylation because of the low organic carbon content in the submerged soils and/or low primary productivity.

Fig. 10.1 Conceptual model of the Hg cycling in primary evolutionary stage reservoirs in Wujiang River Basin, Southwest China

10.3.2 Intermediate Evolutionary Stage of Reservoirs

With the continuous evolution of reservoirs, PD and DF, which are classified as intermediate evolutionary stage, are mesotrophic-eutrophic, which indicates a much higher level of primary productivity compared to the newly constructed reservoirs in the Wujiang River Basin (e.g., YZD, SFY, and HJD are oligotrophic–mesotrophic) (Fig. 10.2). The organic matter content in the sediment was significantly elevated as a result of the continuously increasing autochthonous material. The

Fig. 10.2 Conceptual model of the Hg cycling in intermediate evolutionary stage reservoirs in Wujiang River Basin, Southwest China

active net methylation of inorganic Hg in the sediment occurred in PD and DF (Fig. 10.2). Meanwhile, the concentrations of MeHg in the sediment and the corresponding water column in DF and PD were higher than those in the newly constructed reservoirs (e.g., YZD, HJD, and SFY) in the Wujiang River Basin.

When compared to the newly constructed reservoirs, the intermediate evolutionary stage reservoirs within the Wujiang River Basin (such as DF and PD) changed from MeHg sinks to MeHg sources. DF is characterized as mesotrophic, which is similar to the newly constructed reservoirs (e.g., YZD and SFY). During our sampling campaigns, DF acted as a net sink for THg but a net source for MeHg. These observations suggested that in situ net Hg methylation existed in the intermediate evolutionary stage of the reservoir. More importantly, the Hg methylation processes will be continuously accelerated with increases in the primary productivity of the reservoirs in the Wujiang River Basin in southwestern China.

10.3.3 Advanced Evolutionary Stage of Reservoirs

Over a long-term evolution of the reservoir, the primary productivity of WJD, which is classified as an advanced evolutionary stage, was significantly elevated (Fig. 10.3). WJD was developed as a typical eutrophic reservoir during our sampling periods.

The high-intensity cage aquaculture activity in WJD contributed to the higher primary productivity compared to the newly constructed reservoirs (SFY, HJD, and YZD). Phytoplankton-derived organic matter and the fish feeds and feces were potentially significant sources of organic matter input to the surface sediments of WJD (Fig. 10.3). Even the submerged soils from the WJD contained very low

Fig. 10.3 Conceptual model of the Hg cycling in advanced evolutionary stage reservoirs in Wujiang River Basin, Southwest China

concentrations of organic matter because of the karstic environment in the Wujiang River Basin. However, the organic carbon content in the surface sediment collected from WJD was relatively high in comparison to the sediments from PD and DF (see detail in Chap. 6).

The organic matter in the surface sediment originating from cage aquaculture activities in WJD was easily decomposed by microorganisms. The increased oxygen consumption during fresh organic matter degradation causes progressively more anoxic conditions at the sediment/water interface (see detail in Chap. 6), which leads to the active methylation process. As shown in Chap. 6, WJD (advanced evolutionary stage reservoir) is characterized by a much more active net Hg methylation compared to the primary evolutionary stage reservoirs (such as YZD, SFY, and HJD) and the intermediate evolutionary stage reservoirs (such as PD and DF). The MeHg concentrations in the water column and sediments of WJD were significantly elevated compared to those from the other five reservoirs in the Wujiang River Basin (such as SFY, YZD, HJD, PD, and DF). The current study further demonstrated that both the surface sediment and the hypolimnetic water were the net MeHg sources for the water column in the WJD, which may pose a potential threat to the reservoir system and downstream aquatic ecosystems. The popularity of cage aquaculture appears to be the key factor in the high primary productivity in the reservoirs in the Wujiang River Basin, and this could actually, significantly change the conditions promoting the net MeHg production. Moreover, the contribution of the organic matter to the sediment from cage aquaculture affected the Hg methylation degrees, which would explain the different net MeHg production rates among the different evolutionary stages of reservoirs in the Wujiang River Basin. A hyper state of eutrophication that existed in WJD significantly changed the physical and chemical characteristics of the water column, which resulted in the significantly elevated Hg^0 emission from the surface water to the atmosphere. Therefore, WJD, an advanced evolutionary stage reservoir, acted as a source for both THg and MeHg annually.

References

Abernathy AR, Cumbie PM (1997) Mercury accumulation by largemouth bass (Micropterus salmoides) in recently impounded reservoirs. Bull Environ Contam Toxicol 17:595–602

Avramescu ML, Yumvihoze E, Hintelmann H, Ridal J, Fortin D, Lean DRS (2011) Biogeochemical factors influencing net mercury methylation in contaminated freshwater sediments from the St. Lawrence River in Cornwall, Ontario. Canada Sci Total Environ 409:968–978

Bodaly RA (1997) Bioaccumulation of mercury in the aquatic food chain of newly flooded areas. In: St. Louis VL et al (eds) Metal ions in biological systems. Marcel Dekker, New York, pp 259–287

Bodaly RA, St Louis VL, Paterson MJ, Fudge RJ, Hall BD, Rosenberg DM, Rudd JWM (1997) Bioaccumulation of mercury in the aquatic food chain in newly flooded areas. Met Ions Biol Syst 34:259–287

Brigham ME, Krabbenhoft DP, Olson ML, Dewild JF (2002) Methylmercury in flood-control inpoundment and natural waters of Northwestern Minnesota, 1997–1999. Water Air Soil Pollut 138:61–78

Coelho JP, Pereira ME, Duarte A, Pardal MA (2005) Macroalgae response to a mercury contamination gradient in a temperate coastal lagoon (Ria de Aveiro, Portugal). Estuar Coast Shelf Sci 65:492–500

Cossa D, Gobeil C (2000) Mercury speciation in the lower St. Lawrence estuary. Can J Fish Aquat Sci 57:138–147

Department of environmental protection of Guizhou Province (2003) Water pollution prevention and control report in Hongfeng and Baihua lakes. (in Chinese)

Fearnside PM (2001) Environmental impacts of Brazil's Tucurui Dam: unlearned lessons for hydroelectric development in Amazonia. Environ Manage 27:377–396

Feng YQ, Xia PH, Zhang MS, Li CX, Lin T, Ma JR, Long J, Yang JJ (2011) Analysis on eutrophication of Hongfeng Reservoir on Guizhou plateau. J Guizhou Norm Univ (Nat Sci) 29:29–35. (in Chinese, with English abstract)

Fitzgerald WF, Mason RP (1997) Biogeochemical cycling of mercury in the marine environment. In: Sigel A, Sigel H (eds) Metal ions in biological systems, vol 34-mercury and its effect on environment and biology. Marcel Dekker, New York, NY, pp 53–111

Furutani A, Rudd JW (1980) Measurement of mercury methylation in lake water and sediment samples. Appl Environ Microbiol 40:770–776

Gagnon C, Fisher NS (1997) Bioavailability of sediment-bound methyl and inorganic mercury to a marine bivalve. Environ Sci Technol 31:993–998

Gray JE, Fey DL, Holmes CW, Lasorsa BK (2005) Historical deposition and fluxes of mercury in Narraguinnep reservoir, Southwestern Colorado, USA. Appl Geochem 20:207–220

Gu DX, Shu JH (1988) Water pollution prediction and prevention planning method in the Lake. China Environmental Science Press, Beijing (in Chinese)

Hall BD, St Louis VL, Rolfhus KR, Bodaly RA, Beaty KG, Paterson MJ, Cherewyk KAP (2005) Impact of reservoir creation on the biogeochemical cycling of methyl and total mercury in boreal upland forests. Ecosystems 8:248–266

Hecky RE (1991) Increased methylmercury contamination in fish in newly formed reservoirs. In: Ramsey DJ et al (eds) Advances in mercury toxicology. Plenum Press, New York, pp 33–52

Heyes A, Moore TR, Rudd JWM (1998) Mercury and methylmercury in decomposing vegetation of a pristine and impounded wetland. J Environ Qual 27:591–599

Jin XC, Tu QY (1990) Standard for Lake Eutrophication investigation, 2nd edn. China Environmental Science Press, Beijing (in Chinese)

Kelly CA, Rudd JWM, Bodaly RA, Roulet NP, St Louis VL, Heyes A, Moore TR, Schiff S, Aravena R, Scott KJ, Dyck B, Harris R, Warner B, Edwards G (1997) Increases in fluxes of greenhouse gases and methylmercury following flooding of an experimental reservoir. Environ Sci Technol 31:1334–1344

Larssen T (2010) Mercury in Chinese reservoirs. Environ Pollut 158:24–25

Lin QQ, Han BP, Li T, Wang LM (2001) Reservoir water supply and reservoir eutrophication in Guangdong Province (South China). In: International Lake Environment Committee ed. Proceedings of 9th International Conference on the Conservation and Management of Lakes. Shiga, Japan, Shiga Prefectural Government 3:269–272

Lodenius M, Seppanen A, Herranen M (1983) Accumulation of mercury in fish and man from reservoirs in northern Finland. Water Air Soil Pollut 19:237–246

Lucotte M, Schetagne R, Therien N, Langlois C, Tremblay A (1999) Mercury in the biogeochemical cycle- natural environments and hydroelectric reservoirs of Northern Quebec (Canada). Springer, pp 1–334

Mailman M, Stepnuk L, Cicek N, Bodaly RA (2006) Strategies to lower methyl mercury concentrations in hydroelectric reservoirs and lakes: a review. Sci Total Environ 368:224–235

Mason RP, Sveinsdottir AY (2003) Mercury and methylmercury concentrations in water and largemouth bass in Maryland Reservoirs. Maryland Department of Natural Resources, Annapolis, MD

Ren QF, Chen C, Li L, Wang S, Long SX (2010) Phytoplankton community and its relationship with environmental factors in Hongfeng Lake. Environ Sci Technol 33:59–64. (in Chinese, with English abstract)
Rogers DW, Dickman M, Han X (1995) Stories from old reservoirs: sediment Hg and Hg methylation in Ontario hydroelectric developments. Water Air Soil Pollut 80:829–839
Schetagne R, Doyo JF, Fournier JJ (2000) Export of mercury downstream from reservoirs. Sci Total Environ 260:135–145
St. Louis VL, Rudd JWM, Kelly CA, Bodaly RAD, Paterson MJ, Beaty KG, Hesslein RH, Heyes AW, Majewski AR (2004) The rise and fall of mercury methylation in an experimental reservoir. Environ Sci Technol 38:1348–1358
Tremblay A, Lucotte M, Rheault I (1996) Methylmercury in a benthnic food web of two hydroelectric reservoirs and a natural lake of northern Quebec (Canada). Water Air Soil Pollut 91:255–269
Ullrich SM, Tanton TW, Abdrashitova SA (2001) Mercury in the aquatic environment: a review of factors affecting methylation. Crit Rev Environ Sci Technol 31:241–293
Verdon R, Brouard D, Demers C, Lalumiere R, Laperle M, Schetagne R (1991) Mercury evolution (1978–1988) in fishes of the La Grande hydroelectric complex, Quebec, Canada. Water Air Soil Pollut 56:405–417
Vollenweider RA, Kerekes J (1982) Eutrophication of waters: monitoring assessment and control. Organization for Economic Co-Operation and Development (OECD), Paris
Wang BL, Liu CQ, Wang FS, Yu YX, Zhang LH (2008) The distributions of autumn picoplankton in relation to environmental factors in the reservoirs along the Wujiang River in Guizhou Province, SW China. Hydrobiologia 598:35–45
Wang BL, Liu CQ, Wang FS, Yu YX, Wu YY (2009) Flow cytometric observation of picophytoplankton community structure in the cascade reservoirs along the Wujiang River. SW China. J Limnol 68:53–63
Wei L, Xia T, Yan ZC, Bian XW (2010) Eutrophication in upstream cascade reservoirs of Wujiang River. Water Resour Prot 26:39–43. (in Chinese, with English abstract)
Winfrey MR, Rudd JWM (1990) Environmental factors affecting the formation of methylmercury in low pH lakes. Environ Toxicol Chem 9:853–869
Xia PH, Li QH, Lin T, Hu JW (2011) Limnological characteristics and environmental effects of the Baihua reservoir in Guizhou Plateau, China. Acta Scientiae Circumstantiae 31:1660–1669. (in Chinese, with English abstract)
Xue F, Xia PH, Lin T, Feng YQ, Zhang BX, Jiang Y, Jin CH (2011) Study on the relationship between chlorophyll-a and environmental impact factors in deep reservoir in Karst Areas. J Anhui Agric Sci 39:9811–9814. (in Chinese, with English abstract)
Yan HY, Li QH, Meng B, Wang CP, Feng XB, He TR, Dominik J (2013) Spatial distribution and methylation of mercury in a eutrophic reservoir heavily contaminated by mercury in Southwest China. Appl Geochem 33:182–190
Yan N (2006) The comparison study of the distribution of the phytoplankton and Eutrophication features of two kinds of Karst reservoirs in Guizhou. Guizhou Normal University, Guiyang, China. (in Chinese, with English abstract)
Yang WS, Li XJ, Zhang J (2007) Fuzzy evaluation of water quality and strategies of water pollution prevention in the Aha Lake. South North Water Trans Water Sci Technol 5:61–63. (in Chinese, with English abstract)

MIX
Papier aus verantwortungsvollen Quellen
Paper from responsible sources
FSC® C105338

If you have any concerns about our products,
you can contact us on
ProductSafety@springernature.com

In case Publisher is established outside the EU,
the EU authorized representative is:
Springer Nature Customer Service Center GmbH
Europaplatz 3, 69115 Heidelberg, Germany

Printed by Libri Plureos GmbH
in Hamburg, Germany